Lecture Notes in Computer Science 4538

Commenced Publication in 1973
Founding and Former Series Editors:
Gerhard Goos, Juris Hartmanis, and Jan van Leeuwen

T0223136

Francisco Escolano   Mario Vento (Eds.)

Graph-Based Representations in Pattern Recognition

6th IAPR-TC-15 International Workshop, GbRPR 2007
Alicante, Spain, June 11-13, 2007
Proceedings

 Springer

Volume Editors

Francisco Escolano
DCCIA- Universidad de Alicante
E-03080 Alicante, Spain
E-mail: sco@dccia.ua.es

Mario Vento
DIIE - University of Salerno
1 FISICIANO, Italy
E-mail: mvento@unisa.it

Library of Congress Control Number: 2007927650

CR Subject Classification (1998): I.5, I.4, I.3, I.2.10, G.2.2, F.2.2, E.1

LNCS Sublibrary: SL 6 – Image Processing, Computer Vision, Pattern Recognition, and Graphics

ISSN   0302-9743
ISBN-10   3-540-72902-X Springer Berlin Heidelberg New York
ISBN-13   978-3-540-72902-0 Springer Berlin Heidelberg New York

Springer is a part of Springer Science+Business Media

springer.com

© Springer-Verlag Berlin Heidelberg 2007
Printed in Germany

Typesetting: Camera-ready by author, data conversion by Scientific Publishing Services, Chennai, India
Printed on acid-free paper   SPIN: 12072798   06/3180   5 4 3 2 1 0

Preface

GbR (Graph-Based Representations in Pattern Recognition) is a biennial workshop organized by the TC15 (http://www.greyc.ensicaen.fr/iapr-tc15/) Technical Committee of the IAPR, aimed at encouraging research works in pattern recognition and image analysis within the graph theory framework. This workshop series traditionally provides a forum for presenting and discussing research results and applications at the intersection of pattern recognition, image analysis on one side and graph theory on the other side.

Traditionally, the scientific content of these workshops covers research on problems such as matching, segmentation, object and shape representation, and even image processing, where using graphs is more than interesting approach: all the later problems, and many more, may be attacked from the graph point a view. Sometimes, coarse-to-fine representations such as graph pyramids emerge to deal with some of the latter problems. Topological notions are also interesting in the case of object representation. More recently, translation into the graph (or to the string) domain of data-mining procedures usually designed for vectors (e.g., clustering) and the embedding of graphs in subspaces have opened new intriguing perspectives. Related to the latter problem, the design of proper measures or distances between graphs, trees, or between the nodes of graphs or trees, is an interesting challenge (e.g., kernel design). These elements are contemplated in the current proceedings.

In addition, the avenue of new structural/graphical models and structural criteria (e.g., belief-propagation, specific graphs under the constellation approach, graph-cuts) has impacted the current edition of GbR and some papers, especially those related to segmentation, are also included in these proceedings.

Furthermore, and in connection with data-mining with graphs, the intersection between graph representations and machine learning motivated GbR 2007 (which was celebrated in Alicante, Spain, June 11-13) us to organize in conjunction with the Learning from and with Graphs (http://eurise.univ-st-etienne.fr/GBR_workshop2007/) workshop (June 14) of PASCAL (Pattern Analysis, Statistical Modelling and Computational Learning), which is European Excellence Network. This one-day workshop was within the PASCAL thematic program of Graph Theory Methods in Machine Learning. The aim of this workshop was, thus, to promote the collaboration between the different communities interested in learning with and from graphs.

The papers presented in these proceedings have been reviewed by at least two members of the Program Committee (we sincerely thank all of them and the additional referees for their efforts). There where 54 papers from authors of 14 countries. The Program Committee selected 23 of them oral presentation and 14 as posters. The 37 resulting papers are published in this volume.

Finally, special thanks to Luc Brun, the new president of TC 15, for his coordination efforts.

April 2007   Francisco Escolano
                                                                      Mario Vento

References

1. Luc Brun and Mario Vento (Eds.) *Graph-based Representations in Pattern Recognition, GbRPR 2005*, volume 3434 of LNCS, Poitiers, France, April 2005, IAPR-TC15, Springer. ISBN 3-540-25270-3.
2. Edwin Hancock and Mario Vento (Eds.) *Graph-based Representations in Pattern Recognition, GbRPR 2003*, volume 2726 of LNCS, York, UK, June/July 2003, IAPR-TC15, Springer. ISBN 3-540-40452-X.
3. Jean-Michel Jolion, Walter G. Kropatsch and Mario Vento (Eds.) *Graph-based Representations in Pattern Recognition, GbR 2001* Ischia, Italy, May 2001, IAPR-TC15, CUEN. ISBN 88-7146-579-2.
4. Walter G. Kropatsch and Jean-Michel Jolion (Eds.) *Graph-based Representations in Pattern Recognition, GbR 1999* Haindorf, Austria, May 1999, IAPR-TC15, OCG. ISBN 3-85403-126-2.

Organization

Program Co-chairs

Francisco Escolano
DCCIA - Universidad de Alicante
EPS Campus San Vicente Ap. 99
E-03080 Alacant Spain

Mario Vento
DIIE - University of Salerno
Via Ponte Don Melillo, 1
FISCIANO (SA) Italy

Program Committee

Luc Brun	GREYC ENSICAEN, France
Horst Bunke	University of Bern, Switzerland
Terry Caelli	National ICT Australia
Luigi Pietro Cordella	Università di Napoli, Italia
James Coughlan	SKERI, San Francisco, CA, USA
Sven Dickinson	University of Toronto, Ontario, Canada
Francisco Escolano	University of Alicante, Spain
Edwin R. Hancock	University of York, UK
Jean-Michel Jolion	Lab. LIRIS, INSA Lyon, France
Walter G. Kropatsch	Vienna University of Technology, Austria
Marcello Pelillo	Università Ca' Foscari di Venezia, Italy
Kaleem Siddiqi	McGill University, Canada
Andrea Torsello	Università Ca' Foscari di Venezia, Italy
Mario Vento	University of Salerno, Italy

Referees

I. Bloch (France)	T. Gärtner (Germany)	M.A. Lozano (Spain)
A. Braquelaire (France)	B. Jain (Germany)	J.M. Sáez (Spain)
G. Damiand (France)	O. Lezoray (France)	M. Sebban (France)
A. Deruyver (France)	A. Levinshtein (Canada)	S. Stolpner (Canada)

Sponsoring Institutions

Robot Vision Group, University of Alicante (UA)
Pattern Analysis, Statistical Modelling and Computational Learning (PASCAL)
European Excellence Network
Generalitat Valenciana: Department of Company, University and Science
Office for Extracurricular Activities (UA)
Institute of Informatics Research (UA)
Alicante Convention Bureau, Patronato Municipal de Turismo
Department of Computer Science and Artificial Intelligence (UA)
Office for Research, Developement and Innovation (UA)

Table of Contents

Graph-Based Segmentation and Image Processing

Graph-Based Clustering

Graph Representations

Pyramids, Combinatorial Maps and Homologies

Graph Clustering, Embedding and Learning

Bipartite Graph Matching for Computing the Edit Distance of Graphs

Kaspar Riesen, Michel Neuhaus, and Horst Bunke

Department of Computer Science, University of Bern,
Neubrückstrasse 10, CH-3012 Bern, Switzerland
{riesen,mneuhaus,bunke}@iam.unibe.ch

Abstract. In the field of structural pattern recognition graphs constitute a very common and powerful way of representing patterns. In contrast to string representations, graphs allow us to describe relational information in the patterns under consideration. One of the main drawbacks of graph representations is that the computation of standard graph similarity measures is exponential in the number of involved nodes. Hence, such computations are feasible for rather small graphs only. One of the most flexible error-tolerant graph similarity measures is based on graph edit distance. In this paper we propose an approach for the efficient compuation of edit distance based on bipartite graph matching by means of Munkres' algorithm, sometimes referred to as the Hungarian algorithm. Our proposed algorithm runs in polynomial time, but provides only suboptimal edit distance results. The reason for its suboptimality is that implied edge operations are not considered during the process of finding the optimal node assignment. In experiments on semi-artificial and real data we demonstrate the speedup of our proposed method over a traditional tree search based algorithm for graph edit distance computation. Also we show that classification accuracy remains nearly unaffected.

1 Introduction

Graph matching refers to the process of evaluating the structural similarity of graphs. A large number of methods for graph matching have been proposed in recent years [1]. One idea is to consider the spectral decomposition of graphs rather than the graphs themselves [2,3], which provides for a number of convenient properties. It seems, however, that spectral methods are often rather sensitive to structural errors and sometimes only applicable to unlabeled graphs in a straight-forward manner. In other approaches, relaxation labeling techniques, artificial neural networks, and genetic algorithms have been used to map the nodes of one graph to the nodes of another graph such that the edge structure, and possibly labels attached to nodes and edges, are preserved as accurately as possible [4,5,6]. Such algorithms perform quite efficiently, but they are limited in that they are often applicable to special classes of graphs only. Recently a general framework for graph matching has been proposed [7]. This aproach uses random walk based models to compute topological features for each node and

F. Escolano and M. Vento (Eds.): GbRPR 2007, LNCS 4538, pp. 1–12, 2007.

shows that the adoption of classic bipartite graph matching algorithms offers a straightforward generalization of the algorithm given for graph isomorphism.

One of the most flexible methods for error-tolerant graph matching that is applicable to various kinds of graphs is based on the edit distance of graphs [8,9]. The idea of graph edit distance is to define the dissimilarity of graphs by the amount of distortion that is needed to transform one graph into the other. Using the edit distance, an input graph to be classified can be analyzed by computing its dissimilarity to a number of training graphs. For classification, the resulting distance values may be fed, for instance, into a nearest-neighbor classifier. Alternatively, the edit distance of graphs can also be interpreted as a pattern similarity measure in the context of kernel machines, which makes a large number of powerful methods applicable to graphs [10], including support vector machines for classification and kernel principal component analysis for pattern analysis. The edit distance has proved to be suitable for error-tolerant graph matching in various applications [11,12,13].

The main drawback of graph edit distance is its computational complexity, which is exponential in the number of nodes of the involved graphs. Consequently, the application of edit distance is limited to graphs of rather small size in practice. To render the matching of graphs less computationally demanding, a number of methods have been proposed. In some approaches, the basic idea is to perform a local search to solve the graph matching problem, that is, to optimize local criteria instead of global, or optimal ones [11,14,15]. In [16], a linear programming method for computing the edit distance of graphs with unlabeled edges is proposed. The method can be used to derivelower and upper edit distance bounds in polynomial time.

In this paper, we propose a new efficient algorithm for graph edit distance computation. The method is based on a (globally optimal) fast bipartite optimization procedure mapping nodes, or edges, of one graph to nodes, or edges, of another graph. To this end we make use of Munkres' algorithm [17]. Originally, this algorithm has been proposed to solve the assignment problem in polynomial time. However, in the present paper we adapt the original algorithm such that one can compute graph edit distance. In experiments (on semi-artificial and real-world data) we demonstrate that the proposed method allows us to speed up the computation of graph edit distance substantially, while at the same time recognition accuracy is not much affected. In Section 2, we briefly introduce graph edit distance. In Section 3, Munkres' algrithm for bipartite graph matching and its extension for computing graph edit distance are described. In Section 4, we give some experimental results achieved by Munkres' algorithm and compare them with results from other methods. Finally, in Section 5, we draw conclusions.

2   Graph Edit Distance

The key idea of graph matching is to define a dissimilarity measure for graphs [8,9]. In contrast to statistical pattern recognition, where patterns are described by vectors, graphs do not offer a straightforward distance model like the

Euclidean distance. A common way to define the dissimilarity of two graphs is to determine the minimal amount of distortion that is needed to transform one graph into the other. These distortions are given by insertions, deletions, and substitutions of nodes and edges. Given two graphs – the source graph g_1 and the target graph g_2 – the idea is to delete some nodes and edges from g_1, relabel some of the remaining nodes and edges (substitutions) and possibly insert some nodes and edges, such that g_1 is finally transformed into g_2. A sequence of edit operations that transforms g_1 into g_2 is called an *edit path* between g_1 and g_2. One can introduce cost functions for each edit operation measuring the strength of the given distortion. The idea of such cost functions is that one can define whether or not an edit operation represents a strong modification of the graph. Hence, between two structurally similar graphs, there exists an inexpensive edit path, representing low cost operations, while for structurally different graphs an edit path with high costs is needed. Consequently, the *edit distance* of two graphs is defined by the minimum cost edit path between two graphs. In the following we will denote a graph by $g = (V, E, \alpha, \beta)$, where V denotes a finite set of nodes, $E \subseteq V \times V$ a set of directed edges, $\alpha : V \to L_V$ a node labeling function assigning an attribute from L_V to each node, and $\beta : E \to L_E$ an edge labeling function. The substitution of a node u by a node v is denoted by $u \to v$, the insertion of u by $\varepsilon \to u$, and the deletion of u by $u \to \varepsilon$.

The edit distance can be computed by a tree search algorithm, where possible edit paths are iteratively explored, and the minimum-cost edit path can finally be retrieved from the search tree [8,18]. This method allows us to find the optimal edit path between two graphs. However, its drawback is the exponential time-complexity, which makes the algorithm applicable to small graphs only. In this paper we propose another way of computing graph edit distance based on bipartite graph matching.

3   Munkres' Algorithm for Graph Matching

The process of graph matching can be seen as an assignment problem: How can one assign the nodes of graph g_1 to the nodes of graph g_2, such that the overall edit costs are minimal? For two graphs g_1 and g_2 with n and m nodes respectively, there are $n \cdot (n - 1) \cdot \ldots \cdot (n - m + 1)$ possible assignments of which several may be optimal. The time complexity of a brute force algorithm is therefore $O(n^n)$.

Originally, Munkres' algorithm was proposed to solve an assignment problem in a more sophisticated and efficient way than brute force [17]. The assignment problem is the task of finding the best – that is the minimum-cost – assignment of the elements of a set S_1 to the elements of another set S_2. One can prove that the method of [17] finds the minimum-cost assignment of two given sets in a time-complexity of $O(n^3)$, where $n = \max\{|S_1|, |S_2|\}$. Applied to the problem of graph edit distance computation, the above-mentioned sets are the nodes of the graphs to be matched: $S_1 = V_1$ and $S_2 = V_2$. Hence, Munkres' algorithm can be applied to graph edit distance computation. Despite the optimality of the minimum-cost assignment of the involved nodes, the edit path found by this method is only suboptimal. The reason is that edge operations are always implied by node operations. Munkres' algorithm finds the minimum-cost node assignment, without considering edge edit operations. In our proposed algorithm for graph edit distance computation, the costs of these implied operations are added at the end of the computation. Hence, the node assignments and the implied edge assignments found by the algorithm need not correspond to the optimal edit path.

Formally, given two graphs $g_1 = (V_1, E_1, \alpha_1, \beta_1)$ and $g_2 = (V_2, E_2, \alpha_2, \beta_2)$, where $n = |V_1|$ and $m = |V_2|$, Munkres' algorithm finds the $k = min\{n, m\}$ node assignments with minimal costs. If one does not impose any restrictions on the cost function, it may be that the edit sequence *delete-insert* is less expensive than a simple *substitution* of nodes: $(c(u \to \varepsilon) + c(\varepsilon \to v)) < c(u \to v)$. Whenever this case occurs the costs for the substitution can be replaced by the costs caused by the edit sequence *delete-insert*. In the remainder of this paper we assume in fact that substitution costs are always given by the minimum-cost edit operations: $c(u \to v) = \min\{(c(u \to \varepsilon) + c(\varepsilon \to v)), c(u \to v)\}$.

Munkres' algorithm is given in detail in Alg. 1, and Fig. 1 illustrates the execution of the algorithm by means of an example.

3.1   Munkres' Algorithm

In the description of Munkres' algorithm we will use a cost matrix and distinguish some lines (rows or columns) of the matrix and some zero elements. In the remainder of this paper we will speak of *covered* or *uncovered* lines and *starred* or *primed* zeros, for zeros marked with a star, and zeros marked with a prime, respectively.

In line 1 of Alg. 1 the initial cost-matrix C is generated. Matrix elements represent the costs of a node assignment given by a predefined cost function c: $C_{i,j} = c(v_i \to u_j)$. The smallest element of each row is subtracted from the actual row in line 2. This leads to a matrix in which each line contains at least one zero element. In line 3 all zeros in a row or column with no starred zeros are getting marked with a star. This completes the preliminary steps.

The algorithm enters STEP 1, in which all the columns containing a starred zero are covered. If $k = min\{n, m\}$ columns are covered, the algorithm has found the k minimum-cost node assignments. Assignment pairs are then indicated by the positions of the starred zeros in the matrix. In the example in Fig. 1 only two columns are covered, so that we proceed with STEP 2. In this part of the algorithm it is determined whether the algorithm continues with STEP 3 or 4. Before switching to one of the two subsequent steps, STEP 2 possibly has to be repeated several times. If there exists an uncovered zero in the cost-matrix, and there is no starred zero in its row, the algorithm proceeds with STEP 3. If there is an uncovered zero in the cost-matrix, but there is also a starred zero in its row, some covering- and uncovering-operations have to be performed and STEP 2 is repeated with the modified matrix. If finally the cost-matrix contains no uncovered zeros any more, the smallest element e_{min} is saved and the algorithm continues with STEP 4. In the example given in Fig. 1, STEP 2

Algorithm 1. Computation of the minimum-cost node assignement by Munkres' algorithm

Input:	Non-empty graphs $g_1 = (V_1, E_1, \alpha_1, \beta_1)$ and $g_2 = (V_2, E_2, \alpha_2, \beta_2)$, where $V_1 = \{u_1, \ldots, u_n\}$ and $V_2 = \{v_1, \ldots, v_m\}$
Output:	The minimum-cost node assignment

1: Generate the $n \times m$ cost-matrix C, where each element represents the cost of a single node assignment: $C_{i,j} = c(u_i \rightarrow v_j)$. Initialize $k = min\{n, m\}$.
2: For each row r in C, subtract its smallest element from every element in r
3: For all zeros z_i in C, mark z_i with a star if there is no starred zero in its row or column
4: **STEP 1**:
5: **for** Each column containing a starred zero **do**
6:     cover this column
7: **end for**
8: **if** k columns are covered **then GOTO** DONE **else GOTO** STEP 2 **end if**
9: **STEP 2**:
10: **if** C contains an uncovered zero **then**
11:     Find an arbitrary uncovered zero Z_0 and prime it
12:     **if** There is no starred zero in the row of Z_0 **then**
13:         **GOTO** STEP 3
14:     **else**
15:         Cover this row, and uncover the column containing the starred zero **GOTO** STEP 2.
16:     **end if**
17: **else**
18:     Save the smallest uncovered element e_{min} **GOTO** STEP 4
19: **end if**
20: **STEP 3**: Construct a series S of alternating primed and starred zeros as follows:
21: Insert Z_0 into S
22: **while** In the column of Z_0 exists a starred zero Z_1 **do**
23:     Insert Z_1 into S
24:     Replace Z_0 with the primed zero in the row of Z_1. Insert Z_0 into S
25: **end while**
26: Unstar each starred zero in S and replace all primes with stars. Erase all other primes and uncover every line in C **GOTO** STEP 1
27: **STEP 4**: Add e_{min} to every element in covered rows and subtract it from every element in uncovered columns. **GOTO** STEP 2
28: **DONE**: Assignment pairs are indicated by the positions of starred zeros in the cost-matrix.

has to be performed twice before switching to STEP 4. The effect of STEP 4 is to decrease each uncovered element of the matrix by e_{min}, increase each twice-covered element by e_{min}, and leave each once-covered element unaltered. All primed and starred zeros are once-covered, so that each of these zeros are still zeros after performing STEP 4. Moreover, there is at least one new uncovered zero in the cost-matrix, so that STEP 2 can be repeated. In the example in Fig. 1, STEP 2 reiterates twice before finding an uncovered zero without a starred zero in its row, so that the algorithm proceeds with STEP 3. In this step the algorithm construct a special sequence S of altering primed and starred zeros, the stars in S are deleted, and the primes of S are starred accordingly. The resulting set of starred zeros is larger, by one, than the previous set of starred zeros. The algorithm switches to STEP 1, and it is checked whether the starred zeros describe a complete node assignment or not. After performing STEP 3 and going back to STEP 1 in Fig. 1, the starred elements represent the minimum-cost node assignment and the algorithm terminates. Note that in the example of Fig. 1 the dimensions of the graphs to be matched are the same. In the general case, where $n > m$ or $m > n$ the costs for $\max\{0, n - m\}$ node deletions and $\max\{0, m - n\}$ node insertions have to be added to the minimum-cost node assignment.

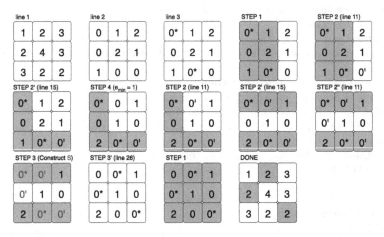

Fig. 1. Finding the minimum-cost node assignment by Munkres' algorithm – shown is the 3×3 cost-matrix C, where each element represents the cost of a single node assignment: $C_{i,j} = c(u_i \to v_j)$. Covered rows and columns are shaded.

3.2 Plain-Munkres and Adjacency-Munkres

Obviously, one can use the algorithm described in Section 3.1 to approximately compute the edit distance of a pair of graphs. In this algorithm only node, but no edge information is taken into account, i.e. no information of the adjacent edges is used for constructing the cost matrix:

$$C_{i,j} = c(v_i \to u_j)$$

In the remainder of this paper we will refer to this version of Munkres' algorithm as PLAIN-MUNKRES. To enhance the algorithm one can use information about the edges adjacent to a node when the initial cost matrix is constructed in line 1. Whenever the cost of a node assignment $c(u_i \to v_j)$ has to be calculated for the cost matrix, one can add the costs resulting from the minimum-cost edge assignment for all edges connected to u_i and v_j. To find the minimal edge assignment Munkres' algorithm can be used again. Instead of the costs for node assignments the costs for edge substitutions are used to generate the cost matrix. So for every entry $C_{i,j}$ in the initial cost matrix one has to compute Munkres' algorithm recursively for all the adjacent edges of u_i and v_j. In the remainder of this paper we will call this method ADJACENCY-MUNKRES, because it uses the information of the adjacent edges of the nodes to be assigned (e_{vi} stands for all edges connected to v_i and e_{uj} for all edges connected to u_j):

$$C_{i,j} = c(v_i \to u_j) + \min\{\sum c(e_{vi} \to e_{uj})\},$$

where $\min\{\sum c(e_{vi} \to e_{uj})\}$ is computed by Munkres' algorithm using a new cost-matrix C_e with entries $C_{i,j} = c(e_{vi} \to e_{uj})$.

4 Experimental Results

The method of Munkres compared to brute force methods is expected to perform significantly faster. Instead of an exponential time-complexity, Munkres' algorithm finds the node assignment in polynomial time. However, one has to keep in mind that the optimal node assignment need not be the optimal edit path. Thus, the computation of graph edit distance by Munkres' algorithm is suboptimal in the sense that only approximate edit distance values are obtained. The crucial question is whether the accuracy of the suboptimal distance remains sufficiently accurate for pattern recognition applications.

We propose a k-nearest-neighbor classifier in conjunction with edit distance to address the classification problems considered in this paper. Given a labeled training set of graphs, an unknown test graph is assigned to the class that occurs most frequently among the k nearest graphs (in terms of edit distance) from the training set. The fundamental idea of k-nearest-neighbor classifiers is based on the assumption that graphs of the same class are similar in their structure. In the experiments, insertion and deletion costs are set to constant values, and substitution costs are set proportional to the Euclidean distance of a pair of labels. First, we optimize these parameters on a validation set, and subsequently, the optimized parameters are applied to the unknown test set. This validation phase is accomplished with the optimal tree-search algorithm proposed in [8,9]. This optimal calculation of the edit distance will serve us as a reference system (REFERENCE METHOD). In [19] two simple, but effective modifications of the standard edit distance algorithm are proposed that allow us to suboptimally compute edit distance in an efficient way. These suboptimal modifications are referred to as BEAMSEARCH and PATHLENGTH. Both versions are controlled by a parameter t, which is used to control the trade-off between suboptimality of the distances and the matching time. The methods proposed in this paper are also compared to these two algorithms.

4.1 Letter Database

The first database consists of graphs representing distorted letter drawings. In this experiment we consider the 15 capital letters that consists of straight lines

Table 1. Letter Database: Classification accuracy and average running time

Method	1-NN	3-NN	5-NN	Time [ms]
REFERENCE METHOD [8,9]	82.0	80.7	82.7	468.0
BEAMSEARCH ($t = 100$) [19]	82.0	80.7	82.0	18.0
PATHLENGTH ($t = 1.1$) [19]	79.3	82.7	84.0	8.0
PLAIN-MUNKRES	81.3	84.0 ∘	83.4	0.2
ADJACENCY-MUNKRES	82.7	84.0 ∘	82.7	2.8

∘ Statistically significantly better than the reference system ($\alpha = 0.05$).

only *(A, E, F, ...)*. For each class, a prototype line drawing is manually constructed. To obtain aribtrarily large sample sets of drawings with arbitrarily strong distortions, distortion operators are applied to the prototype line drawings. This results in randomly shifted, removed, and added lines. These drawings are then converted into graphs in very intuitive manner by representing lines by edges and ending points of lines by nodes. Each node is labeled with a two-dimensional attribute giving its position. The graph database used in our experiments is composed of a training set, a validation set, and a test set, each of size 150. The letter graphs consist of 4.6 nodes and 4.4 edges on the average. In Table 1 we give the classification accuracy of three nearest-neighbor classifiers and the average time it takes to compute a single edit distance of two graphs. It turns out that the speedup of all suboptimal methods is remarkable, while the classification accuracy remains high. As a matter of fact, the accuracy is not statistically significantly worse than that of the reference system. Moreover, both versions of Munkres outperform the reference system with a 3-NN classifier by achieving statistically significantly better results. Note that PLAIN-MUNKRES is more than 2000 times faster than the reference system and 40 times faster than the subopotimal tree search variant PATHLENGTH. The fact that suboptimal methods can outperform exact methods concerning the accuracy is caused by the ability of the suboptimal methods to correct misclassifications by assigning higher costs to pairs of graphs from different classes than the exact algorithm. To understand this phenomenon one has to distinguish between *inter-class distances*, that is edit distances between graphs from different classes, and *intra-class distances*, that is edit distances between graphs from the same class. As above-mentioned, there exists an inexpensive edit path between two structurally similar graphs, while for structurally different graphs a high cost edit path is required. By computing the edit distance by means of Munkres' algorithm the low cost *intra-class distances* remain low, while higher cost *inter-class distances* remain high or even increase. This increasing of *inter-class distances* is the main reason for enhancing the classification accuracy.

4.2 Image Database

For a more thorough evaluation of the classification accuracy, we apply the proposed methods of Munkres to the problem of image classification. Images are

Table 2. Image Database: Classification accuracy and average running time

Method	1-NN	3-NN	5-NN	Time [ms]
REFERENCE METHOD [8,9]	46.3	48.2	44.4	20
BEAMSEARCH($t = 5$) [19]	48.2	50.0	44.4	6
PATHLENGTH($t = 1.1$) [19]	48.2	50.0	46.3	5
PLAIN-MUNKRES	50.0	44.4	42.6	2
ADJACENCY-MUNKES	48.2	44.4	42.6	15

converted into attributed graphs by segmenting them into regions, eliminating regions that are irrelevant for classification, and representing the remaining regions by nodes and the adjacency of regions by edges [20]. The image database consists of 5 classes (*city, countryside, people, snowy, streets*) and is split into a training set, validation set, and test set of size 54 each. The image graphs consist of 2.8 nodes and 2.5 edges on the average. The nearest-neighbor classification performance and the running time of the edit distance computation are given in Table 2. Note that in this application ADJACENCY-MUNKRES is not able to improve the accuracy of PLAIN-MUNKRES and is much slower than all the other suboptimal methods. We observe that the classification accuracy achieved by Munkres' method remains nearly unaffected when compared to the reference method, i.e. the results of the suboptimal methods are not significantly worse than those of the exact computation by REFERENCE METHOD. The speedup of Munkres' algorithm is again remarkable. It is ten times faster than the reference system.

4.3   Fingerprint Database

Finally we apply the proposed algorithms to the difficult problem of fingerprint classification. For this purpose, we construct graphs from fingerprint images of the NIST-4 database by extracting characteristic regions in fingerprints and converting the result to attributed graphs [11]. We use a validation set of size 300 and a training set and a test set of size 500 each. The fingerprint graphs consist of 5.2 nodes and 8.6 edges on the average. In our experiment, we address the 4-class problem (*arch, left loop, right loop, whorl*). Note that for this dataset, the exact edit distance and the distances computed by PATHLENGTH ($t = 1, 2, 3, \ldots$) cannot be computed because the search tree grows too large. So the reference system is given by the method proposed in [11]. The results in Table 3 clearly demonstrate the power of the new algorithms proposed in this paper. One single computation of the edit distance is more than 100 times faster than the reference system, and the classification accuracy remains high and exceeds the results of the reference system with the 3-NN classifiers significantly. Because of specific properties of our predefined cost function for fingerprint graphs, it is useless to run ADJACENCY-MUNKRES. In contrast to other data sets, the edges of the fingerprint graphs yield the crucial information about the structure of a given graph. Node attributes are not relevant and therefore the costs of all node edit operations are set to zero. Hence, the initial entries in the cost matrix C are given by

Table 3. Fingerprint Database: Classification accuracy and average running time

Method	1-NN	3-NN	5-NN	Time [ms]
REFERENCE METHOD [11]	82.6	83.8	84.8	11.0
BEAMSEARCH ($t = 30$) [19]	85.2	87.6 ∘	87.4 ∘	32.0
PATHLENGTH ($t = $ ARBITRARY) [19]	–	–	–	–
PLAIN-MUNKRES	83.6	87.2 ∘	86.6	0.1

∘ Statistically significantly better than the reference system [11] ($\alpha = 0.05$).
– Empty entries indicate computation failure due to lack of memory.

the costs of the edge operations. Consequently, a recursive computation for the adjacent nodes becomes unnecessary and therefore only PLAIN-MUNKRES is used in conjunction with edit distance computation for fingerprint graphs.

Summarizing we conclude that the proposed methods based on Munkres' algorithm are remarkably faster than the exact methods for computing the edit distance. Fortunately on all tested datasets the classification performance is not negatively affected. The reason for good performance is that edit distances that are increased by our suboptimal algorithms are often from graphs from different classes. Therefore the classification accuracy does not decrease. On the contrary, in some cases the suboptimal methods outperforms the exact ones.

5 Conclusion

One of the main problems in graph matching is that standard algorithms for computing the similarity of graphs – e.g. tree search algorithms – are exponential in the number of involved nodes. Hence, such algorithms are applicable to small graphs only. In this paper, we propose a novel procedure based on Munkres' algorithm to compute the edit distance. Since our method considers the implied edge operations only locally, the resulting edit distances are suboptimal. Despite the suboptimality, the proposed method accomplishes the graph matching process much faster and leaves classification rates nearly unaffected. Sometimes, the accuracy is even higher than that of the exact method. This means that the suboptimality mainly leads to an increase of inter-class distances, while intra-class distances, which are more relevant for classification, are not strongly affected. The speedup of the suboptimal algorithm compared to exact edit distance algorithms is massive. We provide an experimental evaluation and demonstrate the usefulness of our method on semi-artificial line drawings, on scenery images, and fingerprints.

Acknowledgements

This work has been supported by the Swiss National Science Foundation NCCR program *Interactive Multimodal Information Management (IM) 2* in the Individual Project *Multimedia Information Access and Content Protection*.

References

1. Conte, D., Foggia, P., Sansone, C., Vento, M.: Thirty years of graph matching in pattern recognition. Int. Journal of Pattern Recognition and Artificial Intelligence 18(3), 265–298 (2004)
2. Umeyama, S.: An eigendecomposition approach to weighted graph matching problems. IEEE Transactions on Pattern Analysis and Machine Intelligence 10(5), 695–703 (1988)
3. Luo, B., Wilson, R., Hancock, E.R.: Spectral embedding of graphs. Pattern Recognition 36(10), 2213–2223 (2003)
4. Christmas, W.J., Kittler, J., Petrou, M.: Structural matching in computer vision using probabilistic relaxation. IEEE Transactions on Pattern Analysis and Machine Intelligence 17(8), 749–764 (1995)
5. Suganthan, P.N., Teoh, E.K., Mital, D.P.: Pattern recognition by graph matching using the potts MFT neural networks. Pattern Recognition 28(7), 997–1009 (1995)
6. Cross, A., Wilson, R., Hancock, E.: Inexact graph matching using genetic search. Pattern Recognition 30(6), 953–970 (1997)
7. Gori, M., Maggini, M., Sarti, L.: Exact and approximate graph matching using random walks. IEEE Transactions on Pattern Analysis and Machine Intelligence 27(7), 1100–1111 (2005)
8. Bunke, H., Allermann, G.: Inexact graph matching for structural pattern recognition. Pattern Recognition Letters 1, 245–253 (1983)
9. Sanfeliu, A., Fu, K.S.: A distance measure between attributed relational graphs for pattern recognition. IEEE Transactions on Systems, Man, and Cybernetics (Part B) 13(3), 353–363 (1983)
10. Neuhaus, M., Bunke, H.: Edit distance based kernel functions for structural pattern classification. Pattern Recognition 39(10), 1852–1863 (2006)
11. Neuhaus, M., Bunke, H.: An error-tolerant approximate matching algorithm for attributed planar graphs and its application to fingerprint classification. In: Fred, A., Caelli, T., Duin, R., Campilho, A., de Ridder, D. (eds.) Structural, Syntactic, and Statistical Pattern Recognition. LNCS, vol. 3138, pp. 180–189. Springer, Heidelberg (2004)
12. Ambauen, R., Fischer, S., Bunke, H.: Graph edit distance with node splitting and merging and its application to diatom identification. In: Hancock, E., Vento, M. (eds.) GbRPR 2003. LNCS, vol. 2726, pp. 95–106. Springer, Heidelberg (2003)
13. Robles-Kelly, A., Hancock, E.R.: Graph edit distance from spectral seriation. IEEE Transactions on Pattern Analysis and Machine Intelligence 27(3), 365–378 (2005)
14. Boeres, M.C., Ribeiro, C.C., Bloch, I.: A randomized heuristic for scene recognition by graph matching. In: Ribeiro, C.C., Martins, S.L. (eds.) WEA 2004. LNCS, vol. 3059, pp. 100–113. Springer, Heidelberg (2004)
15. Sorlin, S., Solnon, C.: Reactive tabu search for measuring graph similarity. In: Brun, L., Vento, M. (eds.) GbRPR 2005. LNCS, vol. 3434, pp. 172–182. Springer, Heidelberg (2005)
16. Justice, D., Hero, A.: A binary linear programming formulation of the graph edit distance. IEEE Trans. on Pattern Analysis ans Machine Intelligence 28(8), 1200–1214 (2006)
17. Munkres, J.: Algorithms for the assignment and transportation problems. Journal of the Society for Industrial and Applied Mathematics 5, 32–38 (1957)
18. Hart, P.E., Nilsson, N.J., Raphael, B.: A formal basis for the heuristic determination of minimum cost paths. IEEE Transactions of Systems, Science, and Cybernetics 4(2), 100–107 (1968)

19. Neuhaus, M., Riesen, K., Bunke, H.: Fast suboptimal algorithms for the computation of graph edit distance. In: Ribeiro, C.C., Martins, S.L. (eds.) WEA 2004. LNCS, vol. 3059, pp. 163–172. Springer, Heidelberg (2005)
20. Le Saux, B., Bunke, H.: Feature selection for graph-based image classifiers. In: Marques, J.S., Pérez de la Blanca, N., Pina, P. (eds.) IbPRIA 2005. LNCS, vol. 3523, pp. 147–154. Springer, Heidelberg (2005)

Matching of Tree Structures
for Registration of Medical Images

Jan Hendrik Metzen[1], Tim Kröger[2], Andrea Schenk[2], Stephan Zidowitz[2],
Heinz-Otto Peitgen[2], and Xiaoyi Jiang[3]

[1] University of Bremen, Faculty of Mathematics and Computer Science,
Robert Hooke Str. 5, 28359 Bremen, Germany
jhm@informatik.uni-bremen.de
[2] MeVis Research GmbH, Universitaetsallee 29, 28359 Bremen, Germany
[3] University of Münster, Faculty of Mathematics and Computer Science,
Einsteinstraße 62, 48149 Münster, Germany

Abstract. Many medical applications require a registration of different
images of the same organ. In many cases, such a registration is accomplished by manually placing landmarks in the images. In this paper we
propose a method which is able to find reasonable landmarks automatically. To achieve this, nodes of the vessel systems, which have been
extracted from the images by a segmentation algorithm, will be assigned
by the so-called association graph method and the coordinates of these
matched nodes can be used as landmarks for a non-rigid registration
algorithm.

1   Introduction

Medical imaging methods like computed tomography (CT) and magnetic resonance imaging (MRI) are able to provide three-dimensional, digital images of
organs like liver or lung. In many medical applications, it is desirable to provide
the user different images of the same organ. For instance, this might be reasonable if a lung shall be examined both in the inhaled and exhaled state or if there
are CT as well as MRI images of the same organ. Another possible application
is the monitoring of an organ over a long time period by regularly scanning the
organ.

Because of respiration, heartbeat etc. it is possible that the position and
shape of an organ might considerably differ between two scannings. That makes
it difficult to detect regions in the images which depict the same part of the
organ. Such a mapping between different images of an organ is called *registration*.
Following Hill [1], registration is the process of transforming different image data
sets into one coordinate system.

In order to allow an automatic registration of image data sets it is necessary
to use properties of the organs which are invariant against respiration, heartbeat
etc. The vessel system of the organs is one possibility for such an invariant feature. The position and extension of these vessel systems might change but their
structure remains (nearly) constant. The identification of corresponding areas

F. Escolano and M. Vento (Eds.): GbRPR 2007, LNCS 4538, pp. 13–24, 2007.
© Springer-Verlag Berlin Heidelberg 2007

in these structures provides structural information, which eases the registration of the image data sets. The vessel systems of liver and lung (e. g. portal vein of the liver and bronchi respectively) are trees and it is therefore possible to apply structural pattern recognition methods to this problem of *matching tree structures.*

2 State of the Art

There are different approaches for matching of tree structures. We briefly sketch four of them:

Pelillo et al. [2,3] used the so-called association graph for detecting maximal subtree isomorphism of rooted and free trees. Possible assignments of tree nodes are represented as nodes of the association graph. Two nodes of the association graph are connected via an edge if the corresponding assignments are consistent. Two assignments are considered to be consistent if the topological relationship between the two involved nodes in both trees is equal. The definition of this topological relationship differs for matching of rooted and free trees; for free trees it is exactly the topological distance[1] of two nodes while for rooted trees the difference of node levels[2] additionally has to be equal. In the derived association graph a maximal clique is detected by applying pay-off monotonic dynamics from evolutionary game theory on a continuous formulation of the problem obtained by the Motzkin-Straus theorem [4].

Bartoli et al. [5] and Pelillo et al. [6] proposed an extension of the association graph approach to achieve many-to-one and many-to-many matchings of attributed trees. Many-to-many matching means that a group of nodes can be assigned (contracted) to a single node in the other tree while many-to-one matching means that this relationship holds only in one direction. The latter might be adequate when matching a tree to a model. For the purpose of many-to-many matching, each node is rated with a value $r \in \mathbb{R}^+$ which depends only on its attributes. A group of nodes can be contracted if the ratings of all but one node fall below a certain threshold.

Tschirren et al. [7] proposed a method for matching of human airway trees. They first perform a pruning step on the trees in order to improve their comparability and subsequently a rigid registration in order to map the trees into the same coordinate system. Thereafter, a hierarchical approach using an association graph is applied to the data to accomplish a matching. While this approach performs well for some input trees, there are two major drawbacks: The method needs robust ways of detecting major branchpoints in the trees and relies on the invariance of the topological distance. The former may be possible in airway trees but proved to be difficult in liver vessel systems. The latter is susceptible to erroneous segmentation due to noise (see Figure 1).

---

[1] The topological distance is defined as the number of edges, which have to be traversed on the unique path from one node to another.

[2] The level of a tree node is defined as the number of edges, which have to be traversed on the unique path from a node to the root of the tree.

Charnoz et al. [8] proposed an algorithm, which performs a parallel depth first search on both trees. During this process, a set of matching hypotheses is generated. All matching hypotheses contained in this set are rated and the global optimal matching is chosen. Because of the exponential number of potential matching hypotheses it is crucial to study only the most promising hypotheses. Hence, on each step of the depth first search only a certain number of those node assignments is considered, whose node attributes are similar. This selection based on local properties is risky since the global optimal matching may contain assignments of nodes whose attributes differ significantly. If one correct assignment of nodes close to their root is missed, the whole generated set of matching hypotheses may be significantly flawed. The approach attained promising results when matching one tree segmented from the Visible Mans liver with a perturbed version of itself. Nevertheless, the applicability to trees segmented from real patient data with significant differing topology and node attributes (see Section 3) remains open.

In this work, we will propose an enhanced version of the association graph approach. The association graph approach proposed by Pelillo et al. [2,3] has been applied successfully to matching of shock trees and shape-axis trees. When applying it to matching of anatomical vessel trees, there are a few additional issues which have to be considered:

- Due to noise and motions of the recorded organs, there will nearly always be errors in the extracted tree structures: Noise, for example, might result in additional branches, which do not exist in the real organ. An additional branch can influence topological distances as well as the level of nodes (see Figure 1).
- Since the resolutions of CT and MRI have increased continuously in the recent years, the tree structures might be quite big, i. e. have up to 1000 nodes.

A many-to-many matching as described by Bartoli and Pelillo [5,6] might in principle deal with the first issue; however, it is not clear how to obtain a rating based on node attributes that has small values for exactly those nodes, which were erroneously detected.

Therefore, the described approaches which use an association graph are not well suited for the purpose of matching of anatomical vessel trees. The main contribution of this work is to enhance the association graph approach so that the method can deal with the issues mentioned above. Consequently, it is adequate for the matching of anatomical vessel trees.

3 Methods

3.1 Association Graph

In this section, we will present our enhanced version of the association graph approach. We start with the definition of the enhanced tree association graph, and explain it in more details later on:

Definition 1 (Tree Association Graph). *Let* $T_1 = (V_1, E_1, w_1)$ *and* $T_2 = (V_2, E_2, w_2)$ *be two rooted trees. We define the* tree association graph $G = (V_A, E_A)$ *of* T_1 *and* T_2 *with respect to a set of unary constraints* C_F *and a set of binary constraints* C_G *as:*

1. $V_A = \{v_a \in V_1 \times V_2 | \sum_{f_i \in C_F} w_i f_i(v_a) \geq 0.5\}, \sum_i w_i = 1, w_i \in [0, 1]$
2. $E_A = \{(v_a, v_b) \in V_A \times V_A | \sum_{g_j \in C_G} v_j g_j(v_a, v_b) \geq 0.5\}, \sum_j v_j = 1, v_i \in [0, 1]$

An interpretation of this definition is as follows: A node $v_a = (v_{a1}, v_{a2}) \in V_A$ represents the potential assignment of the tree nodes v_{a1} and v_{a2}. Consequently, the set V_A is the set of all promising assignments of nodes in V_1 to nodes in V_2. An assignment is considered to be *promising* if it fulfills a set of unary constraints, formalising similarity measures for two nodes, to a certain extent. Each unary constraint measures the similarity of two nodes and rates this similarity with a value between 0 and 1. If the rating is close to 1, the two nodes are nearly indistinguishable for this constraint, while two nodes with rating close to 0 possess only little similarity. Furthermore, each of these constraints has a parameter which determines its selectivity[3]. Since there are different unary constraints and not all of them might be equally decisive, each unary constraints is weighted with a factor $w_i \in [0, 1]$. The sum of all weights has to be 1. An assignment is promising if the average weighted sum of the rating of all unary constraints for this assignment is greater than or equal to 0.5. Hence, in contrast to the original definition [2,3,5,6] of an association graph $V_A \neq V_1 \times V_2$ but $V_A \subsetneq V_1 \times V_2$. The reduction of the cardinality of V_A enables us to apply the association graph approach to trees with a great number of nodes. A collection of unary constraints is introduced in Subsection 3.2.

Two nodes $v_a = (v_{a1}, v_{a2})$ and $v_b = (v_{b1}, v_{b2})$ of an association graph are connected via an edge e iff the assignments $v_{a1} \leftrightarrow v_{a2}$ and $v_{b1} \leftrightarrow v_{b2}$ are *consistent* to each other. Two assignments are considered to be consistent if they fulfill a set of binary constraints, which formalise consistency measures for two assignments, to a certain extent. Analogue to unary constraints, each binary constraint gives a rating between 0 and 1, is controlled by a selectivity parameter, and is weighted with a factor v_j. Two assignments are consistent if the average weighted sum of the ratings of all binary constraints for the corresponding association graph nodes is greater than or equal to 0.5. Some binary constraints are proposed in Section 3.3.

The aim of matching of tree structures is to determine a set of node assignments of maximum cardinality in which each two assignments are pairwise consistent. Such a set corresponds directly to a maximum clique[4] in the association graph. Since detecting a maximum clique of a graph is known to be $\mathcal{NP}$-hard [9],

---

[3] "Selectivity" means how similar two nodes must be to be rated with a value greater than 0.5.

[4] Given an arbitrary undirected graph $G = (V, E)$, a subset of vertices $C \subseteq V$ is called a clique if all its vertices are mutually adjacent; a clique is said to be maximum if there is no other clique with higher node cardinality in the graph.

we have to apply approximate methods. As discussed in Section 2, Pelillo gives a promising approach for approximating the maximum clique based on applying pay-off monotonic dynamics from evolutionary game theory on a continuous formulation of the problem obtained by the Motzkin-Straus theorem [4]. This approach has been successfully adopted to this problem.

3.2 Unary Constraints

Unary constraints detect promising assignments of tree nodes in an early stage of the matching process. Obviously, two nodes which might be assigned should be similar. In this context, similarity means that some local properties of the nodes should differ only to a small amount. Examples of such local properties are:

- The level of a tree node.
- The length (or diameter) of the discharging edge of a node. The discharging edge is the unique incident edge of a node, whose other endpoint is a node with minor level. The length of an edge has been computed during the segmentation process and is defined as the length of the anatomical vessel, which corresponds to the edge.
- The size of the induced subtree of a node. The induced subtree of a node is that part of the tree which is rooted in this node. The size of a subtree is defined as the sum of the length of all edges of this subtree.
- The spatial coordinate of a node.

Unfortunately, all of these properties are perturbed by noise, movements of the organs and resultant errors during the segmentation and extraction of the tree structure. For example, the spatial coordinate of a node in the lung is heavily influenced by respiration. The size of a subtree depends among other things on the resolution of the medical imaging method. Methods with higher resolution are able to detect more subtle parts of the tree structure, which increase the size of a subtree. Noise might cause the erroneous detection of a node, which can influence the level of nodes as well as the length of edges (see Figure 1).

Thus, these local properties proposed above are no good choices for unary constraints (as shown by the results presented in Section 4) . A more reliable property, though not local, is the spatial course of the unique path from a node v to a reference node[5] r of the tree. This path traverses a set of edges, whose spatial course is described by a sequence of skeleton points. Hence, the spatial course of each path can be described as a polyline $S = [v, v_1, ..., v_n, r]$ in $\mathbb{R}^3$ consisting of the concatenation of the skeleton points of the edges. We will now define a similarity measure for two of these polylines. This similarity measure compares the curve progression but is independent of the length of the polylines. Therefor, each polyline will be normalised first by transforming it as follows:

---

[5] A reference node is a tree node, which can be reliably detected in both trees. Possibilities for reference nodes are the root or the first major bifurcation node of a tree.

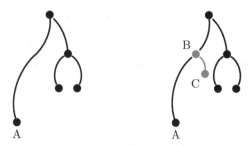

Fig. 1. Erroneously detected nodes influence the level of nodes as well as the length of edges. In this example, the erroneous node C causes an additional bifurcation node B, which splits one edge into two parts. Thus, the level of the leftmost node A has increased by one as well as its topological distance to all other tree nodes. Furthermore the length of its discharging edge has decreased considerably, because this edge has been split into two parts.

1. The polyline will be displaced by $-v$ so that its start point is located in the origin of the coordinate system.
2. Each skeleton point of the polyline will be scaled by $\frac{1}{||r-v||_2}$ so that the start and end point of the polyline will have the euclidean distance 1.
3. The polyline will be rotated in order that its end point will be $x_1 = (1,0,0)^t$. For this purpose, the angle α between the vectors $r - v$ and $x = (1,0,0)^t$ will be computed. Thereafter, the polyline will be rotated by α around the axis, which is orthogonal to $r - v$ and x. This axis is uniquely determined unless $r - v = \pm(1,0,0)^t$. In this case, an axis which is orthogonal to x can be arbitrarily chosen, because α has to be 0 or π.

After this transformation, all polylines $S = [s_0, s_1, ..., s_{n-1}, s_n]$ will begin at $s_0 = (0,0,0)^t$ and end at $s_n = (1,0,0)^t$. We define a partial-polyline as $S_i = [s_0, s_1, ..., s_i]$ and the length of a polyline S_i as $||S_i|| = \sum_{j=1}^i ||s_j - s_{j-1}||_2$. Furthermore, we define a parametrisation f of a polyline as follows: $f : [0; 1] \to \mathbb{R}^3$ with $f(\frac{||S_i||}{||S||}) = s_i$, in particular $f(0) = (0,0,0)^t$ and $f(1) = (1,0,0)^t$. The other values of f are linearly interpolated: For $t \in \left] \frac{||S_i||}{||S||}, \frac{||S_{i+1}||}{||S||} \right[$ let $f(t) = (1-\alpha)s_i + \alpha s_{i+1}$ with $\alpha = \frac{t||S||-||S_i||}{||S_{i+1}||-||S_i||}$.

An obvious similarity measure for two polylines S_1 and S_2 with parametrisation f_1 and f_2 is the integral $\int_0^1 ||f_1(t) - f_2(t)||_2 \, dt$. This integral corresponds to the area between the polylines. However, as shown in Figure 2, this similarity measure is not appropriate since there is still one degree of freedom which affects the value of the integral. Therefore, a better similarity measure is

$$d = \min_{\phi \in [0,2\pi]} d(\phi) \quad \text{with} \quad d(\phi) = \int_0^1 ||f_1(t) - A_\phi f_2(t)||_2 \, dt,$$

where A_ϕ is a matrix which describes a rotation by ϕ degrees around the x-axis. Unfortunately, this optimisation of the angle ϕ is computationally expensive.

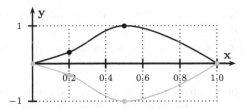

Fig. 2. Two normalized polylines which are to be compared. If the rotational degree of freedom around the x-axis is not considered, even very similar polylines can be rated as very different. In this example, if one of the polylines is rotated by π around the x-axis, the polylines would be identical.

This is critical since we have to apply each unary constraints very often[6], which requires the computation of d each time.

Instead of computing the area between two polylines, the directions of the polylines can be compared as basis for a similarity measure, too. Since the polylines are piecewise linear, the derivative $f'(t)$ exists almost everywhere and f' is piecewise constant. For each $t \in [0, 2\pi]$, $g_i(t) = \frac{f_i'(t)}{||f_i'(t)||_2}$ is a unit vector which describes the direction of the polyline's tangent for the parameter value t. We define the similarity of two unit vectors a and b as their dot product $a^t b$. A property of the dot product of two unit vectors is $-1 \leq a^t b \leq 1$ with $a^t b = 1 \Leftrightarrow a = b$. Hence, the similarity of two polylines can be defined as $\frac{1}{2} \int_0^1 (1 + g_1(t)^t g_2(t)) dt \in [0, 1]$. The greater the value of the integral, the more similar are the two polylines. However, this similarity measure is influenced by the rotational degree of freedom, too. Fortunately, the optimisation problem

$$\max_{\phi \in [0, 2\pi]} \frac{1}{2} \int_0^1 (1 + g_1(t)^t A_\phi g_2(t)) dt \tag{1}$$

with a matrix A_ϕ which describes a rotation around the x-axis can be solved analytically [10]. Using $g_i = (g_{i1}, g_{i2}, g_{i3})^t$ the solution is given by

$$\max_{\phi \in [0, 2\pi]} \int_0^1 (g_1(t)^t A_\phi g_2(t)) dt = C + \sqrt{D^2 + E^2}$$

with $C = b_{11}$, $D = b_{22} + b_{33}$, $E = b_{23} - b_{32}$ and $b_{jk} = \int_0^1 g_{1j}(t) g_{2k}(t) dt$.

3.3 Binary Constraints

Binary constraints determine whether two assignments $v_{a1} \leftrightarrow v_{a2}$ and $v_{b1} \leftrightarrow v_{b2}$ are *consistent* to each other, i.e. whether the corresponding nodes of the association graph shall be connected via an edge. First ideas for binary constraints might be:

---

[6] If the two trees have n and m nodes respectively, a unary constraint has to be applied $O(nm)$ times.

- The *euclidean distance* d_e of both nodes in the two trees should be similar, i. e. $d_e(v_{a1}, v_{b1}) \approx d_e(v_{a2}, v_{b2})$
- The *topological distance* d_t of both nodes in the two trees should be similar, i. e. $d_t(v_{a1}, v_{b1}) \approx d_t(v_{a2}, v_{b2})$

Unfortunately, the same disturbing factors (noise, respiration, heartbeat) which affect the local properties in Section 3.2 influence the mentioned properties, too. Thus, we have to develop more sophisticated similarity measures. It has turned out that the comparison of the directions of two polylines can be easily transferred onto the binary case and results in a robust constraint: Instead of comparing the two polylines which connect the nodes with the reference nodes of their trees we can simply compare the two polylines which connect v_{a1} and v_{b1}, and v_{a2} and v_{b2}, respectively.

Since binary constraints might be applied very often[7] the comparison of the directions of two polylines can be too expensive for large trees. In this case, the comparison of the *length* $l(v_a, v_b)$ of the connecting path[8] of two nodes v_a and v_b is an option, which can be computed very fast and yields acceptable results. Another similarity measure, which is computationally cheap, is the *curvature* of the connecting path. We define the curvature c of a path connecting two nodes v_a and v_b as $c(v_a, v_b) = l(v_a, v_b)/d_e(v_a, v_b)$.

4 Results

The proposed method has been implemented and tested in the *MeVisLab* research and prototyping platform (http://www.mevislab.de/). To provide a basis for the empirical evaluation of the quality of the matchings achieved by the association graph method, we use two matchings as ground truth, which were created manually by human experts. These datasets match a pair of bronchi trees (in inhaleted and exhaleted state) and a pair of portal vein trees (one CT and one MRI image). In each case, both trees have roughly 200 nodes and both ground truths contain 34 assignments of significant nodes distributed all over the trees.

We analyzed the quality of unary and binary constraints. In this section we present mainly the results obtained with the portal vein dataset. Nevertheless, the results for the bronchi tree were similar. First, we present a comparison[9] of

---

[7] If the two trees have n and m nodes respectively, in the worst case a binary constraint has to be applied $O(n^2 m^2)$ times. The worst case occurs, if most of the possible assignments have been rated as promising by the unary constraints.

[8] The length of a path is defined as the sum of the length of its edges.

[9] Since the proposed algorithm can only be tested with a combination of unary and binary constraints, we show here the results of a particular unary constraint in combination with a specific set of binary constraints. Since we used the same set of binary constraints for all tests, we can compare the quality of the unary constraint. Furthermore, it has turned out that the relative quality of the unary constraints does not depend on the choice of the set of binary constraints.

Table 1. Comparison of different unary constraints. For each constraint, the selectivity parameter σ has been chosen empirically in order to optimise the resulting matching. Neither the comparison of the diameter of the discharging edge (*EdgeDiameter*) nor the size of the induced subtree (*SubtreeSize*) of a node yield in promising results. Better results are obtained by the non-local constraints. These constraints require the choice of a reference node. For the results presented here, the root has acted as reference node. The runtime was measured on a Pentium4 3.2GHz.

Constraint	σ	Correct	Error	Runtime
EdgeDiameter	1.5	0	3	0.08 sec
SubtreeSize	1.1	4	2	2.36 sec
PathCurvature	1.08	7	3	0.16 sec
PolylineArea	1.04	11	1	14.79 sec
PolylineDirection	1.15	18	4	2.58 sec

the different unary constraints proposed in Subsection 3.2. We will use the following names for the different types of constraints (regardless if unary or binary) proposed in Section 3:

EdgeDiameter: Compares the diameter of the discharging edge of a node.
SubtreeSize: Compares the size of the subtree induced by a node.
PathLength: Compares the length of the (unique) path connecting two nodes.
PathCurvature: Compares the curvature of the (unique) path connecting two nodes.
PolylineArea: Compares the area between two (normalized) polylines.
PolylineDirection: Compares the direction of two (normalized) polylines.

As can be seen in Table 1, the best results yield from the *PolylineArea* and the *PolylineDirection* constraints. As expected, the runtime of the *PolylineArea* constraint is (because of the expensive optimization of the angle ϕ) significantly larger than the runtime of the other constraints. Further improvements of the matching can be achieved when combining several unary constraints. It has turned out that it is optimal to combine the *PathCurvature* constraint with the *PolylineDirection* constraint in the proportion 1 : 4 (see Table 2).

Similarly, we have analysed the quality of several binary constraints proposed in Subsection 3.3 (in combination with the optimal set of unary constraints, see above). The results are shown in Table 3 and Table 4.

Subsequently, the set of constraints and parameters, which was optimal for the portal vein datasets, has been applied to the bronchi tree datasets to evaluate if the approach performs equally well for datasets originated from other organs. The results are summarised in Table 5, and Figure 3 depicts the matching of the portal vein trees. The quality of both matchings is satisfying, whereas the matching of the portal vein dataset is superior to that of the bronchi trees. Since this discrepancy remains when differing weights and selectivity parameters, the

Table 2. Combination of unary constraints. In parentheses, the weight of the respective constraint is shown. A combination of the *PolylineArea* and the *PolylineDirection* does not improve the quality of the matching. It is likely that this is due to the fact that both constraints assess similar properties and a combination comprises a lot of redundancy. Better results are obtained when combining one of these constraints with the *PathCurvature* constraint. The best result is achieved when combining this constraint with the *PolylineDirection* constraint in the proportion 1 : 4. This proportion is grounded in the fact that the *PolylineDirection* constraint provides better results than the *PathCurvature* constraint when applied solely.

Configuration	Correct	Error
PolylineArea(0.5) : PolylineDirection(0.5)	16	4
PathCurvature(0.5) : PolylineDirection(0.5)	13	2
PathCurvature(0.2) : PolylineDirection(0.8)	20	2
PathCurvature(0.2) : PolylineArea(0.8)	10	4

Table 3. Comparison of binary constraints. For each constraint, the selectivity parameter σ has been chosen empirically in order to optimise the resulting matching. The *PathLength* constraint alone does not yield in good results as it is not distinctive enough. Better results are obtained by the *PathCurvature* and *PolylineDirection* constraints. The *PolylineArea* constraint is omitted here, since its runtime is (due to the large number of applications of a binary constraint) very high and it does not yield in better results than the *PolylineDirection* constraint.

Constraint	σ	Correct	Error	Runtime
PathLength	1.7	2	4	6 sec
PathCurvature	1.3	18	2	8.9 sec
PolylineDirection	1.13	17	2	203.97 sec

Table 4. Combination of binary constraints. The weight of the respective constraint is shown in parentheses. Even though the *PathLength* constraint alone does not seem to be promising, it can improve the results of the *PathCurvature* and *PolylineDirection* constraints. The best results are achieved when combining the *PolylineDirection* with the *PathLength* constraint in the proportion 3:1.

Configuration	Correct	Error
PolylineDirection(0.5) : PathCurvature(0.5)	17	4
PathCurvature(0.75) : PathLength (0.25)	20	2
PolylineDirection(0.75) : PathLength(0.25)	17	0

matching of bronchi trees is apperently intrinsically more complex than matching of portal veins. One possible reason is that bronchi trees are dichotomous structures, which means that they have many subtrees which are very similar to each other. An example for such similar subtrees are the right and the left part of a bronchi tree, which split in the first major branchpoint.

Table 5. Results of the matching process: In both cases, the matching algorithms assigned approximately 80 nodes. In case of the portal vein dataset, 17 of these assignments were covered by the ground truth and none of them was incorrect. In case of the bronchi tree, 25 of the assignments were covered by the ground truth and 4 of them were incorrect. Nevertheless, these erroneous assignments match nodes, which are topological neighbours and geometrically at close quarters and therefore, it was even difficult for humans to determine the correct matching of these nodes.

Dataset	Portal Vein	Bronchi Tree
Correct	17	21
Error	0	4
Runtime	207.35 sec	369.23 sec

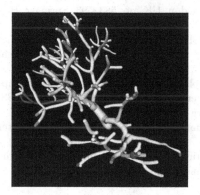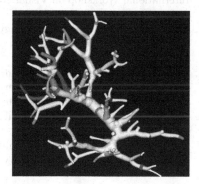

Fig. 3. Matching of two portal veins. Depicted are two portal veins as well as the attained matching. Assigned nodes are dyed in the same colour. If a set of node assignments induces a subtree isomorphism, the whole subtrees are dyed with the same colour.

5   Conclusions

The results indicate that the proposed method is able to achieve good results for typical examples of vessel trees. A significant ratio of tree nodes is assigned in an admissible amount of time. The acquired matching covers most parts of the trees and contains no or only few errors. In our future work, we will analyze if the acquired landmarks are able to improve the registration of the image datasets. Furthermore, it will be examined if the method acquires promising results for harder datasets like images taken during the regeneration of a liver after living liver donation. Also, it will be analyzed if a rigid registration and a hierarchical decomposition of the trees (as described in [7]) can reduce the required computation time.

References

1. Hill, D.L.G., Batchelor, P.G., Holden, M., Hawkes, D.J.: Medical image registration. Physics in Medicine and Biology 46, R1–R45 (2001)
2. Pelillo, M., Siddiqi, K., Zucker, S.W.: Matching hierarchical structures using association graphs. IEEE Trans. Pattern Anal. Mach. Intell. 21(11), 1105–1120 (1999)
3. Pelillo, M.: Matching free trees, maximal cliques, and monotone game dynamics. IEEE Trans. Pattern Anal. Mach. Intell. 24(11), 1535–1541 (2002)
4. Pelillo, M.: Replicator equations, maximal cliques, and graph isomorphism. Neural Comput. 11(9), 1933–1955 (1999)
5. Bartoli, M., Pelillo, M., Siddiqi, K., Zucker, S.W.: Attributed tree homomorphism using association graphs. ICPR 02, 2133–2136 (2000)
6. Pelillo, M., Siddiqi, K., Zucker, S.W.: Many-to-many matching of attributed trees using association graphs and game dynamics. In: IWVF-4. Proceedings of the 4th International Workshop on Visual Form, London, UK, pp. 583–593. Springer, Heidelberg (2001)
7. Tschirren, J., Mclennan, G., Palagyi, K., Hoffman, E.A., Sonka, M.: Matching and anatomical labeling of human airway tree. IEEE Transactions on Medical Imaging 24(12), 1540–1547 (2005)
8. Charnoz, A., Agnus, V., Malandain, G., Soler, L., Tajine, M.: Tree matching applied to vascular system. In: Brun, L., Vento, M. (eds.) GbRPR 2005. LNCS, vol. 3434, pp. 183–192. Springer, Heidelberg (2005)
9. Karp, R.: Reducibility among combinatorial problems. Complexity of Computer Computations, pp. 85–103 (1972)
10. Metzen, J.H.: Matching von Baumstrukturen in der medizinischen Bildverarbeitung. Diploma Thesis, University of Münster (2006)

Graph-Based Methods for Retinal Mosaicing and Vascular Characterization

Wendy Aguilar, M. Elena Martinez-Perez, Yann Frauel[1], Francisco Escolano, Miguel Angel Lozano[2], and Arturo Espinosa-Romero[3]

[1] Department of Computer Science, Insituto de Investigaciones en Matemáticas Aplicadas y en Sistemas, Universidad Nacional Autónoma de México, CP 04500
[2] Robot Vision Group. Universidad de Alicante, Spain
[3] Faculty of Mathematics, Universidad Autónoma de Yucatán. Anillo Periférico Norte. Tablaje Cat. 13615, Colonia Chuburná Hidalgo Inn. Mérida, Yucatán, México

Abstract. In this paper, we propose a highly robust point-matching method (*Graph Transformation Matching* - GTM) relying on finding the consensus graph emerging from putative matches. Such method is a two-phased one in the sense that after finding the consensus graph it tries to complete it as much as possible. We successfully apply GTM to image registration in the context of finding mosaics from retinal images. Feature points are obtained after properly segmenting such images. In addition, we also introduce a novel topological descriptor for quantifying disease by characterizing the arterial/venular trees. Such descriptor relies on diffusion kernels on graphs. Our experiments have showed only statistical significance for the case of arterial trees, which is consistent with previous findings.

1 Introduction

Image registration is a fundamental problem to several image processing and computer vision applications. In the case of medical retinal images, two images taken before and after laser surgery can be registered to detect location of lesions. Two images taken from the same eye but at different times can be registered to quantify the severity of disease and the progression of therapy. A series of images of the same retina can be registered to form a mosaic image giving a complete view of the retina. The quantification of some diseases is commonly made by measuring differences in tree vascular structure between groups. Measurements can be either geometrical or topological [9].

In this paper we present a twofold graph-based method applied to retinal image analysis: 1) a graph-based point-matching algorithm to allow the construction of mosaics in order to have a larger view of the retina, and 2) a tree vascular characterization in order to find structural differences such as those found by other authors [4,5]. The graph-based point-matching algorithm, *Graph Transformation Matching* (GTM), is an efficient method for dealing with high rates of outlying matches, which is the main drawback of some continuation methods for graph matching like *Softassign*, or its kernelized version [7], which

F. Escolano and M. Vento (Eds.): GbRPR 2007, LNCS 4538, pp. 25–36, 2007.
© Springer-Verlag Berlin Heidelberg 2007

optimize quadratic cost functions. In addition to the latter, we propose a new spectral descriptor relying on diffusion kernels for characterizing the topologies of normotensive and hypertensive arterial trees.

The rest of the paper is organized as follows. Section 2 describes the process of segmenting vascular trees (arterial and venular) and extracting the key features from retinal images, which is a critical step for further analysis. Section 3 describes the *Graph Transformation Matching*: (i) the basic approach, (ii) the optimized algorithm, and (iii) the recovery phase. We test the algorithm in the context of retinal image alignment and compare it with *Softassign*. Section 4 is devoted to describe the interpolation method used for the alignment to generate a mosaic view. Section 5 presents our spectral descriptor for vascular characterization and show its adequacy. Finally, in Section 6 we outline our conclusions.

2   Image Feature Extraction

Previous work has been done in order to analyze and extract features from retinal images. The process consists of two main steps: i) the segmentation of blood vessels to generate a binary image, and ii) the analysis of the binary image. Features of interest herein are branching and crossing points as feature points for mosaicing and, extraction of arterial and venous vessel trees for characterization. Blood vessels are segmented based on a multi-scale analysis of the first and second derivatives of the images in combination with a region growing algorithm [10]. Figs. 1(a) and 1(c) show two different views of retinal images and Figs. 1(b) and 1(d) their segmented binary images, respectively. The optic disc region is on the bottom-left, vessels are tracked from this area outwards.

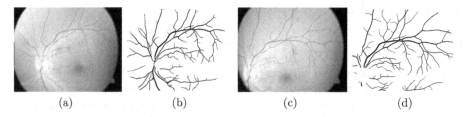

(a)   (b)   (c)   (d)

Fig. 1. Retinal images. (a) *img15* and (c) *img21* are two different views from the same eye-ball and, (b) and (d) are their respective segmented binary images.

A semi-automatic method to measure and quantify geometrical and topological properties of continuous vascular trees on binary retinal images was developed on a previous work [9]. The analysis of the binary image involves: i) labelling each vessel tree, ii) detecting significant points, iii) extracting the vessel tree by a chain code tracking method and iv) measuring geometrical and topological parameters.

Labelling each vessel tree involves thinning the segmented binary image to produce its skeleton. Three types of significant points in the skeleton must be detected: terminal, bifurcation and crossing points. In a first pass, skeleton pixels with only one neighbor in a 3×3 neighborhood are labelled as terminal points and pixels with 3 neighbors are labelled as candidate bifurcation points. Fig. 2(a) shows the skeleton of the tree with the candidate points marked with circles.

Because vessel crossing points appear in the skeleton as two bifurcation points very close to each other, a second pass is made using a fixed size window centered on the candidate bifurcations. The number of intersections of the skeleton with the window frame determine whether the point is a bifurcation or a crossing. After this process a chain code is used to label the rest of the skeleton points in order to track the tree. Fig. 2(b) shows the branching and crossing points marked with circles over the skeleton and the tree tracked on black. Finally, after the tracking process the selected tree is isolated. Fig. 2(c) shows an arterial tree extracted. The user should select the tree to be tracked and decided if it is an arterial or a venous tree.

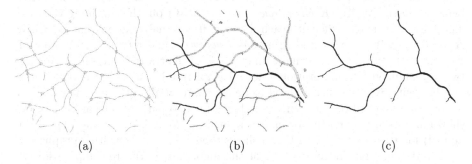

(a)   (b)   (c)

Fig. 2. Binary image analysis: (a) skeleton with candidate points marked with circles, (b) branching and crossing points marked with circles and tracked tree marked in black. (c) arterial tree extracted.

3   Graph Transformation Matching

Once significant points have been detected in both images, we proceed to get an initial matching between them. This can be done by making a cross-correlation around each significant point and matching with the point having the highest correlation value. From this process two sets of corresponding points $P = \{p_i\}$ and $P' = \{p'_i\}$ of size N (where p_i matches p'_i) are found. The *Graph Transformation Matching* (GTM) algorithm consists of two phases: a pruning and a recovering phase. In the pruning phase, a *median K-NN graph* $G_P = (V_P, E_P)$ is computed as follows: vertices $V_P = v_1, ..., v_N$ are given by the positions of the N corresponding points. A non-directed edge (i, j) exists when p_j is one of the K closest neighbors of p_i and also $\|p_i - p_j\| \leq \eta$. Being $\eta = med_{(l,m) \in V_P \times V_P} \|p_l - p_m\|$ the median of all distances between pairs of vertices. The first condition states that a vertex can just validate the structure of its closest neighbors, while the second condition restricts the proximity of validation which filters structural deformations due to outlying points. If there are not K vertices that support the structure of p_i then this vertex is disconnected completely. The graph G_P, which is not necessarily connected, has the $N \times N$ adjacency matrix A_{ij}, where $A_{ij} = 1$ when $(i,j) \in E_P$ and $A_{ij} = 0$ otherwise. Similarly, the graph $G_{P'} = (V_{P'}, E_{P'})$ for points p_i' has adjacency matrix A_{ij}', also of dimension $N \times N$ because of the one-to-one initial matching M.

GTM relies on the hypothesis that outlying matchings in M may be removed, with high probability, by iteratively applying a simple structural criterion [1]. Thus, GTM iterates as follows: (i) selecting an outlying matching; (ii) removing matched features corresponding to the outlying matching, as well as this matching itself, and (iii) recomputing both *median K-NN graphs*. Structural disparity is approximated by computing the residual adjacency matrix $R_{ij} = |A_{ij} - A_{ij}'|$ and selecting column $j^o = argmax_{j=1...N} \sum_{i=1}^{N} R_{ij}$, the one that yields the maximal number of different edges in both graphs. The selected structural outliers are the features forming the pair (p_{j^o}, p_{j^o}'). Thus, we remove v_{j^o} from G_P and v_{j^o}' from $G_{P'}$, and (p_{j^o}, p_{j^o}') from M. Then, after decrementing N, a new iteration begins, and the median *K-NN graphs* are computed from the surviving vertices. The algorithm stops when it reaches the null residual matrix, that is, when $R_{ij} = 0, \forall i, j$. It seeks for finding a *consensus graph* and returns the number of vertices of this graph. Fig. 3 shows an example of the transformation process for two retinal images, from iteration 0 (initial graphs) to iteration 71 (final identical graphs), with $K = 4$ which showed to be adequate in all our experiments.

An example of initial and final matchings for two pairs of retinal images are shown in Figs. 4 (a) and (b) for images named *img15* and *img12*, and Figs. 4 (c) and (d) for *img15* and *img21*. Fig. 5 shows their respective resulting graphs.

Considering that the bottleneck of the algorithm is the re-computation of the graphs, which takes $O(N^2 log N)$ (the same as computing the median at the beginning of the algorithm) and also that the maximum number of iterations

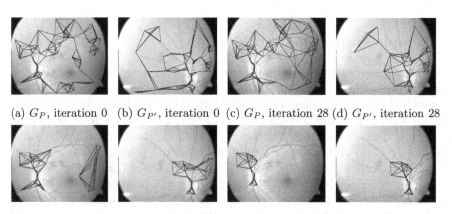

(a) G_P, iteration 0   (b) $G_{P'}$, iteration 0   (c) G_P, iteration 28   (d) $G_{P'}$, iteration 28

(e) G_P, iteration 56   (f) $G_{P'}$, iteration 56   (g) G_P, iteration 71   (h) $G_{P'}$, iteration 71

Fig. 3. Graph Transformation process example. G_P corresponds to the *median K-NN graph* of *img15* and $G_{P'}$ for *img12* during iterations 0, 28, 56 and 71.

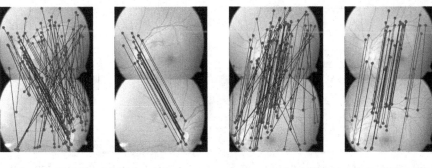

(a) Initial matches,   (b)Final matches,   (c) Initial matches,   (d) Final matches,
img15 with *img12*   *img15* with *img12*   *img15* with *img21*   *img15* with *img21*

Fig. 4. GTM initial and final matches from the pruning phase: (a)-(b) *img15* with *img12* and (c)-(d) *img15* with *img21*

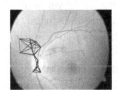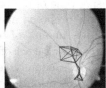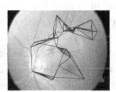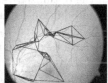

(a) G_P for matching (b) $G_{P'}$ for matching (c) G_P for matching (d) $G_{P'}$ for matching
img15 with *img12*   *img15* with *img12*   *img15* with *img21*   *img15* with *img21*

Fig. 5. Graphs resulting from GTM in its pruning phase: (a)-(b) *image15* with *image12* and (c)-(d) *image15* with *image21*

is N, the worst case complexity is $O(N^3 log N)$. A significant improvement in the reconstruction of the graphs was made. It consists of replacing the graph representation (adjacency matrix) by three new structures: i) a matrix O_F of size $n \times n$ where rows represent output edges for each vertex (ordered by distances smaller than the median and where the first K locations represent the actual output edges for that vertex, the rest are the potential next connections), ii) an array of linked lists I_F of dimension n with input edges for each vertex and iii) an array N_F of dimension n that keeps a reference to the next available edge to connect in O_F (initially with value $K + 1$).

Two implementations were made (in C language), one corresponding to the brute force algorithm and the other to the optimized version. The time required for the algorithm depends on two variables: i) the number of initial vertices (which is directly related to the number of matches) and ii) the number of iterations. To test the significance of the optimization, two experiments were made by fixing one factor at a time. Time results for both implementations (in seconds) are shown in Fig. 6, (a) and (b). These graphs make evident the improvement in time due to the optimization of the algorithm. The difference was

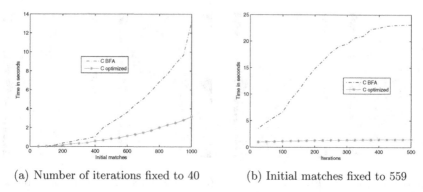

(a) Number of iterations fixed to 40   (b) Initial matches fixed to 559

Fig. 6. Graphs of times reported for GTM algorithms implemented in C: (a) brute force and (b) optimized

more significant in the second case (b) suggesting an almost constant behavior of the optimized version versus the increasing time of the brute force version.

In addition to the efficiency desired for the proposed matching algorithm, it was also wanted to make it robust to a large amount of outliers. To test the latter, an experiment was made consisting of taking a set of 60 correct matches and introduce different percentages of outliers randomly generated, from 10% to 95%. Results showed that in the case of 85% of outliers, the *Graph Transformation* algorithm could recover 48 correct matches (from the 60 available) with just 2 mismatches. In the case of 95% of outliers it recovered 24 with 2 mismatches. Images used in this paper contain at most 74% of natural outliers. For some applications it is wanted to recover high quantities of correct matches

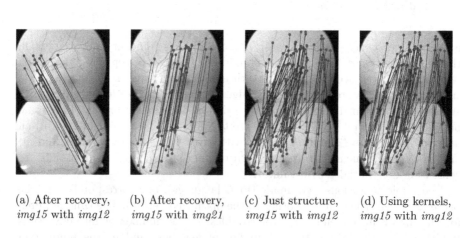

(a) After recovery,   (b) After recovery,   (c) Just structure,   (d) Using kernels,
img15 with *img12*   *img15* with *img21*   *img15* with *img12*   *img15* with *img12*

Fig. 7. (a)-(b) Results from the recovery phase when matching (a) *img15* with *img12* and (b) *img15* with *img21*. (c)-(d) *Softassign* results from matching *img15* with *img12* using (c) just structure and (d) kernels.

tolerating some mismatches. For these cases, here we propose a second phase of the GTM algorithm named *Recovering Phase*. It consists in taking final graphs and matches obtained from the pruning phase and adding iteratively all rejected matches one at a time, recomputing the corresponding *K-NN graphs* and testing the residual matrix condition. If this condition is satisfied, then this match is considered as correct. Otherwise, it is discarded. Figs. 7 (a) and (b) show the resulting matchings after the recovery phase for *img15* with *img12* and *img15* with *img21*. Compare these results with those showed in Figs. 4 (b) and (d).

We compared our results versus those obtained from the *Softassign* algorithm. Figs. 7(a) and (b) present the results obtained from matching *img15* with *img12*, using *Softassign* with (a) just structure and (b) kernels. In the case of using kernels and costs, the resulting matching was exactly the same as the input, suggesting that the costs completely influence the matching process. Figs. 7(c) and (d) show the contribution of the recovery phase.

The algorithm was also tested with other retinal images. Some of the results are shown in Figs. 8 and 9.

img13 with *img15*　　*img13* with *img27*　　*img14* with *img15*　　*img14* with *img27*

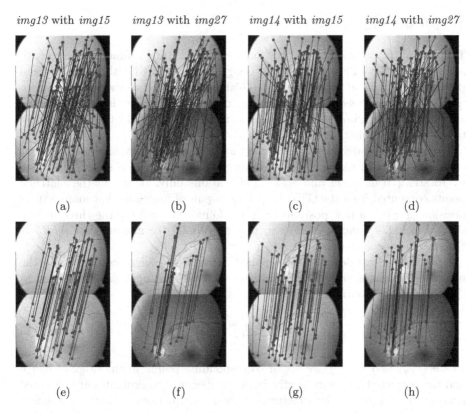

(a)　　　　　　(b)　　　　　　(c)　　　　　　(d)

(e)　　　　　　(f)　　　　　　(g)　　　　　　(h)

Fig. 8. Other matching results. (a)-(d) show the initial matches and (e)-(h) show the corresponding final matches from *GTM* algorithm.

img13 with img15   img13 with img27   img14 with img15   img14 with img27

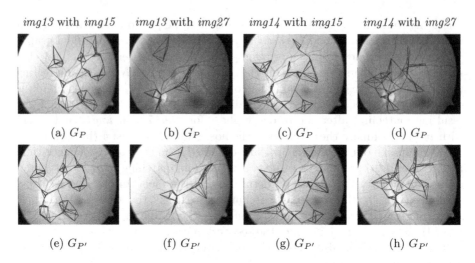

(a) G_P (b) G_P (c) G_P (d) G_P

(e) $G_{P'}$ (f) $G_{P'}$ (g) $G_{P'}$ (h) $G_{P'}$

Fig. 9. Resulting graphs from GTM algorithm of other results

4   Mosaicing

The retinal images used in this section were taken using a fundus camera with a 50° field of view (Zeiss FF 450 IR). Images were acquired with a CCD camera (Sony Power HAD 3CCD Color Video Camera) attached to the fundus camera with 768×576 pixels in size. Since the surface of the retina is curved, almost spherical, the interimage transformation model use to build the mosaic must take this into account.

A quadratic model surface is a good approximation that allows corrections of misalignment of blood vessels that cannot be corrected on spherical surface by rotation, translation and scale modifications only. Based on the matched points computed from the GTM technique, a pair of functions that maps a pixel position (x, y) to a new position (x', y') is found. We will use the three images showed in Fig. 4 (img15, img12 and img21) to build the mosaic using img15 as the reference coordinate system.

A quadratic transformation is applied between the reference image and the images to be changed into its coordinate system. The functions are defined as the polynomial equations:

$$
\begin{aligned}
x' &= \sum_{i=0}^{m} \sum_{j=0}^{m-i} a_{ij} x^i y^j, \\
y' &= \sum_{i=0}^{m} \sum_{j=0}^{m-i} b_{ij} x^i y^j, \quad m = 2
\end{aligned}
\tag{1}
$$

where (x, y) and (x', y') are set of corresponding points in the original image and the corrected image, respectively. A number of N coordinates are collected from both images and, by substituting them in equations 1, a set of N linear equations with respect to the coefficients $a_{i,j}$ and $b_{i,j}$ are obtained. The solution is found by least squares criterion and an approximation of the functions that describe the correct mapping is found. The mapping is expressed in terms of a pair of transformation maps, (M_x, M_y), that record the correspondence between every pixel in the original image $I(x, y)$ and the corrected image $I'(x, y)$:

$$I'(x, y) = I(M_x(x, y), M_y(x, y)) \tag{2}$$

The final image obtained from mosaicing $img12$, $img15$ and $img21$ is shown in Fig. 10.

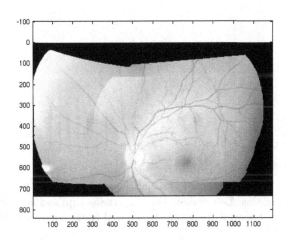

Fig. 10. The result mosaic after matching images $img15$ with $img12$ and with $img21$

5   Spectral Vascular Characterization

Previous approaches [4,9] to quantify changes in vascular trees due to disease are mainly focused on geometric parameters (branching angles, length-to-diameter ratio, diameter, and so on) whereas topological measurements (e.g. asymmetry, number of terminal branches) have been less studied [5,9]. We propose a new topological measure based on the computation of diffusion kernels [2,6] on binary trees. Images used in this section were obtained from [8]. 20 images (10 normotensive and 10 hypertensive) were taken using a fundus camera with 30° field of view (Kowa FX-50R). Photographic negatives were digitized and reduced to 533 × 509 pixels size. Vessel trees were segmented and extracted as described in section 2.

Given a tree $T = (V, G)$ with: m vertices, adjacency matrix A and Laplacian L where: $L_{ij} = -1$ when $(i, j) \in E$, $L_{ii} = \sum_i A_{ij}$, and $L_{ij} = 0$ otherwise, the diffusion kernel K of A is defined by the matrix exponentiation of the Laplacian:

$$K \equiv e^{-\beta L} = I_m + L + \frac{1}{2!}L^2 + \frac{1}{3!}L^3 + \dots, \tag{3}$$

being I_m the $m \times m$ identity matrix. In the particular case of an infinite binary tree, we have that the kernel value between two vertices i and j depends on the length d_{ij} of the unique path connecting them [3,6]:

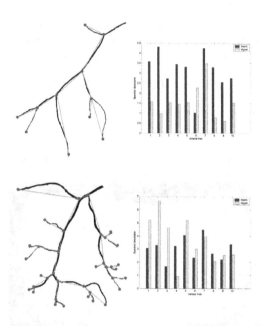

Fig. 11. Characterization. Left: Examples of hypertense (up) and normotensive (down) arterial trees. Right: Spectral Descriptor for arterial (up) and venular (down) trees.

$$K_{ij} = K^2(d_{ij}) = \frac{2}{\pi} \int_0^\pi \frac{e^{-\beta(1-\cos x)}}{4(1-\cos^2 x)} \sin x [\sin(d_{ij}+1)x - \sin(d_{ij}-1)x] dx$$

$$K_{ii} = K^2(0) = \frac{1}{\pi} \int_0^\pi e^{-\beta(1-\cos x)} dx$$

(4)

Equation 4 shows a well known property of kernels on trees. Alike in terms of random walks, K_{ij} can be regarded as the probability that a *lazy* random walk reaches j from i, in binary trees, where there is a unique path between two nodes. Such probability depends on the length of the path, because the branching factor is constant. In the infinite case, K_{ii} is constant but this does not happen in the finite case where K_{ii} is given by $1 - \sum_j K_{ij}$ and the distribution of distances d_{ij} is not uniform. In addition, for the i−th node in the tree, the coefficients $\{K_{ij}, j = 1 \dots m\}$ define a spectral signature, in this case a probability density function specifying the probability of reaching each node in the tree from the i−th one. For the application domain considered in this paper, we propose to quantify the comparison of the $\{K_{ij}\}$ probability distribution functions of vertices from normotensive and hypertensive trees with those associated to nodes in perfect binary trees with the same cardinality, say m. Such comparison will yield path-length invariance when characterizing a given binary tree, but it requires to properly mapping each tree on the perfect one (completely balanced), a sort of simple tree matching. After such matching, if the i−th node in the tree is mapped on i' in the perfect tree, a simple *spectral tree descriptor* is given by:

$$SD(T) = \sum_{i=1}^{m}(H_i^{K_T} - H_{i'}^{K_P})^2 \text{ being } H_i^K = \sum_{j=1}^{m} K_{ij} \log K_{ij}. \qquad (5)$$

where: K_T is the diffusion kernel of the tree and K_P is the one of the perfect binary tree with the same number of nodes; H^{K_T} and H^{K_P} are the entropies of the probability distributions associated to the vertices and induced by the kernels [7]. In Fig. 11 we show our preliminary characterization experiments. We have considered both arterial and venular trees for 10 normotensive and 10 hypertensive subjects. After the Wilcoxon rank test, only statistically significant differences ($p < 0.0211$) where found in arterial trees. However, when comparing descriptors corresponding to venular trees the differences where no significant ($p = 0.3847$). These findings are consistent with the evidence already reported of geometrical and topological parameters [8], and show that the new spectral descriptor may complement existing measures.

6   Conclusions

We have presented both a novel graph-based algorithm applied to the match of correspondence points for retinal mosaicing, and a structural characterization of retinal tree vasculature. On one hand, the results obtained for mosaicing depend on the amount of overlapping regions between images and on the number of matching points. Further work has to be done in order to obtained larger views and more robust results using more matching points to get better mapping approximations. The GTM algorithm has demonstrated to be a fast, robust and reliable method for feature matching. On the other hand, the analysis of purely topological indices of retinal blood vessels made by [8,9], using the same database, showed that arterial and venous trees are asymmetric and indices such as the number of terminal edges not including the root: N_T, the total sum of external path length in the tree: P_e, and the total number of external-internal edges: E_I, can characterize normal arterial trees from those of hypertensive subjects. They displayed no differences in topology for venous trees between groups. However, it is known that the latter topological indices are tree size dependent and thus a normalization factor should be applied. On the contrary, the spectral descriptor presented in this paper, which is consistent with previous findings, is size invariant and it has proved to be effective. Such descriptor can be considered another useful measurement as a complement for the analysis of geometry.

Acknowledgments

Authors would like to thank to *Clinica Lomas Altas* at México, D.F. who provided us with the images used for mosaicing and to *Department of Clinical & Cardiovascular Pharmacology* at St. Mary's Hospital, London for the clinical images use in the vascular characterization.

References

1. Aguilar, W.: Object recognition based on the structural correspondence of local features. Master's thesis, UNAM, Mexico city (2006)
2. Chung, F.R.K.: Spectral graph theory. In: Conference Board of Mathematical Science CBMS, Providence, RI, American Matematical Society, vol. 92 (1997)
3. Chung, F.R.K., Yau, S.-T.: Coverings, heat kernels and spanning trees. Electronic Journal of Combinatorics, 6 (1999)
4. Gelman, R., Martinez-Perez, M.E., Vanderveen, D.K., Moskowitz, A., Fulton, A.: Diagnosis of plus disease in retinopathy of prematurity using retinal image multisacle analysis (risa). Investigative Ophthalmology & Visual Science 46(12), 4734–4738 (2005)
5. Hughes, A.D., Martinez-Perez, M.E., Jabba, A.-S., Hassan, A., Witt, N.W., Mistry, P.D., Chapman, N., Stanton, A.V., Beevers, G., Pedrinelli, T., Parker, K.H., Thom, S.A.M.: Quantification of topological changes in retinal vascular architecture in essential and malignant hypertension. Journal of Hypertension 24(5), 889–894 (2006)
6. Kondor, R., Lafferty, J.: Diffusion kernels on graphs and other discrete input spaces. In: Proc. Intl. Conf. on Machine Learning, Los Altos CA, pp. 315–322 (2002)
7. Lozano, M.A., Escolano, F.: A significant improvement of softassign with diffusion kernels. In: Fred, A., Caelli, T.M., Duin, R.P.W., Campilho, A., de Ridder, D. (eds.) Structural, Syntactic, and Statistical Pattern Recognition. LNCS, vol. 3138, pp. 76–84. Springer, Heidelberg (2004)
8. Martinez-Perez, M.E.: Computer Analysis of the Geometry of the Retinal Vasculature. PhD thesis, Imperial College, London, UK (2001)
9. Martinez-Perez, M.E., Hughes, A.D., Stanton, A.V., Thom, S.A., Chapman, N., Bharath, A.A., Parker, K.H.: Retinal vascular tree morphology: A semi-automatic quantification. IEEE Transactions on Biomedical Engineering 49(8), 912–917 (2002)
10. Martinez-Perez, M.E., Hughes, A.D., Thom, S.A., Bharath, A.A., Parker, K.H.: Segmentation of blood vessels from red-free and fluorescein retinal images. Medical Image Analysis 11(1), 47–61 (2007)

Stereo Vision for Obstacle Detection: A Graph-Based Approach

P. Foggia[2], Jean-Michel Jolion[3], A. Limongiello[1], and M. Vento[1]

[1] Dip. di Ingegneria dell'Informazione ed Ingegneria Elettrica
Università di Salerno, Via Ponte don Melillo, I84084 Fisciano (SA), Italy
{alimongiello,mvento}@unisa.it
[2] Dip. di Informatica e Sistemistica
Università di Napoli, Via Claudio 21, I80125, Napoli, Italy
foggiapa@unina.it
[3] Lyon Research Center for Images and Information Systems, UMR CNRS 5205 Bat. J. Verne
INSA Lyon 69621, Villeurbanne Cedex, France
Jean-Michel.Jolion@insa-lyon.fr

Abstract. We propose a new approach to stereo matching for obstacle detection in the autonomous navigation framework. An accurate but slow reconstruction of the 3D scene is not needed; rather, it is more important to have a fast localization of the obstacles to avoid them. All the methods in the literature, based on a punctual stereo matching, are ineffective in realistic contexts because they are either computationally too expensive, or unable to deal with the presence of uniform patterns, or of perturbations between the left and right images. Our idea is to face the stereo matching problem as a matching between homologous regions. The stereo images are represented as graphs and a graph matching is computed to find homologous regions. Our method is strongly robust in a realistic environment, requires little parameter tuning, and is adequately fast, as experimentally demonstrated in a comparison with the best algorithms in the literature.

Keywords: Stereo vision; Obstacle detection; Graph matching; Autonomous mobile robots; Automated guided vehicles.

1 Introduction

During the last years, the Computer Vision community has shown an increasing interest in applications like Automated Guided Vehicles (AGV) or Autonomous Mobile Robots (AMR). In the literature many approaches have been proposed for Visual Navigation of a mobile platform. DeSouza and Kak [1] provide an excellent survey of Mobile Robot Navigation; Bertozzi, *et al.* [2], and Kastrinaki, *et al.* [3], propose a survey for Vision-based intelligent vehicles. A very challenging task is the so-called *obstacle detection*, that is the detection of an obstacle in an unstructured environment and without prior knowledge of the obstacle appearance. Many authors have expressed their conviction that a robotic vision system should aim at reproducing the human vision system, and so should be based on stereo vision. The greatest advantage of stereo vision with respect to

F. Escolano and M. Vento (Eds.): GbRPR 2007, LNCS 4538, pp. 37–48, 2007.

other techniques (e.g. optical flow, or model-based) is that it produces a full description of the scene, can detect motionless and moving obstacles (without defining a complex obstacle model), and is less sensitive to the environmental changes (the major disadvantage of optical-flow techniques). The stereo vision provides a 3D representation (or at least an approximation like a 2D ½ representation) of the scene, and an interpretation of the structure can produce information about objects in the environment that may obstacle the motion.

A pair of images acquired from a stereo camera implicitly contains depth information about the scene: this is the main assumption of stereo vision, based on the binocular parallax property of the human visual system. The main difficulty is to establish a correspondence between points of the two images representing the same point of the scene; this process is called *disparity matching*. The set of displacements between matched pixels is usually indicated as *disparity map*. All the approaches, in the literature, are based on this punctual definition of the disparity. We propose an extension of that concept, namely we define a disparity value for a whole region of the scene starting from the two homologous views of it in the stereo pair. The main reason of this extension is that a punctual approach is redundant for Autonomous Mobile Robot (AMR) and Automatic Vehicle Guidance (AVG) applications. In fact, in this framework, it is not very important to have a good reconstruction of the surfaces, but it is more important to identify adequately the space occupied by each object in the scene (as soon as possible to avoid collisions), even by just assigning to it a single disparity information. Moreover the punctual approaches are lacking in robustness in some realistic frameworks, especially for video acquired from a mobile platform. Most of the algorithms available in off-the-shelf systems [23] are unable to deal with large uniform regions or with vibration of the cameras. On the other hand, some efforts have been done in the literature to improve the robustness of the algorithms, but at the price of a significant increase of the running time. Our method estimates the average depth of the whole region by an integral measure, and so has fewer problems with uniform regions than other methods have. The estimate of the position of the regions is sufficiently accurate for navigation, also in the mentioned cases, and it is fast enough for real time processing.

This paper is organized as follows: Section 2 presents the related works; Section 3 shows an overview of our approach; Section 4 is devoted to the algorithm. Finally, in Section 5 there is a discussion of experimental results on standard stereo database and also on our stereo video sequence. Conclusions are drawn in Section 6.

2 Related Works

We will present a brief description of the most important methods for stereo matching; for more details, there is a good taxonomy proposed by Scharstein and Szeliski [4], and a survey on stereo vision for mobile robots by Zhang [5]. There are two major types of techniques, in the literature, for disparity matching: the area-based and feature-based techniques. Moreover, the area-based algorithms can be classified in local and global approaches. The local area-based algorithms [6],[7],[8] provide a correspondence for each pixel of the stereo pair. They assume that each pixel is surrounded by a window of pixels having similar disparity; these windows are matched using correlation or a similar technique. They produce a dense disparity map (i.e. a map providing a disparity for each pixel), more detailed than it is needed for AMR aims. Furthermore, they can be quite unreliable, not only in homogeneous regions, but also in textured regions for an inappropriately chosen window size. On the other side, the global area-based approaches (that also yield a dense map) try to propagate disparity information from a pixel to its neighbors [9],[10], or they define and minimize some energy function over the whole disparity map [11],[12],[13]. They have a better performance in homogeneous regions, but they frequently have parameters which are difficult to set, and are highly time-consuming. The feature-based approaches [14],[15],[16] detect and match only "feature" pixels (as corner, edges, etc.). These methods produce accurate and efficient results, but compute sparse disparity maps (disparity is available only in correspondence to the feature points). AMR applications require more details, such as some information about the size of the objects; also a rough shape of the objects is needed for guiding a robot in the environment or for basic recognition tasks (e.g. in industrial applications, or for platooning of robots).

In the literature, there are also a few works that are based on a color segmentation of the stereo pair in order to enforce depth smoothness and delineate sharp depth boundaries [17]. Namely, they are methods that use segmentation to refine the output of a pixel-based algorithm in order to face problems of matching ambiguity in homogeneous color regions.

All the proposed methods, as already said, look for a punctual matching in the stereo pair. Therefore, some constraints on the pixels being considered have been introduced, since the first works on the stereopsis by Marr and Poggio [9],[14] in order to guarantee good results and to reduce the complexity. These constraints are basically of two kinds: *Geometric constraints on imaging system*, such as the *horizontal epipolar line* constraint that reduces the two-dimensional search for correspondence into a one-dimensional one; *Geometric and physical constraints on the scene*, assuming that distance varies slowly almost everywhere (the so-called *continuity* constraint), that a given point from one image can match *no more than one* point from the other image (the so-called *uniqueness* constraint), and that matching pixels have similar intensity values (the so-called *compatibility* constraint). To guarantee these constraints, the stereo pair is supposed to be acquired from a sophisticated system, so that the energy distributions of the two images are as similar as possible. Moreover, a pre-processing phase is needed, before the correspondence finding step, to compensate the hardware setup (*calibration* phase), or to assume an horizontal epipolar line (epipolar *rectification*). Unfortunately, in realistic applications of mobile robots these constraints are not easy to guarantee. The two images of the stereo pair could have a different energy distribution, the motion of the mobile platform on a rough ground could produce mechanical vibrations of the cameras, and consequently local or global perturbations between the two images, that could undermine the initial phases of calibration and rectification. We want to relax some constraints on the input images in order to consider a more realistic acquisition system, and consequently we add some other constraints on our goal. A good representation of the scene must be related to the navigation aim, so that the resolution of the problem has to be chosen in order to permit a good tread-off between efficiency and effectiveness.

3   Overview of the Strategy

The main idea of our approach is to obtain a disparity map looking at the distance between homologous regions (instead of pixels) in the stereo images. Let these regions be called *blobs*. In this way the computation of the disparity map is carried out on a set of pixels having the same spatial and color proprieties, producing a more robust performance with respect to local and global perturbations in the two images.

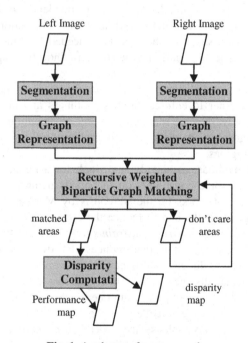

Fig. 1. A scheme of our approach

It should be noted that a blob is not an object; objects are decomposed into several blobs, so the overall shape of the object is however reconstructed, except for uncommon pathological cases. An example of pathological case can be a uniform object almost along the line of sight, but it has been satisfactorily dealt with only by global criteria optimization, which is extremely time consuming.

In our approach (see Fig. 1), the left and right images are segmented and each area identifies a node of a graph. The segmentation process is simple and very fast. In fact, we are not interested in a fine segmentation, because we do not pursue a reconstruction aim. Anyway, we need similar segments between the left and right image in order to correctly find homologous regions. This objective is possible, in fact the stereo images are likely similar because they represent two different view points of the same scene. Moreover, the segmentation process does not influence the rest of algorithm, because a recursive definition of the matching and a performance function guarantee a recovery of some segmentation problems. A bipartite graph matching between the two graphs is computed in order to match each area of the left image with only one area of the right image. This process yields a list of reliably matched areas and a list of so-called *don't care* areas. By calculating a vertical displacement between the corresponding areas, a depth is found for those areas of the reference image (i.e. left image). The list of the don't care areas, instead, could be processed in order to refine our result. As it is clear, this approach is robust even in case of uniform texture and it does not need a strong calibration process because it looks for area correspondence and not pixel correspondence. On the other hand, an effort is required in graph matching to assure real-time requirements. The application time is reduced using some constraints for a quicker computation of the bipartite graph matching.

4   The Algorithm

Segmentation and Graph representation. The first phase of the algorithm is the segmentation of the stereo images and their graph representation. We need a very fast segmentation process that produces similarly segmented areas between the left and right images. We have used a simple multi-threshold segmentation. It is essentially based on the quantization of the histogram in some color ranges (of the same size). The left and right segmentations are very similar, considering an adaptive quantizat-ion for each image according to its lighting condition. A connected component detection procedure is applied on each segmented image to obtain 4-connected areas of the same color. Each connected area (blob) is then represented as a node of an attributed graph. Each node has the following attributes:

- *colMean*: the RGB mean value of the blob (m_r, m_g, m_b);
- *size*: the number of pixels in a connected area;
- *coord*: the coordinates of the box containing the blob (*top, left, bottom, right*);
- *blobMask*: a binary mask for the pixels belonging to the blob.

It is easy to understand that a segmentation yielding many segments can be more accurate but creates lots of nodes, consequently requiring a more expensive graph matching process. On the other hand, a rougher segmentation process generates matching nodes that are very dissimilar in size and shape. As a compromise, we consider a segmentation process tuned to over-segment the image, and subsequently we filter the image in order to discard small noisy areas.

Graph Matching. Formally our matching algorithm can be described in the following way. A number of nodes is identified in each frame (left and right image) and a progressive label is associated to each node (blob). Let $G^L = \{N_0^L, \ldots, N_n^L\}$ and $G^R = \{N_0^R, \ldots, N_m^R\}$ be the two graphs representing the left and right image respectively. The solution of the spatial matching problem, between two stereo frames, is an injective mapping between a subset of G^L and a subset of G^R. The problem at hand can be represented by using a matrix whose rows and columns are respectively used to represent the nodes of the set G^L, and the nodes of the set G^R (correspondence matrix). The element (i,j) of the matrix is 1 if we have a matching between the element N_i^L with the element N_j^R, it is 0 otherwise. Each row contains no more than one value set to 1. If the j-th row or the i-th column contains only zeros, it means that it is a don't care node. The bijective mapping τ: $G^L \rightarrow G^R$ solves a suitable Weighted Bipartite Graph Matching (WBGM) problem. A Bipartite Graph (BG) [19] is a graph where nodes can be divided into two sets such that no edge connects nodes in the same set. In our problem, the first set is G^L, while the second set is G^R. Before the correspondence is determined, each node of the set G^L is connected with each node of the set G^R, thus obtaining a Complete BG. In general, an assignment between two sets G^L and G^R is any subset of $G^L \times G^R$, i.e., any set of ordered pairs whose first elements belongs to G^L and whose second elements belongs to G^R, with the constraint that each node may appear at most once in the set. A maximal assignment, i.e. an assignment containing a maximal number of ordered pairs is known as a matching (BGM) [20]. A cost function is then introduced, so that each edge (N_i^L, N_j^R) of the complete bipartite graph is assigned a cost. This cost takes into account how similar are the two nodes N_i^L and N_j^R. The lower is the cost, the more suitable is that edge. If the cost of an edge is higher than a threshold, the edge is considered unprofitable and is removed from the graph. Let us now introduce the cost function:

$$Cost = \frac{colCost + d\,i\,mCost + posCost}{3}$$

$$colCost = \frac{\displaystyle\sum_{i\in\{m\_r,m\_g,m\_b\}} \left| colMean_i^L - colMean_i^R \right|}{3*256}$$

$$d\,i\,mCost = \frac{\displaystyle\sum_{\substack{i\in\{bottom,right\}\\ j\in\{top,left\}}} \left| (i^L - j^L) - (i^R - j^R) \right|}{width + height}$$

$$posCost = \frac{\displaystyle\sum_{i\in\{bottom,right,top,left\}} \left| i^L - i^R \right|}{2*(width + height)}$$

(1)

The matching with the lowest cost among the ones with maximal cardinality is selected as the best solution. The problem of computing a matching having minimum cost is called Weighted BGM (WBGM). This operation is generally time-consuming; for this reason the search area (that is the subset of possible couples of nodes) is bounded by the **epipolar** and **disparity bands**. These constraints come from stereo vision geometry, but in our case they represent a generalization. The epipolar band is a generalization for epipolar line, that is the maximum horizontal displacement of two corresponding nodes (generally its value can be a few pixel). Disparity band, instead, is a vertical displacement, so a node of the right image can move on the left almost of α*maxdisparity pixels (with α a small integer). These two displacements are computed with respect to the centers of the bounding box of the two blobs.

The graph matching process yields a list of reliably matched areas and a list of so-called *don't care* areas. The matched areas are considered in the following section for the disparity computation. The list of the don't care areas, instead, is processed in order to group adjacent blobs in the left and right image and consequently reduce split and merge artifacts of the segmentation process. Finally, a new matching of these nodes is found. The recursive definition of this phase assures a reduction of the don't care areas in few steps, but sometimes this process is not needed because don't care areas are very small.

Disparity Computation. The disparity computation is faced superimposing the corresponding nodes until the maximum covering occurs. The overlapping is obtained moving the bounding box of the smallest region into the bounding box of the largest one; precisely, the bounding box with the minimum width is moved horizontally into the other box, and the bounding box with the minimum height is moved vertically into the other box. The horizontal displacement, corresponding to the best fitting of the matched nodes, is the disparity value for the node in the reference image (left image).

5   Experimental Results

In the literature, tests are usually performed with standard databases composed of static images, well-calibrated and acquired in uniform lighting. The Middlebury web site [21] is a good reference for some stereo images and to compare some stereovision algorithms. Nowadays, in AMR and AGV applications it is not defined a quantitative measurement for performance evaluation. In [18] it is proposed a quantitative performance evaluation for disparity map, but in case of reconstruction aims. For this reason in this paper, it is also proposed a quantitative method to compare stereo algorithms when the goal is the obstacle detection and no longer the 3D reconstruction of the scene. The following Fig. 2 shows our result on the Tsukuba DB and a comparison with other approaches. We have selected the best methods in the literature: squared differences (SSD) and graph cuts (GC) [21]. The experiments have been performed on a notebook Intel P4 1.5 GHz, 512 Mb RAM, and we have considered a resolution of 384x288 pixel.

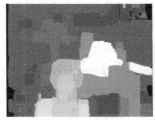  

SSD: Time < 1 sec   Graph Cut: Time 70 sec   **OUR: Time 1.14**

Fig. 2. A comparison with other approaches

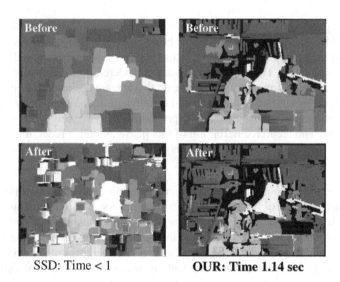

SSD: Time < 1   **OUR: Time 1.14 sec**

Fig. 3. SSD and Our approach after a vertical translation of 2 pixels

In Fig. 3 it is clear the robustness of our approach in relation to the loss of the horizontal epipolar constraint. In order to consider a quantitative comparison of the algorithms for obstacle detection aim, we define a simple module that detects the obstacles from the disparity map. Each 4-connected region with the same disparity value is identified with a bounding box and its distance from the observer. We select the obstacles as the connected regions that belong to a chosen range of distances, in fact an obstacle is an object so close to the mobile platform to forbid the navigation. Two performance index are defined in order to valuate: the capability of the algorithm to identify adequately the space occupied by each obstacle (*occupancy performance*); the correctness of depth computation (*distance performance*). For each frame of the video sequence acquired from the platform, let R_G be the real obstacle regions (*Ground Truth*), let R_D be the obstacle regions detected by the algorithm, and let R_I be the subset of regions correctly detected as obstacles by the algorithm ($R_I = R_G \cap R_D$). The occupancy performance is evaluated with the measures of *precision* and *recall*:

$$recall = \frac{R_I}{R_G}$$
$$precision = \frac{R_I}{R_D} \tag{2}$$

The distance performance is evaluated with the following *relative distance error*:

$$relative\ distance\ error\ (rde) = \frac{|detected\ distance - real\ distance|}{real\ distance} \tag{3}$$

The distance of an obstacle is related to its disparity value following the relation:

$$distance = k_{px/m} \frac{baseline \cdot focal\ lenght}{disparity} \tag{4}$$

It should be noted that for each real obstacle (*Ground Truth*) could be more than one overlapped obstacle regions detected by the algorithm. The detected distance for that obstacle is supposed to be a weighted mean distance of all the overlapped regions. The weights are set up to the sizes of each overlapping area. We report some results obtained on a realistic video acquired from our mobile platform. The video sequence (100 frames) is characterized by camera vibration, light changing, uniform obstacles.

Fig. 4. Some frames of the video sequence

The proposed method is compared with the Small Vision System (SVS) by Konolige [23,24], that is the most popular system in off-the-shelf systems. Namely, the SSD stereo matching algorithm has been implemented in SVS, taking care the real-time requirement and filtering the solution to reject false stereo matches. We consider two different version of that algorithm: *SSD* and *SSD multi-scale*.

Fig. 5. Disparity Map Results: On the left side our method, on the center side the SSD, and on the right side the SSD multi-scale

Table 1. Precision and Recall

algorithm	recall	precision	Relative distance error
our method	**0.886**	**0.439**	**0.046**
SSD	0.209	0.478	0.191
SSD multi-scale	0.469	0.349	0.180

The results in the previous table show that our method is much better than the other two, especially for the occupancy performance. In fact, as it is clear from Fig. 6 and Fig. 7, our approach can better overlap the space occupied by the real obstacles, as like the SSD multi-scale algorithm has big opening areas inside the obstacles.

Fig. 6. Some results of our obstacle detection algorithm

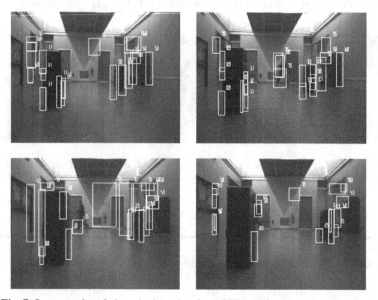

Fig. 7. Some results of obstacle detection from SSD multi-scale stereo algorithm

6 Conclusions

We have presented a stereo matching algorithm that is especially oriented towards AMR and AGV applications, providing a fast and robust detection of object positions instead of a detailed but slow reconstruction of the 3D scene. The algorithm has been experimentally validated showing an encouraging performance when compared to the most commonly used matching algorithms, especially on real-world images. Future works are oriented to test our method in outdoor environment and to develop a temporal coherence of the solution in the video sequence.

References

1. De Souza, G.N., Kak, A.C.: Vision for Mobile Robot Navigation: A Survey. IEEE Transactions on Pattern Analysis and Machine Intelligence 24(2), 237–267 (2002)
2. Bertozzi, M., Broggi, A., Fascioli, A.: Vision-based intelligent vehicles: State of the art and perspectives. Robotics and Autonomous Systems 32, 1–16 (2000)
3. Kastrinaki, V., Zervakis, M., Kalaitzakis, K.: A survey of video processing techniques for traffic applications. Image and Vision Computing 21, 359–381 (2003)
4. Scharstein, D., Szeliski, R.: A Taxonomy and Evaluation of Dense Two-Frame Stereo Correspondence Algorithms. International Journal of Computer Vision 47(1), 7–42 (2002)
5. Zhang, C.: A Survey on Stereo Vision for Mobile Robots. Technical report, Dept. of Electrical and Computer Engineering, Carnegie Mellon University, Pittsburgh, PA, 15213, USA (2002)
6. Kanade, T., Okutomi, M.: A stereo matching algorithm with an adaptive window: Theory and experiment. IEEE Transaction on Pattern Analysis and Machine Intelligence 16(9), 920–932 (1994)
7. Fusiello, A., Roberto, V.: Efficient stereo with multiple windowing. In: Proceedings of IEEE Conference on Computer Vision and Pattern Recognition, Puerto Rico, pp. 858–863 (1997)
8. Veksler, O.: Stereo matching by compact windows via minimum ratio cycle. In: Proceedings of the International Conference on Computer Vision, Vancouver, Canada, vol. I, pp. 540–547 (2001)
9. Marr, D., Poggio, T.A.: Cooperative computation of stereo disparity. Science 194(4262), 283–287 (1976)
10. Zitnick, C.L., Kanade, T.: A cooperative algorithm for stereo matching and occlusion detection. IEEE Transaction on Pattern Analysis and Machine Intelligence 22(7), 675–684 (2000)
11. Geiger, D., Ladendorf, B., Yuille, A.: Occlusions and binocular stereo. International Journal of Computer Vision 14, 211–226 (1995)
12. Roy, S.: Stereo without epipolar lines: A maximum-flow formulation. International Journal of Computer Vision 34(2/3), 1–15 (1999)
13. Boykov, Y., Veksler, O., Zabih, R.: Fast approximate energy minimization via graph cuts. IEEE Transaction on Pattern Analysis and Machine Intelligence 23(11), 1222–1239 (2001)
14. Marr, D., Poggio, T.A.: A computational theory of human stereo vision. RoyalP B(204), 301–328 (1979)
15. Grimson, W.E.L.: A computer implementation of a theory of human stereo vision. Royal B(292), 217–253 (1981)
16. Candocia, F., Adjouadi, M.: A similarity measure for stereo feature matching. IEEE Transaction on Image Processing 6, 1460–1464 (1997)
17. Tao, H., Sawhney, H.S.: Global matching criterion and color segmentation based stereo. In: WACV00, pp. 246–253 (2000)

18. Scharstein, D., Szeliski, R.: High-accuracy stereo depth maps using structured light. In: Proceedings of IEEE Computer Society Conference on Computer Vision and Pattern Recognition, Madison, WI, vol. 1, pp. 195–202 (2003)
19. Baier, H., Lucchesi, C.L.: Matching Algorithms for Bipartite Graphs. Technical Report DCC-03/93, DCC-IMECC-UNICAMP, Brazil (1993)
20. Kuhn, H.W.: The Hungarian Method for the Assignment Problem. Naval Research Logistics Quarterly 2, 83–97 (1955)
21. http://cat.middlebury.edu/stereo/
22. http://www.videredesign.com/
23. Konolige, K.: (2006) Web site. http://www.ai.sri. com/software/SVS
24. Konolige, K.: Small vision systems: hardware and implementation. Intl. Symp. On Robotics Research, pp. 111–116 (1997)

Graph Based Shapes Representation and Recognition

Rashid Jalal Qureshi, Jean-Yves Ramel, and Hubert Cardot

Université François-Rabelais de Tours
Laboratoire d'Informatique (EA 2101)
64, Avenue Jean Portalis, 37200 Tours – France
{rashid.qureshi,jean-yves.ramel,hubert.cardot}@univ-tours.fr

Abstract. In this paper, we propose to represent shapes by graphs. Based on graphic primitives extracted from the binary images, attributed relational graphs were generated. Thus, the nodes of the graph represent shape primitives like vectors and quadrilaterals while arcs describing the mutual primitives relations. To be invariant to transformations such as rotation and scaling, relative geometric features extracted from primitives are associated to nodes and edges as attributes. Concerning graph matching, due to the fact of NP-completeness of graph-subgraph isomorphism, a considerable attention is given to different strategies of inexact graph matching. We also present a new scoring function to compute a similarity score between two graphs, using the numerical values associated to the nodes and edges of the graphs. The adaptation of a greedy graph matching algorithm with the new scoring function demonstrates significant performance improvements over traditional exhaustive searches of graph matching.

Keywords: Inexact graph matching, graph based representation, shape matching.

1 Introduction

Graphs are flexible and powerful representations that have been successfully applied in computer vision, pattern recognition and related areas. In pattern recognition, graphs have been proved to be effective for representation purposes; the nodes typically represent objects or parts of objects, while the edges describe relations between objects or objects parts [1]. Concerning representation of shapes, most of the time, a classical technique is used that include representing shapes by thinnest representation of the original shape that preserves the topology. The set of idealised lines obtained by thinning [2] is called the skeleton or medial axis. The skeletons do not hold information about local thickness which can play a vital role to distinguish solid shapes and linear shapes. In addition to that, all thinning techniques generally introduce rough branches at the crossing and junction of lines (figure.1a). It is evident from figure.1b that due to boundary irregularity the skeleton disturbed a lot. Therefore the graphs build using skeletons by considering end points and junctions as vertices and branches as edges, are not reliable as small changes in the boundary can cause serious changes in the skeleton

F. Escolano and M. Vento (Eds.): GbRPR 2007, LNCS 4538, pp. 49–60, 2007.
© Springer-Verlag Berlin Heidelberg 2007

and thus ultimately can influence the generated graphs in term of extra or missing nodes and/or edges. An improvement of the topological skeleton representation of a binary shape is presented in [3] by applying parametric morphological pruning transformations to eliminate short branches on a skeleton. But more complex processing is required in the following stages which are not always effective [4].

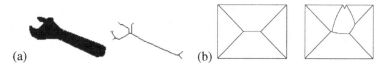

(a)   (b)

Fig. 1. The sensitivity to noise of the medial axis: small changes in the boundary may induce significant changes in the medial axis

On the other hand, a great deal of effort has been devoted over the past decades to devise efficient and robust algorithms for the fundamental problem of graph matching. An update on recent development is presented in [5]. Basically, Graph matching is the process of finding correspondence between the nodes and the edges of two graphs that satisfies some constraints ensuring that similar substructures in one graph are mapped to similar substructure in the other.

Although simple, the basic idea of graph matching suffers from a number of drawbacks. Computational complexity is an inherent difficulty of the graph matching problem. The subgraph isomorphism is proven to be NP-complete [6]. The detection of a maximum common subgraph, as well as the computation of graph edit distance is also known to be NP-hard problems [7] i.e., any of the known algorithms requires an exponential number of computational steps in the worst case.

A comparison of frequently used algorithms i.e., best backtracking methods [8] [9], a group theory based Naughty algorithm [10], VF algorithm and VF2 algorithm [11] is given in [12]. The author concludes that, it does not exit an algorithm that is definitely better than all others and it depend on graph type e.g., Naughty algorithm perform better for randomly connected graphs of quite large size, while for sparse graphs or more regular graphs like 2D meshes, VF2 is the best algorithm[13].

In this work we had tried to explore the power of graphs as a tool for structural representation as well as for the purpose of classification. Concerning structural representation of shapes, we propose a contour based approach which has a dual nature and is capable of representing both filled and linear shapes. We suggest improvements in graph matching methods to avoid exhaustive searches and we believe that the matching step should involve quantifiable similarities rather than simply "yes" or "no" type responses. A match is not merely a correspondence, but a correspondence that has been quantified according to its "goodness".

Therefore, we propose a novel method to compute a similarity measure between graphs. It can be viewed as two steps process, first, choice of a mapping, and second, computation of a similarity score for that particular mapping between the nodes and arcs of the two graphs. Finally, selecting the best mapping that yield highest similarity score as compare to others. The attractive feature of this technique is its capability to restrict the search space by using a partial similarity measure and its robustness to noise and distortion. The method is invariant to affine transformations and has a polynomial time complexity.

The remainder of this paper is organized as follow: In section 2, we introduce the structural representation construction steps. The proposed similarity measure for shape graphs matching is presented in section 3. In Section 4, we present results of experiments done in this regards. Finally, in Section 5, we draw conclusion from this work and discuss future works.

2 Graph Based Representation of Shapes

Due to their representational power, graphs are widely used to represent complex structures in computer vision and pattern recognition applications. The main aim of the representation phase is to reduce the amount of data, to outline the image component, and to represent them in such a way that insignificant differences among various instances are smoothed and to highlight the significant ones. In order to have a compact and transformation invariant representation of various shapes, we present a dual nature graph. That's to say, if the shape is solid (see figure. 2), a vectorization procedure will generate a list of vectors representing the contours of the shape and each vector will become node of the graph with their spatial relations as edges.

On the other hand if the shape is a linear one, which is composed of lines and arcs (see figure. 3), for such shapes we defined another graphic primitive-the quadrilaterals obtained by matching opposite contours vectors. The graph nodes in this case will represent quadrilaterals extracted from the shape and edges shows the spatial relationships between their neighbours. The detail steps are presented in the following subsections.

2.1 Raster to Vectors Conversion

Starting with a binary image, we describe shapes by their contours using vector primitives. For this purpose a polygonal approximation of contours of the shapes was done using Wall and Danielson iterative algorithm [14]. This step provides a sequence of vectors, segments between two consecutive control points, with its attributes like initial point $P_1(x_1,y_1)$, final point $P_2(x_2,y_2)$, length (ℓ), and angle(θ) which are stored in a chained list. The method only needs a single threshold i.e., the ratio between the algebraic surface and the length of the segments which makes this linear time algorithm fast and efficient.

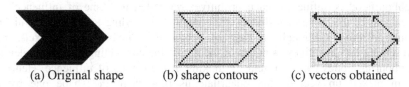

(a) Original shape   (b) shape contours   (c) vectors obtained

Fig. 2. Extraction of vectors from binary shape

2.2 Vectors to Quadrilaterals

For the representation of thin shapes or linear shapes, we used another graphic primitive - the quadrilateral. For this purpose contours are extracted from the image and vectorized. Then a matching algorithm starts that try to match opposite close vectors having similar slope and parallelism criteria for generating quadrilaterals out of them. The detail construction steps are given in [15].

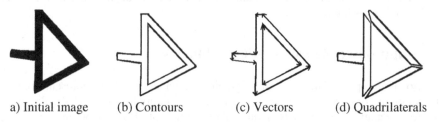

a) Initial image   (b) Contours   (c) Vectors   (d) Quadrilaterals

Fig. 3. Construction of quadrilaterals

2.3 Graph Generation

In our attributed relational graph, the graphic primitives extracted from shape are represented by graph nodes and the relationships between these graphic primitives are represented by arcs between such nodes. Both nodes and arcs are associated with attributes corresponding to properties (features) of primitives and their mutual relationships respectively.

We defined our attributed relational graph G as a 4-tuple : $G = (V, E, \alpha, \beta)$, where

- V is the finite set of vertices,
- $E \subseteq V \times V$ is the set of edges,
- $\alpha : V \rightarrow A_V^i$ function assigning attributes to vertices,
- $\beta : E \rightarrow A_E^j$ function assigning attributes to edges.

Here, A_V and A_E denote sets of vertex and edge attributes, respectively, i is varying from 1 to δ and j is varying from 1 to Ω. While δ and Ω represent the number of attributes associated to a vertex and an edge of the graph respectively.

To describe binary shape in a simple way, actually, both nodes and edges of the graph contain only a single numerical value as an attribute.

To search the connecting edges between nodes of the graph, a region of possible neighbourhood is define for each primitive; we called it "Zone of influence". The dimensions of this zone of influence are computed according to length and thickness of the primitive (although in case of vector the thickness is 1 pixel). Thus, representing the primitives as nodes of the graph, the connecting edges between their neighbouring nodes can be found by looking at their zone of influence. Hence, the system picked up vectors and quadrilaterals present in the shape one by one, generating its zone of influence; spotting and storing all other vectors and quadrilaterals that fall in this particular zone as neighbours (see fig. 4). The relative length (λ_i) computed as the ratio of the length of the primitive to the length of the longest primitive found in that particular shape i.e., ($\lambda_i = \ell_i / \ell_{max}$) is associated to nodes. While relative angle between the primitives i.e., $\varphi_{ij} = |\theta_i - \theta_j|$ (θ_i, θ_j are the angles with horizontal axis) is associated to edge attribute.

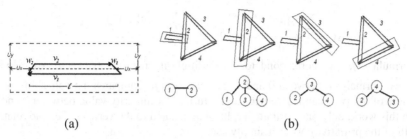

(a)   (b)

Fig. 4. (a) Zone of influence of a quadrilateral (b) Influence Zone of the quadrilaterals and their corresponding sub-graphs respectively

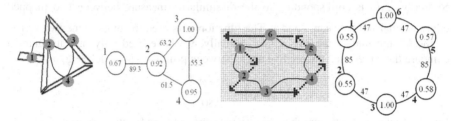

Fig. 5. Attributed Relational Graphs (ARGs), relative length as node attribute and relative angle as edge attribute

3 Recognition Phase

We believe that comparing graphs using flexible selection and matching criteria of nodes is the key to effective and significant performance improvement. We propose a new similarity measure to test the effectiveness of a given mapping and used a new similarity measure coupled with a greedy algorithm [16] for searching possible node pairs in the two graphs to avoid exhaustive complete search.

3.1 The Similarity Score Computation

To compare two graphs with numerical attributes on vertices and edges. We propose a new distance based measure which is capable of computing vertex-to-vertex similarity and edge-to-edge similarity present in a particular mapping. We propose to calculate the similarity score of a given mapping (Mp) as:

$$Sc_{Mp} = \left[\sum_{i=1}^{m}(1 - \Delta V_i) + \sum_{j=1}^{n}(1 - \Delta E_j) - \left(\sum_{i=1}^{k} \omega_i + \sum_{j=1}^{\ell} \omega'_j \right) \right] \tag{1}$$

Where, m is total number of mapped vertices in a mapping and n is the total number of edges in between them. The ω_i and ω'_j are the weights associated to the split of the ith vertex and the jth edge in the given mapping (M_p) respectively. The splits are the association of a vertex (or an edge) in one graph to more than one vertices (or edges) in the other graph. The weight of the split depends on the number of attributes associated to the vertices or edges.

$$\Delta V_i = \frac{\sum\limits_{k=1}^{\delta} f_k\left(A_V^k, A_{V'}^k\right)}{\delta} \quad \text{and} \quad \Delta E_j = \frac{\sum\limits_{k=1}^{\Omega} g_k\left(A_E^k, A_{E'}^k\right)}{\Omega} \tag{2}$$

In formula (1), ΔV_i correspond to the dissimilarity measure between two mapped vertices, normalized between 0 and 1. The function f_k compares the values of the kth attribute of the two mapped vertices and return a dissimilarity value between 0 and 1. As in this work, only one attribute has been associated to the vertices (i.e., the relative length of the primitive), we can simply use

$$f_1\left(A_V^1, A_{V'}^1\right) = \left|\lambda_i - \lambda_{i'}\right| \tag{3}$$

Similarly, ΔE_j is corresponding to the dissimilarity measure between two mapped edges normalised between 0 and 1. The function g_k is used to compare the two values of the kth attribute of the two edges. Actually, we have used only one attribute to compare the edges (i.e., the relative angle between primitives), we define g_1 as :

$$g_1\left(A_E^1, A_{E'}^1\right) = \frac{\left|\varphi_{ij} - \varphi_{i'j'}\right|}{180} \tag{4}$$

Finally, to normalize the graph similarity measure between 0 and 1, we use:

$$\text{Sim}(G, G') = \frac{Sc_{Mp}}{\left[\delta \times (C(V) + C(V')) + \Omega \times (C(E) + C(E'))\right]} \tag{5}$$

Here Sc_{Mp} is the score of the mapping computed according to new similarity formula given in eq. 1. While C is cardinality function that return the number of vertices or edges in a graph. While δ and Ω represent the number of attributes associated to the a vertex and an edge of the graph respectively.

3.2 Splits: Multiple Associations of Nodes and Arcs

The splits are the association of a vertex (or an edge) in one graph to more than one vertex (or edges) in the other graph. For example, in the two graphs given below, node "2" has been associated to node "B" as well as node "C", thus causing the split of node "2" and reducing the score of the mapping.

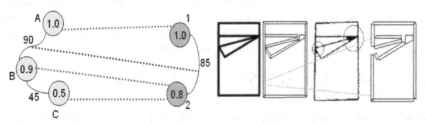

We allowed this phenomenon because due to noise and distortion the quadrilaterals or vectors can be either broken or unwanted parasitic quadrilaterals and vectors can appear. Hence, by using splits, on one hand association of nodes in the two graphs is a bit more flexible, while on the other hand, by considering it as a penalty the score of matching is controlled.

3.3   The Graph Matching Routine (Simgraph)

The simgraph routine takes two graphs G_1 and G_2 as input and return best mapping of nodes and its score. It is an iterative incremental algorithm. Beginning with an empty mapping set. In each iteration possible nodes pairs are pointed out, and are allowed to enter into the current mapping with the assumption of increase in score. If the node pair increases the over all score of the mapping, it is kept in the current mapping and is saved as best mapping so far. On contrary if it does not bring any improvement to the score, the node pair is rejected and the system search for another possible candidate pair (combination of nodes from the two graphs excluding those which are already present in the current mapping). This process continues till the score stopped increasing and there are no more virgin edges. The $Edges_{virgins}$ are those connecting edges of a node which are not yet part of the current mapping; this concept is used to break ties in a situation when there are more than one node pairs which can equally increase the score function. The detail algorithm is presented below.

Simgraph Algorithm

Input: Two attributed graphs G_1 and G_2
Output: Best mapping between nodes in G_1 and G_2, and a
             similarity score

Initialization:
1. Current Mapping($M_{current} \leftarrow \phi$)
2. Best Mapping ($M_{best} \leftarrow \phi$)
3. Maximum Score (Sc_{max} = 0)
4. **Repeat**
5. Select a nodes pair $(v_i , v_j) \notin M_{Current}$
6. $M' = M_{Current} \cup (v_i , v_j)$
7. Sc = **Call** $Mapping\_Score(M')$
8.   **If** ($Sc > Sc_{max}$) **then**
9.   Candidates $\leftarrow (v_i , v_j)$
10.   **End if**
11. $\forall (v_i , v_j) \in$ Candidates, Select (v_i, v_j) having maximum
       $Edges_{virgin}$ (edges not present in the current mapping)
12. $M_{Current} \leftarrow M_{Current} \cup (v_i , v_j)$
13. **If** $Mapping\_Score(M_{Current}) > Mapping\_Score(M_{best})$ **then**
14. $M_{best} \leftarrow M_{Current}$
15. $Sc_{max} = Mapping\_Score(M_{best})$
16. **End if**
17. **Until**
18. $Edges_{virgin}$ = 0 $\wedge$ $Sc \leq Sc_{max}$

For example, two different possible mappings are shown in the figure 6, When node "3" is mapped to node "$2'$" and node "4" to "$3'$", the score obtained is **5.9**, while in another combination where node "3" is mapped to node "$3'$" and node "4" to node "$2'$", the score improved to **5.94**. Thus the system takes the decision based on the score of the mapping and will keep this mapping as the best possible mapping so far.

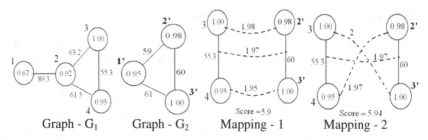

Fig. 6. Two possible mapping and their similarity score

4   Results

To test the proposed method, three datasets were used. First, a large database of closed binary shapes collected by the LEMS Vision Group at Brown University [17], second an online Encyclopedia of western signs and ideograms [18] containing 2500 symbols, and third the linear graphic symbol database proposed for the symbol recognition contest (GREC'03) [19]. The tests were launched on different subsets of these shapes dataset. For the sake of conciseness, only the results of some prototypes are shown in table 1 and 2. The low similarity scores between false neighbours is a clear indication of discriminating power of the proposed approach. The table.3 shows the images retrieved based on their similarity with respect to the query shapes in a certain range (here it's up to 70% similarity).

Table 1. Score of similarity between shapes belonging to different groups

	🕊	🦋	🐈	👗	⚱	🐟	🐕
🕊	1.000	0.535	0.445	0.566	0.517	0.645	0.529
🦋		1.000	0.630	0.563	0.527	0.554	0.630
🐈			1.000	0.492	0.398	0.502	0.579
👗				1.000	0.652	0.673	0.532
⚱					1.000	0.615	0.522
🐟						1.000	0.605
🐕							1.000

Table 2. Score of similarity between different instances of the same group

	1.00	0.62	0.75	0.83	0.67	0.79	0.65
		1.00	0.58	0.63	0.72	0.66	0.57
			1.00	0.70	0.64	0.75	0.63
				1.00	0.67	0.74	0.79
					1.00	0.69	0.62
						1.00	0.74
							1.00

Table 3. Score of similarity and shape retrieved based on query shape

Query shape	Total shapes	Rank-1	Rank-2	Rank-3	Rank-4	Rank-5
Q_1	43	1.000 bd-007	0.925 bd-030	0.888 bd-019	0.784 bd-024	0.744 bd-042
Q_2	100	1.000 bf-001	0.872 bf-003	0.826 bf-026	0.792 bf-009	0.786 bf-011
Q_3	100	1.000 f-026	0.801 f-075	0.793 f-044	0.774 f-065	0.733 f-046
Q_4	45	1.000 v-004	0.849 v-018	0.814 V-017	0.754 v-038	0.678 v-044

The score of similarity degrade gracefully depending on the changes in the shape boundaries. Similarly table 4 shows the top 5 ideograms found that were close to the query sign based on their mutual similarity measures. For query ideograms Q3 in table.4 we can see that the first 4 neighbour are the instances of the same ideograms at different orientations, the similarity score of 1.00 computed by the system showed that our proposed system is invariant of rotation as well. To test the system robustness regarding vectorial distortion, a hand drawn dataset based on the graphic symbols of GREC-database was synthetically generated (fig. 7).

The results obtained were quite satisfactory. However, it would be intresting to see the performance of the system on different levels of degradation with respect to noise.

Table 4. Score of similarity and shape retrieved based on query shape

Query shape	Rank-1	Rank-2	Rank-3	Rank-4	Rank-5
Q_1	1.000 C-1106	0.903 C-1112	0.859 C-1113	0.831 C-1111	0.789 C-1107
Q_2	1.000 C-0304	0.860 C-0305	0.819 C-0306	0.781 C-0308	0.709 C-0307
Q_3	1.000 C-0206a	1.000 C-0206b	1.000 C-0206c	1.000 C-0206d	0.534 C-0202a
Q_4	1.000 C-430b	0.909 C-0419	0.641 C-0431b	0.588 C-0431a	0.562 C-0421

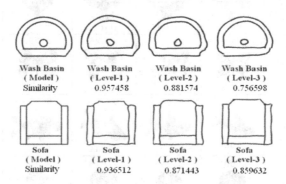

Fig. 7. The three level of synthetic hand-drawn symbols

5 Conclusion

In this work we deal with the recognition of thin and filled shapes. We propose a general methodology which uses an efficient representation of shapes based on "Attributed Relational Graphs" (ARGs). We believe that the proposed similarity measure play a vital role in evaluating the appropriateness of the nodes and arcs mapping between the two graphs. In addition, the use of greedy algorithm improves the efficiency of the search techniques and bound it to polynomial time solution.

The proposed graph based representations generated using graphic primitives like vectors and quadrilaterals are understandable, highly invariant to scaling and rotation, and are insensitive to small changes in shape as well. However, the results can be further improved by adding more attributed to nodes and edges of the graphs. Future works include, having a combined graph of quadrilaterals and vectors to test the recognition of mixed shapes (partially filled and partially linear) based on proposed similarity measure.

References

1. Bunke, H.: Error Correcting Graph Matching: On the Influence of the Underlying Cost Function. IEEE transactions on Pattern Analysis and Machine Intelligence 21, 917–922 (1999)
2. Lam, L., Lee, S.W., Suen, C.Y.: Thinning Methodologies-A Comprehensive Survey. IEEE Transactions on Pattern Analysis and Machine Intelligence 14, 869–885 (1992)
3. Ruberto, C.D., Rodriguez, G., Casta, L.: Recognition of shapes by morphological attributed relational graphs. citeseer.ist.psu.edu/535355.html (2002)
4. Cordella, L.P., Vento, M.: Symbol Recognition in Documents: A Collection of Techniques. International Journal of Document Analysis and Recognition 3, 73–88 (2000)
5. Bunke, H.: Recent developments in graph matching. In: The Proc. of 15th Int. Conf. Pattern Recognition, vol. 2, pp. 117–124 (2000)
6. Mehlhorn, K.: Graph Algorithms and NP-Completeness, vol. 2. Springer-Verlag, Berlin Heidelberg (1984)
7. Dickinson, P.J., Bunke, H., Dadej, A., Kraetzl, M.: On Graphs with Unique Node Labels. In: Hancock, E.R., Vento, M. (eds.) GbRPR 2003. LNCS, vol. 2726, pp. 13–23. Springer, Heidelberg (2003)
8. Ullman, J.R.: An Algorithm for Subgraph Isomorphism. Journal of the Association for Computing Machinery 23, 31–42 (1976)
9. Schmidt, D.C., Druffel, L.E.: A Fast Backtracking Algorithm to Test Directed Graphs for Isomorphism Using Distance Matrices. Journal of the Association for Computing Machinery 23, 433–445 (1976)
10. McKay, B.D.: Practical graph isomorphism. Congr. Numerantium 30, 45–87 (1981)
11. Cordella, L.P., Foggia, P., Sansone, C., Vento, M.: An improved algorithm for matching large graphs. In: Proc. 3rd IAPR –TC15 Workshop Graph Based Representations in Pattern Recognition, pp. 149-159 (2001)
12. Foggia, P., Sansone, C., Vento, M.: A performance comparison of five algorithms for graph isomorphism. In: The 3rd IAPR-TC15 Workshop on Graph-Based Representations in Pattern Recognition, Cuen, 188–199 (2001)
13. Conte, D., Foggia, P., Sansone, C., Vento, M.: Thirty Years of Graph Matching in Pattern Recognition. The. International Journal of Pattern Recognition and Artificial Intelligence 18, 265–298 (2004)
14. Wall, K., Danielsson, P.: A fast sequential method for polygonal approximation of digitized curves. Computer Vision, Graphics and Image Processing 28, 220–221 (1984)
15. Ramel, J.Y., Vincent, N., Emptoz, H.: A structural representation for understanding line - drawing images. International Journal on Document Analysis and Recognition 3, 58–66 (2000)

16. Champin, P.A., Solnon, C.: Measuring the Similarity of Labelled Graphs. In: Proceedings of the 5th International Conference on Case-Based. 5th International Conference on Case-Based, pp. 80–95. Springer-Verlag, Berlin Heidelberg (2003)
17. http://www.lems.brown.edu/~dmc/
18. http://www.symbols.com/
19. GREC'03 sample images http://www.cvc.uab.es/grec2003/SymRecContest/

A Continuous-Based Approach for Partial Clique Enumeration

Samuel Rota Bulò, Andrea Torsello, and Marcello Pelillo

"Ca' Foscari" University of Venice
{srotabul,torsello,pelillo}@dsi.unive.it

Abstract. In many applications of computer vision and pattern recognition which use graph-based knowledge representation, it is of great interest to be able to extract the K largest cliques in a graph, but most methods are geared either towards extracting the single clique of maximum size, or enumerating all cliques, without following any particular order. In this paper we present a novel approach for partial clique enumeration, that is, the extraction of the K largest cliques of a graph. Our approach is based on a continuous formulation of the clique problem developed by Motzkin and Straus, and is able to avoid extracting the same clique multiple times. This is done by casting the problem into a game-theoretic framework and iteratively rendering unstable the solutions that have already been extracted.

1 Introduction

Many applications of computer vision and pattern recognition which use graph-based knowledge representation have to deal with the problem of finding complete subgraphs (cliques) of their structural descriptions. Examples of problems that have successfully been reduced to a clique-finding problem range from matching [2], to category learning and knowledge discovery [17,9], to clustering [1,18], to stereo matching [13], to name just a few. Furthermore, clique finding is also linked with the learning of graphical structure by the Hammersley-Clifford theorem [11].

The maximum clique problem (MCP) deals with the challenge of finding the largest complete subgraph of an undirected and unweighted graph. It falls in the crucial class of NP-Complete problems, whose intractability forces us to fall back on approximation methods. Unfortunately, even approximating the MCP is intractable [12]. Due to this pessimistic state of affairs much attention has gone into developing efficient heuristics for the MCP, for which no formal guarantee of performance may be provided, but are nevertheless useful in practical applications. We refer to Bomze et al. [5] for a survey concerning algorithms, applications, and complexity issues of this important problem.

In a recent series of papers [19,10,7] we find approaches that are centered around a classical result from graph theory due to Motzkin and Straus [16], that allows us to formulate the MCP as a continuous quadratic optimization problem with simplex constraints. This program is typically solved by the replicator

F. Escolano and M. Vento (Eds.): GbRPR 2007, LNCS 4538, pp. 61–70, 2007.
© Springer-Verlag Berlin Heidelberg 2007

dynamics, well-known continuous- and discrete-time dynamical systems, developed and studied in the field of evolutionary game theory and for which it can be shown that there exists a one-to-one correspondence between stable points and maximal cliques of the corresponding graph.

In several contexts, it is of great interest to have an approach that can extract several large cliques, in particular, we would like to be able to efficiently extract the K largest cliques in a graph. For example in knowledge discovery, where categories are abstracted in terms of cliques, each element can belong to multiple categories, and hence we are interested in discovering more than one category [17,9]. In a completely different domain, Horaud and Skordas [13] use the largest cliques to find stereo correspondences in image pairs. While exact search-based enumerative algorithms are guaranteed to generate every maximal clique, in general they cannot guarantee a specific order in which these are found, in particular they give no guarantee about the relative size of the clique obtained at each step.

In this paper we present an approach which uses a continuous formulation to enumerate a user-defined number of large cliques. Ideally, we would like to obtain the K largest maximal cliques after a small number of enumerations. Clearly, the actual size of the extracted cliques depends on the effectiveness of the continuous formulation, but, experimental evidence tells us that the approach performs fairly well [19].

The basis of this approach rests on the fact that under a certain family of quadratic problems, there is a bijection between asymptotically stable points of the replicator dynamics and maximal cliques. Once we have extracted a maximal clique, we would like to avoid that the dynamics converge to the same clique. Intuitively, what our method does is to render unstable the associated rest point. To do this, we deal with directed graphs, and apply a particular asymmetric graph-extension for every maximal clique we want to render unstable. By iterating this extension process, we progressively reduce the set of asymptotically stable points of the replicator dynamics, and, hence, we obtain a continuous-based enumerative algorithm.

2   A Family of Quadratic Programs for Maximum Clique

Let $G = (V, E)$ be an undirected graph without self-loops, where $V = \{1, 2, \ldots, n\}$ is the set of vertices and $E \subseteq V \times V$ the set of edges. Two vertices $u, v \in V$ are *adjacent* if $(u, v) \in E$. A subset C of vertices in G is called a *clique* if all its vertices are mutually adjacent. It is a *maximal clique* if it is not subset of other cliques in G. It is a *maximum clique* if no other cliques of G have a strictly greater cardinality. The cardinality of a maximum clique of G is also called *clique number* and denoted by $\omega(G)$.

The *adjacency matrix* of G is the $n \times n$ symmetric matrix $A_G = (a_{ij})$ where $a_{ij} = \chi_E((i, j))$. Here, $\chi_A(i)$ represents the indicator function that returns 1 if $i \in A$, 0 otherwise.

Consider the following constrained quadratic program.

$$\text{maximize} \quad f_\alpha(\boldsymbol{x}) = \boldsymbol{x}'(A_G + \alpha I)\boldsymbol{x} \quad \text{s.t.} \quad \boldsymbol{x} \in \Delta \subset \mathbb{R}^n, \tag{1}$$

where n is the order of G, I the identity matrix, and α is a real parameter. In 1965 Motzkin-Straus [16] established a connection between the maximum clique problem and the program in (1) with $\alpha = 0$; they related the clique number of G to global solutions $\boldsymbol{x}^*$ of the program through the formula $\omega(G) = (1 - f_0(\boldsymbol{x}^*))^{-1}$, and showed that a subset of vertices C with cardinality $|C|$ is a maximum clique of G if and only if [1] its *characteristic vector* $\boldsymbol{x}^C \in \Delta$, where $x_i^C = \chi_C(i)|C|^{-1}$, is a global maximizer of f_0 on Δ. Gibbons, Hearn, Pardalos and Ramana [10], and Pelillo and Jagota [20], extended the Motzkin-Straus theorem by providing a characterization of maximal cliques in terms of local maximizers of f_0 in Δ.

A drawback of the original Motzkin-Straus formulation is the existence of "spurious" solutions, i.e., maximizers of f_0 that are not in the form of characteristic vectors. Bomze et al.[6] proved that for $0 < \alpha < 1$ all local maximizer of (1) are strict and are in one-to-one relation with the characteristic vectors of the maximal cliques of G, hence, overcoming the problem.

In order to find the maxima of (1) we cast the problem in a game-theoretic setting and use the replicator dynamics, a well-known formalization of the selection process. In the next section we will review some concepts from evolutionary game theory that will be useful throughout the paper and provide the link between game theory and maximal cliques.

3   A Game-Theoretic Perspective

Let $O = \{1, 2, \ldots, n\}$ be the set of *pure strategies* available to the players and $A = (a_{ij})$ the $n \times n$ payoff or utility matrix [23] where a_{ij} is the payoff that a player gains when playing the strategy i against an opponent playing j. In biological contexts, payoff are typically measured in terms of Darwinian fitness or reproductive success whereas in economics applications, they usually represent firms' profits or consumers' utilities.

A *mixed strategy* is a probability distribution $\boldsymbol{x} = (x_1, x_2, \ldots, x_n)'$ over the available strategies in O. Mixed strategies clearly lie in the standard simplex of the n-dimensional Euclidean space $\Delta = \{\boldsymbol{x} \in \mathbb{R}^n : \boldsymbol{e}'\boldsymbol{x} = 1, \quad \boldsymbol{x} \geq 0\}$ where $\boldsymbol{e}$ is the vector with all components equal to 1.

The *support* of a mixed strategy $\boldsymbol{x} \in \Delta$, denoted by $\sigma(\boldsymbol{x})$, defines the set of elements with non-zero probability: $\sigma(\boldsymbol{x}) = \{i \in O : x_i > 0\}$

The expected payoff that a player obtains by playing the element i against an opponent playing a mixed strategy $\boldsymbol{x}$ is $u(\boldsymbol{e}^i, \boldsymbol{x}) = (A\boldsymbol{x})_i = \sum_j a_{ij} x_j$, where $\boldsymbol{e}^i$ is the vector with all components equal zero except for the ith-component which is equal to 1. Hence, the expected payoff received by adopting a mixed strategy $\boldsymbol{y}$ is $u(\boldsymbol{y}, \boldsymbol{x}) = \boldsymbol{y}'A\boldsymbol{x}$.

---

[1] In the original paper Motzkin-Straus proved the "only-if" part of this theorem. The converse however is a straightforward consequence of their result (Pelillo & Jagota, 1995) [20].

Evolutionary game theory considers an idealized scenario wherein pairs of individuals are repeatedly drawn from a large population to play a two-player symmetric game. Each player is not supposed to behave rationally or have a complete knowledge of the details of the game, but he acts according to a pre-programmed pure strategy. This dynamic activates some selection process that results in the evolution of the fittest strategies.

A well-known formalization of the selection process is given by the replicator equations [23]: $\dot{x}_i = x_i(u(e^i, x) - u(x, x))$.

If the payoff matrix is symmetric then $x'Ax$ is strictly increasing along any non-constant trajectory of any payoff-monotonic dynamics [23]. This result allows us to establish a bijective relation between the local solutions of program (1), namely characteristic vectors of maximal cliques of G, and asymptotically stable points of the replicator dynamics with payoff matrix $A_G + \alpha I$ and $0 < \alpha < 1$.

In order to obtain enumeration of maximal cliques through a continuous formulation we move from undirected graphs to directed graphs, or, in other words, from symmetric payoff matrices to asymmetric payoff matrices. If we loosen the symmetry constraint, then all the results that bind local solutions to asymptotically stable points and maximal cliques do not hold any longer, and $x'Ax$ is not a Lyapunov function for the dynamics.

The *best replies* against a mixed strategy x is the set of mixed strategies $\beta(x) = \{y \in \Delta : u(y, x) = \max_z u(z, x)\}$.

A mixed strategy x is a *Nash equilibrium* if it is a best reply to itself, i.e. $\forall y \in \Delta, u(y, x) \leq u(x, x)$. This implies that for all $i \in \sigma(x)$, $u(e^i, x) = u(x, x)$, hence the payoff of every strategy in the support of x is constant, while all strategies outside the support of x earn a payoff that is less than or equal $u(x, x)$.

A strategy x is said to be an *evolutionary stable strategy* (ESS) if it is a Nash equilibrium and for all $y \in \Delta$ such that $u(y, x) = u(x, x)$ we have that $u(x, y) > u(y, y)$. Intuitively, *ESS* are strategies such that any small deviation from them will lead to an inferior payoff.

Consider the following quadratic program

$$\text{maximize} \quad \pi(x) = x'Ax \quad \text{s.t.} \quad x \in \Delta \subset \mathbb{R}^n, \tag{2}$$

where A is a symmetric matrix. We have that x is a Nash equilibrium of a two-player game with payoff matrix A, if and only if it satisfies the Karush-Kuhn-Tucker (KKT) conditions for (2). In fact the KKT conditions can be written as

$$u(e^i, x) = (Ax)_i \begin{cases} = \lambda & \text{if } i \in \sigma(x) \\ \leq \lambda & \text{if } i \notin \sigma(x) \end{cases}$$

for some real λ. However it is clear that $\lambda = x'Ax = u(x, x)$ and what we obtain is exactly the definition of a Nash equilibrium. Hence local solution of (2) are indeed Nash equilibria, but the converse does not necessarily hold.

A two-player symmetric game where the payoff matrix is also symmetric is called *doubly-symmetric* game. Loser and Akin [15] showed that for all doubly symmetric games the average payoff $u(x, x)$ increases along every non-stationary solution path to the replicator dynamics.

If we consider program (1) with $0 < \alpha < 1$, we have that the set of ESS is equivalent to the set of maximal cliques of the related graph. We refer to [8] and [6] for a deeper insight of the relation between ESS and maximal cliques.

Through this change in perspective, we can move from a constrained maximization problem, to a game-theoretic setting. Instead of finding local solutions of a quadratic program, we look for ESS of a doubly symmetric game. The advantage of this new approach is that we can generalize the Motzkin-Straus result to non symmetric payoff matrices and, hence, directed graphs.

Let $G = (V, E)$ be a directed graph. A *doubly-linked clique* of G is a set $S \subseteq V$ such that for all $u, v \in S$, $(u, v) \in E$ implies $(v, u) \in E$. The clique is *saturated* if there is no $t \in V \setminus S$ such that for all $s \in S$, $(s, t) \in E$.

In [22] we find the following result.

Theorem 1. *Let $G = (V, E)$ be a directed graph with adjacency matrix A, $S \subseteq V$ is a saturated doubly-linked clique of G if and only if x^S is an ESS for a two-player game with payoff matrix $B = A' + \alpha I$, where $0.5 < \alpha < 1$.*

We have already seen that if we consider an undirected graph G and the payoff matrix $A_G + \alpha I$ with $0 < \alpha < 1$, then the ESSs of the related two-player game are in one-to-one correspondence with maximal cliques of G. However if we strengthen the constraint on α to lay between 0.5 and 1, then we can see that the concept of saturated doubly-linked clique is a direct generalization to the asymmetric case of the concept of maximal clique, i.e. ESSs are in one-to-one correspondence with saturated doubly-linked cliques.

4   Continuous-Based Enumeration

In this section we will present our continuous-based enumeration approach and prove its correctness. In order to render unstable a given ESS x it is enough to drop the Nash condition for x. A simple way to do it without affecting other equilibria, is to add a new strategy z that is a best reply to x, but to no other ESS. This way, x will no longer be asymptotically stable.

Let $G = (V, E)$ be an undirected graph and $G' = (V, E')$ be its directed version where for all $(u, v) \in E$, $(u, v), (v, u) \in E'$. Given a set Σ of maximal cliques of G, we extend G' obtaining the Σ-*extension* G^Σ of G. The extension is as follows. For each clique $S \in \Sigma$, we create a new vertex v, called Σ-*vertex*, and put edges from v to each vertex in S and from each vertex not in S to v. After this operation, each Σ-vertex v dominates a particular clique S of Σ. Further, each vertex not in S dominates the Σ-vertex v so that it cannot be part of a new asymptotically stable strategy.

Theorem 2. *Let $G = (V, E)$ be an undirected graph, Σ be a set of maximal cliques of G and A be the adjacency matrix of the Σ-extension G_Σ of G. Let Φ be a two person symmetric game with payoff matrix $A + \alpha I$ with $0.5 \leq \alpha < 1$. Then x is an ESS equilibrium of $\mathcal{G}$ if and only if it is the characteristic vector of a maximal clique of G not in Σ.*

Proof. ($\Rightarrow$) By (1) if x is an ESS of Φ then it is the characteristic vector of a saturated doubly-linked clique S of G_Σ. By construction of G_Σ, the only possible doubly-linked cliques are subsets of V, therefore S is a clique of G. It is also maximal and not in Σ because otherwise it would not be saturated.

($\Leftarrow$) Consider $S \notin \Sigma$ a maximal clique of G. Then by construction of G^Σ, it is a saturated doubly-linked clique of G^Σ and hence by [22] x^S is an ESS equilibrium of $\mathcal{G}$.

The continuous-based enumerative algorithm uses this result in the following way. We iteratively find an asymptotically stable point through the replicator dynamics. If we have an ESS, then we have found a new maximal clique[2]. After that, we extend the graph by adding the newly extracted clique to Σ, hence rendering its associated strategy unstable, and reiterate the procedure until we have enumerated the selected number of maximal cliques.

The space complexity of this algorithm is $O\{(n + K)^2\}$, where n is the graph order and K is the number of enumerated cliques, while the time complexity is $O\{\gamma K (n + K)^2\}$, where γ is the average number of iterations that the replicator dynamics require to converge (in the experiments we present in the next section we have that $\gamma < 15$).

5   Experimental Results

In this section we asses the ability of our continuous-based enumerative heuristic (CEH) to extract large cliques. To this end we apply the enumeration to the extraction of the maximum clique from the DIMACS benchmark graphs. For each graph, we run the method 20 times and took for each run, the maximum between the first 300 enumerated maximal cliques.

In order to extract the maximal clique from a characteristic vector, we avoid the standard thresholding technique on the value of each component of the characteristic vector, but rather we use the values of each component as indicators for a New-Best-In heuristic [14]. This is a sequential greedy heuristic that, starting from an empty set of vertices, iteratively constructs a maximal clique by inserting the clique-preserving vertex v that maximizes $w_v + \sum_{j \in S} \chi_E((v, j)) w_j$ where E is the set of edges of the graph, S is the set of clique-preserving vertices and $w = (w_1, \ldots, w_n)$ is a weight vector, in our case the mixed strategy obtained through the replicator dynamics. An added advantage of this approach is that we can stop the dynamics before the dominated strategies where driven to a hard zero, and still be able to extract the associated maximal clique. This can significantly improve the speed of the approach as a lower number of iterations are needed to extract each clique. the method.

In figure (1) we show the results obtained by enumerating about 450 maximal cliques of a random graph of order 100 and density 0.25. For each enumeration the graph plots the average size of the last 40 cliques in order to clarify the

---

[2] We have never experienced an AS point that was not an ESS, so we strongly believe that theorem (2) can be generalized to asymptotically stable points.

Table 1. Comparative results on DIMACS benchmark graphs

Name	#	ρ	BR	Min	Clique size Avg.(S.Dev.)	Max	K	Avg. time	IHN	AIH	CBH	QMS	RLS
brock200_1	200	0.75	21	20	20.050 (0.224)	21	156	7.85s	-	20	20	21	21
brock200_2	200	0.50	12	10	10.400 (0.503)	11	24	7.25s	-	10	12	12	12
brock200_3	200	0.61	15	13	13.750 (0.444)	14	19	7.40s	-	13	14	15	15
brock200_4	200	0.66	17	15	15.850 (0.587)	17	2	7.50s	-	16	16	17	17
brock400_1	400	0.75	27	23	23.800 (0.410)	24	49	19.90s	-	24	23	27	25
brock400_2	400	0.75	29	23	23.450 (0.510)	24	24	20.00s	-	24	24	29	29
brock400_3	400	0.75	31	23	23.700 (0.657)	25	10	19.90s	-	24	23	31	25
brock400_4	400	0.75	33	23	23.900 (0.641)	25	77	19.90s	-	23	24	33	33
brock800_1	800	0.65	23	19	19.600 (0.503)	20	4	51.35s	-	20	20	23	21
brock800_2	800	0.65	24	19	19.900 (0.447)	21	3	51.60s	-	18	19	24	21
brock800_3	800	0.65	25	19	19.750 (0.550)	21	245	51.30s	-	19	20	25	22
brock800_4	800	0.65	26	19	19.550 (0.510)	20	17	51.25s	-	19	19	26	21
c-fat200-1	200	0.08	12	12	12 (0)	12	1	7.40s	12	12	12	12	12
c-fat200-2	200	0.16	24	24	24 (0)	24	1	7.95s	24	24	24	24	24
c-fat200-5	200	0.43	58	58	58 (0)	58	1	18.80s	58	58	58	58	58
c-fat500-1	500	0.04	14	14	14 (0)	14	1	24.70s	14	14	14	14	14
c-fat500-2	500	0.07	26	26	26 (0)	26	1	28.90s	26	26	26	26	26
c-fat500-5	500	0.19	64	64	64 (0)	64	1	41.50s	64	64	64	64	64
c-fat500-10	500	0.37	126	126	126 (0)	126	1	62.75s	-	126	126	126	126
hamming6-2	64	0.90	32	32	32 (0)	32	1	2.79s	32	32	32	32	32
hamming6-4	64	0.35	4	4	4 (0)	4	1	2.23s	4	4	4	4	4
hamming8-2	256	0.97	128	128	128 (0)	128	1	9.90s	128	128	128	128	128
hamming8-4	256	0.64	16	16	16 (0)	16	1	10.15s	16	16	16	16	16
johnson8-2-4	28	0.56	4	4	4 (0)	4	1	1.11s	4	4	4	4	4
johnson8-4-4	70	0.77	14	14	14 (0)	14	1	2.68s	14	14	14	14	14
johnson16-2-4	120	0.76	8	8	8 (0)	8	1	4.28s	8	8	8	8	8
johnson32-2-4	496	0.88	16	16	16 (0)	16	1	25.50s	16	16	16	16	16
keller4	171	0.65	11	11	11 (0)	11	1	2.20s	-	9	10	11	11
keller5	776	0.75	27	25	26.600 (0.681)	27	5	28.55s	-	16	21	26	27
keller6	3361	0.82	≥59	51	52.250 (0.910)	54	45	761.75s	-	31	-	53	59
MANN_a9	45	0.927	16	16	16 (0)	16	1	1.87s	-	16	16	16	16
MANN_a27	378	0.990	126	125	125.100 (0.308)	126	124	36.30s	-	117	121	125	126
MANN_a45	1035	0.996	345	341	342.100 (0.641)	343	85	528.00s	-	-	-	342	345
p_hat300-1	300	0.24	8	8	8 (0)	8	1	11.15s	8	8	8	8	8
p_hat300-2	300	0.49	25	25	25 (0)	25	9	12.80s	25	25	25	25	25
p_hat300-3	300	0.74	36	34	34.550 (0.605)	36	218	13.75s	36	36	36	35	36
p_hat500-1	500	0.25	9	9	9 (0)	9	1	22.65s	9	9	9	9	9
p_hat500-2	500	0.50	36	34	35.300 (0.571)	36	36	31.05s	36	36	35	36	36
p_hat500-3	500	0.75	50	48	48.500 (0.510)	49	7	35.40s	49	49	49	48	50
p_hat700-1	700	0.25	11	9	10.700 (0.571)	11	2	35.45s	11	9	11	11	11
p_hat700-2	700	0.50	44	43	43.400 (0.503)	44	1	56.30s	44	44	44	44	44
p_hat700-3	700	0.75	62	60	60.500 (0.607)	62	1	67.25s	61	60	60	62	62
p_hat1000-1	1000	0.25	10	10	10 (0)	10	1	60.55s	10	-	-	10	10
p_hat1000-2	1000	0.50	46	44	45.250 (0.550)	46	26	104.00s	46	-	-	45	46
p_hat1000-3	1000	0.75	68	63	63.900 (0.718)	65	50	127.80s	68	-	-	65	68
p_hat1500-1	1500	0.25	12	11	11 (0)	11	1	114.95s	-	10	11	12	12
p_hat1500-2	1500	0.50	65	62	63.150 (0.745)	64	51	255.60s	-	64	63	64	65
p_hat1500-3	1500	0.75	94	88	89.750 (1.333)	92	178	326.75s	-	92	94	91	94
san200_0.7_1	200	0.70	30	19	29.050 (2.964)	30	11	7.70s	30	15	15	30	30
san200_0.7_2	200	0.70	18	13	13 (0)	13	1	7.45s	15	12	12	18	18
san200_0.9_1	200	0.90	70	70	70 (0)	70	2	9.60s	70	46	46	70	70
san200_0.9_2	200	0.90	60	57	59.800 (0.696)	60	2	8.95s	41	39	36	60	60
san200_0.9_3	200	0.90	44	36	39.800 (2.375)	44	85	8.70s	-	35	30	40	44
san400_0.5_1	400	0.50	13	7	7.900 (0.308)	8	20	16.55s	-	7	8	13	13
san400_0.7_1	400	0.70	40	40	40 (0)	40	2	20.05s	40	20	20	40	40
san400_0.7_2	400	0.70	30	18	21.250 (4.315)	30	32	18.75s	30	15	15	30	30
san400_0.7_3	400	0.70	22	15	15.800 (0.410)	16	7	18.15s	-	12	14	18	22
san400_0.9_1	400	0.90	100	100	100 (0)	100	1	27.40s	100	51	50	100	100
san1000	1000	0.50	10	8	8.400 (0.503)	9	30	59.90s	10	8	8	15	15
sanr200_0.7	200	0.70	18	17	17.850 (0.366)	18	29	7.60s	17	18	18	18	18
sanr200_0.9	200	0.90	42	39	40.550 (0.686)	42	243	8.80s	41	41	41	41	42
sanr400_0.5	400	0.50	13	12	12.750 (0.444)	13	30	17.25s	12	13	12	13	13
sanr400_0.7	400	0.70	21	19	20.250 (0.550)	21	59	19.10s	21	21	20	20	21

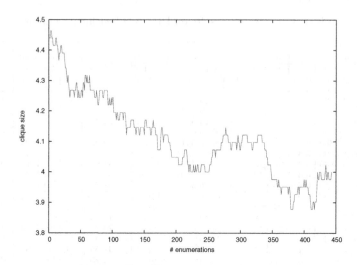

Fig. 1. Average size of the extracted clique over the number of extractions

descending tendency. As it can be seen, the approach enumerates the clique in approximately decreasing order of size.

Table (1) shows the results obtained with CEH on the DIMACS benchmark. We compared our approach with a neural-network-based heuristic, Inverted Neural Network (IHN) [4] and with other Motzkin-Straus -based heuristics for MCP, i.e. Annealed Imitation Heuristic (AIH) [21], Continuous Based Heuristic (CBH) [10] and Qualex Motzkin-Straus (QMS) [7]. Furthermore, we also compare the approach with Reactive Local Search (RLS) [3], a state-of-the-art heuristic search-based algorithm for MCP.

The table includes the name of the DIMACS graph (Name), the number of vertices ($\#$), the graph density (ρ), the optimum size (BR) . In the second part we find the results obtained with CEH: the minimum (Min), the average size and standard deviation (Avg), and the maximum size (Max) obtained among 20 runs of CEH, each enumerating 300 cliques. The column labeled with K provides the number of enumerations required before the maximum was found. The running times are referred to an unoptimized C implementation on 64-bit PC with a 2 GHz AMD Opteron Processor and 1 Gb RAM. The computation times of the other methods can be found in their respective papers, however they are not comparable because they refer to experiments conducted with different hardware and software settings.

The *c-fat, hamming* and *johnson* families were the easiest to solve, in fact all algorithms find the global optima.

Though CEH, AIH and CBH use the same continuous-based technique, CEH outperforms both algorithms on all DIMACS graphs. The comparison with IHN is not so meaningful because it has been tested on few graph instances, but we can notice that for all families except *sanr* the approaches are comparable, while on the *sanr* graphs CEH is the best performer.

QMS seems to be particularly good on the *brock* family, where it outperforms all other approaches. However, CEH outperforms QMS on *MANN*, *keller* and *sanr* families and performs slightly better on the *p_hat* family, while QMS performs slightly better on the *san* family.

We can see that RLS provides the best performance on almost all DIMACS benchmarks, with the exception of the *brock* family, where QMS is indeed the best. It is worth reminding that RLS is a search based-approach while all the other are continuous-based.

The column K of the tables represent the minimum number of enumerations before the best clique size for the algorithm has been reached. It is in some sense a measure of the action of the enumeration in order to achieve the maximum result. We see that the easy instances of the benchmark are solved within the first enumeration, while more difficult ones, for example *brock*, *san*, *sanr*, require a higher number of enumerations.

6   Conclusions

In this paper we developed a partial clique enumeration algorithm based on the Motzkin-Straus formulation. In order to perform the enumeration, we deal with a directed form of the clique problem and we deal with an asymmetric extension. This way we lose the original connection with the quadratic problem, but, by casting the problem into a game-theoretic framework, we are able to prove a relationship between the evolutionary stable strategies and maximal cliques that have not yet been enumerated. In order to asses the usefulness of the approach we compared it with several state-of-the-art approaches on the problem of extracting the maximum clique from the DIMACS benchmark graphs. The approach proved to be superior to other continuous-based approaches and competitive with the state of the art search heuristics.

References

1. Agrawal, R., Gehrke, J., Gunopulos, D., Raghavan, P.: Automatic subspace clustering of high dimensional data for data mining applications. In: Proc. 1998 ACM-SIGMOD Int. Conf. Management of Data, Seattle, Washington (June 1998)
2. Barrow, H.G., Burstall, R.M.: Subgraph Isomorphism, Matching Relational Structures, and Maximal Cliques. Information Processing Letters 4(4), 83–84 (1976)
3. Battiti, R., Protasi, M.: Reactive local search for the maximum clique problem, Technical Report TR-95-052, International Computer Science Institute, Berkeley, CA (1995)
4. Bertoni, A., Campadelli, P., Grossi, G.: A neural algorithm for the maximum clique problem: analysis, experiments and circuit implementation. Algorithmica 33(1), 71–88 (2002)
5. Bomze, I.M., Budinich, M., Pardalos, P.M., Pelillo, M.: The maximum clique problem. In: Du, D.Z., Pardalos, P.M. (eds.) Handbook of Combinatorial Optimization, Supplement vol. A, pp. 1–74. Kluwer Academic Publishers, Boston, MA (1999)

6. Bomze, I.M., Pelillo, M., Giacomini, R.: Evolutionary approach to the maximum clique problem: Empirical evidence on a larger scale. In: Bomze, I.M., Csendes, T., Horst, R., Pardalos, P.M. (eds.) Developments in Global Optimization, pp. 95–108. Kluwer, Dordrecht, The Netherlands (1997)
7. Busygin, S.: A new trust region technique for the maximum weight clique problem (2002)
8. Cannings, C., Veckers, G.: Patterns of ess's. J. Theor. Biol., p. 132
9. Dmitry, D., Ari, R.: Efficient Unsupervised Discovery of Word Categories Using Symmetric Patterns and High Frequency Words. In: Proceedings of the 21st International Conference on Computational Linguistics and 44th Annual Meeting of the ACL, Association for Computational Linguistics, pp. 297–304 (2006)
10. Gibbons, L.E., Hearn, D.W., Pardalos, P.M., Ramana, M.V.: Continuous characterizations of the maximum clique problem. Math. Oper. Res. 22, 754–768 (1997)
11. Hammersley, J., Clifford, P.: Markov fields on finite graphs and lattices, unpublished manuscript (1971)
12. Hastad, J.: Clique is hard to approximate within $n^{1-\varepsilon}$. In: Proc. 37th Ann. Symp. Found. Comput. Sci., pp. 627–636 (1996)
13. Horaud, R., Skordas, T.: Stereo correspondence through feature grouping and maximal cliques. IEEE Trans. Pattern Anal. Mach. Intell. 11(11), 1168–1180 (1989)
14. Kopf, R., Ruhe, G.: A computational study of the weighted independent set problem for general graphs. Found. Control Engin. 12, 167–180 (1987)
15. Losert, V., Akin, E.: Dynamics of games and genes: Discrete versus continuous time. J. Math. Biol. 17, 241–251 (1983)
16. Motzkin, T.S., Straus, E.G.: Maxima for graphs and a new proof of a theorem of Turán. Canad. J. Math. 17, 533–540 (1965)
17. Nina, M., Dana, R., Ram, S.: A New Conceptual Clustering Framework. Machine Learning 56, 115–151 (2004)
18. Pavan, M., Pelillo, M.: Dominant sets and pairwise clustering. IEEE Transactions on Pattern Analysis and Machine Intelligence 29(1), 167–172 (2007)
19. Pelillo, M.: Replicator equations, maximal cliques, and graph isomorphism. Neural Computation 11(8), 2023–2045 (1999)
20. Pelillo, M., Jagota, A.: Feasible and infeasible maxima in a quadratic program for maximum clique. J. Artif. Neural Networks 2, 411–420 (1995)
21. Pelillo, M., Torsello, A.: Payoff-Monotonic Game Dynamics and the Maximum Clique Problem. Neural Computation 18, 1215–1258 (2006)
22. Torsello, A., Bulò, S.R., Pelillo, M.: Grouping with asymmetric affinities: A game-theoretic perspective. In: CVPR '06: Proceedings of the 2006 IEEE Computer Society Conference on Computer Vision and Pattern Recognition, pp. 292–299. IEEE Computer Society, Los Alamitos, CA (2006)
23. Weibull, J.W.: Evolutionary Game Theory. MIT Press, Cambridge, MA (1995)

A Bound for Non-subgraph Isomorphism

Christian Schellewald*

School of Computing, Dublin City University, Dublin 9, Ireland
Christian.Schellewald@computing.dcu.ie
http://www.computing.dcu.ie/~cschellewald/

Abstract. In this paper we propose a new lower bound to a subgraph isomorphism problem. This bound can provide a proof that no subgraph isomorphism between two graphs can be found. The computation is based on the SDP relaxation of a – to the best of our knowledge – new combinatorial optimisation formulation for subgraph isomorphism. We consider problem instances where only the structures of the two graph instances are given and therefore we deal with simple graphs in the first place. The idea is based on the fact that a subgraph isomorphism for such problem instances always leads to 0 as lowest possible optimal objective value for our combinatorial optimisation problem formulation. Therefore, a lower bound that is larger than 0 represents a proof that a subgraph isomorphism don't exist in the problem instance. But note that conversely, a negative lower bound does not imply that a subgraph isomorphism must be present and only indicates that a subgraph isomorphism is still possible.

1 Introduction

The graph isomorphism problem is a well known problem in computer science and usually involves also the problem of finding the appropriate matching. Therefore it is also of interest in computer vision. If an object is represented by a graph the object could be identified as subgraph within a possibly larger scene graph. *Error-correcting* graph matching [1] – also known as *error-tolerant* graph matching – is a quite general and appropriate approach to calculate an assignment between the nodes of two graphs. It is based on the minimisation of so called *graph edit costs* which appear when one graph is turned into the other by some predefined edit operations. Commonly introduced graph edit operations are deletion, insertion, and substitution of nodes and edges. Each graph edit operation has a cost assigned which is application dependent. The minimal graph edit cost defines the so called *edit distance* between two graphs. The idea to define the edit distance for graph matching goes back to Sanfeliu and Fu [2] in 1983. Before that the edit distance was mainly used for string matching. Several algorithms for error correcting graph matching have been proposed that are based on different methods like tree search [3], genetic algorithms [4] and others (see e.g. [1]).

---

* This research was supported by Marie Curie Intra-European Fellowships within the 6th European Community Framework Programme.

F. Escolano and M. Vento (Eds.): GbRPR 2007, LNCS 4538, pp. 71–80, 2007.

In this paper we first propose an combinatorial optimisation formulation for the subgraph isomorphism problem that can be seen as a error-correcting graph matching approach. The integer optimisation problem we end up with is generally an indefinite quadratic integer optimisation problem which is NP-hard [5]. For example Pardalos and Vavasis showed in that indefinite quadratic programs are NP-hard problems, even if the quadratic program is very simple (see [6]). Then we compute a (convex) SDP relaxation of the combinatorial problem to obtain a lower bound to the subgraph isomorphism problem. The bound can be computed with standard methods for semidefinite programs. Finally we show that the bound can indeed be used to proof that no subgraph isomorphism between two graphs can be found.

Several approaches have been proposed to tackle the subgraph isomorphism problem [7,8,3,9]. Our approach differs to a more recent proposed approach that is based on a reformulation to a largest clique problem [10,11]. Our approach intends to find the full first graph as an subgraph isomorphism in the second graph where the largest clique represents the largest common subgraph isomorphism.

2   Preliminaries

In this work we consider simple graphs $G = (V, E)$ with nodes $V = \{1, \ldots, n\}$ and edges $E \subset V \times V$. We denote the first possibly smaller graph with G_K and the second graph with G_L. The corresponding sets V_K and V_L contain $K = |V_K|$ and $L = |V_L|$ nodes respectively. We assume that $L \geq K$. We make extensive use of the direct product $C = A \otimes B$, which is also known as Kronecker product [12]. It is the product of every matrix element A_{ij} of $A \in \mathbb{R}^{n \times m}$ with the whole matrix $B \in \mathbb{R}^{p \times q}$ resulting in the larger matrix $C \in \mathbb{R}^{np \times mq}$.

The subgraph isomorphism is a mapping $m : V_K \mapsto V \subset V_L$ of all nodes in the graph G_K to a subset V of V_L with K nodes of the graph G_L such that the structure is preserved. That means that any two nodes i and j from G_K that are adjacent must be mapped to nodes $m(i)$ and $m(j)$ in G_L that are adjacent too. The same has to be true for the inverse mapping $m^{-1} : V \mapsto V_K$ which maps the nodes V of the subgraph to nodes V_K of G_K.

3   Combinatorial Objective Function

In this section we propose and proof a formulation of the combinatorial problem of finding a sub-graph isomorphism. The general idea is to find a bipartite matching between the set of nodes from the smaller graph to the set of nodes of the larger graph. The bipartite matching is evaluated by an objective function that can be interpreted as a comparison of the structure between all possible node pairs in the first graph and the structure of the node pairs to which the nodes are matched in the second graph. A matching that leads to no structural differences has no costs and represents a sub-graph isomorphism. Mathematically the evaluation can be performed by a simple quadratic objective function $x^\top Q x$. The full task of finding a sub-graph isomorphism results in the following combinatorial quadratic optimisation problem, which details are explained below:

$$\min_x x^\top Q x$$

$$\text{s.t. } A_K x = e_K , \ A_L x \le e_L \tag{1}$$

$$x \in \{0,1\}^{KL}$$

The constraints that make use of the matrices $A_K = I_K \otimes e_L^\top$ and $A_L = e_K^\top \otimes I_L$ ensure that the vector x is a 0,1-indicator vector which represents a bipartite matching between the two node sets of the graphs. Here $e_n \in \mathbb{R}^n$ represents a vector with all elements 1 and $I_n \in \mathbb{R}^{n \times n}$ denotes the unit matrix. A vector element $x_{ji} = 1$ indicates that the node i of the first set V_K is matched to the node j in the second set V_L otherwise $x_{ji} = 0$. The elements of the indicator vector $x \in \{1,0\}^{KL}$ are ordered as follows:

$$x = (x_{11}, \cdots, x_{L1}, x_{12}, \cdots, x_{L2}, \cdots, x_{1K}, \cdots, x_{LK})^\top. \tag{2}$$

We illustrate such an indicator vector in figure 1 where a bipartite matching between two sets of nodes and the corresponding indicator vector are shown. The matrix Q within the objective function $x^\top Q x$ of the optimisation problem

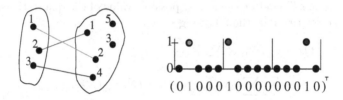

$$(0\,1\,0\,0\,0\,1\,0\,0\,0\,0\,0\,0\,0\,1\,0)^\top$$

Fig. 1. The illustration of the 0, 1-indicator vector on the right side is a representation of the matching which is shown on the left side of this figure

(1) can be written in a short form using the Kronecker product:

Definition 1. *Relational Structure Matrix*

$$Q = N_K \otimes \bar{N}_L + \bar{N}_K \otimes N_L \tag{3}$$

Here N_K and N_L are the 0, 1-adjacency matrices of the two graphs. The matrices $\bar{N}_K$ and $\bar{N}_L$ represent the complementary adjacency matrices which are computed as follows:

Definition 2. *Complementary Adjacency Matrices*

$$\bar{N}_L = E_{LL} - N_L - I_L \qquad \bar{N}_K = E_{KK} - N_K - I_K$$

These complementary adjacency matrices can be interpreted as 0, 1-indicator matrices for *non-adjacent* nodes. They have the element $(\bar{N})_{ij} = 1$ if the

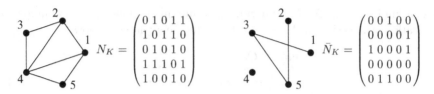

Fig. 2. An example graph and its adjacency matrix N_K along with its complementary adjacency matrix $\bar{N}_K$

corresponding nodes i and j are not directly connected in the graph. The adjacency matrix N_K for a small graph along with its complementary adjacency matrix are shown in figure 2. In the following we show that a 0,1-solution vector x^* of the optimisation problem (1) with an optimal objective value of zero represents a subgraph isomorphism. We first show that zero is the smallest possible value and than we show that every derivation of a subgraph isomorphism results in an objective value > 0.

Proposition 1. *The minimal value of the combinatorial optimisation problem (1) is zero.*

Proof. The elements of Q and x are all non-negative. In fact the elements are either zero or one. Therefore the lowest possible value of the quadratic cost term which can be rewritten as the following sum

$$x^\top Q x = x^\top (N_K \otimes \bar{N}_L + \bar{N}_K \otimes N_L)x$$

$$= \sum_{a,r}^{K,L} \sum_{b,s}^{K,L} [(N_K)_{ab}(\bar{N}_L)_{rs} + (\bar{N}_K)_{ab}(N_L)_{rs}]x_{ra}x_{sb} \qquad (4)$$

is zero. □

Proposition 2. *A solution with the minimal value of zero of the quadratic optimisation problem (1) represents a sub-graph isomorphism.*

To proof this we consider the term within the sum and show it leads only to a cost > 0 if the considered matching violates the condition for a subgraph isomorphism.

Proof. Only if the product $x_{ra}x_{sb}$ is one the term within the sum (4) can be different from zero and the part $[(N_K)_{ab}(\bar{N}_L)_{rs} + (\bar{N}_K)_{ab}(N_L)_{rs}]$ must be considered. In the following we refer to this part of the term also as *structure comparison term*. There are two cases that lead to $x_{ra}x_{sb} = 1$:

 – Case A: The node a and node b in G_K represent the same node ($a = b$). But as the diagonals of N_K and $\bar{N}_K$ are zero one obtains that $(N_K)_{aa} = 0$ and $(\bar{N}_K)_{aa} = 0$. In this case the term $[(N_K)_{aa}(\bar{N}_L)_{rr} + (\bar{N}_K)_{aa}(N_L)_{rs}]x_{ra}x_{ra}$ is always zero and does not contribute to the sum.

- Case B: The nodes a and b in G_K represent different nodes $(a \neq b)$ in G_K and due to the bipartite matching constraint a value $x_{ar} x_{bs} = 1$ represents the situation $x_{ar} = 1$ and $x_{bs} = 1$ which means that the nodes a and b are mapped to two different nodes r and s in the second graph G_L, respectively. Considering now the term $[(N_K)_{ab}(\bar{N}_L)_{rs} + (\bar{N}_K)_{ab}(N_L)_{rs}]$ we observe that all four possible structural cases between two pairs of nodes in the two graphs are valued with a cost of zero or one.

All these sub-cases from case B that could lead to a non-zero value in the structure comparison term and therefore in the sum are listed in the table 1. In the following we summarise the meaning of the cases and we will see that costs are only added for every difference between the structure of graph G_K and the considered subgraph of the second graph G_L.

Table 1. List of all outcomes of the structure comparison term between two different nodes a and b of graph G_K that are mapped to two different nodes r and s in the second graph G_L. Only in case I and IV the structure is preserved and can lead to an isomorphism. No cost is added in this cases. The other cases (II and III) don't preserve the structure and lead to an total cost > 0. For details see the text.

case	configuration	$(N_K)_{ab}$	$(\bar{N}_L)_{rs}$	$(\bar{N}_K)_{ab}$	$(N_L)_{rs}$	cost
I	a,b adjacent; r,s adjacent	1	0	0	1	0
II	a,b adjacent; r,s not adjacent	1	1	0	0	1
III	a,b not adjacent; r,s adjacent	0	0	1	1	1
IV	a,b not adjacent; r,s not adjacent	0	1	1	0	0

- I: If the two nodes a and b in the first graph are neighbours, $(N_K)_{ab} = 1$, then no cost is added in (4) if the nodes r and s in the scene graph are neighbours, too: $(\bar{N}_L)_{rs} = 0$.
- II: Otherwise if a and b are neighbours in G_K and the corresponding nodes r and s are no neighbours in the second graph, $(\bar{N}_L)_{rs} = 1$, then a cost of 1 is added.

The configurations I and II are visualised in figure 3.

- III: Analogously, the structure comparison term penalises assignments where pairs of nodes (a and b) in the graph G_K become neighbours in the second graph G_L which were not adjacent before.
- IV: Finally if a and b are not adjacent in the first graph G_K and the nodes r and s in G_L are also not adjacent, no cost is added.

Figure 4 illustrates situation III and IV in detail.

This shows that only mappings that lead to a change in the structure are penelised with a cost. Structure preserving mappings which are compatible with a subgraph isomorphism are without costs.

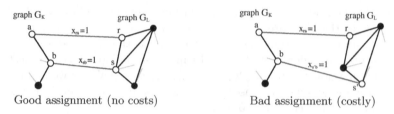

Good assignment (no costs)   Bad assignment (costly)

Fig. 3. Left: Adjacent nodes a and b in the graph G_K are assigned to adjacent nodes r and s in the graph G_L. **Right.** Adjacent nodes a and b are no longer adjacent in the graph G_L after the assignment. The left assignment leads to no additional costs while the right undesired assignment adds 1 to the total cost.

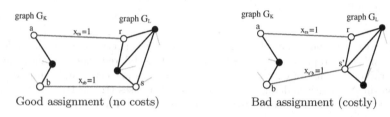

Good assignment (no costs)   Bad assignment (costly)

Fig. 4. Left: Nodes a and b which are not adjacent in the object graph G_K are assigned to nodes which are also not adjacent in the scene graph G_L. **Right:** A pair of nodes a and b become neighbours r and s' after assignment. The left assignment is associated with no additional costs in (4). The undesired assignment on the right side adds 1 to these costs.

Note that due to the symmetry of the adjacency matrices the quadratic cost term $x^\top Q x$ is symmetric too and every difference in the compared structures of the two graphs is considered twice resulting in a cost of 2 for every difference in the structure.

Finally the sum (4) and therefore the objective function $x^\top Q x$ considers all possible combinations of node pairs a and b that are mapped to r and s, respectively. And only for matchings which lead to no difference in the mapped sub-structure and vice versa all the terms within the sum (4) are zero. In this case the bipartite matching represents a subgraph isomorphism.   □

We wish to emphasise that the minimisation of (1) represents the search for a bipartite matching which has the smallest possible structural deviation between G_K and the considered subgraph of G_L. Therefore (1) can be seen as a graph edit distance with a cost of 2 for each addition or removal of an edge that is needed to turn the first graph into the considered subgraph of the other graph.

4   Convex Problem Relaxation

The combinatorial isomorphism approach (1) can be relaxed to a (convex) semidefinite program (SDP) which has the following standard form:

$$\min \; \mathrm{Tr}\left[\tilde{Q}X\right]$$

$$\text{s.t. } \mathrm{Tr}[A_i X] = c_i \quad \text{for} \quad i = 1, \ldots, m \tag{5}$$

$$X \succeq 0$$

The constraint $X \succeq 0$ means that X has to be positive semidefinite. This convex optimization problem can be solved with standard methods like interior point algorithms (see e.g. [13]). Note that the solution of the relaxation (5) provides a lower bound to (1). Below, we describe how we derive such a semidefinite program from (1). For more information on semidefinite programming we refer to [14].

5   Convex Relaxation

The convex relaxation in this section follows the relaxation explained in detail in [15]. In order to obtain an appropriate SDP relaxation for the combinatorial subgraph matching problem, we start with the reformulation of the objective function of (1)

$$f(x) = x^\top Q x \;=\; \mathrm{Tr}\left[x^\top Q x\right] \;=\; \mathrm{Tr}\left[Q x x^\top\right] \;=\; \mathrm{Tr}\left[\tilde{Q}X\right], \tag{6}$$

We take into account the following summarised constraints of the form $\mathrm{Tr}[A_i X] = c_i$ which intend to include the original bipartite matching constrainst in a suitable way. In particular we describe the constraint marices A_i. The equality constraint $\sum_{j=1}^{L} x_{ij} = 1, i = 1, \ldots, K$, which are part of the bipartite matching constraints represent the constraint that each node of the smaller graph is mapped to exactly one node of the scene graph. We define K constraint matrices $^{\mathrm{sum}}A^j \in \mathbb{R}^{(KL+1)\times(KL+1)}, j = 1, \ldots, K$ which ensure (taking the order of the diagonal elements into account) that the sum of the appropriate portion of the diagonal elements of X is 1. As we deal with the diagonal elements of X we exploit also the fact that $x_i = x_i^2$ holds true for 0/1-variables. The matrix elements for the j-th constraint matrix $^{\mathrm{sum}}A^j$ can be expressed as follows:

$$^{\mathrm{sum}}A^j_{kl} = \sum_{i=(j-1)L+1}^{jL+1} \delta_{ik}\delta_{il} \quad \text{for} \quad k,l = 1, \ldots, KL+1$$

For these constraints the constants c_j are: $c_j, j = 1, \ldots, K$. As all integer solutions $\tilde{X} = xx^\top \in \mathbb{R}^{KL \times KL}$, where x represents a bipartite matching, have zero-values at those matrix positions where $I_K \otimes (E_{LL} - I_L)$ and $(E_{KK} - I_K) \otimes I_L$ have non-zero elements we want to force the corresponding elements in $X \in \mathbb{R}^{KL \times KL}$ to be zero. The matrices $E_{LL} \in \mathbb{R}^{L \times L}$ and $E_{KK} \in \mathbb{R}^{R \times R}$ are matrices where all elements are 1. The matrices $I_{nn} \mathbb{R}^{n \times n}$ represent the unit matrices. This can be achieved with the constraint matrices $A^{ars}, A^{\hat{s}\hat{a}\hat{b}} \in \mathbb{R}^{KL \times KL}$ which are determined by the indices a, r, s and $\hat{s}, \hat{a}, \hat{b}$. They have the following matrix elements

$$A^{ars}_{kl} = \delta_{k,(aL+r)}\delta_{l,(aL+s)} + \delta_{k,(aL+s)}\delta_{l,(aL+r)}, \tag{7}$$

$$A^{\hat{s}\hat{a}\hat{b}}_{kl} = \delta_{k,(\hat{s}K+\hat{b})}\delta_{l,(\hat{s}K+\hat{a})} + \delta_{k,(\hat{s}K+\hat{a})}\delta_{l,(\hat{s}K+\hat{b})}, \tag{8}$$

where $k, l = 1, \ldots, KL$.

The indices a, r, s and $\hat{s}, \hat{a}, \hat{b}$ attain all valid combinations of the following triples where $s > r$ and $\hat{b} > \hat{a}$:

$$(a, r, s) : \quad a = 1, \ldots, K; \; r = 1, \ldots, L; \; s = (r + 1), \ldots, L$$
$$(\hat{s}, \hat{a}, \hat{b}) : \quad \hat{s} = 1, \ldots, L; \; \hat{a} = 1, \ldots, K; \; \hat{b} = (\hat{a} + 1), \ldots, K$$

For this constraints the constant c has to be zero. With this we define $(LL - L)K/2 + (KK - K)L/2$ additional constraints that ensure zero-values at the corresponding matrix positions of X.

6   Early Results to the Non-isomorphism Bound

For the early results presented in this section we used our implementation described in [15] where we had to set the similarity vector to a zero vector. Furthermore we introduced a parameter $\alpha > 0$ which is just a scaling parameter for the objective function and should not have a influence on the solution other than a scaling. An illustrative example for a subgraph isomorphism problem is depicted in figure 5. For this example we compute a lower bound > 0 using the SDP relaxation (5), which proves that a subgraph isomorphism does not exist in this problem instance. Note that we did not eliminate mappings that could not lead to an subgraph isomorphism.

The possible objective values of (1) are restricted to discrete values as the quadratic term $\alpha x^\top Q x$ can only reach values which are multiples of 2α. The discrete distribution of the objective values for the subgraph isomorphism problem shown in figure 5 is depicted in figure 6 where we have set $\alpha = 0.3$. For a first preliminary investigation of this bound we created 1000 small subgraph matching problem instances for which we have chosen the size of the two graphs G_K and G_L to be $K = 7$ and $L = 15$, respectively. The edge probability of the graph G_K was set to 0.5 and the probability for an edge in the second graph was set to 0.2. The results for this experiment series reveal that for various problem instances it is indeed possible to conclude that no subgraph isomorphism exist. We have obtained 388 problem instances with

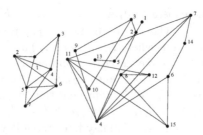

Fig. 5. Example for a randomly created subgraph problem. Is there a subgraph isomorphism ? For the shown problem instance we can compute a lower bound > 0 for (1) which proves that no subgraph isomorphism is present.

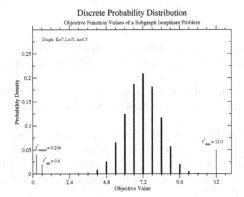

Fig. 6. The distribution of the objective values for the subgraph isomorphism problem which is shown in figure 5. The objective values are restricted to discrete values, as the quadratic term $\alpha x^\top Q x$ can only attain values which are multiples of 2α. Here we have set α arbitrarily to 0.3. The optimal objective value is 0.6 and the obtained lower bound is $0.204 > 0.0$, which is a non-isomorphism proof for this problem instance.

a lower bound > 0.0 which proves that no subgraph isomorphism can occur in this problem instances. The other 612 problem instances have a lower bound ≤ 0.0. For 436 ($\approx 71\%$) of these problem instances the combinatorial optimum is > 0.0 indicating that the relaxation is not tight enough to detect that no subgraph isomorphism can occur.

7 Discussion

We proposed a bound to the subgraph isomorphism problem and showed that the bound is not only of theoretical interest but also applies to several instances of subgraph matching problems. It would be interesting to investigate which criteria a subgraph matching problem has to fullfill to result in a tight relaxation. Such insights could be usefull in the process of creating or obtaining object graphs from images for object recognition tasks.

The tightness and therefore the lower bound can be improved by reducing the dimension of the problem size. For example one can eliminate a mapping $i \mapsto j$ if the degree (The number of incident edges.) of an node i is larger than the degree of node j in the second graph. Such a mapping cannot lead to a subgraph isomorphism. An other improvement could be expected when also inequalities are included in the SDP relaxation. None of these improvements are used for the presented results. However, for increasing problem instances the relaxation will probably get less tight and a lower bound ≤ 0.0 becomes more likely. But note that even less tight solutions still lead to good integer solutions (see e.g. [15]).

References

1. Bunke, H.: Error correcting graph matching: On the influence of the underlying cost function. IEEE Trans. Pattern Analysis and Machine Intelligence 21(9), 917–922 (1999)
2. Sanfeliu, A., Fu, K.S.: A distance measure between attributed relational graphs for pattern recognition. IEEE Transaction on Systems, Man and Cybernetics 13(3), 353–362 (1983)
3. Messmer, B.T., Bunke, H.: A new algorithm for error-tolerant subgraph isomorphism detection. IEEE Trans. Patt. Anal. Mach. Intell. 20(5), 493–504 (1998)
4. Wang, Y.-K., Fan, K.-C., Horng, J.-T.: Genetic-based search for error-correcting graph isomorphism. IEEETSMC: IEEE Transactions on Systems, Man, and Cybernetics, 27 (1997)
5. Garey, M.R., Johnson, D.S.: Computers and Intractability, a Guide to the Theory of NP-Completeness. W. H. Freeman and Company, New York (1991)
6. Pardalos, P.M., Vavasis, S.A.: Quadratic programming with one negative eigenvalue is np-hard. J. Global Optim. 1, 15–22 (1991)
7. Ullmann, J.R.: An algorithm for subgraph isomorphism. Journal of the ACM 23(1), 31–42 (1976)
8. Barrow, H.G., Burstall, R.M.: Subgraph isomorphism, matching relational structures and maximal cliques. Information Processing Letters 4(4), 83–84 (1976)
9. Eppstein, D.: Subgraph isomorphism in planar graphs and related problems. Journal of Graph Algorithms and Applications 3(3), 1–27 (1999)
10. Bomze, I., Budinich, M., Pardalos, P., Pelillo, M.: The maximum clique problem. In: Du, D.-Z., Pardalos, P.M. (eds.) Handbook of Combinatorial Optimization, vol. 4, Kluwer Academic Publishers, Boston, MA (1999)
11. Pelillo, M.: Replicator equations, maximal cliques, and graph isomorphism. Neural Computation 11(8), 1933–1955 (1999)
12. Graham, A.: Kronecker Products and Matrix Calculus with Applications. Ellis Horwood Limited and John Wiley and Sons (1981)
13. Borchers, B.: CSDP: A C library for semidefinite programming. Optimization Methods and Software 11(1), 613–623 (1999)
14. Wolkowicz, H., Saigal, R., Vandenberghe, L. (eds.): Handbook of Semidefinite Programming. Kluwer Academic Publishers, Boston (2000)
15. Schellewald, C., Schnörr, C.: Probabilistic subgraph matching based on convex relaxation. In: Rangarajan, A., Vemuri, B., Yuille, A.L. (eds.) EMMCVPR 2005. LNCS, vol. 3757, pp. 171–186. Springer, Heidelberg (2005)

A Correspondence Measure for Graph Matching Using the Discrete Quantum Walk

David Emms, Edwin R. Hancock, and Richard C. Wilson

Department of Computer Science
University of York
YO10 5DD, UK

Abstract. In this paper we consider how coined quantum walks can be applied to graph matching problems. The matching problem is abstracted using an auxiliary graph that connects pairs of vertices from the graphs to be matched by way of auxiliary vertices. A coined quantum walk is simulated on this auxiliary graph and the quantum interference on the auxiliary vertices indicates possible matches. When dealing with graphs for which there is no exact match, the interference amplitudes together with edge consistencies are used to define a consistency measure. We have tested the algorithm on graphs derived from the NCI molecule database and found it to significantly reduce the space of possible matchings thereby allowing the graphs to be matched directly. An analysis of the quantum walk in the presence of structural errors between graphs is used as the basis of the consistency measure. We test the performance of this measure on graphs derived from images in the COIL-100 database.

1 Introduction

Quantum algorithms have recently attracted considerable attention in the theoretical computer science community. This is primarily because they offer considerable speed-up over classical algorithms. For instance, Grover's [8] search method is polynomially faster than its classical counterpart, and Shor's factorization method is exponentially faster than classical methods. However, quantum algorithms also have a richer structure than their classical counterparts since they use qubits rather than bits as the basic representational unit [14]. Consequentially, an n qubit quantum computer would manipulate a state in $\mathbf{C}^{2^n}$ as opposed to $\mathbf{Z}_2^n$, which is the case classically. For instance, this structure is exploited in Shor's algorithm where the Fourier transform is used to locate prime factors.

It is this issue of richer representations that is the subject of this paper. We are interested in how the idea of quantum walks can be applied to the problem of graph matching. From a practical perspective, there have been a number of useful applications of random walks. One of the most important of these is the analysis of routing problems in network and circuit theory. Of more recent interest is the use of ideas from random walks to define the page-rank index for internet search engines such as Googlebot [4].

F. Escolano and M. Vento (Eds.): GbRPR 2007, LNCS 4538, pp. 81–91, 2007.
© Springer-Verlag Berlin Heidelberg 2007

In the pattern recognition community there have been several attempts to use random walks for graph matching. These include the work of Robles-Kelly and Hancock which has used both a standard spectral method and a more sophisticated one based on ideas from graph seriation to convert graphs to strings, so that string matching methods may be used [15]. Meila and Shi use a random walk based on pairwise similarities between image pixels to carry out clustering and thus segmentation of images [11]. Gori, Maggini and Sarti [7] on the other hand, have used ideas borrowed from page-rank to associate a spectral index with graph nodes and have then used standard subgraph isomorphism methods for matching the resulting attributed graphs. In addition, Nadler, Lafon and Coifman [12] used random walks to define a diffusion distance between data points in order to carry out clustering and dimensionality reduction.

Quantum walks have been introduced as quantum counterparts of random walks [10] and posses a number of interesting properties not exhibited by classical random walks. The paths of the coined quantum walk have been used to define a matrix representation of graphs that is able to lift the cospectrality of certain classes of graphs that are typically hard to distinguish [6,5].

In this paper, we present a novel auxiliary graph structure, based on a pair of graphs to be matched, and simulate a coined quantum walk on this structure. The auxiliary structure contains auxiliary vertices connecting each pair of vertices from the two graphs. It is on these auxiliary vertices that the two walks interfere, and by identifying where this interference is exact we are able to identify matches between the graphs. To test the algorithms effectiveness at finding isomorphisms, we carry out experiments using graphs from the US National Cancer Institute database of molecules [1]. In addition, we carry out a sensitivity analysis in order to investigate its behaviour in the presence of structural errors. This allows us to define a 'consistency measure' between graphs which utilizes probabilities that pairs of vertices from the two graphs match. We use this consistency measure to cluster graphs derived from real-world data.

2   The Coined Quantum Walk

In what follows, we present a brief overview of the coined quantum walk. Let $G = (V_G, E_G)$ be a graph with vertex set V_G and edge set, $E_G = \{\{u, v\} | u, v \in V_G$, u adjacent to $v\}$. The degree of a vertex $u \in V_G$, denoted $d(u)$, is the number of vertices adjacent to u. Quantum processes are reversible, and in order to make the walk reversible a particular state of the walk must give both the current location of the walk and its previous location [2]. To this effect each edge $\{u, v\} \in E$ is replaced by a pair of directed arcs (u, v) and (v, u) and the set of these arcs is denoted D_G. The basis states for the quantum walk are vectors in a Hilbert space, $\mathcal{H} \cong \mathbb{C}^{|D_G|}$, and are denoted, using Dirac's bra-ket notation, as $|uv\rangle$, where $(u, v) \in D_G$. Such a state is interpreted as the walk being at vertex v having been at u. A general state for the walk is written as a 'superposition'

$$|\psi\rangle = \sum_{(u,v)\in D_G} \alpha_{uv}|uv\rangle$$

where $\alpha_{uv} \in \mathbb{C}$.

The probability that a walk is in a particular state is given by, $P(|uv\rangle) = \alpha_{uv}\alpha_{uv}^*$ where α_{uv}^* is the complex conjugate of α_{uv}. Thus a state is normalized such that $\sum_{(u,v)\in D_G} \alpha_{uv}\alpha_{uv}^* = 1$. For the purpose of this work the amplitudes will be real, albeit negative as well as positive. The fact that states can have negative amplitudes is of key importance as it allows various paths to cancel out (destructive interference) and this is utilized by our algorithm.

The evolution is linear and conserves probabilities, in addition it respects the connectivity structure of the graph and transitions are only allowed between adjacent vertices. Consider a state $|\psi\rangle = |uv\rangle$ where the degree of v, $d(v) = r+1$. That is, v is adjacent to u and a further r vertices, $w_1, w_2, \ldots, w_r$. For this state one step of the walk is such that

$$|uv\rangle \rightarrow a|vu\rangle + b\sum_{i=1}^{r} |vw_i\rangle \quad a, b \in \mathbb{C}.$$

Two separate amplitudes, a and b, can be used since the transitions from $|uv\rangle$ to $|vu\rangle$ can be distinguished from those to $|vw_i\rangle$ without reference to any (arbitrary) labellings of the vertices or edges. Since probability must be conserved, $a^2 + rb^2 = 1$. It is usual to use the 'Grover diffusion operators' [8] for the walk, which are such that $a = 2/d(v) - 1$ and $b = 2/d(v)$ (for transitions from vertex v) since these provide the transition operator that is furthest from the identity.

For a general graph, with edges replaced by arcs as described above, the real-orthogonal matrix, U_G, governing the evolution of the walk can be written as

$$U_G((u,v),(w,x)) = \begin{cases} \frac{2}{d(v)} - \delta_{u,x} & \text{if } v = w; \\ 0 & \text{otherwise.} \end{cases}$$

for all $(u,v),(w,x) \in D_G$, where $\delta_{u,x}$ is the Kronecker delta.

3   Exact Matching

Given two graphs, G and H, the basis of our approach is to create an auxiliary graph, $\Gamma(G,H)$ on which the talk walks can interfere. This graph is symmetric with respect to interchanging its two arguments, and is such that if the two graphs are isomorphic, deconstructive quantum interference will take place. Furthermore, if they are isomorphic, the deconstructive interference will be exact, in the sense that the quantum amplitudes will be zero on certain special states. Moreover, these states directly indicate pairs of matching vertices.

Given a pair of isomorphic graphs $G = (V_G, E_G)$ and $H = (V_H, E_H)$ we would like to find a mapping, $\phi : V_G \rightarrow V_H$ such that $\{\phi(u), \phi(v)\} \in E_H$ if and only if $\{u, v\} \in E_G$. The algorithm operates by taking the two graphs that are to be matched and connecting all pairs of vertices where one vertex is from G and one is from H. The coined quantum walk is then simulated on this auxiliary graph. The intermediate vertices provide states on which quantum interference takes place between the two walks, the final step is to simulate this interference to give a set of quantum amplitudes indicating possible matches between pairs of vertices.

More precisely, we form a new graph $\Gamma = (V_\Gamma, E_\Gamma)$ whose vertex and edge sets are given by $V_\Gamma = V_G \cup V_H \cup V_A$ and $E_\Gamma = E_G \cup E_H \cup E_A$, where V_A is the set of auxiliary vertices and E_A the set if edges connecting the two graphs by way of these auxiliary vertices. That is

$$V_A = \{v_{\{g_i,h_j\}} | g_i \in V_G, h_j \in V_H\}$$
$$E_A = \{\{g_i, v_{\{g_i,h_j\}}\}, \{h_j, v_{\{g_i,h_j\}}\} | g_i \in V_G, h_j \in V_H\}.$$

Thus the vertices $g_i \in V_G$ and $h_j \in V_H$ are linked via an auxiliary vertex, denoted $v_{\{g_i,h_j\}}$. The structure of the auxiliary graph is shown in Figure 1. The auxiliary graph is similar to the association graph [3]. However, information about the structure of the two graphs comes from incorporating the original graphs themselves rather than through the connections between the auxiliary vertices, as is done in the association graph.

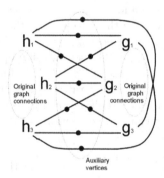

Fig. 1. The auxiliary graph, $\Gamma(G,H)$, showing the vertices $g_1, g_2, g_3 \in V_G$ and $h_1, h_2, h_3 \in V_H$ connected by way of auxiliary vertices

The walk evolves in discrete steps according to the rule $|\psi^{t+1}\rangle = U_\Gamma |\psi^t\rangle$ from a starting state $|\psi^0\rangle$ with initial amplitudes,

$$\alpha_{uv}^0 = \begin{cases} 1 \text{ if } \{u,v\} \subset V_G \text{ or } \{u,v\} \subset V_H; \\ 0 \text{ otherwise.} \end{cases}$$

After the walk has been evolved for a given number of steps, T, an interference operator, R, acts on the state, giving the final state $|\psi'\rangle = R|\psi^T\rangle$. The interference operator is such that, for all pairs of vertices, $g \in V_G$ and $h \in V_H$, the difference between the amplitudes of the corresponding states $|gv_{\{g,h\}}\rangle$ and $|hv_{\{g,h\}}\rangle$ determines the final amplitude of the state $|hv_{\{g,h\}}\rangle$. For these pairs of states, in the basis $(|gv_{\{g,h\}}\rangle, |hv_{\{g,h\}}\rangle)$, the step corresponds to the application of the Hadamard operator

$$\begin{pmatrix} \alpha'_{gv_{\{g,h\}}} \\ \alpha'_{hv_{\{g,h\}}} \end{pmatrix} = \frac{1}{\sqrt{2}} \begin{pmatrix} 1 & 1 \\ 1 & -1 \end{pmatrix} \begin{pmatrix} \alpha^T_{gv_{\{g,h\}}} \\ \alpha^T_{hv_{\{g,h\}}} \end{pmatrix}.$$

Consider two graphs, G and H, such that there is an isomorphism $\phi : V_G \to V_H$ between them. If $g \in V_G$ and $h \in V_H$ are two vertices such that $\phi(g) = h$ then, as a result of the symmetry of the auxiliary graph and starting state, $\alpha^t_{gv_{\{g,h\}}} = \alpha^t_{hv_{\{g,h\}}}$ for all times, t. Consequently, in the final state, the amplitude $\alpha'_{hv_{\{g,h\}}} = 0$ whenever $\phi(g) = h$. Thus we can use the amplitudes of these states to identify possible isomorphisms between the two graphs.

4 Structural Errors

As described above, if the two graphs are isomorphic then the amplitudes on the interference states for matching pairs of vertices will all be zero. In many situations, however, there will exist structural errors and so no complete isomorphism will exist. In this case it is not guaranteed that the amplitudes for the 'correct' matches between vertices will be exactly zero.

In order to investigate the robustness of the method in the presence of such errors we generated a random graph together with a partner that differed from it by a set number of edges. We found that the amplitudes for false matches could be modelled as a Gaussian with standard deviation, σ_f, and zero mean. The distribution for true matches on the other hand was much better modelled by the more strongly peaked double-exponential distribution with standard deviation σ_t and zero mean where $\sigma_t < \sigma_f$ (Fig. 2):

$$\text{Gaussian: } p(\alpha|f) = \frac{e^{-\frac{\alpha^2}{2\sigma_f^2}}}{\sqrt{2\pi}\sigma_f}, \quad \text{Double exponential: } p(\alpha|t) = \frac{e^{-\frac{\sqrt{2}|\alpha|}{\sigma_t}}}{\sigma_t\sqrt{2}}$$

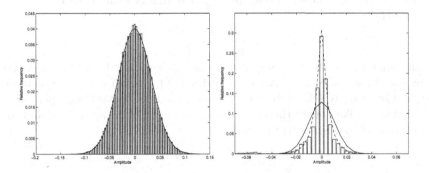

Fig. 2. The distribution of amplitudes for 'non-matching' vertices (left) and 'matching' vertices (right) for pairs of graphs on 15 vertices, differing by 2 edges, for 50 pairs of graphs. A Gaussian distribution has been fitted to the non-matching vertices and a Gaussian (solid line) and double-exponential (dashed line) distributions for the matching vertices.

Since the distribution for corresponding vertices is far more strongly peaked, we hope to still be able to distinguish between the distributions even in the presence of noise. Calculating the probability of a true match given the amplitude, α, can be achieved by applying Bayes' rule and this information can be combined with structural constraints for inexact graph matching tasks.

Given some observed interference amplitude, $\alpha_{hv_{\{g,h\}}}$, (which we will henceforth write simply as α), we wish to know the probability that the pair (g, h) is a true match. We model the probability distribution for α as the sum of the two distributions, the probability it originates from true match and the probability it originates from a false match. Hence,

$$p(\alpha) = p(\alpha|t)p(t) + p(\alpha|f)p(f)$$
$$= \frac{p(t)}{\sigma_t\sqrt{2}}e^{-\frac{\sqrt{2}|\alpha|}{\sigma_t}} + \frac{p(f)}{\sqrt{2\pi}\sigma_f}e^{-\frac{\alpha^2}{2\sigma_f^2}},$$

where $p(t) = 1 - p(f) = \frac{1}{n}$ and $n = |V_G| = |V_H|$. By applying Bayes' rule, we can find the probability of a true match given α using

$$p(t|\alpha) = \frac{p(\alpha|t)p(t)}{p(\alpha)}$$

$$= \frac{\frac{1}{\sigma_t n\sqrt{2}}e^{-\frac{\sqrt{2}|\alpha|}{\sigma_t}}}{\frac{1}{\sigma_t n\sqrt{2}}e^{-\frac{\sqrt{2}|\alpha|}{\sigma_t}} + \left(\frac{n-1}{n}\right)\frac{1}{\sqrt{2\pi}\sigma_f}e^{-\frac{\alpha^2}{2\sigma_f^2}}}.$$

Note that, to calculate $p(t|\alpha)$ it is necessary to estimate the variances of the two distributions, σ_t^2 and σ_f^2. Let X be the set of n^2 interference amplitudes between the vertices of G and those of H. The set X is the union of two disjoint sets; X_t, the set of interference amplitudes for the true matches and X_f, the set of interference amplitudes for the false matches. The variance of X, denoted σ^2, can be measured directly and this is related to the the other variances by

$$\sigma^2 = \frac{\sigma_t^2}{n} + \frac{(n-1)\sigma_f^2}{n}.$$

Clearly another equation linking these variances is required, and for this higher moments of the distribution X can be utilized. The fourth central moment of a distribution (closely related to the kurtosis) measures how strongly peaked the distribution is. We denote the forth central moments by μ, μ_t and μ_f, and they can be related to one another by

$$\mu = \frac{\mu_t}{n} + \frac{(n-1)\mu_f}{n}.$$

Since the fourth central moment of the Gaussian distribution is given by $\mu_f = 3\sigma_f^4$ and that for the double exponential distribution by $\mu_t = 6\sigma_t^4$, we have that

$$\mu = \frac{6\sigma_t^4}{n} + \frac{3(n-1)\sigma_f^4}{n}.$$

Since μ and σ^2 are known, the equations expressing these in terms of σ_t^2 and σ_f^2 can be solved simultaneously to give variances for the matching and non-matching amplitudes. Thus, given a set of interference amplitudes for the possible pairings of the vertices of G with the vertices of H, we are able to calculate the probability that a particular pairing is part of a correct match.

5   A Correspondence Measure from the Matching Probabilities

The probabilities for matches between pairs of vertices from the two graphs can be combined with structural information in order to give a 'correspondence measure', which quantifies the quality of the match between the two graphs. Consider two pairs of vertices (g_1, h_1) and (g_2, h_2), where $g_1, g_2 \in V_G$ and $h_1, h_2 \in V_H$, corresponding to two possible matches with probabilities $p(g_1, h_1)$ and $p(g_1, h_2)$ respectively. We could calculate a purely local correspondence measure between the graphs by summing the number of times that for such pairs (g_1, h_1) and (g_2, h_2), are edge consistent. That is, either $\{g_1, g_2\} \in E_G$ and $\{h_1, h_2\} \in E_H$, or $\{g_1, g_2\} \notin E_G$ and $\{h_1, h_2\} \notin E_H$. In general, as structural errors are introduced this measure will decrease. However, we can also introduce the global information encoded in the probabilities of matches derived from the quantum walk to improve this measure. The quantity $p(g_1, h_1)p(g_2, h_2)$ will in general be larger when both (g_1, h_1) and (g_2, h_2) are correct matches than if either, or neither, of them are. Thus we can define a correspondence measure by

$$M_E(G, H) = \frac{1}{T} \sum_{\{g_1, g_2\} \in E_G} \sum_{\{h_1, h_2\} \in E_H} p(g_1, h_1)p(g_2, h_2)$$

Alternatively, a correspondence measure can be given by

$$M_{\overline{E}}(G, H) = \frac{1}{T} \sum_{\{g_1, g_2\} \notin E_G} \sum_{\{h_1, h_2\} \notin E_H} p(g_1, h_1)p(g_2, h_2),$$

where $T = \sum_{g_1, g_2 \in V_G} \sum_{h_1, h_2 \in V_H} p(g_1, h_1)p(g_2, h_2)$ is used to normalize the measures. Thus, we include in the sum the term $p(g_1, h_1)p(g_2, h_2)$ if the match is consistent in terms of their edge connectivity.

Which of the two measures, $M_E(G, H)$ or $M_{\overline{E}}(G, H)$, gives the better results depends on the edge density of the graphs being compared. Let $|E_H| \approx |E_G| = |E|$, we have that the maximum number of edges for a graph on n vertices is, $E_{\max} = \frac{n(n-1)}{2}$. We have found using sets of randomly generated graphs with structural errors introduced that if $|E| < 0.5E_{\max}$ then the presence of an edge between a pair of vertices is more significant than the absence of an edge and so $M_E(G, H)$ gives a better measure. If, on the other hand, $|E| > 0.5E_{\max}$, the absence of an edge between a pair of vertices is more significant and hence $M_{\overline{E}}(G, H)$ gives a better measure (if $|E| \approx 0.5E_{\max}$ then both measure work well). So as to utilize both these measures when they are at their strongest we define the consistency measure that we use for all graphs to be $M(G, H) = aM_E(G, H) + bM_{\overline{E}}(G, H)$ where $b = |E|/E_{\max}$ and $a = 1 - b$. This measure gives the strengths of the two separate measures for both low edge-density and high-density graphs.

6   Experiments

6.1   Graph Matching

We present experiments carried out on graphs representing the structure of a subset of the molecules from the NCI database of molecules [1]. In the database, a particular molecule is represented as a graph with vertex attributes giving the type of atom and edges representing bonds. We disregard the type of atoms and only make use of the bond structure thereby giving a set of non-attributed graphs, a number of which are isomorphic, the goal being to identify these. The following experiments deal with a set of approximately 1400 of these on up to 28 vertices.

For a particular pair of graphs we use our algorithm to prune the space of possible matches and check the matches returned. If there are multiple matches we simply search through the possible matches in order. To avoid the possibility that this processes is aided by the ordering of the vertices we carry out a random permutation of the nodes of each graph. In some cases too many possible matches remain and so if after checking 1000 possible matches more remain then we class the question of whether two graphs are isomorphic as undecided. Some upper limit is needed using this method although the choice of 1000 is essentially arbitrary. The problem of large numbers of matches remaining could also be addressed by using a more sophisticated method of searching through these although we do not consider this here.

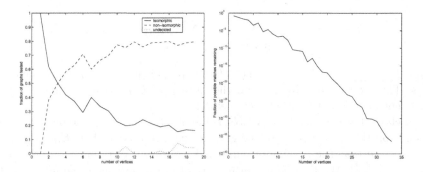

Fig. 3. Left: The proportion of graphs that are isomorphic, non-isomorphic or undecided as a function of the number of vertices. Right: Log plot of number of matches searched as a fraction of the total number of permutations in order to identify the correct permutations when isomorphic graphs are compared (undecided graphs excluded).

For the graphs tested Figure 3 shows the fraction of isomorphic graphs, non-isomorphic graphs and those that are undecided as a function of the number of vertices. We see that the algorithm is able to to reconstruct the match or identify that the graphs are non-isomorphic for these graphs with a low percentage undecided, this percentage can be lowered further by allowing more than 1000 matches to be checked.

In the second experiment, for each graph we carried out a random permutation of its vertices and then attempted to recover this permutation using our algorithm. In order to analyse the performance of the algorithm we considered the average number of matches that needed to be checked– once the search space had been pruned– in order to reconstruct the permutation, and recorded the fraction of graphs for which we did not recover the permutation (Figure 3 and Table 1). As can be seen, the quantum walk is able to significantly prune the size of the search space, and hence make the problem of finding matches for graphs on large numbers of vertices significantly easier.

6.2   Graph Clustering

Our second set of experiments test the ability of the consistency measure to cluster similar graphs. The graphs were derived from images in the COIL-100 database, a set of images of objects viewed from a series of angles [13]. To derive graphs from these images Harris and Stephen's corner detector [9] was used to detect feature points for the graph and the Delaunay triangulation of these points were taken. We use the Delaunay graphs derived in this way since they incorporate important structural information from the images.

Table 1. The fraction of the graphs tested for which more than 1000 matches still remain after the interference step of the walk

$|V|$	> 1000 matches	$|V|$	> 1000 matches
≤18	0	26	0.42
19	0.25	27	0.4
20	0.08	28	0.67
21	0.17	29	0.42
22	0.17	30	0.71
23	0.17	31	0.58
24	0.42	32	0.60
25	0.25		

For each object we took 9 graphs derived from a viewing the objects at equally space intervals of $10°$. We obtain graphs with on average 30 vertices and for each pair of graphs we calculate the consistency measure. Figure 4 shows a PCA embedding of the graphs using the consistency measure, each represented by a thumbnail of the object from which the are derived. Although some graphs representing different objects are grouped together, the clusters are mostly of just one object.

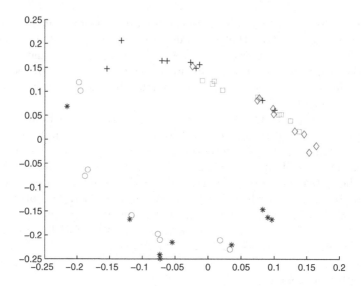

Fig. 4. A PCA embedding of the Images from the COIL database using the consistency measure

7   Conclusion

In this paper we have looked at one of the ways in which the richer structure inherent in quantum processes can be utilized classically. We have described an auxiliary graph that can be used for the purpose of graph matching. By simulating a discrete quantum walk on this graph, quantum interference can be used to compare the two graphs. The walk gives rise to a set of amplitudes corresponding to possible pairings of the vertices of the two graphs. If the graphs are isomorphic then the states for which the interference is exact are used to significantly prune the space of possible mappings between the graphs allowing us to recover the isomorphism. We have tested the algorithm on graphs representing molecular structures and found that it reduces the space of matches sufficiently for us to match the graphs directly. We have analysed how the algorithm behaves in the presence of structural errors and used this analysis to develop a consistency measure that can be used for the clustering of graphs. We carried out experiments on real world data for which such errors exist and demonstrated the effectiveness of the consistency measure for clustering. As further work we would like to look at the normalisation of the consistency measure since the measure, as currently defined, can be effected unduly by the edge densities of the graphs and minimizing this is likely to improve the performance of the approach.

References

1. US National Cancer Institute Database (2006) http://resresources.nci.nih.gov/database.cfm?id=1231
2. Aharonov, D., Ambainis, A., Kempe, J., Vazirani, U.: A fast quantum mechanical algorithm for database search. In: Proc. 28th ACM Symp on Theory of Computation, pp. 50–59. ACM Press, New York (1996)
3. Barrow, H.G., Burstall, R.M.: Subgraph isomorphism, matching relational structures and maximal cliques. Information Processing Letters 4(4), 83–84 (1976)
4. Brin, S., Page, L.: The anatomy of a large-scale hypertextual Web search engine. Computer Networks and ISDN Systems 30(1–7), 107–117 (1998)
5. Emms, D., Hancock, E., Severini, S., Wilson, R.C.: A matrix representation of graphs and its spectrum as a graph invariantn. Electronic Journal of Combinatorics 13(1), R34 (2006)
6. Emms, D., Severini, S., Wilson, R.C., Hancock, E.: Coined quantum walks lift the co-spectrality of graphs and trees. In: Rangarajan, A., Vemuri, B., Yuille, A.L. (eds.) EMMCVPR 2005. LNCS, vol. 3757, pp. 332–345. Springer, Heidelberg (2005)
7. Gori, M., Maggini, M., Sarti, L.: Graph matching using random walks. In: IEEE 17th ICPR, August 2004 (2004)
8. Grover, L.: A fast quantum mechanical algorithm for database search. In: STOC '96: Proc. 28th ACM Theory of computing, pp. 212–219. ACM Press, New York (1996)
9. Harris, C., Stephens, M.: A combined corner and edge detector. In: Proc. of 4th Alvey Vision Conference, Manchester, vol. 15, pp. 147–151 (1988)
10. Kempe, J.: Quantum random walks – an introductory overview. Contemporary Physics 44(4), 307–327 (2003)
11. Meila, M., Shi, J.: A random walks view of spectral segmentation (2001)
12. Nadler, B., Lafon, S., Coifman, R., Kevrekidis, I.: Diffusion maps, spectral clustering and eigenfunctions of fokker-planck operators. In: Advances in Neural Information Processing Systems 18, MIT Press, Cambridge, MA (2006)
13. Nene, S.A., Nayar, S.K., Murase, H.: Columbia object image library (coil-100) (1996)
14. Nielson, M., Chuang, I.: Quantum Computing and Quantum Information. Cambridge University Press, Cambridge (2000)
15. Robles-Kelly, A., Hancock, E.: Graph edit distance from spectral seriation. IEEE Transactions on Pattern Analysis and Machine Intelligence 27, 365–378 (2005)

A Quadratic Programming Approach to the Graph Edit Distance Problem

Michel Neuhaus[1] and Horst Bunke[2]

[1] LIP6, Université Pierre et Marie Curie
104 avenue du Président Kennedy, F-75016 Paris, France
`mneuhaus@iam.unibe.ch`
[2] Institute of Computer Science, University of Bern
Neubrückstrasse 10, CH-3012 Bern, Switzerland
`bunke@iam.unibe.ch`

Abstract. In this paper we propose a quadratic programming approach to computing the edit distance of graphs. Whereas the standard edit distance is defined with respect to a minimum-cost edit path between graphs, we introduce the notion of fuzzy edit paths between graphs and provide a quadratic programming formulation for the minimization of fuzzy edit costs. Experiments on real-world graph data demonstrate that our proposed method is able to outperform the standard edit distance method in terms of recognition accuracy on two out of three data sets.

1 Introduction

In structural pattern recognition, the edit distance measure has been widely used for error-tolerant graph matching. The successful application of graph edit distance is mainly due to its intuitive and universal definition. Based on a node and edge distortion model, the edit distance is defined as the minimum amount of distortion that is needed to transform a given graph into another one [1,2], which follows the intuitive understanding that the more dissimilar two graphs are, the more transformation operations have to be performed. Graph edit distance is applicable to arbitrarily labeled and arbitrarily structured graphs — and other data structures such as strings [3], trees [4], and hyper-graphs [5] — and can therefore be considered a universal matching scheme for complex patterns. In practice, the flexibility of graph edit distance, which allows us to assign weights to individual distortion operations based on the type of distortion and the involved nodes and edges, renders edit distance applicable to various practical graph matching tasks.

Computing the edit distance of two graphs results in a time and space complexity that is exponential in the number of nodes of the two graphs. Particularly in the presence of large graphs, the edit distance problem is computationally very demanding. In recent years, a number of methods have been proposed to render the computation of graph edit distance feasible. In [6], an approximate edit distance algorithm for planarly embedded nodes is introduced. The algorithm

F. Escolano and M. Vento (Eds.): GbRPR 2007, LNCS 4538, pp. 92–102, 2007.

exploits the position node information that is available in many graph representations in pattern recognition. The approximate edit distance is computed in an iterative procedure by successively optimizing local matching criteria. Two fast suboptimal variants of a standard edit distance algorithms are proposed in [7]. The idea is to restrict the matching process to promising candidates by applying a technique for search tree pruning and a re-weighting of edit costs. These approaches have in common that they attempt to refine the standard tree search algorithm for edit distance to speed up the computation.

In the present paper, we propose to circumvent the standard inefficient algorithm altogether by addressing the edit distance problem by means of quadratic programming. The basic idea is to formulate the minimum-cost optimization problem of edit distance in the well-known mathematical framework of quadratic programming [8], which allows us to tackle the complex graph matching problem using standard optimization methods. In the longer term, it would be desirable to develop fast (possibly suboptimal) optimizers for the particular edit distance quadratic programming formulation, which is not covered in this paper. Our main contribution is an alternative method for the computation of edit distance.

The quadratic programming approach leads us to the notion of fuzzy edit paths. The result of our method is either a minimum-cost fuzzy edit path or, after defuzzification, a standard edit path between two graphs. In this respect, the method we propose in this paper is loosely related to relaxation labeling techniques for graph matching [9,10], where the idea is to define the matching problem as a node labeling problem and to apply iterative procedures refining the labeling until a sufficiently accurate matching is obtained. Unlike these relaxation labeling techniques, which are sometimes defined for numerically labeled or weighted graphs only, the method we propose is applicable to arbitrarily labeled graphs and is closely related to the standard edit distance measure. In [11], a linear programming method for computing the edit distance of graphs with unlabeled edges that is somehow related to our approach is introduced.

This paper is structured as follows. In Section 2, we briefly introduce graph edit distance. The proposed quadratic programming formulation of edit distance is described in Section 3. Experimental results on three real-world graph data sets are given in Section 4. Finally, in Section 5 a few summarizing conclusions are drawn.

2   Graph Edit Distance

Graph edit distance is an error-tolerant dissimilarity measure on graphs. The edit distance method is applicable to arbitrarily labeled graphs, that is, graphs with any kind of labels attached to nodes and edges. A graph is commonly defined by a four-tuple $g = (V, E, \mu, \nu)$, where V denotes a finite set of nodes, $E \subseteq V \times V$ is a set of directed edges, $\mu : V \to L$ is a node labeling function assigning each node a label from alphabet L, and $\nu : E \to L$ is an edge labeling function. Note that in practical applications, numerical labels (attribute vectors) usually prevail. The idea of edit distance is to define a set of basic graph distortion operations, or edit operations, and define the dissimilarity of two given graphs by the minimal amount of edit operations that are needed to transform one graph into the other one [1,2]. While the edit distance concept theoretically allows for a wide range of edit operations, for most applications it is sufficient to consider the insertion, deletion, and substitution of nodes and edges only. A node deletion operation, for instance, refers to the removal of a node and its adjacent edges, and an edge substitution operation is equivalent to changing the label of an edge. The edit distance method can be tailored to specific application by assigning each edit operation a cost value reflecting the strength of the corresponding distortion. For instance, changing an edge label by a small amount might often be considered a weaker distortion than the removal of a node together with all edges connected to this node. In this particular case, the edge substitution would be assigned a lower cost than the node deletion. The total edit costs of a given sequence of edit operations transforming one graph into another one, or edit path between the two graphs, is obtained by summing up the costs of the individual edit operations. Finally, the edit distance of two graphs is defined as the minimum cost edit path between them, that is, the least expensive way to edit one graph into the other one, given an edit operation model and an edit cost function. If we denote by $P(g, g')$ the set of edit paths transforming a graph g into a graph g' and by C the function assigning costs to edit operations, the edit distance of g and g' is defined by

$$d(g, g') = \min_{(w_1, \ldots, w_k) \in P(g,g')} \sum_{i=1}^{k} C(w_i) \ , \tag{1}$$

where $(w_1, \ldots, w_k)$ represents an edit path consisting of k edit operations.

The simplest way to compute edit distance is obviously to generate all edit paths between two graphs and determine the one with minimum costs. In more sophisticated approaches, lookahead techniques or heuristics are used to determine which edit paths seem to be promising candidates for exploration. A standard edit distance computation algorithm is based on an A* tree search algorithm with efficient heuristics [1,7,12]. The idea is to systematically explore all relevant edit paths by traversing, in a best-first fashion, a search tree with inner nodes representing partial edit paths and leaf nodes representing complete edit paths. The flexibility of edit distance, potentially allowing any node of one graph to be mapped to any node of the second graph, results in exponential computational costs in terms of time and space complexity. That is, the edit distance of graphs is typically tractable for graphs with up to about a dozen of nodes only.

3   Quadratic Programming for Graph Edit Distance

Quadratic programming is a particular type of mathematical optimization problem [8]. It turns out that the graph edit distance problem needs only a few slight adaptations to fit into the quadratic programming framework, which makes a new class of algorithms available for the computation of graph edit distance.

3.1   Quadratic Programming

Quadratic programming refers to a range of optimization problems satisfying a general mathematical form. In the following, the quadratic programming problem will be described and briefly discussed. First, let the set of real matrices of dimension $a \times b$ be denoted by $\mathbb{R}^{a \times b}$. For a given dimension $n \geq 1$, let us assume that a symmetric matrix $Q \in \mathbb{R}^{n \times n}$ and a vector $c \in \mathbb{R}^n$ are given. Furthermore for $l, m \geq 1$, let matrices $R \in \mathbb{R}^{l \times n}$ and $S \in \mathbb{R}^{m \times n}$ as well as vectors $u \in \mathbb{R}^l$ and $v \in \mathbb{R}^m$ be given. The general quadratic programming problem can then be formulated as [8]

$$\text{Minimize } f(x) = \frac{1}{2}x'Qx + c'x \qquad \text{for } x \in \mathbb{R}^n \tag{2}$$

$$\text{such that}$$

$$Rx = u$$
$$Sx \geq v \ .$$

Note that the vector inequality constraint in the last line means that all components of the two vectors must satisfy the inequality. Solving the quadratic programming problem consists of finding an $x \in \mathbb{R}^n$ that minimizes $f(x)$ such that the given equality and inequality conditions are satisfied. The expression *quadratic programming* is due to the fact that the target function $f(x)$ is a quadratic function of the argument x. The equality constraint can be seen as a compact representation of l independent equality conditions (one per line of matrix R), and similarly the inequality constraint is equivalent to m inequality conditions.

Quadratic programming problems can always be solved, or shown to be unfeasible, in a finite amount of time. However, the actual complexity of the computation depends strongly on the characteristics of the problem, in particular on the matrix Q and the number of relevant inequality constraints [8]. If Q is positive definite, for instance, the quadratic programming problem can typically be solved as efficiently as linear programming problems. Furthermore, it is also known in this case that there exists a globally optimal solution, provided that the equality and inequality constraints are satisfied for at least one vector. The methods commonly used to solve quadratic programming problems can roughly be divided into interior point methods, active set methods, and conjugate gradient methods [8]. In our experiments, we use the interior point algorithm from the Computational Optimization Program Library [13].

A classic example of a quadratic programming problem is the management of investment portfolios [8]. The idea is to model the tradeoff between risk and expected return for a collection of investments. Quadratic programming can be used to derive an investment strategy that predicts high returns with low variance. The popular support vector machine method for classification and regression is another example. The maximum-margin hyperplane separating two classes can be found by solving a quadratic programming problem [14], namely by minimizing the squared norm of the hyperplane weight vector given a number of linear constraints. In the following, we will apply quadratic programming to the graph edit distance problem.

3.2   Fuzzy Edit Path

The standard graph edit distance is defined by the minimum-cost edit path between two graphs. A common interpretation of substitutions in an optimal edit path is that they indicate which parts of one graph can be identified in the other graph. That is, a set of node substitutions can be seen as a mapping of nodes of one graph to nodes of another graph. Analogously, deleted (or inserted) nodes and edges can be interpreted as those nodes and edges of the first graph (second graph) that cannot be matched, with sufficient accuracy, to nodes and edges of the second graph (first graph). Hence, given an edit path between two graphs, each node and edge is either substituted with another node and edge, or deleted or inserted.

The basic idea of fuzzy edit paths is to allow nodes and edges of one graph to be simultaneously assigned to several nodes and edges of another graph. In the following, let us assume that two graphs $g = (V, E, \mu, \nu)$ and $g' = (V', E', \mu', \nu')$ with $|V| = n$ and $|V'| = n'$ are given. Clearly, there exist $n \cdot n'$ distinct substitutions of a node $u \in V$ with a node $v \in V'$, such a substitution being denoted by $u \rightarrow v$. A fuzzy edit path is defined by assigning a weight to each possible node substitution. Formally, a fuzzy edit path between g and g' is a function $w : V \times V' \rightarrow [0, 1]$ satisfying the conditions

$$\sum_{v \in V'} w(u, v) = 1 \text{ for each } u \in V \quad \text{and} \quad \sum_{u \in V} w(u, v) = 1 \text{ for each } v \in V' \ . \quad (3)$$

This weighting function w can be understood as a kind of membership function reflecting how well a node substitution conforms to, or how strongly it violates, the structure and labels of the two graphs. The interpretation of a fuzzy edit path that is optimal with respect to some matching criterion is that two nodes u, v with a large value of $w(u, v)$ are likely to correspond to a good structural match, while nodes with small values of $w(u, v)$ should rather be considered unmatchable. The advantage of fuzzy edit paths over standard edit paths is that they allow us to integrate ambiguity directly in the definition of edit paths, instead of being forced to settle for one edit transformation for each node and edge.

In order to construct a standard edit path from a fuzzy edit path, a defuzzification procedure can be carried out. A straight-forward defuzzification method consists in selecting from all fuzzy node substitutions those with large fuzzy weights. The first node substitution to be inserted into the standard edit path is obtained by selecting from all fuzzy node substitutions the one with largest fuzzy weight, say the substitution $u \rightarrow v$. In the following steps, all fuzzy node substitutions involving u and v will no longer be considered. The second node substitution of the standard edit path is obtained by selecting from the remaining fuzzy node substitutions the one with largest fuzzy weight. Again, all fuzzy node substitutions containing either one of the two nodes of the selected substitution are ignored in successive steps. This iterative procedure is continued until no more node substitutions can be extracted. The remaining nodes are considered equivalent to node deletions and insertions. Finally, edge operations are inferred from node operations. Note that in the computation of fuzzy edit paths, edge edit operation costs will be included in the definition of fuzzy weights attached to node substitutions. That is, not only the substitution of nodes, but also the edge structure plays a role in the defuzzification procedure outlined above.

3.3   Quadratic Programming Formulation

In the preceding paragraphs, fuzzy edit paths have been introduced as an extension to standard edit paths. The remaining question is how to compute a fuzzy edit path between two graphs that is optimal with respect to some node and edge matching criterion. The method we propose in this paper is based on a quadratic programming formulation of the graph matching problem. The basic idea is to encode node and edge edit costs in a cost matrix and minimize the overall costs corresponding to a fuzzy edit path.

Again, let the two graphs under consideration be denoted by $g = (V, E, \mu, \nu)$ and $g' = (V', E', \mu', \nu')$, and let $|V| = n$ and $|V'| = n'$. It is clear that there exist $n \cdot n'$ substitutions between g and g'. In view of this, we construct a real matrix $Q \in \mathbb{R}^{nn' \times nn'}$ where rows and columns are indexed by substitutions $u \to v$, where $u \in V, v \in V'$. That is, each row, and the corresponding column, of the matrix is associated with one distinct node substitution. The matrix Q is then constructed in such a way that diagonal entries hold the costs of node substitutions, while off-diagonal entries correspond to edge edit costs. The entry at position $(u \to v, u \to v)$ is set to the node substitution costs of $u \to v$; the entry at position $(u \to v, p \to q)$ is set to the edge edit costs resulting from substituting $u \to v$ and $p \to q$, depending on the existence of edges between u and p as well as between v and q. It should be noted that edit costs can be defined for any kind of node and edge labels, including symbols from a finite alphabet and complex labels such as strings. The proposed approach is thus not limited to graphs with numerical labels, but applicable to arbitrarily labeled graphs, which is one of the strengths of graph edit distance.

An example of two graphs with $n = n' = 3$ is provided in Fig. 1, where nodes are labeled with a two-dimensional position attribute and edges are unlabeled. It is clear that in this example the nine possible distinct node substitutions are $A \to a, A \to b, A \to c, B \to a, B \to b, B \to c, C \to a, C \to b, C \to c$. When constructing the 9×9 matrix Q, each row and column is associated with one of these substitutions. In Fig. 2, an example cost matrix Q is shown for the two graphs in Fig. 1. Note that in this example, node substitution costs are set equal to the squared Euclidean distance of the two node labels, and node and edge insertion and deletion costs are set to a constant value of 10. The substitution of unlabeled edges can be carried out for free. For example, since node A is labeled with $(1, 1)$ and node a with $(1, 6)$, the substitution $A \to a$ results in costs $Q_{A \to a, A \to a} = 25$. Since there exists an edge between A and B as well as

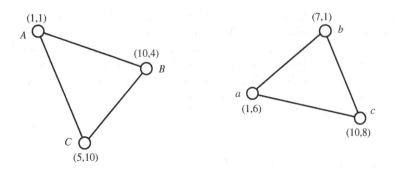

Fig. 1. Two example graphs g (left) and g' (right)

an edge between a and b, the substitutions $A \rightarrow a$ and $B \rightarrow b$ involve no edge operations costs, hence $Q_{A \rightarrow a, B \rightarrow b} = 0$. As node A is not connected to itself by an edge, the substitutions $A \rightarrow a$ and $A \rightarrow b$ involve the insertion of an edge, which leads to $Q_{A \rightarrow a, A \rightarrow b} = 10$.

Recall that fuzzy edit paths are defined in Sect. 3.2 as functions assigning weights to all possible node substitutions between two graphs. Also, the matrix Q consists of one row, and column, per node substitution. In view of this, we define a fuzzy cost function assigning each row of Q a weight according to the rules stated in Sect. 3.2. That is, each row, and the corresponding column, is associated with a node substitution and a fuzzy weight. It should be noted that these fuzzy weights are not pre-defined, but to be determined in the optimization process. Hence, the idea of the reformulated graph matching problem is to find fuzzy weights that satisfy the conditions of a fuzzy edit path and minimize the structural error. To this end, we propose to minimize the expression $x'Qx$, where x denotes the $n \cdot n'$-dimensional vector of fuzzy weights, one for each row of Q. The minimization is carried out over all fuzzy weights x satisfying the conditions defined in Sect. 3.2. In this optimization formulation, the weight associated with a node substitution $u \rightarrow v$ will influence not only the weighting of the node substitution costs of $u \rightarrow v$ (in the diagonal entry $Q_{u \rightarrow v, u \rightarrow v}$), but also all edge edit costs involving the substitution $u \rightarrow v$ (in off-diagonal entries $Q_{u \rightarrow v, p \rightarrow q}$ and $Q_{p \rightarrow q, u \rightarrow v}$). Clearly, this optimization process aims at assigning large weights to node substitutions that involve low node and edge costs, and assigning small weights to node substitutions that result in high costs. Note that this optimization principle is not identical to the minimum cost edit path concept in standard graph edit distance, but the intuitive interpretation of minimizing penalty costs for structural errors is comparable.

The optimization problem described above can be formulated in the standard quadratic programming framework. To this end, the matrix Q in Eq. 2 is defined as the cost matrix Q described above, the solution vector x in Eq. 2 is the weight vector x mentioned above, and vector c in Eq. 2 is the zero vector. Furthermore, it is easy to see that the conditions of consistent fuzzy weights can be formulated in terms of equality and inequality conditions — $Rx = u$ stating that fuzzy weights sum up to 1 as shown in Eq. 3, and $Sx \geq v$ restricting considerations to fuzzy

$$Q = \begin{pmatrix} 25 & 10 & 10 & 10 & 0 & 0 & 10 & 0 & 0 \\ 10 & 36 & 10 & 0 & 10 & 0 & 0 & 10 & 0 \\ 10 & 10 & 130 & 0 & 0 & 10 & 0 & 0 & 10 \\ 10 & 0 & 0 & 85 & 10 & 10 & 10 & 0 & 0 \\ 0 & 10 & 0 & 10 & 18 & 10 & 0 & 10 & 0 \\ 0 & 0 & 10 & 10 & 10 & 16 & 0 & 0 & 10 \\ 10 & 0 & 0 & 10 & 0 & 0 & 32 & 10 & 10 \\ 0 & 10 & 0 & 0 & 10 & 0 & 10 & 85 & 10 \\ 0 & 0 & 10 & 0 & 0 & 10 & 10 & 10 & 29 \end{pmatrix}$$

	Solution
$A \to a$	0.662
$A \to b$	0.297
$A \to c$	0.041
$B \to a$	0.000
$B \to b$	0.619
$B \to c$	0.381
$C \to a$	0.338
$C \to b$	0.084
$C \to c$	0.578

		Constraints satisfied
$A \to \ldots$:	$0.662 + 0.297 + 0.041 = 1$	Constraints satisfied
$B \to \ldots$:	$0.000 + 0.619 + 0.381 = 1$	
$C \to \ldots$:	$0.338 + 0.084 + 0.578 = 1$	
$\ldots \to a$:	$0.662 + 0.000 + 0.338 = 1$	
$\ldots \to b$:	$0.297 + 0.619 + 0.084 = 1$	
$\ldots \to c$:	$0.041 + 0.381 + 0.578 = 1$	

Fig. 2. Example quadratic programming problem matrix Q (corresponding to the graphs in Fig. 1) and solution weight vector satisfying fuzzy edit path constraints

weights between 0 and 1. In Fig. 2, the result of the quadratic programming approach to matching the two graphs in Fig. 1 is shown. The solution vector clearly satisfies the fuzzy edit path constraints. After defuzzification, we obtain the same optimal edit path as the standard edit distance algorithm, $\{A \to a, B \to b, C \to c\}$. Note that from the solution vector, it is not only possible to extract the most likely edit path, but also other edit paths that seem to be rather likely, such as the one with edit operations $\{A \to b, B \to c, C \to a\}$ in our case.

4   Experimental Results

In this section, we evaluate how the proposed method performs in comparison to the standard edit distance method on three graph data sets representing letters, images, and diatoms. These data sets are considered difficult because of non-compact and overlapping classes.

The letter data set consists of line drawings of 15 capital letters. Nodes are labeled with a position attribute, and edges are unlabeled. The graphs are split into a training set and validation set each of size 150 and a test set of size 750. The image data set consists of 5 classes of region adjacency graphs representing images after processing and filtering [15]. Nodes contain a color histogram attribute, and edges contain a region adjacency attribute. The diatom data set contains a total of 162 patterns split into a training set, validation set, and test set of equal size. The diatom data set is derived from microscopic images of 22 diatom classes. These images first undergo a segmentation process and are then transformed into graphs [16]. Nodes contain attributes describing region

Table 1. Recognition accuracy on validation set (VS) and test set (TS)

Data set	Method	Accuracy VS	Accuracy TS	Running time
Letter	Standard	67.3	69.3	12.4'
	Proposed	76.7	74.9 •	19.5'
Image	Standard	64.8	48.1	9s
	Proposed	72.2	59.3 •	18s
Diatoms	Standard	86.5	66.7 •	8s
	Proposed	54.1	47.2	15s

• Improvement over other method statistically significant ($\alpha = 0.05$).

features, and edges contain a common boundary attribute. This data set is split into a training set and validation set each of size 37 and a test set of size 36.

The recognition accuracy of a k-nearest-neighbor classifier based on the standard edit distance method [1,6] and the quadratic programming method proposed in this paper are given in Table 1. Note that the relevant classification accuracy is the one on the test set. In two out of three cases, the proposed method outperforms the standard method significantly, while on the third data set, the proposed method is clearly inferior. Note that the test set results marked with a dot are significantly better than the other ones on a statistical significance level of $\alpha = 0.05$. These classification results show that the proposed method based on quadratic programming constitutes a viable alternative to the standard edit distance method on certain data sets. As far as the running time is concerned, the proposed method seems to require typically twice the running time of the standard algorithm. It should be noted, however, that the efficiency of the proposed method heavily depends on the implementation of the quadratic programming algorithm at hand.

5   Conclusions

In this paper we propose a novel approach to computing graph edit distance. The idea is based on fuzzy edit paths between graphs. In contrast to standard edit distance, the result of the graph matching process is not a transformation of one graph into the other one (an edit path), but rather for each possible node substitution a computed fuzzy weight. The higher this fuzzy weight, the less the corresponding node substitution violates the node and edge structure and labels of the two graphs. For the computation of fuzzy edit paths from standard edit operation costs, a quadratic programming algorithm can be used. The aim is to compute a fuzzy edit path that minimizes the structural error in a manner similar to the minimization of edit costs in graph edit distance. The resulting fuzzy edit path can then easily be turned into a standard edit path by means of a defuzzification procedure. An experimental evaluation on graphs representing line drawings, images, and diatoms demonstrates that the proposed method significantly outperforms the standard edit distance method on two out of three data sets, although the standard edit distance algorithm is typically twice as fast.

In the future, we intend to further study the applicability of quadratic programming principles to graph matching. We would like to investigate whether it may be advantageous to use other quadratic programming formulations of the minimum-cost edit path problem than the one presented in this paper. Also, while applying a defuzzification procedure to fuzzy edit path turns out to be advantageous to directly using fuzzy edit paths for classification, there does not exist a unique way to turn fuzzy edit paths into standard edit paths. For instance, applying error-minimization techniques such as Munkres' algorithm for defuzzification might be a viable alternative to our proposed iterative procedure.

Acknowledgments

The first author was supported by the Swiss National Science Foundation under grant no. PBBE2-113362.

References

1. Bunke, H., Allermann, G.: Inexact graph matching for structural pattern recognition. Pattern Recognition Letters 1, 245–253 (1983)
2. Sanfeliu, A., Fu, K.: A distance measure between attributed relational graphs for pattern recognition. IEEE Transactions on Systems, Man, and Cybernetics (Part B) 13(3), 353–363 (1983)
3. Wagner, R., Fischer, M.: The string-to-string correction problem. Journal of the Association for Computing Machinery 21(1), 168–173 (1974)
4. Selkow, S.: The tree-to-tree editing problem. Information Processing Letters 6(6), 184–186 (1977)
5. Bunke, H., Dickinson, P., Kraetzl, M.: Theoretical and algorithmic framework for hypergraph matching. In: Roli, F., Vitulano, S. (eds.) ICIAP 2005. LNCS, vol. 3617, pp. 463–470. Springer, Heidelberg (2005)
6. Neuhaus, M., Bunke, H.: An error-tolerant approximate matching algorithm for attributed planar graphs and its application to fingerprint classification. In: Fred, A., Caelli, T.M., Duin, R.P.W., Campilho, A., de Ridder, D. (eds.) Structural, Syntactic, and Statistical Pattern Recognition. LNCS, vol. 3138, pp. 180–189. Springer, Heidelberg (2004)
7. Neuhaus, M., Riesen, K., Bunke, H.: Fast suboptimal algorithms for the computation of graph edit distance. In: Yeung, D.-Y., Kwok, J.T., Fred, A., Roli, F., de Ridder, D. (eds.) Structural, Syntactic, and Statistical Pattern Recognition. LNCS, vol. 4109, pp. 163–172. Springer, Heidelberg (2006)
8. Nocedal, J., Wright, S.: Numerical Optimization. Springer, Heidelberg (2000)
9. Christmas, W., Kittler, J., Petrou, M.: Structural matching in computer vision using probabilistic relaxation. IEEE Transactions on Pattern Analysis and Machine Intelligence 17(8), 749–764 (1995)
10. Wilson, R., Hancock, E.: Structural matching by discrete relaxation. IEEE Transactions on Pattern Analysis and Machine Intelligence 19(6), 634–648 (1997)

11. Justice, D., Hero, A.: A binary linear programming formulation of the graph edit distance. IEEE Trans. on Pattern Analysis and Machine Intelligence 28(8), 1200–1214 (2006)
12. Tsai, W., Fu, K.: Error-correcting isomorphism of attributed relational graphs for pattern analysis. IEEE Transactions on Systems, Man, and Cybernetics (Part B) 9(12), 757–768 (1979)
13. Zhang, X., Ye, Y.: Computational Optimization Program Library: Convex Quadratic Programming. University of Iowa (1998)
14. Schölkopf, B., Smola, A.: Learning with Kernels. MIT Press, Cambridge, MA (2002)
15. Le Saux, B., Bunke, H.: Feature selection for graph-based image classifiers. In: Marques, J.S., Pérez de la Blanca, N., Pina, P. (eds.) IbPRIA 2005. LNCS, vol. 3523, pp. 147–154. Springer, Heidelberg (2005)
16. Ambauen, R., Fischer, S., Bunke, H.: Graph edit distance with node splitting and merging and its application to diatom identification. In: Hancock, E., Vento, M. (eds.) GbRPR 2003. LNCS, vol. 2726, pp. 95–106. Springer, Heidelberg (2003)

Image Classification Using Marginalized Kernels for Graphs

Emanuel Aldea[1], Jamal Atif[2,*], and Isabelle Bloch[1]

[1] ENST (GET - Telecom) Paris, Dept. TSI, CNRS UMR 5141 LTCI
46 rue Barrault, 75634 Paris Cedex 13, France
{aldea,bloch}@enst.fr
[2] Groupe de Recherche sur les Energies Renouvelables (GRER)
Université des Antilles et de la Guyane, Campus de St Denis, 97 300 Cayenne, France
jamal.atif@guyane.univ-ag.fr

Abstract. We propose in this article an image classification technique based on kernel methods and graphs. Our work explores the possibility of applying marginalized kernels to image processing. In machine learning, performant algorithms have been developed for data organized as real valued arrays; these algorithms are used for various purposes like classification or regression. However, they are inappropriate for direct use on complex data sets. Our work consists of two distinct parts. In the first one we model the images by graphs to be able to represent their structural properties and inherent attributes. In the second one, we use kernel functions to project the graphs in a mathematical space that allows the use of performant classification algorithms. Experiments are performed on medical images acquired with various modalities and concerning different parts of the body.

1 Introduction

Most of the traditional machine learning techniques ultimately cope with basic numeric features given in the form of arrays [1]. Such input information is processed for various purposes, like classification or regression.

Nevertheless, it has become clear recently that machine learning should be able to cope equally with more complex input data, such as images, molecules, graphs or hypergraphs. The attributes that one can use to describe the input information are complex and very often inaccurate. In this context, classical learning methods do not provide a generic solution to the problem of processing complex input data.

Instead of changing the classical machine learning algorithms, our choice is to go in the opposite direction and to adapt the input for classification purposes so as to decrease structural complexity and at the same time preserve the attributes that allow assigning data to distinct classes.

As these complex structures started to emerge from various scientific areas (computer science, chemistry, biology, geography), one possible approach that we also employ in the current work has been to add a supplementary preprocessing step

---

* This work has been partially funded by GET and ANR grants during J. Atif's post-doctoral position at Telecom Paris.

F. Escolano and M. Vento (Eds.): GbRPR 2007, LNCS 4538, pp. 103–113, 2007.

involving structure and attribute extraction. In this way, we can return to the vectorial case by projecting a complex structure x belonging to a certain general space X into the n-dimensional real vector space $\mathbb{R}^n$ or in an infinite-dimensional Hilbert space. Different approaches have been used to project images in classifiable spaces. In one of them, images are treated as indivisible objects [2] and only global attributes are extracted. Another strategy has been to treat images as "bags" that contain indivisible objects [3], and interpret them by indexing these objects and their attributes, but ignoring whatsoever the relationships among them. However, the novel strategy that best defines our approach is to interpret images as organized sets of objects [4,5,6] and extract at the same time object attributes and structural information.

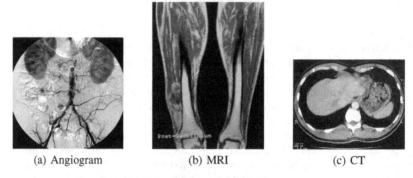

(a) Angiogram   (b) MRI   (c) CT

Fig. 1. Examples of medical images

We intend to use our work for the classification of 2D gray level medical images acquired using different techniques and concerning different parts of the human body. More precisely, the test base includes angiograms (Fig. 1(a)), sonograms, MRIs - magnetic resonance images (Fig. 1(b)), X-rays, CTs - computed tomograms (Fig. 1(c)), acquired with different imaging systems and following different protocols (even in each class). Most images also present small annotations, intended for the human reviewers. A good classification technique for this family of images should be able to cope with the generality factor due to the variety of classes and of acquisition techniques and equally with that due to noise (annotations, arrows) which are placed by instruments or by technicians and which are supposed to facilitate the work of medical teams.

This article starts with a brief presentation of support vector machines in Section 2 and of kernel methods for graphs in Section 3, emphasizing the one that represents the starting point for our work. Afterwards we describe in Section 4 our graph model for images, based on a generic, non-supervised segmentation followed by attribute extraction on the resulting structure. In Section 5, we explain where the difficulty of working with these attributes resides, and we propose a classification method adapted for image-issued graphs. This constitutes indeed the main contribution of the paper. Preliminary results on medical images are discussed in Section 6.

2 SVM Classifiers and Kernel Machines

In its basic form, a Support Vector Machine (SVM) classifier uses two sets of discriminative examples for training; these examples belong to a vector space endowed with a dot product. The main advantage of this classifier is the fact that it minimizes the classification error while maximizing the distance from the training examples to the separating hyperplane. It also allows the definition of a soft margin to prevent the mislabeled examples from perturbing too much the classification. Although SVMs have been originally designed as linear classifiers, they have been extended to perform non-linear discrimination [7] by using a "kernel trick", that replaces the dot product needed in computation by a non-linear positive definite kernel function. As a consequence, the examples are projected into a Hilbert space of higher dimension, called the feature space, which allows the construction of a linear classifier that is not necessarily linear in the initial space.

An important observation is that the classifier only needs the value of the kernel function between the examples. An additional advantage of this approch is that it allows classifying elements issued from spaces which are not naturally endowed with inner products (such as graph, tree or string spaces), as long as we use a valid kernel function.

3 Marginalized Graph Kernels

In this section, we briefly describe the marginalized kernel for labeled graphs.

We perform feature extraction on an undirected graph G, whose set of vertices is $\mathbb{X}$. The graph is labeled using the functions $v : \mathbb{X} \to \mathbb{S}_v$ for its vertices and $e : \mathbb{X} \times \mathbb{X} \to \mathbb{S}_e$ for its edges, $\mathbb{S}_v$ and $\mathbb{S}_e$ being two label sets. For the sake of clarity, we note $v(x)$ by v_x and $e(x_1, x_2)$ by $e_{x_1 x_2}$.

Feature extraction is carried out by first creating a set of random walks [8,9]. The first element of the walk is a vertex x_1 given by a certain probability distribution over $\mathbb{X}$. At a subsequent moment during the generation, the walk will end at the current vertex x_i with a fixed (small) probability or it will continue by visiting a neighboring vertex x_{i+i}.

For each walk $h = (x_1, x_2, \ldots, x_n)$, labeled as $l_h = (v_{x_1}, e_{x_1 x_2}, v_{x_2}, \ldots, v_{x_n})$, the probability to obtain it may be expressed as:

$$p(h|G) = p_s(x_1) \prod_{i=2}^{n} p_t(x_i|x_{i-1}) \tag{1}$$

in which p_s and p_t have to be chosen in order to build $p(h|G)$ as a probability distribution in the random walk space $\mathbb{X}^* = \cup_{i=1}^{\infty} \mathbb{X}^i$, the union of all random walk spaces of a certain finite length i. One proposal for p_s and p_t, that we have also adopted for our model, is given for example in [9].

The kernel between two graphs G and G' measures the similarity of all the possible random walk labels, weighted by their probabilities of apparition:

$$K(G, G') = \sum_{h} \sum_{h'} k(h, h') p(h|G) p(h'|G') \tag{2}$$

As for the kernel between two random walk labels, a natural option is to define it as 0 if the walks have different lengths, and the product of all the kernels for their corresponding constituent parts otherwise:

$$k(h, h') = k^v(v_{x_1}, v_{x'_1}) \prod_{i=2}^{n} k^e(e_{x_{i-1}x_i}, e_{x'_{i-1}x'_i}) k^v(v_{x_i}, v_{x'_i}) \tag{3}$$

where k^v and k^e denote the kernel functions used for computing vertex and edge similarity, respectively. In computational chemistry, where this kernel has been extensively and successfully used, label functions have a limited range and therefore an appropriate kernel for assessing vertex or edge label similarity is the Dirac kernel:

$$k_\delta(z, t) = \begin{cases} 1 & , \ if \ z = t \\ 0 & , \ otherwise \end{cases} \tag{4}$$

Even so, computing the marginalized kernel for two graphs is difficult in the absence of two supplementary variables [10]. The first one $\Pi_s = ((\pi_s(x, x'))_{(x,x')\in\mathbb{X}^2}$ is a $|\mathbb{X}||\mathbb{X}'|$ vector containing the joint start probabilities of two vertices $x \in \mathbb{X}$ and $x' \in \mathbb{X}'$ if they have the same label, and 0 otherwise. The second variable needed for the kernel computation is $\Pi_t = ((\pi_t((x_1, x'_1)|(x_2, x'_2))))_{(x_1,x'_1),(x_2,x'_2)\in\mathbb{X}^2}$ is a $|\mathbb{X}||\mathbb{X}'| \times |\mathbb{X}||\mathbb{X}'|$ square matrix whose elements assess the joint transition probability between two pairs of vertices belonging to the first and to the second graph, if and only if these vertex pairs and the corresponding edge pair are identically labeled (otherwise the probability is null):

$$\begin{cases} \pi_s(x, x') = p_s^G(x)p_s^{G'}(x') \\ \pi_t((x_1, x'_1)|(x_2, x'_2)) = p_t^G(x_1|x_2)p_t^{G'}(x'_1|x'_2) \end{cases} \tag{5}$$

Using these new variables and $\mathbb{1}$ - the vector with all its values equal to 1, the kernel can be evaluated as:

$$K(G, G') = \Pi_s^T(I - \Pi_t)^{-1}\mathbb{1} \tag{6}$$

Due to the inversion of Π_t which dominates the computation cost, the problem has an order of complexity of $O((|\mathbb{X}||\mathbb{X}'|)^3)$. However, one may take advantage of the sparsity of Π_t, as well as of other methods [10,11], in order to boost the performance of the algorithm. Many of these improvements are conditioned by a small range of labels and a low degree of vertex connectivity.

4   Graph Models of Images

The first step of our method consists in extracting and modeling image information. In order to achieve this, we use a labeled graph support (vertices are labeled as well as edges). The graph is obtained by first segmenting the initial image into regions, which allow us to describe its structure and to facilitate the information extraction step. Distinct regions correspond to vertices, while edges model the spatial relationships between regions. Beside this information brought by the structural expressivity of the graph, we integrate in the labeling relevant intrinsic information that we describe in detail later. Therefore, the interest of using a graph structure goes beyond the structural expressivity and is due to the possibility that it offers to save various data and link them to particular components.

Unsupervised segmentation, as the low-level processing stage of our classification system, is an important and at the same time difficult task. Good results of a classifier with no prior information on the elements to be classified imply the use of a segmentation method that works reasonably well for any input image type.

For our processing stage we adopt a generic hierarchical image segmentation paradigm [12,13,14]. We suppose that the image is divided into components that may be further divided into subcomponents. This decomposition may be represented by a tree whose root node is the whole image and whose leaf nodes represent a partition of tiny regions built at the beginning of the processing step. This partition may be for example the set of pixels of the image. The advantage of employing a hierarchical segmentation method is that changes are gradual, unlike for other methods where the variation of one parameter may induce a completely different segmentation map. This aspect is relevant because medical images which have been acquired using different protocols but show the same body parts are sensitive to segmentation methods that use absolute thresholds. As opposed to that, hierarchical segmentation gives emphasis to relative relationships between image subconstituents.

To generate the leaf node partition of the tree, instead of employing each pixel as a terminal node in a tree, we use a watershed over-segmentation that leaves us however with a very large number of small regions. At this point we start climbing in the tree structure by merging neighboring regions that have the closest average gray levels:

$$dif_g(r_1, r_2) = \left| av_g(r_1) - av_g(r_2) \right| \tag{7}$$

where $av_g(r)$ denotes the average gray level in the region r.

Concerning the stopping condition for the fusion process, we have chosen to set a dynamic threshold t_f. If the smallest gray level difference between two neighboring regions is higher than the threshold, we decide that the regions are not similar enough for the fusion to be performed and we stop. The threshold is dynamic because we compute it at each step as a (fixed) fraction f of the difference between the highest and the lowest region gray levels that exist in the image:

$$t_f = f \cdot (\max_{r \in Im} av_g(r) - \min_{r \in Im} av_g(r)) \tag{8}$$

As an example, we present a typical mammography in Fig. 2(a), along with one of the best possible human-assisted watershed based segmentations (Fig. 2(b)) and the result of the unsupervised method presented above (Fig. 2(c)).

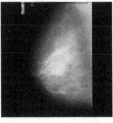 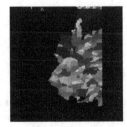 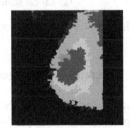

(a) Initial image   (b) Direct segmentation   (c) Region fusion

Fig. 2. Mammography segmentation

Once the fusion has ended, we compute the following attributes for the resulting regions, encoded as vertices:

- region surface in pixels,
- relative surface, a real value that represents the percent of the image covered by the concerned region,
- average gray value of the region,
- relative average gray value, corresponding to an affine transform with respect to the highest and lowest average gray values in the image, g_{min} et g_{max}:

$$gray_{rel}(r) = \frac{g - g_{min}}{g_{max} - g_{min}}, \qquad (9)$$

- region perimeter,
- region compacity, in $[0, 1/(4\pi)]$ and defined as the ratio between its surface and its squared perimeter,
- number of neighboring regions.

For the time being, the only relationship encoded by the edges (implicitely) is the neighborhood.

In Fig. 3, we present how the region number evolves when we modify the fusion threshold. Regions in images with a stronger initial over-segmentation tend to merge faster, so that for a fraction $f > 0.1$, results will start to be similar enough to those of images that presented a medium and low over-segmentation due to a smaller size or to a lower contrast, for example. This is an interesting result of our approach.

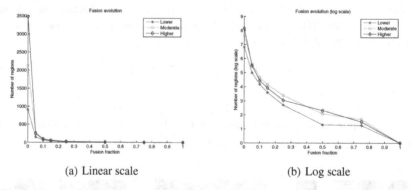

(a) Linear scale   (b) Log scale

Fig. 3. Threshold fusion fraction influence on fusion results in terms of number of regions

5   A Kernel for Image-Based Graphs

For image-based graphs, we propose a marginalized kernel different of that used in computational chemistry, which is able to better cope with specific image attributes.

A major structural difference in image-based graphs concerns the connectivity. While it is uncommon that atoms present more than four links towards the rest of the structure they belong to, this changes dramatically in the case of image regions, where there is no limit for the number of neighbors a region might possess. Potential optimizations based on graph sparsity become useless and, at the same time, the region neighborhood relationship has a lower importance than it has in a chemical compound, where the number of vertex neighbors $n(x)$ could be a used by a Dirac type kernel, as in Eq. (4). In our case however, the function $n(x)$ is unreliable as it is heavily influenced by the segmentation step, and cannot be helpful in building a vertex kernel or a significant part of it.

Another major difference in image-based graphs concerns the labeling. In the initial approach, vertices and edges are labeled using a small set of chemical symbols and possible bindings, and much information is given by the existence of the edge. The fundamental modification in the case of image-based graphs is that the labeling variable space becomes continuous and multi-dimensional, and a significant part of the information migrates from the graph structure to the labeling of its constituent parts.

The marginalized kernel presented in Section 3 employs a Dirac kernel for vertices and edges, which is useful for assessing structural similarities but is not adapted for a graph whose labeling is a major source of information. Under these circumstances, we have tried to adapt the vertex and edge kernels in order to define a proper similarity estimate for them.

The original graph kernel $K(G, G')$ defined in Eq. (2) is estimated by summing the similarities of all pairs of random walks of equal length. For a certain pair of such random walks (h, h'), let us suppose that we get simultaneously to a pair of corresponding vertices (x_i, x_i'). At this point we analyze the next transition in each walk; if the labels of the next two edges $(e_{x_i x_{i+1}}, e_{x_i' x_{i+1}'})$ and the labels of the next two vertices $(v_{x_{i+1}}, v_{x_{i+1}'})$ are not identical, the similarity brought by these walks will be null and we start analyzing another pair of walks. Otherwise we multiply the current similarity of the walks by the probabilities for the two transitions occurring in each walk. This leaves us with a probability of getting these random walks from start to end of:

$$p(h, h') = \left(p_s^G(v) \prod_{i=2}^{n} p_t^G(v_i | v_{i-1}) \right) \cdot \left(p_s^{G'}(v_1') \prod_{i=2}^{n} p_t^{G'}(v_i' | v_{i-1}') \right) \tag{10}$$

in which we suppose implicitely that for the walks (h, h'), the labels of all the constituent parts are identical. Using a Dirac similarity function for vertices and edges, it is obvious that random walk kernels in Eq. (3) will be also Dirac functions, so the graph kernel in Eq. (2) is reduced to the direct sum of all the probabilities $p(h, h')$ as in Eq. (10) computed for identically labeled random walks.

This strategy works for discrete ranged kernel functions, but in the case of region attributes like gray level or surface, we need a less discriminative kernel. Possible solutions to this problem are the Gaussian radial basis function (RBF) kernel and the triangular kernel [15]:

$$K^{RBF}(x, y) = exp\left(-\frac{\|x - y\|^2}{2\sigma^2} \right)$$

$$K^{\Delta}(x, y) = \begin{cases} \frac{C - \|x - y\|}{C}, & if \ \|x - y\| \leq C, \\ 0, & otherwise \end{cases} \tag{11}$$

The advantage of the first kernel over the second is that it offers a smoother, Gaussian discrimination compared to the uniform discrimination of the triangular kernel. However, beside an increase in computation time, the disadvantage of K^{RBF} is that it does not vanish at finite bounds, while the triangular kernel has a compact support.

We are entitled to use any of these kernels in the place of the Dirac kernel because they are also known to be positive definite and their use inside the graph kernel respects the closure properties of the family of kernel functions.

The next step is to integrate these values in the graph kernel computation. If we employ in Eq. (2) the joint probability from Eq. (10) and we replace the generic value $k(h, h')$ with that of Eq. (3), we get:

$$K(G, G') = \sum_h \sum_{h'} \left[k^v(v_{x_1}, v_{x'_1}) \prod_{i=2}^{n} \left(k^e(e_{x_{i-1}x_i}, e_{x'_{i-1}x'_i}) \cdot k^v(v_{x_i}, v_{x'_i}) \right) \right.$$

$$\left. \times p_s^G(x_1) \cdot p_s^{G'}(x'_1) \cdot \prod_{i=2}^{n} \left(p_t^G(x_i|x_{i-1}) p_t^{G'}(x'_i|x'_{i-1}) \right) \right] \quad (12)$$

By comparing the kernel equation Eq. (2) with its revised form Eq. (12), we can notice the adaptation of the variables from Eq. (5) that we must perform in order to use the same method for the computation of the new kernel function:

$$\begin{cases} \pi_s(x, x') = p_s^G(x) p_s^{G'}(x') \cdot k^v(v_{x_1}, v_{x'_1}) \\ \pi_t((x_1, x'_1)|(x_2, x'_2)) = p_t^G(x_1|x_2) p_t^{G'}(x'_1|x'_2) \cdot k^e(e_{x_{i-1}x_i}, e_{x'_{i-1}x'_i}) k^v(v_{x_i}, v_{x'_i}) \end{cases} \quad (13)$$

The vertex and edge kernel functions appear in this model as probability multipliers along transitions, which penalize paths with respect to their constituent dissimilarities. Using the revised variables π_s and π_t from Eq. (13), we can now employ Eq. (6) to compute the revised graph kernel from Eq. (12).

In the general case of an attribute set $A = \{a_1, \ldots, a_n\}$ associated to a graph component, the kernel function will be extended in order to take into account all the elements of A. Kernel functions related to these various attributes allow us to treat them in a unified way, merging them in a unified similarity estimate [16]. As each kernel provides us with a partial description of data properties, we are interested in building a parameterized combination that employs each attribute according to its relevance. In our work, we have employed a linear combination of base kernels:

$$K_A = \sum_{i=1}^{|A|} \lambda_i K_{a_i} \quad (14)$$

where the multipliers $\lambda_i \geq 0$ satisfy $\sum_{i=1}^{n} \lambda_i = 1$. This time too, the weighted sum of definite positive functions preserves the key property of definite positiviness of the result.

6   Experimental Results

Based on this adapted marginalized kernel, we have conducted some preliminary experiments, whose purpose is to assess its viability and the impact of different graph

Table 1. Recognition rates based on the relative surface attribute s_{rel}

Recognition rate	$C = 0.05$ $\sigma = 0.0167$	$C = 0.15$ $\sigma = 0.0500$	$C = 0.2$ $\sigma = 0.0667$	$C = 0.5$ $\sigma = 0.1667$	$C = 0.6$ $\sigma = 0.2000$	$C = 1$ $\sigma = 0.3333$
K^{Δ}	0.81	0.74	0.83	0.83	0.83	0.86
K^{RBF}	0.93	0.95	0.93	0.86	0.86	0.86

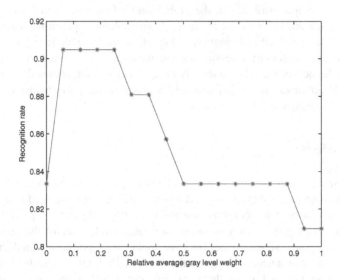

Fig. 4. Performance of a linear combination between a triangular kernel for relative surface (C=0.2) and a triangular kernel for relative average gray level (C=0.5)

attributes on its performance. As training examples, we have used ten head X-rays (coronal view) for the first class and ten mammographies (sagittal view) for the second one. For the moment, edges are not, beside their implicit structural importance, taken into account; therefore, we consider them as having the same label and we concentrate on the richer vertex attributes. We have particularly analyzed two of them which are adjusting to global image content: the relative surface s_{rel} with respect to the image surface and the relative average gray value $gray_{rel}$ defined in Eq. (9). They are less prone to perturbations, rescaling, contrast or brightness variations, etc.

In a first phase of our experiment, we have compared the performances of K^{RBF} and K^{Δ} in Eq. (11) for the s_{rel} attribute and for different parameterizations of C and respectively σ, on a testing sample of 42 images. For obvious statistics reasons, results for the two kernels are directly comparable in the situations where the value of C is at the 3-sigma level: $C = 3\sigma$.

Results in Table 1 show that the RBF kernel performs well in the case of a strong discrimination (i.e. if region areas differ by more than one tenth of the image surface, the kernel returns a very small similarity value). While simplifying the discrimination function, the triangular kernel does not manage to discriminate as efficiently as the RBF kernel in the initial range of the surface attribute.

In a second step, we have built a linear kernel as in Eq. (14) based on both s_{rel} and $gray_{rel}$, in order to analyze the classification performance as a function of the individual kernel multipliers. The tests are performed, as before, on the sample of 42 images. The discrimination thresholds are fixed at 0.2 and 0.5 for the surface and gray level attributes respectively. Gray level weight is gradually increased from 0 to 1 in the unified kernel equation.

The graph shown in Fig. 4 proves that performance may be improved drastically by combining multiple attributes in the global kernel function. Even for the limited use of two vertex attributes in the absence of edge labeling, preliminary results are encouraging. The weighted combination of kernels should be able to use information from multiple data sources by assessing the relative importance of each of them.

Triangular kernels prove to be noticeably faster than Gaussian ones and we hope that further weight optimization [16,17] will help us increase the performance of a linear kernel based on triangular subcomponents.

7   Conclusions

We have presented a new version of marginalized graph kernel which extends the one being used in computational chemistry and which allows the processing of image-based graphs. This new approach incorporates in the similarity computation specific properties of image-based graphs, such as image attributes, irrelevance of the numbers of neighbors of a segmented region, etc. We have applied this approach to medical image classification, based on a generic segmentation method. Preliminary results validate this model and further work will be needed in investigating which of the possible attributes are relevant for graph-based image representation and classification. We are also interested in labeling edges with relationship attributes which go beyond planar neighborhood and which are essential for expressing globally image content. In the same direction, we could try to use some results concerning the kernel integration theory in order to find the most suitable multipliers for a certain attribute set that we consider relevant.

References

1. Caruana, R., Niculescu-Mizil, A.: An empirical comparison of supervised learning algorithms. In: Proc. 23rd Int. Conf. on Machine Learning (2006)
2. Chapelle, O., Haffner, P., Vapnik, V.: Svms for histogram-based image classification. IEEE Transactions on Neural Networks, special issue on Support Vectors (1999)
3. Sivic, J., Russell, B.C., Efros, A.A., Zisserman, A., Freeman, W.T.: Discovering object categories in image collections. In: Proc. IEEE Int. Conf. on Computer Vision (ICCV), (2005)
4. Ros, J., Laurent, C., Jolion, J.M., Simand, I.: Comparing string representations and distances in a natural images classification task. In: GbR'05, 5th IAPR-TC-15 workshop on graph-based representations, pp. 72–81 (2005)
5. Neuhaus, M., Bunke, H.: Edit distance based kernel functions for attributed graph matching. In: Brun, L., Vento, M. (eds.) GbRPR 2005. LNCS, vol. 3434, pp. 352–361. Springer, Heidelberg (2005)

6. Neuhaus, M., Bunke, H.: A random walk kernel derived from graph edit distance. In: SSPR/SPR, pp. 191–199 (2006)
7. Vapnik, V.: Statistical Learning Theory. Wiley-Interscience, Chichester (1998)
8. Gaertner, T., Flach, P., Wrobel, S.: On graph kernels: Hardness results and efficient alternatives. In: Proc. 16th Annual Conf. on Computational Learning Theory, pp. 129–143 (2003)
9. Kashima, H., Tsuda, K., Inokuchi, A.: Marginalized kernels between labeled graphs. In: Proc. 20st Int. Conf. on Machine Learning, pp. 321–328 (2003)
10. Mahé, P., Ueda, N., Akutsu, T., Perret, J.L., Vert, J.P.: Extensions of marginalized graph kernels. In: ICML '04: Proc. 21st Int. Conf. on Machine Learning (2004)
11. Borgwardt, K., Vishwanathan, S., Schraudolph, N., Kriegel, H.P.: Protein function prediction via faster graph kernels. In: NIPS Bioinformatics Workshop (2005)
12. Haris, K., Estradiadis, S.N., Maglaveras, N., Katsaggelos, A.K.: Hybrid image segmentation using watersheds and fast region merging. IEEE Transactions on Image Processing 7, 1684–1699 (1998)
13. Beaulieu, J.M., Goldberg, M.: Hierarchy in picture segmentation: A stepwise optimization approach. IEEE Trans. Pattern Anal. Mach. Intell. 11, 150–163 (1989)
14. Brun, L., Mokhtari, M., Meyer, F.: Hierarchical watersheds within the combinatorial pyramid framework. In: Kuba, A., Nyúl, L.G., Palágyi, K. (eds.) DGCI 2006. LNCS, vol. 4245, pp. 34–44. Springer, Heidelberg (2006)
15. Mahé, P., Ralaivola, L., Stoven, V., Vert, J.P.: The pharmacophore kernel for virtual screening with support vector machines. J. Chem. Inf. Model. 46, 2003–2014 (2006)
16. Schlkopf, B., Tsuda, K., Vert, J.P.: Kernel Methods in Computational Biology. The MIT Press, Cambridge, Massachussetts (2004)
17. Boyd, S., Vandenberghe, L.: Convex Optimization. Cambridge University Press, Cambridge (2004)

Comparing Sets of 3D Digital Shapes Through Topological Structures

Laura Paraboschi, Silvia Biasotti, and Bianca Falcidieno

Istituto di Matematica Applicata e Tecnologie Informatiche - CNR
Via De Marini 6 - 16149 Genova, Italy

Abstract. New technologies for shape acquisition and rendering of digital shapes have simplified the process of creating virtual scenes; nonetheless, shape annotation, recognition and manipulation of both the complete virtual scenes and even of subparts of them are still open problems.

Once the main components of a virtual scene are represented by structural descriptions, this paper deals with the problem of comparing two (or more) sets of 3D objects, where each model is represented by an attributed graph. We will define a new distance to estimate the possible similarities among the sets of graphs and we will validate our work using a shape graph [1].

Keywords: graph-matching, scene comparison, structural descriptor, shape retrieval.

1 Introduction

Object recognition is one of the main tasks of Computer Vision and Graphics. In particular, there is a growing consensus that shapes are mentally represented and coded in terms of relevant parts and their spatial configuration, or structure [2]. This fact suggests that *describing* a shape through structural descriptors requires a limited amount of information, focused, for instance, on the shape topology and, if attributed, on some of its geometric features.

An important point is that the similarity between two shapes is assessed not only in terms of identical global matching but it is based on the contribution of common features compared to those distinguishing them. A distance measure, in particular a metric, translates the intuitive concept of closeness in a mathematical environment, considering as similar two objects that are close. In our work, 3D objects are coded using structural graphs and then compared through a distance measure defined over attributed graphs.

In the last decades several approaches have been developed for graph matching, and especially in defining distances between them.

The matching problem is solvable in polynomial time when dealing with trees. An interesting contribution has been proposed by Torsello et al. [3], where the authors present four distance metrics for attributed trees based on the notion of a maximum similarity subtree isomorphism. Since many problems deal with

F. Escolano and M. Vento (Eds.): GbRPR 2007, LNCS 4538, pp. 114–125, 2007.
© Springer-Verlag Berlin Heidelberg 2007

graphs instead of trees, a possible solution is to reduce the graph representation into an attributed tree obtained through edit operations (removing, adding or replacing a vertex or an edge). Unfortunately these techniques discard a lot of information about the original structure of the object and the tree obtained is intrinsically non-unique.

As far as graph matching, several approaches are based on maximum common subgraph MCS or minimum common supergraph mcs, like those proposed by Bunke [4] and Fernández and Valiente [5]. In both cases the distances proposed are metrics between non-attributed graphs, therefore the attributes associated to graphs are not taken into account. Since the problem of finding the MCS or the mcs is NP-complete, some authors have proposed algorithms for approximating their computation (see Marini et al. [6]), obviously providing a lower bound of the graph distance.

Another large family of methods is based on linear algebra techniques, exploiting the adjacency, incident or Laplacian matrices of a graph. For example, Shokoufandeh et al. [7] present a signature based on a combination of the spectral properties of the adjacency matrix that provides an effective mechanism for indexing large databases of graphs. The method described by Robles-Kelly and Hancock [8] adopts a brushfire search procedure using the rank-order of the coefficients of the leading eigenvector of the adjacency matrix. Similarly, Wilson et al. [9] analyse the spectrum of the Laplacian matrix of the graph. Since we have identified this approach as appropriate for our purposes, we describe it in Section 2. Finally the method of Bapat [10] introduces the concept of a tree with attached graphs and defines a distance matrix on it.

Last, the need of having flexible graph matching frameworks that admit the mapping of many nodes into many others has led to the definition of the so-called "many-to-many" approach. For example Demirci et al. [11] translate the problem into a many-to-many geometric points matching task, for which the Earth Mover's Distance algorithm (see [12]) is well suited.

Often in real applications it is important to compare sets of objects simultaneously present in different scenes. For example, this is relevant when a global view of the elements in a scene database is required, maybe for indexing or filtering purposes, before handling a fine recognition analysis. In fact, when massive volumes of data are provided, a fast and high-level analysis of the single scenes could significantly increase the number of computations. Moreover our approach can be adopted for comparing objects made of sets of single parts, and, therefore, overcomes the limitation, typical when using structural descriptors, of comparing objects made of a single connected component. In our context this problem is translated in the comparison of sets of objects represented by their structural graphs; therefore, we are interested in developing techniques able to treat sets of graphs. In this contribution we define a distance between two of these sets, providing a result able to establish a similarity, if it exists, among the scenes, which is our primary goal. Then two different normalizations of our distance are proposed: the first one is useful for ranking scene databases, while the second one is better for comparing just two sets of graphs, and corresponds to the measure proposed in Paraboschi et al. [13]. In this paper the focus is mainly on the first one, which has been tested on a database of single objects and on two databases of scenes.

The remainder of this paper is organized as follows. In Section 2 we describe our approach focusing on the properties of our distance and discussing the relation among the spectrum of the Laplacian matrix related to the scene graph and the structural descriptor of the single scene components. Then, experiments on datasets of scenes are discussed in Section 4. Conclusions, discussions and future developments end the paper.

2   Graph-Based Scene Comparison

Starting from the work of Wilson et al. [9], that converts graphs into pattern vectors by using the spectral decomposition of the Laplacian matrix and basis sets of symmetric polynomials, we extend that approach to scene comparison defining a novel scene graph. Moreover, two new distances are introduced to compare two scenes and the relation between the Laplacian spectrum of the scene graph and those of the components is discussed.

A graph $G \in \mathcal{G}$ is usually defined as a couple (V, E), where $\mathcal{G}$ is a set of graphs, V is the node set ($n = |V|$), and E the edge set. If a function $w : V($ or $E) \to \mathbb{R}$ is given, G is said to be attributed. Defining the Laplacian matrix L of a graph as [9] (we suppose from now that the spectrum of every Laplacian matrix is in increasing order), let $\mathbf{e}_i$ be an eigenvector ($\|\mathbf{e}_i\|_2 = 1 \; \forall i$) and λ_i be its eigenvalue. If $\Phi = (\Phi_{i,j})_{i,j=1,\dots,n} = (\sqrt{\lambda_1}\mathbf{e}_1, \dots, \sqrt{\lambda_n}\mathbf{e}_n)$, then $L = \Phi\Phi^t$.

The graph topology is invariant for any permutation of node labels (see [9] and [14]); it follows that the eigenvalues of L are invariant for any of these permutations and they can be used as a spectral representation of a graph, i.e.,

$$\lambda_j = \sum_{i=1}^{n} \Phi_{i,j}^2 \; . \tag{1}$$

Equation (1) is a symmetric polynomial in the components of eigenvectors $\mathbf{e}_i$. To measure invariant features of graphs, the authors propose to consider the set of elementary symmetric polynomials $S_j(v_1, \dots, v_n) = \sum_{i_1 < \dots < i_j} v_{i_1} v_{i_2} \cdots v_{i_j}$, $j = 1, \dots, n$.

Then, a matrix $F = (f_{i,j})_{i,j=1,\dots,n}$, $f_{i,j} = \text{sign}(S_j(\Phi_{1,i}, \dots, \Phi_{n,i})) \ln(1 + |S_j(\Phi_{1,i}, \dots, \Phi_{n,i})|)$ is introduced to define the so-called **feature vector** of the graph G as $\mathbf{B} = (f_{1,1}, \dots, f_{1,n}, \dots, f_{n,1}, \dots, f_{n,n})^t$.

Finally, given two graphs $G_1, G_2 \in \mathcal{G}$ whose feature vectors B_i, $i = 1, 2$, are known, a possible metric on $\mathcal{G}$ is given by

$$d(G_1, G_2) = \|\mathbf{B}_1 - \mathbf{B}_2\|_2 \; . \tag{2}$$

2.1   Comparing Sets of Graphs

In order to extend the previous technique to match sets of graphs, a virtual node X without attributes is introduced in every set (in analogy to VRML files, see for example the definition of grouping nodes at http://www.agocg.ac.uk/brief/vrml.htm), and it is joined to one of the vertices of minimum degree for each graph $G_i \in \mathcal{G}$ (see Fig.1). Since the Laplacian spectrum is invariant for node label permutations, the choice of the vertex to which X is joined is not relevant for the extraction of the eigenvalues of L.

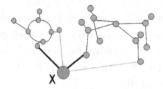

Fig. 1. Two graphs joined by a virtual node X in a scene graph. Provided that for each graph X is connected to a node of minimum degree, the choice of that node is irrelevant (X can be indifferently joined to the two graphs by the black edges or by the grey ones).

A single and connected graph is obtained, whose Laplacian matrix has the form represented in (3). If the set is composed by m graphs $G_i = (V_i, E_i)$, $|V_i| = n_i$, with $i = 1, \ldots, m$, it is

$$L = \begin{pmatrix} m & -1 & 0 & \cdots & -1 & & 0 \\ -1 & l_{1,1}^1 + 1 & \cdots & l_{1,n_1}^1 & & & \\ & \cdots & \cdots & \cdots & & & \\ 0 & l_{n_1,1}^1 & \cdots & l_{n_1,n_1}^1 & & & \\ \vdots & & & & \ddots & & \\ -1 & & & & l_{1,1}^m + 1 & \cdots & l_{1,n_m}^m \\ & & & & \cdots & \cdots & \cdots \\ 0 & & & & l_{n_m,1}^m & \cdots & l_{n_m,n_m}^m \end{pmatrix}. \tag{3}$$

$L(G_h) = L^h = (l_{i,j}^h)_{i,j=1,\ldots,n_h}$ is the Laplacian matrix of the h-th graph, and the element added to $l_{1,1}^h \; \forall h$ stresses the existence of the virtual node joined with a component vertex (we suppose that, after a node label permutation, X is joined to the first node of the component), whose degree increases of 1.

When dealing with attributed graphs, the same procedure described in (3) is used to modify the expression of the attributed Laplacian matrix [9], that takes into account both node and edge attributes.

2.2   A New Pseudo-metric and Two Possible Normalizations

To compare two sets of graphs, we introduce a new distance:

$$D(G_1, G_2) := \left| \|\mathbf{B}_1\|_2 - \|\mathbf{B}_2\|_2 \right| . \tag{4}$$

D is a pseudo-metric: it satisfies positivity, symmetry and triangle inequality; identity is not verified ($D(G_1, G_2) = 0 \nRightarrow G_1 \simeq G_2$). Between D in (4) and d in (2) the following relation holds:

$$D(G_1, G_2) = \left| \|\mathbf{B}_1\|_2 - \|\mathbf{B}_2\|_2 \right| \le \|\mathbf{B}_1 - \mathbf{B}_2\|_2 = d(G_1, G_2) \ .$$

In our experiments we use a normalized version of (4), which is useful to arrange a database with m sets; let it be $\|\mathbf{B}_{\bar{\imath}}\|_2 = \max_{i=1,\dots,m} \|\mathbf{B}_i\|_2$; then

$$D_m(G_1, G_2) = \frac{\left| \|\mathbf{B}_1\|_2 - \|\mathbf{B}_2\|_2 \right|}{\|\mathbf{B}_{\bar{\imath}}\|_2} \ . \tag{5}$$

(5) has the same properties of D, that is, it is a pseudo-metric.

We arrange the experiments in a scene repository as follows: first of all we extract the descriptors of the scene components, and then estimate the distance between two scenes using (5): the smaller D_m is, the more the two scenes, or better, their structures, are similar.

Finally, we propose also another normalization of D (see [13]): $D_N(G_1, G_2) = \frac{\left| \|\mathbf{B}_1\|_2 - \|\mathbf{B}_2\|_2 \right|}{\max(\|\mathbf{B}_1\|_2, \|\mathbf{B}_2\|_2)}$; D_N is a pseudo-semi-metric, that is, it satisfies neither identity nor the triangle inequality, and it is suitable to compare just two scene graphs. Moreover, if we define $d_N(G_1, G_2) = \frac{\|\mathbf{B}_1 - \mathbf{B}_2\|_2}{2\max(\|\mathbf{B}_1\|_2, \|\mathbf{B}_2\|_2)}$ as a normalization of (2), it follows that $D_N(G_1, G_2) \le 2d_N(G_1, G_2)$, and therefore our measure is a lower bound of the distance proposed in [9].

2.3   A Property of Laplacian Eigenvalues of the Scene

In this Section we analyse the relation between the scene graph we have defined and the descriptors of the single scene components. In fact, from successive inequalities, we can find both an upper and a lower bound for the spectrum of L in (3) which depend on the eigenvalues of the Laplacian matrices of the single scene components. Because these graphs are described also by Laplacian eigenvalues, the property we are going to describe guarantees that the scene graph is comparable with those of singular components. Therefore these results justify the addition of a virtual node, instead of separately considering each scene components.

The introduction of X makes the set of graphs $\mathbf{G}$ connected, and it adds a non-zero eigenvalue to its Laplacian matrix L: this change keeps L symmetric and diagonally dominant, so positive definite.

Let us consider a set with just two graphs G_1 and G_2, whose Laplacian matrices are respectively L^1 and L^2. Denoting $\lambda_1(L^1), \dots, \lambda_{n_1}(L^1)$, $\lambda_1(L^2), \dots, \lambda_{n_2}(L^2)$ their eigenvalues, let $\tilde{L}^1 = (\tilde{l}^1_{i,j})_{i,j=1,\dots,n}$ be such that $\tilde{l}^1_{i,j} = \begin{cases} l_{1,1} + 1 & i = j = 1 \\ l_{i,j} & \text{otherwise} \end{cases}$ and similarly $\tilde{L}^2$; it follows that $\sum_i \lambda_i(\tilde{L}^j) = \sum_i \lambda_i(L^j) + 1$, because the sum of the eigenvalues is equal to the matrix trace. Moreover, we can factor the Laplacian matrix of the scene L in (3) as $L = A_1 + A_2$, such that

$$A_1 = \begin{pmatrix} 1 & -1 & & & \\ -1 & l^1_{1,1}+1 & \dots & l^1_{1,n_1} & \\ & \dots & \dots & \dots & \\ & l^1_{n_1,1} & \dots & l^1_{n_1,n_1} & \\ & & & & \mathbf{0}_{n_2} \end{pmatrix}, \quad A_2 = \begin{pmatrix} 1 & & & -1 & \\ & \mathbf{0}_{n_1} & & & \\ -1 & & l^2_{1,1}+1 & \dots & l^2_{1,n_2} \\ & & \dots & \dots & \dots \\ & & l^2_{n_2,1} & \dots & l^2_{n_2,n_2} \end{pmatrix} ;$$

here $\mathbf{0}_{n_i}$ means a square matrix $n_i \times n_i$ with all 0 entries.

Now let G_3 be the graph G_1 with a node v and an edge e added ($E_3 = E_1 + e$, $V_3 = V_1 + v$, $v \notin V_1$, $e = (v, w)$, $w \in V_1$), from the interlace theorem [15] it follows:

$$0 = \lambda_1(L^3) = \lambda_1(L^1) \le \lambda_2(L^3) \le \lambda_2(L^1) \le \ldots \le \lambda_{n_1}(L^3) \le \lambda_{n_1}(L^1) \le \lambda_{n_1+1}(L^3) \ ,$$

where L^3 denotes the Laplacian matrix of G_3. As a consequence, the non-zero eigenvalues of A_1 interlace those of $\tilde{L}^1$, and the same is valid for A_2 and $\tilde{L}^2$.

Denoting $\lambda_N(L)$, $N = n_1 + n_2 + 1$, the maximum eigenvalue of L, Weyl's inequality (see for example [16]) implies the following relation: $\lambda_N(A_i) \le \lambda_N(L) \le \lambda_N(A_1) + \lambda_N(A_2)$, for $i = 1, 2$, since both A_1 and A_2 are positive semi-definite. Therefore it follows:

$$\max\{\lambda_N(A_1), \lambda_N(A_2)\} \le \lambda_N(L) \le \lambda_N(A_1) + \lambda_N(A_2) \ .$$

In addition we observe that, even if the eigenvalues of L are always non-zero (L positive definite), the eigenvalues of the matrices $A_{i=1,2}$ factorizing L verify $\lambda_1(A_i) = \ldots = \lambda_{N-1-n_i}(A_i) = 0$, $i = 1, 2$. Moreover, Weyl's inequality gives a meaningful result when the index i is $\max\{n_1 + 1, n_2 + 1\}$; if $n_1 > n_2$ then

$$\lambda_{n_1+1}(L) \ge \max\{\lambda_{n_1+1}(A_1), \lambda_{n_1+1}(A_2)\} \ .$$

Analogous results hold for scenes with m elements, that is $L = A_1 + \ldots + A_m$, and for attributed graphs.

3   Shape Graph and Comparison

The construction of the descriptor of every scene component relies on the shape graph defined in Biasotti et al. [1]: given a shape represented by a regular triangle mesh M, the co-domain $[f_{min}, f_{max}]$ of $f : M \to \mathbb{R}$ is subdivided considering nv regular values of f, $f_i \in [f_{min}, f_{max}]$, $i = 1, \ldots, nv$. The level sets of f that correspond to these values partition the mesh M into regions (see Fig.2(b)). Hence all points belonging to a region of a contour are identified and represented as nodes and edges of a traditional graph (see Fig.2(c)).

Three different mapping functions f are considered in our framework, namely the distance from the barycentre, the height function (with respect to the z axis) and the integral geodesic distance.

For every node $v \in V$, corresponding to a region R, it is possible to associate a property characterizing the region R or its boundary B_R; in our context, they are the minimum, maximum and average distance of the barycentre of R from the region vertices, the superior and inferior pseudo-cone lateral areas for each component of R, the superior and inferior B_R lengths, the surface area of R, the percentual area of R with respect to the whole object area, and the value of f in every vertex in V.

Concerning the computational complexity of the method, the extraction of each graph requires $O(\max(v \log v, m))$, where v is the number of vertices in the

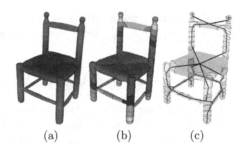

(a)   (b)   (c)

Fig. 2. (a): Evaluation of the distance from the barycenter, (b): the mesh segmentation, (c): the resulting skeleton

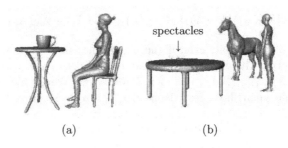

spectacles

(a)   (b)

Fig. 3. (a): Scene 1, and (b): scene 2

original mesh, and m are the vertices in the mesh after partitioning (see [1]). When comparing graphs, the method requires $O(n^3)$ operations, where n is the number of scene nodes, because of the eigendecomposition of the Laplacian matrix. The computation of the symmetric polynomials requires $O(n^2)$ operations, while the comparison of the pattern vectors of G_1 and G_2 requires $O(k^2)$ operations, where $k = \max\{|V_1|, |V_2|\}$.

We show a simple example to explain how the distance D_N behaves when comparing two sets of objects (for more details, see [13]). The two scenes, and their subscenes, in Fig.3 (a) and (b), are analysed comparing the scene graphs provided by the barycentre distance function. In the first scene, there is a woman sitting on a chair in front of a table, on which there is a cup, while in the second one there are a table and a pair of spectacles on it, and next a woman and a horse. When the sum of the pseudo-cone lateral areas is considered as node attribute, the distance is $D_N = 0.2276$, which is a quite high result. On the contrary, if we focus on the woman and the table in both scenes, the distance $D_N = 0.0074$ indicates a possible, maybe strong similarity. Again, when the table and the cup in (a) are compared with the table and the spectacles in (b), $D_N = 0.1617$ refers to less similar subscenes. Finally, if we analyze together the table and the woman in (a) with the table and the horse in (b), $D_N = 0.0894$: this result is an example of *false positive* (that is, a false result of similarity), and is explainable with the fact that the structural descriptor of the chair resembles to that of the horse.

4   Experimental Results

We have evaluated our method, that is, the distance D_m, using closed triangle meshes from various public databases: the AIM@SHAPE repository (http://shapes.aim-at-shape.net), the CAESAR Data Samples (http://www.hec.afrl.af.mil/HECP/Card1b.shtml#caesar samples), the McGill 3D Shape Benchmark (http://www.cim.mcgill.ca/~shape/benchMark/), and the National Design Repository at Drexel University (http://www.designrepository.org). In our experiments we have selected various classes of models, such as tables, chairs, cups, teddy bears and humans.

To analyse the behaviour of D_m, we adopt the *precision-recall* diagrams. Precision and recall are the basic measures used in evaluating search strategies; recall is the ratio of the number of relevant records retrieved to the total number of relevant records in the database, while precision is the ratio of the number of relevant records retrieved to the total number of records retrieved. A few examples are proposed in Fig.4 and they refer to a database of 120 heterogeneous models (see [1]). Supposing that every object is itself a separate scene, and we compare our results with those obtained by Chen et al. [17] and by Kazhdan et al. [18]. It is worth noticing that shape graphs well suit articulated models (see http://www. cim.mcgill.ca/~shape/benchMark/), such as humans and spectacles (Fig.4 (a) and (b)), while both Chen et al. [17] and Kazhdan et al. [18] methods work better on four limbs animals (Fig.4 (c)).

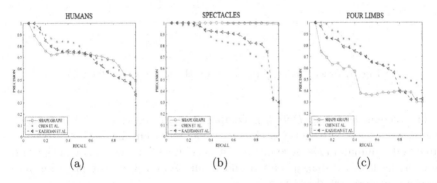

(a)   (b)   (c)

Fig. 4. Comparison between Chen et al., Kazahdan et al. and our shape graph

We finally evaluate our new distance on various scenes. We have generated and preclassified two databases considering in the same class two scenes if they have the same number of objects and also the same kind of components (for example, in a class there are scenes with a human and an animal, in another one a human next to a chair). According to these assumptions, *Data1* consists of 4 scene classes each of them with 20 scenes: in the first class there are a table and a chair, in the second a chair and a human, in the third a table and a human, and in the last a table and a cup. In Fig.5 we present the best precision-recall diagram for each class. This choice is due to the extremely large possibilities that we have when comparing graphs. In this sense, let us remember that, in our context, each object graph is described by 3 measuring functions, and again for each function 10 different attributes are associated to them. Totally, we can choose among 30 different diagrams, and we show here just the best result, which is the diagram with the biggest bounded area.

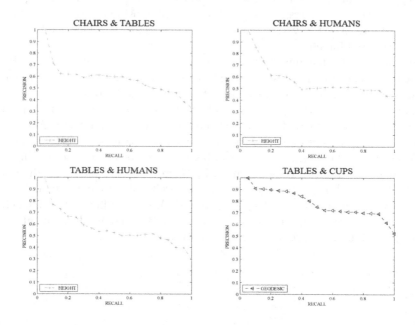

Fig. 5. Precision-recall diagrams on the four classes of *Data1*

The second database (*Data2*) is composed by 80 scenes divided in 4 classes, where, respectively, there are a chair, a table and a cup in the first, a chair, a table and a human in the second, a chair, a table and a teddy bear in the third, and, in the last, a table, a human and a cup. Again, the best precision-recall diagrams are presented in Fig.6.

There are not large differences between the two database diagrams, that is the method performs more or less in the same way on both of them. Moreover, we can observe that in most of the cases the height function used for extracting graphs is the most performing, because all the models are oriented (in the Euclidean space) in the same way; in fact in our experiments we try to deal with scenes from the real world, and consequently with the same orientation. Finally, it is worth noticing that, in *Data2*, the method performs better on the class of scenes with a chair, a table and a human (see Fig.6(b)) than the ones with a chair, a table and a cup (see Fig.6(d)): this is due to the fact that human graphs are richer of

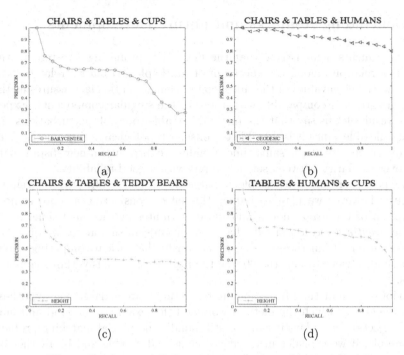

(a)   (b)

(c)   (d)

Fig. 6. Precision-recall diagrams on the four classes of *Data2*

features (articulations) than cup ones, and consequently the shape graph of the whole scene is more characteristic, simplifying their identification.

4.1 Numerical Properties of the Distance

Stability: $\|\mathbf{B}\|_2$, and consequently D_m, is well-conditioned. In fact, it is stable with respect to graph perturbations. For example, if a class of the database used is chosen and each model attribute is slightly perturbed (1% or 2%), the retrieval performance of the method is unaltered. Moreover, if the distance between a graph and its perturbed one is considered, it happens that D_m is lower than 0.08 in the 90% of cases.

Robustness: An inherent numerical error appears in computing symmetric polynomials, that is $S_j(\Phi_{1,i}, \ldots, \Phi_{n,i}) = \lambda_i^{\frac{j}{2}} \sum_{k_1 < \ldots < k_j} V_{k_1,i} V_{k_2,i} \cdots V_{k_j,i}$. Actually it is typical when working with Laplacian matrices that the most significant information on graphs is related to the first eigenvalues, and a common practice is to eliminate some of the last ones. In our context, to guarantee coherent results, it is necessary discarding some eigenvalues: since the last ones are significantly bigger than the previous, a numerical error appears and distorts results. Since the growth of the matrix spectrum has been shown to be almost linear ([19]), the biggest eigenvalues can be automatically removed analysing the increase of the spectrum and discarding the values that diverge from linearity.

5   Concluding Remarks and Future Developments

Our experiments demonstrate that the method is promising. We remark that there is a relation among the spectrum of the Laplacian matrix related to the scene and the eigenvalues of the single components, and this fact assures us that the scene graph is comparable with those of the singular components. Experimental results let us say that the method is stable to graph perturbations, and it is also flexible, since it is possible to consider topological and geometrical features of the objects at the same time. Finally, we introduce a new distance able to recover similarity between sets of graphs with a good reliability.

False positive results are possible, in particular when two objects have a similar structure; however we plan to reduce this effect considering a more complex description of the graph nodes and more attributes at the same time. As far as *false negatives* are concerned, that is when objects similar to the query are not recognized within the same class, we verify that, when comparing scenes, the method retrieves about the 60% of the right class models within the first 30 scenes.

Obviously, the method has some limits. Actually, we make strong hypotheses about the classes which the method works on (for example, the kind and number of objects). Nowadays it can not still handle the partial matching problem, for example, if we consider in a scene a chair and a table and in another one the same chair and table and other 10 objects, the method can not recognize similar subparts, because it is mainly devoted to give a global answer of the scene similarity. However, future developments might be able to improve partial identification. Moreover, we are willing to continue the analysis of the method with respect to the physical and mutual position of the objects, and, if possible, to compare simultaneously graphs extracted with different functions and resolutions, in order to work every time with the most performing descriptor.

Acknowledgments

This work is partially supported by the EU Newtwork of Excellence "AIM@SHAPE" (contract n. 506766) and the CNR ICT Department.

References

1. Biasotti, S., Giorgi, D., Spagnuolo, M., Falcidieno, B.: Size functions for 3D shape retrieval. In: Eurographics Symposium on Geometry Processing, pp. 239–242 (2006)
2. Falcidieno, B., Spagnuolo, M.: A shape abstraction paradigm for modeling geometry and semantics. In: Computer Graphics International, pp. 646–657 (1998)
3. Torsello, A., Hidovic-Rowe, D., Pelillo, M.: Polynomial-time metrics for attributed trees. IEEE Trans. on Pattern Anlysis and Machine Intelligence 27(7), 1087–1099 (2005)
4. Bunke, H., Shearer, K.: A graph distance metric based on the maximal common subgraph. Pattern Recognition Letters 19, 255–259 (1998)

5. Fernández, M.L., Valiente, G.: A graph distance metric combining maximum common subgraph and minimum common supergraph. Pattern Recognition Letters 22, 753–758 (2001)
6. Marini, S., Spagnuolo, M., Falcidieno, B.: From exact to approximate maximum common subgraph. In: Brun, L., Vento, M. (eds.) GbRPR 2005. LNCS, vol. 3434, pp. 263–272. Springer, Heidelberg (2005)
7. Shokoufandeh, A., Macrini, D., Dickinson, S., Siddiqi, K., Zucker, S.W.: Indexing hierarchical structures using graph spectra. IEEE Trans. on Pattern Anlysis and Machine Intelligence 27(7), 1125–1140 (2005)
8. Robles-Kelly, A., Hancock, E.R.: Graph matching using adjacency matrix Markov chains. In: The British Machine Vision Conference, University of Manchester, Manchester, UK, pp. 383–390 (2001)
9. Wilson, R.C., Hancock, E.R., Luo, B.: Pattern vectors from algebraic graph theory. IEEE Trans. on Pattern Anlysis and Machine Intelligence 27, 1112–1124 (2005)
10. Bapat, R.B.: Distance matrix and Laplacian of a tree with attached graphs. Linear Algebra and its Applications 411, 295–308 (2005)
11. Demirci, M.F., Shokoufandeh, A., Keselman, Y., Bretzner, L., Dickinson, S.: Object recognition as many-to-many feature matching. International Journal of Computer Vision 69(2), 203–222 (2006)
12. Rubner, Y., Tomasi, C., Guibas, L.J.: The earth mover's distance as a metric for image retrieval. International Journal of Computer Vision 40(2), 99–121 (2000)
13. Paraboschi, L., Biasotti, S., Falcidieno, B.: 3d scene comparison using topological graphs. In: Eurographics Italian Chapter, Trento (Italy), pp. 87–93 (2007)
14. Horaud, R., Sossa, H.: Polyhedral object recognition by indexing. Pattern Recognition 28(12), 1855–1870 (1995)
15. Mohar, B.: The Laplacian spectrum of graphs. Graph Theory, Combinatorics, and Applications 2, 871–898 (1991)
16. Franklin, J.N.: Matrix theory. Dover Publications, Mineola, NY (1993)
17. Chen, D., Ouhyoung, M., Tian, X., Shen, Y.: On visual similarity based 3D model retrieval. Computer Graphics Forum 22, 223–232 (2003)
18. Kazhdan, M., Funkhouser, T., Rusinkiewicz, S.: Rotation invariant spherical harmonic representation of 3D shape descriptors. In: Eurographics Symposium on Geometry Processing, pp. 167–175 (2003)
19. Dong, S., Bremer, P.T., Garland, M., Pascucci, V., Hart, J.C.: Spectral surface quadrangulation. ACM Trans. on Graphics 25(3), 1057–1066 (2006)

Hierarchy Construction Schemes Within the Scale Set Framework

Jean-Hugues Pruvot and Luc Brun

GREYC Laboratory, Image Team
CNRS UMR 6072
6, Boulevard Maréchal Juin
14050 CAEN cedex France
jhpruvot@greyc.ensicaen.fr

Abstract. Segmentation algorithms based on an energy minimisation framework often depend on a scale parameter which balances a fit to data and a regularising term. Irregular pyramids are defined as a stack of graphs successively reduced. Within this framework, the scale is often defined implicitly as the height in the pyramid. However, each level of an irregular pyramid can not usually be readily associated to the global optimum of an energy or a global criterion on the base level graph. This last drawback is addressed by the scale set framework designed by Guigues. The methods designed by this author allow to build a hierarchy and to design cuts within this hierarchy which globally minimise an energy. This paper studies the influence of the construction scheme of the initial hierarchy on the resulting optimal cuts. We propose one sequential and one parallel method with two variations within both. Our sequential methods provide partitions near an energy lower bound defined in this paper. Parallel methods require less execution times than the sequential method of Guigues even on sequential machines.

1 Introduction

Despite much efforts and significant progresses in recent years, image segmentation remains a notoriously challenging computer vision problem. It's usually a preliminary step towards image interpretation and plays a major role in many applications.

The use of an energy minimisation scheme within the region based segmentation framework allows to define criteria which should be globally optimised over a partition. Several types of methods such as the Level set [1], the Bayesian [2], the minimum description length [3] and the minimal cut [4] frameworks are based on this approach. Within these frameworks the energy of a partition P is usually defined as $E_\lambda(P) = D(P) + \lambda C(P)$ where D and C denote respectively the fit to data and the regularising term. The energy $E_\lambda(P)$ corresponds to the Lagrangian of the constraint problem: minimise $D(P)$ subject to $C(P) \leq \epsilon$. Where ϵ is a function of λ. Under large assumptions, minimising $E_\lambda(P)$ is also equivalent to the dual problem: minimise $C(P)$ subject to $D(P) \leq \epsilon'$, where ϵ' is also a function of λ. Therefore λ may be interpreted as the amount of freedom allowed to minimise D ($D(P) \leq \epsilon'$) while keeping C as low as possible. Since ϵ' is a growing function of λ, as λ is growing, the constraint on D is more

F. Escolano and M. Vento (Eds.): GbRPR 2007, LNCS 4538, pp. 126–137, 2007.

and more relaxed while the importance of the term C is getting more and more important. This parameter λ may thus be interpreted as a *scale parameter* which represents the relative weighting between the two energy terms.

In many approaches the parameter λ is fixed experimentally and a minimisation algorithm determines for a value of λ a locally optimal partition from the set $\mathbb{P}$ of all the possible partitions on image I. A sequence of λ may also be defined a priori in order to compute the optimal partition on each sampled value of λ [5].

The scale set framework proposed by Guigues [5] is based on a different approach. Instead of performing the minimisation scheme on the whole set $\mathbb{P}$ of possible partitions of an image I, Guigues proposes to restrict the search on a hierarchy H. The advantages of this approach are twofold: firstly as shown by Guigues the globally optimal partition on H may be found efficiently while the search on the whole set $\mathbb{P}$ of partitions only provides local minima. Secondly, Guigues shown that if the energy satisfies some basic properties, the whole set of solutions on H when λ describes $\mathbb{R}+$ corresponds to a sequence of increasing cuts within the hierarchy H hereby providing a contiguous representation of the solutions for the parameter λ. A method to build the hierarchy H has been proposed by Guigues. Since the research space used by Guigues is restricted to the initial hierarchy H the construction scheme of this hierarchy is of crucial importance for the optimal partitions *within* H built in the second step.

This paper explores different heuristics to build the initial hierarchy. These heuristics represent different compromises between the energy of the final partitions and the execution times. We first present in Section 2 the scale set framework. The different heuristics are then presented in Section 3. These heuristics are evaluated and compared to the method of Guigues in Section 4.

2   The Scale Set Framework

Given an image I and two partitions P and Q on I, we will say that P is *finer* than Q (or Q is coarser then P) iff Q may be deduced from P by merging operations. This relationship is denoted by $P \unlhd Q$. Let us now consider a theoretic segmentation algorithm P_λ parametrised by λ. We will say that P is an *unbiased multi-scale segmentation* algorithm iff for any couple (λ_1, λ_2) such that $\lambda_1 \leq \lambda_2$, and any image I, $P_{\lambda_1}(I) \unlhd P_{\lambda_2}(I)$. If P_λ is an unbiased multi-scale segmentation algorithm, $P_\lambda(I)$ increases according to λ and the set $H = \bigcup_{\lambda \in \mathbb{R}+} P_\lambda(I)$ defines a hierarchy as an union of nested partitions. Note that the set $\mathbb{P}$ of partitions on I being finite, H must be also finite.

Unbiased multi-scale segmentation algorithms follow a well known causal principal: increasing the scale of observation should not create new information. In other words any phenomenon observed at one scale should be caused by objects defined at finer scales. In our framework, increasing the scale should not create new contours.

The family of energies considered by Guigues corresponds to the set of Affine Separable Energies (ASE) which can be written for any partition P of I in n regions $\{R_1, \ldots, R_n\}$ as:

$$E(P) = D(P) + \lambda C(P) = \sum_{i=1}^{n} D(R_i) + \lambda \sum_{i=1}^{n} C(R_i) = \sum_{i=1}^{n} D(R_i) + \lambda C(R_i)$$

Let us consider a hierarchy H and the sequence $(C_\lambda^*(H))_{\lambda \in \mathbb{R}+}$ of optimal cuts within H. The approach of Guigues is based on the following result: If $E_\lambda(P)$ is an ASE and if $C_\lambda(P)$ is decreasing within $\mathbb{P}$:

$$\forall (P, Q) \in \mathbb{P} \quad P \triangleleft Q \Rightarrow C(P) > C(Q)$$

then the sequence $(C_\lambda^*(H))_{\lambda \in \mathbb{R}+}$ is an unbiased multi-scale segmentation. The union of all $(C_\lambda^*(H))_{\lambda \in \mathbb{R}+}$ defines thus a new hierarchy within H. The tree corresponding to the hierarchical structure of $\bigcup_{\lambda \in \mathbb{R}+} C_\lambda^*(H)$ may be deduced from H by merging with their fathers all the nodes which do not belong to any optimal cuts. Note that an equivalent result may be obtained if no condition is imposed to C but if D is increasing according to λ.

The restriction by Guigues of the research space to a hierarchy may thus be justified by the fact that the set of partitions produced by any unbiased multi-scale segmentation algorithm describes a hierarchy. Conversely, given a hierarchy H, if the energy E_λ is an ASE with a decreasing term C the sequence of optimal cuts of H according to E_λ: $(C_\lambda^*(H))_{\lambda \in \mathbb{R}+}$ is an unbiased multi-scale segmentation algorithm.

Given a partition $P \in \mathbb{P}$, the decrease of C may be equivalently expressed as a sub-additivity relationship:

$$\forall (R, R') \in P \mid R \text{ is adjacent to } R' \quad C(R \cup R') < C(R) + C(R') \tag{1}$$

Note that the sub-additivity of the regularising term C in common is many applications. For example, if C is proportional to some quantity summed up along contours, C is sub-additive due to the removal of the common boundaries between the two merged regions. Moreover, the term C may be interpreted within the Minimum Description Length framework [3] as the amount of information required to encode a partition. Therefore, one can expect C to decrease when the partition gets coarser.

Given a hierarchy H, the sequence of optimal cuts $C_\lambda^*(H)$ within H has to be computed. Let us consider one region R at the second level of the hierarchy (computed from the base) and its set of sons $S_1, \ldots, S_n$. Let us additionally consider the tree $H(R)$ rooted at R within H (Fig. 1(a)). Since R is a level 2 node, the hierarchy $H(R)$ allows only two cuts: one encoding the partition P_1 made of the sons of R whose energy is equal to $E_\lambda(P_1) = \sum_{i=1}^n D(S_i) + \lambda \sum_{i=1}^n C(S_i)$ and one encoding the partition P_2 reduced to the single region R. The energy of P_2 is equal to $E_\lambda(P_2) = D(R) + \lambda C(R)$. Due to the sub additivity of C we have $\sum_{i=1}^n C(R_i) > C(R)$. Therefore, using the linear expression of $E_\lambda(P_1)$ and $E_\lambda(P_2)$ in λ, if $\sum_{i=1}^n D(S_i) < D(R)$ the line $E_\lambda(P_1) = \sum_{i=1}^n D(S_i) + \lambda \sum_{i=1}^n C(R_i)$ is below the line $E_\lambda(P_2) = D(R) + \lambda C(R)$ until a value $\lambda^+(R)$ of λ for which the two lines cross(Fig. 1(b)). If $\sum_{i=1}^n D(S_i) \geq D(R)$, $E_\lambda(P_2)$ is always greater or equal to $E_\lambda(P_1)$ in which case we set $\lambda^+(R)$ to 0. Therefore, in both cases the partition P_1 is associated to a lower energy than P_2 for $\lambda = 0$ until $\lambda = \lambda^+(R)$. Above this value the partition P_2 is associated to the lowest energy. In terms of optimal cuts, P_1 corresponds to the optimal cut of $H(R)$ until $\lambda^+(R)$ and P_2 is the optimal cut above this value(Fig. 1(c)). The value $\lambda^+(R)$ is called the *scale of appearance* of the region R.

Guigues shown that the above process may be generalised to the whole tree. Each node of H is then valuated by a scale of appearance. Some of the nodes of H may get a greater scale of appearance than their father. Such nodes do not belong to any

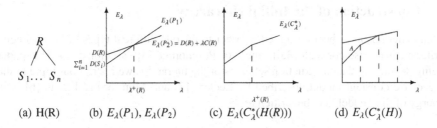

Fig. 1. (a) a node R of the hierarchy whose sons $\{S_1, \ldots, S_n\}$ correspond to initial regions. (b) the energies of the partitions associated to R and $\{S_1, \ldots, S_n\}$ plotted as functions of λ. (c) the energy of the optimal cuts within $H(R)$ (a). (d) an example of concave piecewise linear function encoding the energy of the optimal cuts within a global hierarchy H.

optimal cut and are removed from H during a cleaning step which merges them with their fathers. Each node R of the resulting hierarchy belongs to an optimal cut from $\lambda = \lambda^+(R)$ until the scale of appearance of its father $\lambda^+(\mathcal{F}(R))$, where $\mathcal{F}(R)$ denotes the father of R in H. The value $\lambda^+(R)$ may be set for each node of the tree using a bottom-up process. The optimal cut $C_\lambda^*(H)$ for a given value of λ may then be determined using a top-down process which selects in each branch of the tree the first node with a scale of appearance lower than λ. The set of selected nodes constitutes a cut of H which is optimal by construction according to E_λ. The function $E_\lambda(C_\lambda^*(H))$ corresponds to a concave piecewise linear function whose each linear interval corresponds to the energy of an optimal cut within H (Fig. 1(d)).

Given a hierarchy H and the function $E_\lambda(C_\lambda^*(H))$ encoding the energy of the sequence of optimal cuts, the optimality of H may be measured as the area under the curve $E_\lambda(C_\lambda^*(H))$ for a given range of scales or as the area of the surface A (Fig. 1(d)) between $E_\lambda(C_\lambda^*(H))$ and the energy of the coarsest cut $E_\lambda(P_{max})$. Where P_{max} denote the partition composed of a single region encoding the whole image. We propose in Section 4 an alternative measure of the quality of a hierarchy which allows to reduce the influence of the initial image.

Guigues proposed to build a hierarchy H by using an initial partition P_0 and a strategy called the *scale climbing*. This strategy merges at each step the two adjacent regions R and R' such that:

$$\lambda^+(R \cup R') = \frac{D(R \cup R') - D(R) - D(R')}{C(R) + C(R') - C(R \cup R')} = \min_{(R_1,R_2) \in P^2, R_1 \sim R_2} \frac{D(R_1 \cup R_2) - D(R_1) - D(R_2)}{C(R_1) + C(R_2) - C(R_1 \cup R_2)}$$
(2)

where P denotes the current partition and $R_1 \sim R_2$ indicates that R_1 and R_2 are adjacent in P.

This process merges thus at each step the two regions whose union would appear at the lowest scale. Such a construction scheme is coherent with the further processes applied on the hierarchy. However, there is no evidence that the resulting hierarchy may be optimal according to any of the previously mentioned criteria. We indeed show in the next section that other construction schemes of a hierarchy may lead to lower energies.

3   Construction of the Initial Hierarchy

Many energies have been designed in order to encode different types of homogeneity criteria (piecewise constant [3,6], linear or Polynomial [3] variations,...). This paper being devoted to the construction schemes of the hierarchy, we restrict our topic to the piecewise constant model described by Leclerc [3] and Mumford and Shah [6]. The energy of this model may be written as:

$$E_\lambda(P) = D(P) + \lambda C(P) = \sum_{i=1}^{n} SE(R_i) + \lambda |\delta(R_i)| \qquad (3)$$

where $P = \{R_1, \ldots, R_n\}$ represents the partition of the image, $SE(R_i) = \sum_{p \in R} \|c_p - \mu_R\|^2$ is the squared error of region R_i and $|\delta(R_i)|$ is the total length of its boundaries.

Within the Minimum Description Length framework, $SE(R_i)$ may be understood as the amount of information required to encode the deviation of the data against the model, while $|\delta(R_i)|$ is proportional to the amount of information required to encode the shape of the model. Within the statistical framework, the squared error may also be understood as the log of the probability that the region satisfies the model (i.e. is constant) using a Gaussian assumption while $|\delta(R_i)|$ is a regularising term.

Our approach follows the scale climbing strategy proposed by Guigues (equation 2). Given a set W of regions within a partition P we thus consider the scale of appearance of the region R defined as the union of the regions in W. The heuristics below use this basic approach but differ on the sets W which are considered and on the ordering of the merge operations.

3.1   Sequential Merging

Given a current partition P, let us consider for each region R of P, its set $V(R)$ defined as $\{R\}$ union its set of neighbours and the set $\mathcal{P}^*(V(R))$ of all possible subsets of $V(R)$ including R. Each subset $W \in \mathcal{P}^*(V(R))$ encodes a possible merging of the region R with at least one of its neighbour. Let us denote by $R^W = \bigcup_{R' \in W} R'$ the region formed by the union of the regions in W. Note that the region R^W is connected since R belongs to W and all the regions of W are adjacent to R. Let us additionally consider the two partitions of R^W: $P_{R^W} = \{R^W\}$ and $P_W = W$. The energies associated to these partitions are respectively equal to $E_\lambda(P_{R^W}) = D(R^W) + \lambda C(R^W)$ and:

$$E_\lambda(P_W) = D(W) + \lambda C(W) = \sum_{R' \in W} D(R') + \lambda \sum_{R' \in W} C(R')$$

where $D(W)$ and $C(W)$ denote respectively the fit to data and the regularising terms of the partition P_W.

Since C is sub additive (equation 1) we have $C(W) > C(R^W)$. The energy $E_\lambda(P_W)$ is thus lower than $E_\lambda(P_{R^W})$ until a value $\lambda^+(R^W)$ called the scale of appearance of R^W (Section 2). Using the scale climbing principle, our sequential merging algorithm computes for each region R of the partition the minimal scale of appearance of a region R^W:

$$\lambda_{min}^+(R) = \arg \min_{W \in \mathcal{P}^*(V(R))} \frac{D(R^W) - D(W)}{C(W) - C(R^W)}$$

the set $W \in \mathcal{P}^*(V(R))$ which realises the min is denoted $W_{min}(R)$.

Given the quantities $\lambda^+_{min}(R)$ and $W_{min}(R)$, our sequential algorithm iterates the following steps:

1. Let P denotes the current partition initialised with an initial partition P_0,
2. For each region R of P compute $\lambda^+_{min}(R)$ and $W_{min}(R)$
3. Compute $R_{min} = \arg\,min_{R\in P}\lambda^+_{min}(R)$ and merge all the regions of $W_{min}(R_{min})$.
4. If more than one region remains go to step 2,
5. Output the final hierarchy H encoding the sequence of merge operations.

This algorithm performs thus one merge operation at each step of the algorithm. Note that all the regions of $W_{min}(R_{min})$ are adjacent to R_{min}. Therefore, within the irregular pyramid framework, the merge operation may be encoded by a contraction kernel of depth one composed of a single tree whose root is equal to R_{min}. The computation of $\lambda^+_{min}(R)$ for each region R of the partition requires to traverse $\mathcal{P}^*(V(R))$ whose cardinal is equal to $2^{|V(R)|-1}$. Therefore, if the partition is encoded by a graph $G = (V, E)$, the complexity of each step of our algorithm is bounded by $O(|V|2^k)$ where $|V|$ denotes the number of vertices (i.e. the number of regions) and k represents the maximal vertices's degree of G. The cardinal of V is decreased by $|W_{min}(R_{min})| - 1$ at each iteration. Since $|W_{min}(R_{min})|$ is at least equal to 2, the cardinal of V decreases by at least 1. The computation of $\lambda^+_{min}(R)$ for each region R of the partition may induce important execution times when the degree of the vertices of the graph is important. However, experiments presented in Section 4 show that the cardinal of the subsets $W \in \mathcal{P}^*(R)$ may be bounded without altering significantly the energy of the optimal cuts. Let us finally note that this algorithm includes the scale climbing approach proposed by Guigues. Indeed, the merge operations studied by Guigues (Section 2) correspond to the subsets $W \in \mathcal{P}^*(V(R))$ with $|W| = 2$ which are considered by our algorithm.

3.2 Parallel Merge Algorithm

Our parallel merge algorithm is based on the notion of maximal matching. A set of edges M of a graph $G = (V, E)$ is called a maximal matching if each vertex of G is incident to at most one edge of M and if M is maximal according to this property. Moreover, we would like to design a maximal matching M such that the scale of appearance of the regions produced by the contraction of M is as low as possible. Let us denote by $\iota(e)$, the two vertices incident to e. Using the same approach as in Section 3.1 we associate to each edge e of the graph the scale of appearance $\lambda^+(\iota(e))$ (equation 2) of the region $R^{\iota(e)}$ defined as the union of the regions encoded by the two vertices incident to e. Following, the same approach as Haxhimusa [7] we define our maximal matching as a Maximal Independent Set on the set of edges of the graph. The iterative process which builds the maximal independent set selects at each step edges whose scale of appearance is locally minimal. This process may be formulated thanks to two boolean variables p and q attached to each edge such that:

$$\begin{cases} p^1_e = \lambda^+(e) = min_{e'\in\Gamma(e)}\{\lambda^+(e')\} \\ q^1_e = \bigwedge_{e'\in\Gamma(e)} \overline{p^1_{e'}} \end{cases} \text{ and } \begin{cases} p^{k+1}_e = p^k_e \vee \left(q^k_e \wedge \lambda^+(e) = min_{e'\in\Gamma(e)\,|\,q^k_{e'}}\{\lambda^+(e')\}\right) \\ q^{k+1}_e = \bigwedge_{e'\in\Gamma(e)} \overline{p^{k+1}_{e'}} \end{cases}$$

$$(4)$$

where $\Gamma(e)$ denotes the neighbourhood of the edge e and is defined as $\Gamma(e) = \{e\} \cup \{e' \in E | \iota(e) \cap \iota(e') \neq \emptyset\}$.

This iterative process stops when no change occurs between two iterations. If n denotes the final iteration, the set of edges such that p_e^n is true defines a maximal matching [7] M which encodes the set of edges to be contracted. Moreover, the set of selected edges corresponds to local minima according to the scale of appearance $\lambda^+(e)$. Roughly speaking if $\lambda^+(e)$ is understood as a merge score, one edge between two vertices will be marked ($p_e^k = true$) at iteration k, if among all the remaining possible merge operations involving these two vertices, the one involving them is the one with the best merge score. Note that the construction of a maximal matching is only the first step of the method of Haxhimusa which completes this maximal matching in order to get a decimation ratio of order 2. The restriction of our method to a maximal matching allows to restrict the merge operations to edges which become locally optimal at a given iteration. We thus favour the energy criterion against the reduction factor. As shown by Bield [8], the reduction factor in terms of edges induced by the use of a maximal matching is a least equal to $2\frac{k-1}{2k-1}$ where k is the maximal vertex's degree of the graph. The edge's decimation ratio may thus be very low for graphs with important vertices's degrees. Nevertheless, experiments performed on 100 natural images of the Berkeley database[1] shown that the mean vertex's decimation ratio between levels on this database is equal to 1.73 which is comparable to the 2.0 decimation ratio obtained by Haxhimusa.

The local minima selected in equation 4 are computed on decreasing sets along the iterations in order to complete the maximal matching. We can thus consider that the detected minima are less and less significants as the iterations progress. We thus propose an alternative solution which consists in contracting at each step only the edges selected at the first iteration ($p_e^1 = true$). These edges correspond to minima computed on the whole neighbourhood of each edge. This method may be understood as a combination of the method proposed by Haxhimusa [7] and the stochastic decimation process of Jolion [9] which consists in merging immediately vertices corresponding to local minima.

4   Experiments

The different heuristics presented in this paper have been evaluated on the Berkeley database. The evaluated heuristics include our parallel merge heuristic based on a maximal matching (MM) and the variation of this method(MM^1) which merges at each step the edges selected during the first iteration (Section 3.2). We also evaluated our sequential method (SM) and two variations of this method: the first variation (SM^2), considers for each region R of the partition the subsets of cardinal 2 of $V(R)$. This method corresponds to the heuristic proposed by Guigues. We also evaluated an intermediate method (SM^5) which restricts the cardinal of the subsets of $V(R)$ including R to an upper threshold fixed to five in these experiments. All the experiments have used an initial partition obtained by a Watershed algorithm [10].

Fig. 2 shows 5 optimal cuts obtained for increasing values of λ on the Mushroom and Fisherman images of the Berkeley database[2]. The heuristics used to build the

---

[1] Available at http://www.eecs.berkeley.edu/Research/Projects/CS/vision/bsds/

[2] Color plates are available at the following url: http://www.greyc.ensicaen.fr/~jhpruvot/Cut/

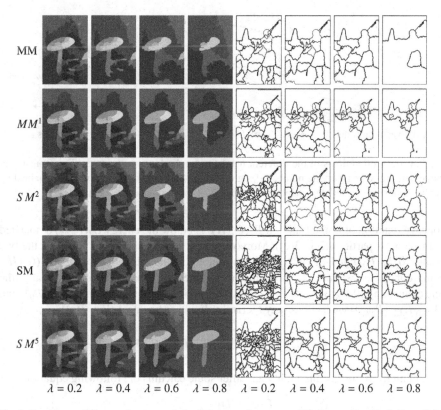

Fig. 2. Partitions of the mushroom and the fisherman images at different scales. Each line of the array corresponds to an heuristic whose acronym is indicated on the first column.

hierarchies are displayed on the first column of Fig. 2. The original images are displayed in Fig. 4(a).

Fig. 3(a) shows the influence of the number of initial regions on the execution time. These curves have been obtained on the Mushroom image with different initial partitions obtained by varying the smoothing parameter of the gradient within our Watershed algorithm.

Fig. 3(b) allows to compare the performance of each heuristic on the whole Berkeley database. However, a direct comparison of the energies obtained by the different heuristics on different images would be meaningless since the shape of the function $E_\lambda(C^*_\lambda(H))$ depends both of the intrinsic performances of the heuristic used to build H and of the image I on which H has been built. We have thus to normalise the energies $E_\lambda(C^*_\lambda(H))$ produced by the different heuristics before any comparison.

Given a hierarchy H, since $C^*_\lambda(H)$ is an unbiased multi-scale segmentation (Section 2), the hierarchy H obtained by each of our methods may be associated to a value λ^H_{max} above which the optimal partition P_{max} is reduced to a single region encoding the whole image. The energy of P_{max} is defined as: $E_\lambda(P_{max}) = D_I + \lambda C_I$ where $D_I = SE(I)$ denotes the global image's squared error and $C_I = |\delta(I)|$ the perimeter of the image. Since the energy of the optimal cuts $E_\lambda(C^*_\lambda(H))$ of a hierarchy H is a piecewise linear

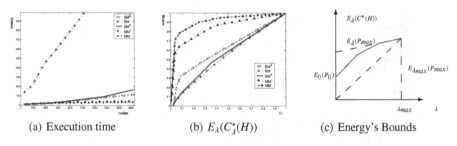

| (a) Execution time | (b) $E_\lambda(C_\lambda^*(H))$ | (c) Energy's Bounds |

Fig. 3. (a) execution times of the different heuristics on the Mushroom image (Fig. 2) using an initial partition with a varying number of regions. (b) mean energies of optimal cuts obtained by our heuristics on the Berkeley database. (c) bounds of the optimal cut's energies.

concave function of λ, the function $E_\lambda(C_\lambda^*(H))$ is below the energy $E_\lambda(P_{max})$ associated to the coarser partition(Fig. 3(c)). Moreover, if P_0 denotes the initial partition, the two points $(0, E_0(P_0))$ and $(\lambda_{max}, E_{\lambda_{max}}(P_{max}))$ belong to the curve. Therefore, $E_\lambda(C_\lambda^*(H))$ being concave, it should be above the line connecting these two points. Finally, the line connecting $(0, 0)$ to $(\lambda_{max}, E_{\lambda_{max}}(P_{max}))$ being below the line joining $(0, E_0(P_0))$ and $(\lambda_{max}, E_{\lambda_{max}}(P_{max}))$ we have for any hierarchy H and any scale λ (Fig. 3(c)):

$$\frac{\lambda}{\lambda_{max}} E_{\lambda_{max}}(P_{max}) \leq E_\lambda(C_\lambda^*(H)) \leq E_\lambda(P_{max})$$

We obtain from this last inequality and after some calculus the following equation:

$$\forall \lambda \in \mathbb{R}+ \quad x_\lambda \leq 1 + \frac{x_\lambda - 1}{1 + x_\lambda E_I} \leq \frac{E_\lambda(C_\lambda^*(H))}{E_\lambda(P_{max})} \leq 1 \text{ with } x_\lambda = \frac{\lambda}{\lambda_{max}} \text{ and } E_I = \frac{\lambda_{max} C_I}{D_I} \quad (5)$$

Therefore, using the normalised energy, $\frac{E_\lambda(C_\lambda^*(H))}{E_\lambda(P_{max})}$ and the normalised scale $x_\lambda = \frac{\lambda}{\lambda_{max}}$, any curve $\frac{E_\lambda(C_\lambda^*(H))}{E_\lambda(P_{max})}$ lies in the upper left part of the unit cube $[0, 1]^2$. Note that this result is valid for any hierarchy H and thus any heuristic.

Using our piecewise constant model (equation 3), the energy $E_\lambda(P_{max})$ is roughly equal to the squared error of the image for small values of λ and may be interpreted as the global variation of the image. The normalised energy allows thus to reduce the influence of the global variation of the images on the energy and to compare energies computed with a same heuristic but on different images. Note however, that the use of the normalised scale $x_\lambda = \frac{\lambda}{\lambda_{max}}$ discards the absolute value of λ_{max}. We thus do not take into account the range of scales for which the optimal cut is not reduced to the trivial partition P_{max}. However, the absolute value of λ_{max} varies according to each image and each heuristics. The normalised scale allows thus to remove the influence of the image. Moreover, our experiments shown thus that for each image, our different heuristics obtain close λ_{max} values.

Fig. 3(b) represents for each value of x_λ and each heuristic, the mean value of the normalised energy $\frac{E_\lambda(C_\lambda^*(H))}{E_\lambda(P_{max})}$ computed on the whole set of images of the Berckley database.

As shown in Fig 3(b) the energy of the optimal cuts obtained by the heuristic MM^1 ($-\blacktriangle-$) is lower than the one obtained by the maximal matching heuristic ($- \bullet -$). This

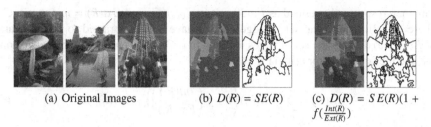

(a) Original Images   (b) $D(R) = SE(R)$ (c) $D(R) = SE(R)(1 + f(\frac{Int(R)}{Ext(R)}))$

Fig. 4. (a) Original images. (b) and (c), partitions of the tower image built with a same heuristic(SM) at a same normalised scale ($x_\lambda = .8$) but with energies defined using two different fit to data terms. (b) is defined using the squared error $D(R) = SE(R)$ while (c) is defined using the formula defined by equation 6.

result is confirmed by Fig. 2 (lines MM and MM^1) where the heuristic MM removes more details of the mushroom at a given scale. This result is connected to the greater decimation ratio of the MM heuristic. The MM heuristic merges at each step regions with important scale of appearance without considering regions which may appear at further steps. The algorithms MM and MM^1 induce equivalent execution times on a sequential machine. The execution times of the method MM^1 ($-\blacktriangle-$) are overlayed by the ones of the method MM ($-\bullet-$) in Fig. 3(a) due to the vertical scale of this figure.

The subjective quality of the partitions obtained by the heuristics MM^1 and SM^2 (Fig. 2) seems roughly similar. We can notice that the heuristic MM^1 seems to produce slightly coarser partitions at each scale. However, considering Fig. 3(b), the optimal energy obtained by the heuristic SM^2 ($-\square-$) are lower than the one obtained by MM^1 ($-\blacktriangle-$). Note that the heuristic MM^1 produces lower execution times than SM^2 even on a sequential machine(Fig. 3(a)).

As shown by Fig. 3(b) the optimal energies produced by the heuristic SM ($-+-$) are always below the one produced by the heuristic SM^2 ($-\square-$). Note that, the curve ($-+-$) is close to the diagonal of the square $[0, 1]^2$. This last point indicates that on most of the images of the Berkeley database the hierarchies produced by the SM heuristic provide optimal cuts whose normalised energy is closed from the lower bound of the optimal cut's energies (equation 5). This result is confirmed by Fig. 2 where the heuristic SM preserves more details of the image at each scale. However, the heuristic SM is the one which requires the more important execution times on a sequential machine (Fig. 3(a)).

The heuristic SM^5 may be understood as a compromise between SM^2 and SM. As shown by Fig. 3(b) the optimal energies obtained by the heuristic SM^5 ($\blacksquare$) are close to the one obtain by $SM(-+-)$ and below the one obtained by $SM^2(-\square-)$. Moreover, as shown by Fig. 3(a), the execution times required by SM^5 are between the one required by the heuristics SM^2 and SM. Finally, the partitions obtained by the SM^5 heuristic in Fig. 2 are closed from the one obtained by the heuristic SM.

Fig. 4 shows results obtained using an other fit to data criterion based on the intuitive notion of contrast. The basic idea of this criterion [11] states that a region should have a higher contrast with its neighbours (called external contrast) than within its eventual subparts (called internal contrast). Let us denote by G_e the mean gradient computed along the contour associated to an edge e. The internal and external contrasts of a region

R are then respectively defined as $Int(R) = max_{e \in CC(R)} G_e$ and $Ext(R) = min_{e \in E | v \in \iota(e)} G_e$. Where $CC(R)$ denotes the set of edges which have been contracted to define R and $e \in E | v \in \iota(e)$ denotes the set of edges incident to v. Our new energy combines the contrast and the squared error criteria as follows:

$$E_\lambda(P) = \sum_{i=1}^{n} S E(R_i) \left(1 + f \left(\frac{Int(R_i)}{Ext(R_i)} \right) \right) + \lambda |\delta(R_i)| \tag{6}$$

where $f()$ denotes a sigmoid function.

A contrasted region will thus have a low ratio between its internal and external contrast. Conversely, a poorly contrasted region may have a fit to data term close to twice its squared error. As shown by Fig. 3(b) and (c) this energy favours highly contrasted regions. For example, the cloud merged with the sky in Fig. 3(b) remains in Fig. 3(c). Moreover, experiments not reported here, shown us that the same type of discussion about the advantages and drawbacks of the different heuristics may be conducted on this new energy with the same conclusions.

5 Conclusion

The Scale Set framework is based on two steps: the determination of a hierarchy according to an energy criterion and the determination of optimal cuts within this hierarchy. We have presented in this article parallel and sequential heuristics to build such hierarchies. The normalised energy of the optimal cuts, associated with these hierarchy are bounded bellow by the diagonal of the unit square $[0, 1]^2$. Our experimental results suggest that our sequential heuristic SM provides hierarchies whose normalised energies are closed from this lower bound. This methods may however require important execution times. We thus propose an alternative heuristic providing lower execution time at the price of generally slightly higher optimal cut's energies. Our parallel methods provide greater energies than the one produced by Guigues's heuristic. However, these methods require less execution times even on sequential machine.

Hierarchies encoding a sequence of optimal cuts are usually composed of a lower number of levels and regions than the initial hierarchies built by our merge heuristics. In the future, we would like to use these hierarchies of optimal cuts in order to match two hierarchies encoding the content of two images sharing a significant part of a same scene.

References

1. Lecellier, F., Jehan-Besson, S., Fadili, M., Aubert, G., Revenu, M., Saloux, E.: Region-based active contours with noise and shape priors. In: proceedings of ICIP'2006, pp. 1649–1652 (2006)
2. Geman, S., Geman, D.: Stochastic relaxation, gibbs distribution, and the bayesian restoration of images. IEEE Transactions on PAMI 6(6), 721–741 (1984)
3. Leclerc, Y.G.: Constructing simple stable descriptions for image partitioning. International Journal of Computer Vision 3(1), 73–102 (1989)

4. Boykov, Y., Kolmogorov, V.: An experimental comparison of min-cut/max-flow algorithms for energy minimization in vision. IEEE Transaction on PAMI 26(9), 1124–1137 (2004)
5. Guigues, L., Cocquerez, J.P., Men, H.: Scale-sets image analysis. Int. J. Comput. Vision 68(3), 289–317 (2006)
6. Mumford, D., Shah, J.: Optimal approximation by piecewise smooth functions and associated variational problems. Communications on Pure. Applied Mathematics 42, 577–685 (1989)
7. Haxhimusa, Y., Glantz, R., Kropatsch, W.: Constructing stochastic pyramids by mides - maximal independent directed edge set. In: Hancock, E.R., Vento, M. (eds.) GbRPR 2003. LNCS, vol. 2726, pp. 35–46. Springer, Heidelberg (2003)
8. Biedl, T., Demaine, E.D., Duncan, C.A., Fleischer, R., Kobourov, S.G.: Tight bounds on maximal and maximum matching. Discrete Mathematics 285(1-3), 7–15 (2004)
9. Jolion, J.M.: Data driven decimation of graphs. In: Jolion, J.M., Kropatsch, W., Vento, M. (eds.) Proceedings of 3rd IAPR-TC15 Workshop on Graph based Representation in Pattern Recognition, Ischia-Italy, pp. 105–114 (2001)
10. Brun, L., Mokhtari, M., Meyer, F.: Hierarchical watersheds within the combinatorial pyramid framework. In: Andrès, É., Damiand, G., Lienhardt, P. (eds.) DGCI 2005. LNCS, vol. 3429, pp. 34–44. Springer, Heidelberg (2005)
11. Felzenszwalb, P., Huttenlocher, D.: Image segmentation using local variation. In: Proceedings of IEEE Conference on CVPR, Santa Barbara, CA, pp. 98–104 (1998)

Local Reasoning in Fuzzy Attribute Graphs for Optimizing Sequential Segmentation

Geoffroy Fouquier[1,2], Jamal Atif[3,*], and Isabelle Bloch[1]

[1] ENST (GET - Telecom Paris), Dept. TSI, CNRS UMR 5141 LTCI
46 rue Barrault, 75634 Paris Cedex 13, France
Geoffroy.Fouquier@enst.fr
[2] EPITA Research and Development Laboratory (LRDE)
14-16, rue Voltaire F-94276 Le Kremlin Bicêtre, France
[3] Groupe de Recherche sur les Energies Renouvelables (GRER)
Université des Antilles et de la Guyane Campus de St Denis 97 300 Cayenne

Abstract. Spatial relations play a crucial role in model-based image recognition and interpretation due to their stability compared to many other image appearance characteristics. Graphs are well adapted to represent such information. Sequential methods for knowledge-based recognition of structures require to define in which order the structures have to be recognized. We propose to address this problem of order definition by developing algorithms that automatically deduce sequential segmentation paths from fuzzy spatial attribute graphs. As an illustration, these algorithms are applied on brain image understanding.

1 Introduction

Knowledge on the spatial organization of a scene carries important information for analyzing and interpreting images of this scene. Spatial relations play a crucial role in this context, since they are less prone to variability than object appearance or shape. Using this knowledge, often represented in symbolic forms, in high reasoning processes requires to link semantic knowledge with low level information extracted from images. Graph representations are well adapted to solve this semantic gap problem.

In [1], spatial relations and graph-based representations have been used for recognizing structures in a progressive way: the recognition of a structure is driven by its relations to previously recognized structures; these relations are encoded in a graph representing generic knowledge. This allows recognizing "difficult" structures at later stages, once more information has been accumulated. In this work, the order in which structures are recognized is defined in a supervised way. Figure 1 shows some segmentation results obtained with a manually defined order.

In this paper, we propose to automate this step, and to infer automatically segmentation paths using reasoning algorithms in the graph. The idea is to start

---

* This work has been partially funded by GET and ANR grants during J. Atif's postdoctoral position at Telecom Paris.

F. Escolano and M. Vento (Eds.): GbRPR 2007, LNCS 4538, pp. 138–147, 2007.

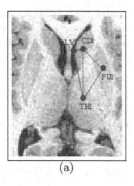
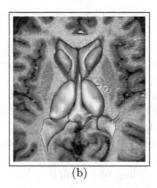
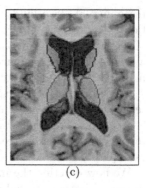

(a)   (b)   (c)

Fig. 1. (a) Slice of a 3D brain magnetic resonance image (MRI). Marked structures are: LVl lateral ventricle, CDl caudate nucleus, THl Thalamus and PUl Putamen. (b, c) Segmentation results from [1].

from a structure, represented as a node in the graph, which is known for being easy to segment and recognize in the images, and to automatically deduce an ordered sequence of structures to be recognized.

A typical application is brain image interpretation, where the domain knowledge involves intensively spatial relations, as acknowledged by neuro-anatomy textbooks [2]. These relations are relatively stable, and exhibit less inter-individual variability than characteristics of the anatomical structures. Graph representations have been used in particular to drive specific recognition procedures (see e.g. [1,3,4] among others). However, in pathological cases, generic knowledge is not always valid and information about the pathology has to be used in order to adapt the reasoning process.

The structure of this paper is as follows. We first describe in Section 2 the graph model, specifically for representing anatomical brain knowledge, along with the fuzzy attributes of edges representing spatial relations. Our contribution on graph-based reasoning is presented in Section 3 for the healthy case. Preliminary results are discussed in Section 4. Some hints towards adaptation of the proposed approach to pathological cases are provided in Section 5.

2   Graph Model

In this paper, we follow the same approach as in [1], and we propose an original method to determine automatically the order in which structures should be segmented, using the spatial relations represented as edge attributes of a graph (nodes represent individual objects, such as anatomical structures in the brain example). Note that this way of using the graph is very different from classical graph matching approaches, widely developed for structural recognition. Let us now summarize the adopted formalism for representing spatial relations.

Fuzzy representations are appropriate to model the intrinsic imprecision of several relations (such as "close to", "behind", etc.), the potential variability (even if it is reduced in normal cases) and the necessary flexibility for spatial

reasoning [5]. Two kinds of questions are raised when coping with spatial relations: (i) given two objects (possibly fuzzy), determine the degree of satisfaction
of a relation; (ii) given one reference object, define the region of space in which
a relation to this reference is satisfied (to some degree). In this paper, we deal
mainly with the second question.

Therefore we rely on spatial representations of the spatial relations: a fuzzy
set in the spatial domain $\mathcal{S}$ defines a region in which a relation to a given object
is satisfied. The membership degree of each point to this fuzzy set corresponds
to the satisfaction degree of the relation at this point [5]. Figure 2 depicts an
example.

We now describe the modeling of the main relations that we use: adjacency,
distances and directional relative positions.

A **distance** relation can be defined as a fuzzy interval f of trapezoidal shape
on $\mathbb{R}^+$, as illustrated in Figure 2. A fuzzy subset μ_d of the image space $\mathcal{S}$ can then
be derived by combining f with a distance map d_A to the reference object A:
$\forall x \in \mathcal{S}$, $\mu_d(x) = f(d_A(x))$, where $d_A(x) = \inf_{y \in A} d(x, y)$.

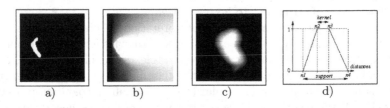

a)   b)   c)   d)

Fig. 2. (a) 2D view of a 3D binary lateral ventricle. (b) Fuzzy spatial representation
of "Right of the lateral ventricle". (c) Fuzzy subset corresponding to "Near the lateral
ventricle". (d) Trapezoidal fuzzy interval.

Directional relations are represented using the "fuzzy landscape approach"
[6]. A morphological dilation δ_{ν_α} by a fuzzy structuring element ν_α representing
the semantics of the relation "in direction α" is applied to the reference object
A: $\mu_\alpha = \delta_{\nu_\alpha}(A)$, where ν_α is defined, for x in $\mathcal{S}$ given in polar coordinates (ρ, θ),
as: $\nu_\alpha(x) = g(|\theta - \alpha|)$, where g is a decreasing function from $[0, \pi]$ to $[0, 1]$, and
$|\theta - \alpha|$ is defined modulo π. This definition extends to 3D by using two angles
to define a direction. The example in Figure 2 has been obtained using this
definition.

Adjacency is a relation that is highly sensitive to the segmentation of the
objects and whether it is satisfied or not may depend on one point only. Therefore
we choose a more flexible definition of adjacency, interpreted as "very close to".
It can then be defined as a function of the distance between two sets, leading to a
degree of adjacency instead of a Boolean value: $\mu_{adj}(A, B) = h(d(A, B))$ where
$d(A, B)$ denotes the minimal distance between points of A and B: $d(A, B) =
\inf_{x \in A, y \in B} d(x, y)$, and h is a decreasing function of d, from $\mathbb{R}^+$ into $[0, 1]$. We
assume that $A \cap B = \emptyset$.

3   Graph-Based Reasoning in Normal Cases

In this section, we deal with normal cases. The aim of the reasoning in the graph is to select the "best" path between a reference structure and a target structure to be segmented and recognized in an image, by exploiting the information encoded in the graph. Note that the number of simple paths (without loops) between two structures is finite. Path extraction is known as an intractable task but we limit our experiments to small graphs. Extensions to larger graphs require to address this issue.

The reference structure, in the case of MRI images of the brain, can typically be the lateral ventricles, which are easy to segment in such images. The notion of "best" path refers to the constraints of the segmentation process: it should allow segmenting a structure in the path based on relations to previous structures in the path, as done in [1] (based on manually defined paths). We propose two methods:

- the first one is based on the evaluation of the relevance of each spatial relation between two structures independently, and on the optimization of the path according to a criterion involving this relevance measure;
- in the second one, we estimate each path globally and select the best one according to another criterion.

3.1   Evaluating Edge Relevance

In this part, we present a criterion of relevance as well as two different methods for path selection.

In the following, $G = (V, E)$ is an attributed relational graph, with V the set of nodes and E the set of edges. An edge interpretor associates to each edge e a fuzzy set μ_{Rel}, defined in the spatial domain, representing the spatial relation carried by this edge to a reference structure as defined in [6]. Similarly a fuzzy set μ_{Obj} is attached to each node.

Relevance Criterion. The relevance of a spatial relation should represent the adequation between μ_{Rel} and μ_{Obj}, i.e. the degree to which the target object fits in the region where the relation to the reference object is satisfied. The comparison measures and their classification according to [7] provide an appropriate formal framework for this purpose.

For both the reference structure, used to compute μ_{Rel}, and the target object, used for μ_{Obj}, we need an a priori knowledge from an anatomical atlas or from a set of pre-segmented images.

M-measure of satisfiability: We use a M-measure of satisfiability [7] defined as:

$$f(Rel, Obj) = \frac{\sum_{x \in S} \min(\mu_{Rel}(x), \mu_{Obj}(x))}{\sum_{x \in S} \mu_{Obj}(x)}. \tag{1}$$

where S denotes the spatial domain. It measures the precision of the position of the object in the region where the relation is satisfied and is maximal if the whole object is included in the kernel of μ_{Rel}. But the size of the region where the relation is satisfied is not restricted and could be the whole image space. Note that if the object is crisp, this measure reduces to $\frac{\sum_{x \in Obj} \mu_{Rel}(x)}{\sum_{x \in S} \mu_{Obj}(x)}$.

Path Selection. Once every edge has been valued with the proposed relevance measure, path selection is achieved with classical algorithms, such as shortest path or maximal flow. Nevertheless, these algorithms have to be adapted to our purpose.

Shortest path: The shortest path algorithm leads to a global optimization, but does not account for potential disparities between edges. A globally satisfactory path can include an edge with a low relevance value. Moreover, this algorithm favors paths with a reduced number of nodes, hence leading to less segmented structures. The adaptation we propose consists in normalizing the cost of each path by its length (in terms of number of nodes).

Let $\mathcal{F}$ denote the set of the fuzzy sets over the spatial domain. Let $f : \mathcal{F} \times \mathcal{F} \rightarrow \mathbb{R}$ be a real valued cost function, here a satisfiability measure. The shortest path between two nodes v and v' is the path $\hat{p}$ solution of:

$$\min_{p \in \mathcal{P}} \left(\frac{\sum_{e \in p} (1 - f(\mu_{Rel}, \mu_{Obj}))}{card(p)} \right) \tag{2}$$

where e is an edge in the path p, $\mathcal{P}$ is the set of paths from v to v', μ_{Obj} is the target node of edge e, μ_{Rel} is the fuzzy set derived from e and $card(p)$ is the number of edges in p.

Maximal flow: We adapt the classical maximal flow notion [8] in order to take the weakest edges into account without penalizing the most informative paths. This is expressed as the maximization of the minimal value along the path:

$$\max_{p \in \mathcal{P}} \left(\min_{e \in p} (f(\mu_{Rel}, \mu_{Obj})) \right) \tag{3}$$

where f is again a satisfiability measure. This formulation allows avoiding paths including relations which are not well satisfied.

3.2   Globally Evaluating Path Relevance

Instead of evaluating the relevance for each edge, we propose in a second method to evaluate the relevance of a whole path by merging spatial knowledge along this path.

Merging Spatial Knowledge. In this approach, we combine information along the path with prior knowledge derived from an anatomical atlas, as illustrated in Figure 3. For each structure in the atlas and each spatial relation encoded in the graph we compute the fuzzy set representing the region where the relation to this structure is satisfied, as previously. Note that it is relevant to merge different relations (distance and direction for example) since all relations use the same

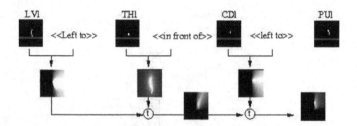

Fig. 3. Merging spatial relations. For each relation carried by an edge on a path, we compute its representation, using a priori knowledge for the structure. Representations of all relations along the path are then merged with a t-norm (here a minimum).

representation framework i.e. fuzzy sets in the spatial domain. The fuzzy sets obtained for all pairs structure/relation along the path p are combined using a t-norm (a conjonctive fusion operator):

$$\mu_p = t[\mu_{Rel_i^p}, i = 1...N^p] \tag{4}$$

where t is a t-norm and p a path composed of N^p relations. In our experiments, we use the minimum t-norm.

Path Evaluation Using Entropy. In this approach, the path selection method we propose relies on a a fuzziness measure, in order to choose the "less fuzzy" path. As a fuzziness measure, we choose the fuzzy entropy measure [9]:

$$H(\mu_p) = -K\left(\sum_{x_i \in S} \mu_p(x_i)\log\mu_p(x_i) + \sum_{x_i \in S}(1 - \mu_p(x_i))\log(1 - \mu_p(x_i))\right) \tag{5}$$

where μ_p is the fuzzy set resulting from the combination of all relations along p and k is a normalizing constant.

The best path $\hat{p}$ is then the path which achieves the minimum of fuzzy entropy:

$$\hat{p} = arg\min_{p\in\mathcal{P}}(H(\mu_p)). \tag{6}$$

Note that this measure is meaningful for representations of relations that are more fuzzy if they are less focused. It is actually the case with our model of relations. For instance, it would be useless to apply this criterion on large crisp regions which would lead to a zero entropy value even if these regions are very extended and of limited help to constrain the segmentation.

4   Results and Discussion

Experiments have been carried out on a small graph presented in Figure 4 containing four cerebral structures: the lateral ventricle (taken as the reference structure), the caudate nucleus, the thalamus and the putamen (the target structure

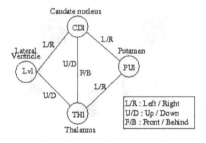

Fig. 4. Small graph used in the experiments

in our experiments). All these structures exist in both brain hemispheres, but only the left side is considered in the reported experiments. Note that the extraction of the structures is supposed to exhibit the same difficulty level.

The edges encode only information about directional relative position in these preliminary experiments. Extending our approach to other binary spatial relations can be achieved in a straightforward manner.

4.1   Edge Valuation

Measures of satisfiability obtained for each edge are presented in Figure 5. The best path according to the satisfiability criterion with normalized shortest path and flow measure is: LVl "left of" CDl "behind" THl "left of" PUl. This path is exactly the one that was previously defined by hand in [1] and that led to the results shown in Figure 1.

Another path with the highest score is: LVl "down of" THl "left of" PUl. This path is less intuitive since it involves a few number of structures. For practical purposes, if several paths exhibit the same global score, the longest path (in terms of number of nodes) is retained.

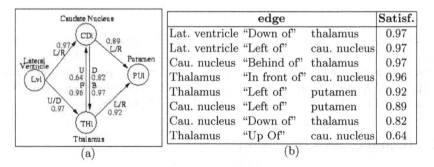

edge			Satisf.
Lat. ventricle	"Down of"	thalamus	0.97
Lat. ventricle	"Left of"	cau. nucleus	0.97
Cau. nucleus	"Behind of"	thalamus	0.97
Thalamus	"In front of"	cau. nucleus	0.96
Thalamus	"Left of"	putamen	0.92
Cau. nucleus	"Left of"	putamen	0.89
Cau. nucleus	"Down of"	thalamus	0.82
Thalamus	"Up Of"	cau. nucleus	0.64

(a)   (b)

Fig. 5. (a) Edges valuation with a measure of satisfiability. (b) Edge ranking according to this measure.

4.2 Merging of Spatial Relations

The best path according to the entropy criterion is: LVl "down of" THl "up of" CDl "left of" PUl. Figure 6 shows a view of the resulting representation of this path. This path contains several changes in direction which explain the strongly focused resulting fuzzy region. More generally, paths with several changes in direction get low entropy while simpler paths get high entropy.

Fig. 6. 2D slice of 3D representation for path LVl "down of" THl "up of" CDl "left of" PUl after merging all spatial relations

5 Graph-Based Reasoning in Pathological Cases

The approaches introduced in Section 3 are not directly applicable in the case of the presence of a pathology and require some adaptation. For instance, the presence of a tumor may induce an important alteration of the appearance and morphometric characteristics of the structure. Although spatial relations are more stable, still modifications of the structural information may occur. Figure 7 presents an example of a pathology in a MRI brain image, illustrating the impact of the tumor on the surrounding structures.

It has been shown in [10] that some spatial relations are more stable than others. A pathology-dependent paradigm has been introduced to adapt a generic reasoning process to specific cases by addressing the fundamental question: given a pathology, which spatial relations do remain stable and to which extent? For this purpose, we designed a computational framework for learning spatial relation stability from a database constituted of healthy and pathological cases, where the main anatomical structures were manually segmented. The degree of stability is inferred from the comparison (using a M-measure of resemblance) between the learned spatial relations for pathological cases and for healthy ones.

In this work we exploit the degree of stability concept to adapt the reasoning approaches designed for healthy cases to pathological ones. This can be achieved in several ways.

The initial graph is filtered so that the spatial relations with a low degree of stability are removed. Then the proposed methods are applied on the filtered graph instead of the initial one. This approach is very severe and does not leave significant place to flexibility, an important property in reasoning and decision making paradigms.

In the second method, the degree of stability is taken into account as an edge attribute and is considered in the cost calculation of the proposed approaches. This approach is a direct extension of the methods proposed for healthy case, and

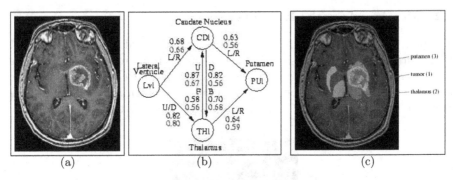

Fig. 7. (a) Axial view of MRI with a tumor close to the lateral ventricle and grey nuclei. (b) Degree of stability learned with a class of similar tumor (in blue). Resulting weighted satisfiability measures (in red). With the sortest path method, the selected path becomes Ventricle "Down Of" thalamus "Left Of" Putamen. (c) Segmentation of the putamen. The tumor is first extract then the thalamus and finally the putamen.

its implementation is straightforward. The integration of the degree of stability must be achieved in a way so that the paths involving pathological or altered structures are penalized. For instance, when using the shortest path method, a weight proportional to the degree of stability is assigned to $f(\mu_{Rel}, \mu_{Obj})$. Figure 7 b) presents the degree of satisfiabity (in blue) learned for each edge in the case of the tumor like the one presented in Figure 7 a) and the weighted measures in red. In this case, the selected path becomes ventricle "Down Of" thalamus "Left Of" Putamen. Figure 7 c) presents a segmentation of the putamen with the same order.

In the global approach, the influence of a relation is decreased by extending its spatial extension (for instance using a fuzzy dilation), so as to increase the resulting degree of fuzziness, and thus unfavoring paths including this relation.

This last idea, of extending the fuzzy representation, is the basis of the third method we propose. Since the fuzzy representation of spatial relations presents the advantage of being flexible in the way they can be constructed, this construction could be correlated to the degree of stability. For instance, the definition of "near the lateral ventricles", as explained in Section 2, is modified by extending the fuzzy interval according to the degree of stability (the less the stability, the more the extension and the more fuzzy). This induces both a lower resemblance and more fuzziness, hence decreasing the relevance of paths including this relation in both approaches.

These approaches are currently being tested on different pathological cases.

6   Conclusion

The main contribution of this paper is to show that the order of structures in a sequential segmentation process can be deduced automatically using graph-based reasoning. We proposed relevance measures of segmentation paths based on fuzzy representations of spatial relations. As an illustration, we applied our method on a small graph representing brain structures. The results are promising since the best path actually allows driving the recognition and segmentation procedure in 3D MRI brain images.

Extensions to the pathological cases are proposed, based on the impact of the pathology on the spatial relations. This part will be further investigated in future work. Applications on larger graphs will also be carried out, which may require to address potential combinatory optimization issues.

References

1. Colliot, O., Camara, O., Bloch, I.: Integration of Fuzzy Spatial Relations in Deformable Models - Application to Brain MRI Segmentation. Pattern Recognition 39, 1401–1414 (2006)
2. Waxman, S.: Correlative neuroanatomy. McGraw-Hill, New York (2000)
3. Mangin, J.F., Frouin, V., Régis, J., Bloch, I., Belin, P., Samson, Y.: Towards better management of cortical anatomy in multi-modal multi-individual brain studies. Physica Medica 12, 103–107 (1996)
4. Hodé, Y., Leammer, E., Jolion, J.M.: Adaptive pyramid and semantic graph: Knowledge driven segmentation. In: Brun, L., Vento, M. (eds.) GbRPR 2005. LNCS, vol. 3434, pp. 213–222. Springer, Heidelberg (2005)
5. Bloch, I.: Fuzzy Spatial Relationships for Image Processing and Interpretation: A Review. Image and Vision Computing 23, 89–110 (2005)
6. Bloch, I.: Fuzzy Relative Position between Objects in Image Processing: a Morphological Approach. IEEE Transactions on Pattern Analysis and Machine Intelligence 21, 657–664 (1999)
7. Bouchon-Meunier, B., Rifqi, M., Bothorel, S.: Towards general measures of comparison of objects. Fuzzy sets and Systems 84(2), 143–153 (1996)
8. Boykov, Y., Kolmogorov, V.: An experimental comparison of min-cut/max- flow algorithms for energy minimization in vision. IEEE Transactions on Pattern Analysis and Machine Intelligence 26, 1124–1137 (2004)
9. Luca, A.D., Termini, S.: A definition of non-probabilistic entropy in the setting of fussy set theory. Information and Control 20, 301–312 (1972)
10. Atif, J., Khotanlou, H., Angelini, E., Duffau, H., Bloch, I.: Segmentation of Internal Brain Structures in the Presence of a Tumor. In: MICCAI Workshop on Clinical Oncology, Copenhagen, pp. 61–68 (2006)

Graph-Based Perceptual Segmentation of Stereo Vision 3D Images at Multiple Abstraction Levels

Rodrigo Moreno[1], Miguel Angel Garcia[2], and Domenec Puig[1,*]

[1] Intelligent Robotics and Computer Vision Group
Department of Computer Science and Mathematics
Rovira i Virgili University
Av. Països Catalans 26, 43007 Tarragona, Spain
{rodrigo.moreno,domenec.puig}@urv.cat
[2] Department of Informatics Engineering
Autonomous University of Madrid
Cra. Colmenar Viejo Km 15, 28049 Madrid, Spain
miguelangel.garcia@uam.es

Abstract. This paper presents a new technique based on perceptual information for the robust segmentation of noisy 3D scenes acquired by stereo vision. A low-pass geometric filter is first applied to the given cloud of 3D points to remove noise. The tensor voting algorithm is then applied in order to extract perceptual geometric information. Finally, a graph-based segmenter is utilized for extracting the different geometric structures present in the scene through a region-growing procedure that is applied hierarchically. The proposed algorithm is evaluated on real 3D scenes acquired with a trinocular camera.

1 Introduction

Segmentation is one of the most important stages in computer vision as a preliminary step towards further analysis and recognition stages. Its goal is to partition a given image into a set of non-overlapping homogeneous regions that likely correspond to the different objects or geometric structures that may be perceived in the scene. When segmentation is applied to 3D images acquired through stereo vision, an additional problem that appears is the high presence of noise.

Although different approaches have been proposed (e.g. [3] reviews seventeen methods), robust segmentation of noisy scenes is still a challenging problem. In this scope, Medioni *et al.* proposed the tensor voting framework [6] as an adequate scheme for extracting perceptual information from noisy 3D images. Their approach recovers the shape of surfaces, edges and junctions present in a given 3D image through tensors and a variant of the marching cubes algorithm [5]. That approach has proven robust even for 3D images constituted by strongly noisy clouds of points.

---

* This work has been partially supported by the Spanish Ministry of Education and Science under project DPI2004-07993-C03-03, by the Department of Universities, Research and Society of Catalonia's Government and by European Social Funds.

F. Escolano and M. Vento (Eds.): GbRPR 2007, LNCS 4538, pp. 148–157, 2007.
© Springer-Verlag Berlin Heidelberg 2007

The proposed algorithm applies the tensor voting but uses a fast graph-based segmenter instead of the marching cubes algorithm, so in that way, the new technique can be useful when it is preferred fast execution over getting shape information of surfaces. Additionally, the proposed technique generates multiresolution segmentations. This feature can be interesting when a different degree of detail in the segmentation is necessary at different areas of the space.

The proposed segmenter is introduced in the next section. Experimental results are described and discussed in section 3. Finally, conclusions and further research lines are given in section 4.

2   3D Image Segmentation

Several technologies can be applied to capture 3D images. One of them is based on the use of stereo cameras, which have numerous advantages over other range systems, such as affordable price, speed of acquisition or simplicity of usage in various applications, both outdoor and indoor. However, there are some problems with this technology that must be taken into account: (a) the accuracy of the range estimation depends on the distance from the camera: points too close or too far lead to wrong estimations, (b) the accuracy of the range estimation decreases with non-textured surfaces, (c) 3D images have often "holes" in regions where range cannot be estimated, (d) the obtained data are quite noisy, (e) quality depends on ambient factors, such as light positions, light amount, material and size of objects, etc., (f) accuracy decreases in areas with large depth discontinuities.

The proposed segmentation algorithm in this paper tries to overcome some of these problems. In the following subsections the overall algorithm is described and the main techniques utilized are introduced.

2.1   Overview of the Algorithm

The proposed algorithm consists of an iterative procedure in which a segmentation is obtained at every iteration. Each segmentation trends to be coarser than the ones obtained in previous iterations, creating in this way, a set of segmentations at different perceptual abstraction levels.

The algorithm has the following steps: firstly, the neighborhood size is estimated, this being a necessary parameter for the rest of the process. Afterwards, an iterative procedure is run with four stages being executed at every iteration: (a) filtering, (b) tensor voting, (c) graph creation and (d) graph segmentation. Every iteration leads to a segmentation of the input image at a progressively higher abstraction level. This iteration is run a given number of times m.

The algorithm returns the set of all calculated segmentations. That set can be used in applications where is interesting to get different levels of detail in different zones of the image.

Even though the quantity of regions has a decreasing trend as iterations increase, the segmentation set does not conform a pyramid. One of the reasons of that is the merging effect: a region can be temporarily divided into two regions in the process of being joint with another one. In the end, the generated segmentations can be seen as a pseudo-pyramid segmentation.

In the following subsections, the main stages of the algorithm are presented.

2.2   Geometric Low-Pass Filters

As said before, 3D images acquired through stereo vision are very noisy. Hence, it is necessary to apply a geometric low-pass filter in order to remove that noise. In this work, local filters have been preferred to global ones since every different region in an image has its own specific features, in particular, point density and amount of noise. Local filtering algorithms are based on local processing of a neighborhood N^B around each 3D point B belonging to the given cloud of 3D points $\mathbf{P}$. N^B may include B itself or not.

Local filters based on either averaging or function-based averaging (e.g., using a Gaussian function to give more weight to nearest neighbors) have been discarded as they do not work well in 3D images with variable density of points, such as the ones obtained through stereo vision. The application of that kind of filters to those 3D images would lead to regions with high densities of points becoming even denser and to low density regions becoming more scattered, creating thus holes in the 3D image or making previous holes bigger.

An alternative way to filter points is to apply 2D linear regression by projecting those points onto the plane (or onto another desired surface) that minimize the squared error in the neighborhood. This technique is known as Moving Least Squares Projection (MLS-Projection) [4]. A Gaussian can be used as a weighting function in order to give more importance to nearest neighbors in the least squares calculations.

The MLS-Projection filter prevents the point density modification problem described above since, by projecting the points onto a plane (or surface), it does not move them towards a specific area of neighbors. The parameters of this filter are the standard deviation of the weighting Gaussian function, σ, and the number of neighbors that constitute a neighborhood, ρ. In the new approach, MLS-projection is used for filtering with a plane as the reference surface and a Gaussian as a weighting function.

2.3   Tensor Voting Framework

The proposed technique uses the tensor voting framework to obtain perceptual information from noisy clouds of 3D points. This framework is based on the aggregation and propagation of local geometric data encoded as tensors. The main steps of tensor voting are described in the next subsections.

Information Encoding. The first step of tensor voting encodes the geometric information associated with every input 3D point from the given cloud as a tensor. This method usually utilizes second order tensors represented by symmetric semidefinite positive matrices.

In order to extract surface likeliness from the cloud of points, the tensor $\mathbf{T}$ of a point is defined according to the next three cases: (a) if the only available information is the 3D coordinates of the point, then $\mathbf{T}$ is defined as the unitary ball tensor represented as an identity matrix, $\mathbf{T} = \mathbf{I}$; (b) if a normal vector $\mathbf{v}$ is already known at that point, then $\mathbf{T} = \mathbf{v}\,\mathbf{v}^T$; and (c) if a tensor $\mathbf{W}$ is already known at that point, then $\mathbf{T} = \mathbf{W}$. It is important to notice that vectors encoded into tensors lose their sign because of the squares involved.

In 3D, tensors can be represented by means of ellipsoids whose shape and orientation are conveyed by their eigenvalues λ_i and their corresponding eigenvectors $\mathbf{e}_i$. By convention, eigenvalues are sorted in descending order, with λ_1 being the largest one. There are three degenerated cases for those ellipsoids:

- a "stick" when $\lambda_1 > 0$, $\lambda_2 = 0$ and $\lambda_3 = 0$
- a "plate" when $\lambda_1 = \lambda_2 > 0$ and $\lambda_3 = 0$
- a "ball" when $\lambda_1 = \lambda_2 = \lambda_3 > 0$

Tensor Voting. In the second step of tensor voting, the tensor associated with every point is propagated to the point's neighborhood through a convolution-like process. In order to apply this convolution, it is necessary to separate the information encoded in tensor form into three components. Thus, a tensor $\mathbf{T}$ can be written as:

$$\mathbf{T} = \lambda_1 \mathbf{e}_1 \mathbf{e}_1^T + \lambda_2 \mathbf{e}_2 \mathbf{e}_2^T + \lambda_3 \mathbf{e}_3 \mathbf{e}_3^T \ . \tag{1}$$

This can be rewritten as:

$$\mathbf{T} = (\lambda_1 - \lambda_2)\mathbf{e}_1 \mathbf{e}_1^T + (\lambda_2 - \lambda_3)(\mathbf{e}_2 \mathbf{e}_2^T + \mathbf{e}_1 \mathbf{e}_1^T) + \lambda_3(\mathbf{e}_1 \mathbf{e}_1^T + \mathbf{e}_2 \mathbf{e}_2^T + \mathbf{e}_3 \mathbf{e}_3^T) \ . \tag{2}$$

Let $s_1 = \lambda_1 - \lambda_2$, $s_2 = \lambda_2 - \lambda_3$ and $s_3 = \lambda_3$ be the saliencies 1 to 3, $\mathbf{ST} = \mathbf{e}_1 \mathbf{e}_1^T$ be the stick tensor, $\mathbf{PT} = \mathbf{e}_1 \mathbf{e}_1^T + \mathbf{e}_2 \mathbf{e}_2^T$ be the plate tensor and $\mathbf{BT} = \mathbf{e}_1 \mathbf{e}_1^T + \mathbf{e}_2 \mathbf{e}_2^T + \mathbf{e}_3 \mathbf{e}_3^T$ be the ball tensor, then:

$$\mathbf{T} = s_1 \, \mathbf{ST} + s_2 \, \mathbf{PT} + s_3 \, \mathbf{BT} \ . \tag{3}$$

Next, it is necessary to define appropriate voting fields, which are equivalent to kernels in classical convolution, for the $\mathbf{ST}$, $\mathbf{PT}$ and $\mathbf{BT}$ components. In the proposed algorithm, the stick, plate and ball voting fields defined in [6] are used since they have a good performance in propagating surface likeliness.

A stick vote at B received from A, $\mathbf{SV}_A^B$, is calculated as the value obtained from the stick voting field at B weighted by saliency 1 at A, s_1^A, when that voting field is centered and oriented at A; the plate vote at B received from A, $\mathbf{PV}_A^B$ is calculated in the same way using the plate voting field and saliency 2 at point A, s_2^A, instead of the stick voting field and s_1^A; and the ball vote $\mathbf{BV}_A^B$ uses the ball voting field and saliency 3 at point A, s_3^A, replacing the stick voting field and s_1^A respectively [6].

Finally, the total vote received at B, $\mathbf{T}^B$, is given by:

$$\mathbf{T}^B = \sum_{A \in N^B} \mathbf{SV}_A^B + \sum_{A \in N^B} \mathbf{PV}_A^B + \sum_{A \in N^B} \mathbf{BV}_A^B \tag{4}$$

where sum of tensors is defined as sum of matrices.

The parameters of the algorithm are σ, which denotes the standard deviation of the Gaussian decay function in the voting fields utilized, and ρ, the number of neighbors that receive a vote [6]. These parameters were selected to be the same as the ones used in filtering.

Analysis of Tensor Voting. After the tensor voting stage, it is necessary to obtain eigenvectors and eigenvalues at every point in order to compute the three aforementioned saliency measures, which allow to determine whether that point belongs to a surface, an edge or a junction. The point likely belongs to a smooth surface if a high saliency s_1 is obtained. In that case, the estimated surface normal is given by $\pm\mathbf{e}_1$. In turn, a high saliency s_2 indicates that the point belongs to an edge whose direction is given by $\pm\mathbf{e}_3$. Finally, a high saliency s_3 is typical for junctions. If the three saliencies are low, the point is likely to belong to noise.

In our approach, tensor voting is applied twice: first, it is applied to the given cloud of 3D points $\mathbf{P}$ by encoding unitary ball tensors at every 3D point in $\mathbf{P}$. Afterwards, a second pass applies tensor voting again, but this time considering the tensors obtained after the first pass at every point, instead of unitary ball tensors.

Finally, it is necessary a normalization of the sum of saliencies after applying the tensor voting algorithm. This is done by calculating $\bar{s}_i$ such that $\sum_{i=1}^{n} \bar{s}_i = 1$. That operation can be done by means of the equation $\bar{\lambda}_i = \frac{\lambda_i}{\lambda_1}$ (i.e., dividing every eigenvalue by the largest one) and recalculating $\bar{s}_i$ using the scaled eigenvalues on the equations given in the previous subsection to calculate saliencies.

2.4   Graph Creation and Segmentation

At this point, every 3D point from the given 3D image is associated with three saliency measures that denote the geometric structure (surface, edge or junction) to which the point likely belongs to. The goal now is to group neighboring points that are likely to belong to the same geometric structure. For that purpose, the graph-based segmenter proposed by Felzenswalb *et al.* [2] is applied, since it has been proven to be fast and to provide good results in image segmentation. Although that segmenter was originally conceived for intensity image segmentation, it can be easily extended to 3D image segmentation as its starting point is a graph that can be created in a variety of ways.

Graph Creation. Given the saliencies obtained in the tensor voting step, a graph is created as described in the following paragraphs. Let ϵ be a small constant, σ be the standard deviation used in the filtering and tensor voting stages and let the predicate: $neighbors(A, B) \Leftrightarrow e^{-\frac{||A-B||^2}{\sigma^2}} > \epsilon$, a graph is built using the following two rules: (a) every point in the cloud $\mathbf{P}$ defines a graph vertex, (b) the vertices corresponding to any pair of 3D points A and B belonging to $\mathbf{P}$ are connected through an edge provided that $neighbors(A, B)$ is *true*.

Every edge connecting neighboring points A and B is associated with a weight w defined according to the following method: let $\mathbf{e}_i^A$ and $\mathbf{e}_i^B$ be the ith eigenvector of the tensors calculated through tensor voting at points A and B respectively, $\bar{s}_i{}^A$ and $\bar{s}_i{}^B$ be the ith normalized saliency of the tensors calculated at points A

and B and $d_i^{AB} = 1 - \frac{2}{\pi} \arccos(|\mathbf{e}_i^A \, \mathbf{e}_i^B|)$ be 1 minus the normalized angle between the ith eigenvectors of A and B, then $w = 1 - (\bar{s}_1^A \, \bar{s}_1^B \, d_1^{AB} + \bar{s}_2^A \, \bar{s}_2^B \, d_3^{AB} + \bar{s}_3^A \, \bar{s}_3^B)$. In this way, this method uses all available saliencies and angles and constraint $0 \le w \le 1$ is guaranteed because $0 \le \bar{s}_i \le 1$.

Graph-Based Segmentation. The segmentation technique is based on a region-growing approach where a function $MInt$ is used as a measure of distance between every pair of neighboring regions C_i and C_j in the graph. This function is given by:

$$MInt(C_i, C_j) = \min(Int(C_i) + \tau(C_i), Int(C_j) + \tau(C_j)) \qquad (5)$$

where $Int(C)$ is the internal difference, defined as the largest weight of the minimum spanning tree for region C and $\tau(C_i)$ is an arbitrary function. In [2], parameter k is proposed to calculate $\tau(C)$ using the formula: $\tau(C) = k/|C|$. The segmenter has the next stages: (a) edges from the graph are sorted in ascending order of weight; (b) a different region is created for every vertex in the graph; (c) for each sorted edge E the following merging rule is applied: let C_i and C_j be the neighboring regions connected in the graph by means of E and w its weight, if $w \le MInt(C_i, C_j)$, then both regions are merged; and finally (d) $Int(C)$ is updated every time another region is merged to C in step (c). The success of this technique basically depends on the graph creation step and on parameter k.

Due to the noisy nature of data, this algorithm requires a post-processing stage to avoid oversegmentation. The basic idea consists of iterating the same segmenter in a hierarchical way. This process is described in Algorithm 1 where "CreateUpperGraph" creates a new graph based on the previous segmentation, with its vertices being the regions found in the previous segmentation and its edges being created using the method described before, but using only one tensor that represents all the tensors in the region.

Algorithm 1. Segmentation Algorithm

1: **function** RESEGMENTATE(V_0, E_0)   ▷ Vertices and Edges
2:     $S_0 \leftarrow Segmentate(V_0, E_0)$
3:     **repeat**
4:         $[V_{i+1}, E_{i+1}] \leftarrow CreateUpperGraph(V_i, E_i, S_i)$
5:         $S_{i+1} \leftarrow Segmentate(V_{i+1}, E_{i+1})$
6:     **until** $NumSegments(S_{i+1}) = NumSegments(S_i)$
7:     **return** S_n
8: **end function**

The tensor of a region is computed by adding the tensors of the elements that constitute that region and normalizing the sum of saliencies. Edges between regions are only calculated if in the previous iteration there is an edge connecting an element from a region to another element from the other region. In this way, the resulting segmentation in an iteration will not have more regions than the segmentations obtained in previous iterations.

2.5   Neighborhood Selection

In order to apply the filtering, tensor voting and graph creation stages, it is necessary to define the size of the neighborhood associated with any given point from the input cloud $\mathbf{P}$. Let ρ be the number of points belonging to that neighborhood. This parameter is estimated from other parameters associated with the algorithm as follows. Let $\mathbf{R}$ be a set of points randomly sampled from $\mathbf{P}$, σ be the standard deviation of the Gaussian decay function used in the filtering and tensor voting stages and *neighbors* be the predicate defined in Sect. 2.4; ρ can be estimated as the average quantity of neighbors of all points $A \in \mathbf{R}$. With this, only those points that give relevant information to their neighbors in the filtering and tensor voting stages are taken into account.

Filtering, tensor voting and graph creation stages use an ANN k-d tree with 3D Euclidian metric to retrieve the neighborhood of a given 3D point, using the aforementioned parameter ρ.

3   Experimental Results

A Digiclops trinocular stereo vision camera and the Digiclops SDK and Triclops SDK libraries [7] have been used to capture the 3D images upon which the proposed technique has been tested. The accuracy of the estimated depth measurements, z, depends on the depth itself, so points farther away than $5m$ are discarded, since their accuracy is above $8.32cm$ [7]. Points at less than $1m$ have also been discarded in being unable to do range estimation at those distances.

The aforementioned libraries only calculate depth when it is possible to do it with a reasonable error. For example, Fig. 1(a) shows a picture taken by the right camera, and Fig. 1(b) depicts in black all pixels whose estimated depth could not be calculated or are too close or too far from the camera. In Fig. 1(c) a 3D view of the test 3D scene is displayed and in Fig. 1(d) is shown a ground truth segmentation calculated by hand using 2D information. In this example, the libraries were able to estimate depth for 45,589 pixels from a total of 76,800, but those numbers can change dramatically (upwards or downwards) in other examples as stereo algorithms are very sensitive to scene conditions. Even though some comparison frameworks have been proposed for range image segmentation algorithms, e.g. Hoover *et al.* [3], none of them have been used in this work because they were designed to compare performance on range images taken by means of laser and structured light scanners and not on noisy images. Instead, ground truth segmentations calculated by hand (using 2D information) were used, taking into account that, as shown in Fig. 1(c) and Fig. 1(d), it is not possible to obtain a perfect segmentation, if results are compared with ground truth segmentations created using 2D information, as misplacing points error of the stereo algorithm increases near big discontinuities and distant zones of the scene.

In order to compare the use of tensors in the proposed algorithm, a non tensor-based technique that uses the same filtering, graph creation and segmenter processes was also implemented, changing only the way in which every edge's weight is calculated: let $\mathbf{n}^A$ and $\mathbf{n}^B$ be the estimated normals at points A and B

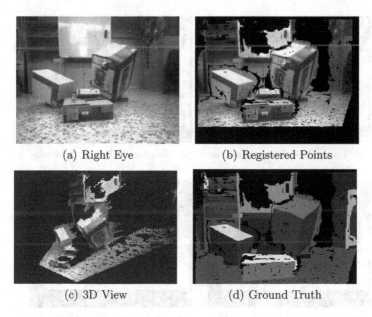

(a) Right Eye   (b) Registered Points

(c) 3D View   (d) Ground Truth

Fig. 1. Example of 3D image

belonging to the cloud $\mathbf{P}$ by means of MLS-Projection with a Gaussian weighting function applied to their neighborhoods, then the weight of the edge that connect them in the graph is calculated as $w = 1 - \mathbf{n}^A \, \mathbf{n}^B$.

The segmentation results obtained from execution of both methods with $\sigma = 20mm$ (filtering and tensor voting parameter) and $k = 0.2$ (segmenter parameter) after five iterations are shown in Fig. 2(a) and Fig. 2(b). The following local metrics are used to evaluate performance of the tested methods: oversegmentation metric on a specific region in the ground truth, calculated as the average of the squared relative areas of intersecting regions in the segmented image and undersegmentation metric, calculated in the same way by exchanging ground truth and the segmented image. Global metrics can be calculated by averaging each local metric weighted by the size of each region. Figure 2(c) and Fig. 2(d) show in gray the local metrics calculated on each region for the proposed algorithm where white means that the segmenter has a perfect performance in that region. These metrics are calculated for all the scene (A), for the 50% nearer points to the camera (N) and for the 50% farther points from the camera (F). The results of the application of both algorithms to two 3D test scenes are depicted in Table 1.

As expected, the best results for oversegmentation were obtained using the tensor-based approach. The global oversegmentation metric is around 60% (66% and 53%), what means that in average a region in the ground truth is segmented approximately into 1.66 regions by the algorithm. However, it is necessary to remark that the segmentation results were suitable in zones near the camera where accuracy is better (86% in average). For far points this metric is around 35% (44% and 26%), i.e., a ground truth zone is divided in average into almost

(a) Tensor-based segmentation

(b) Non tensor-based segmentation

(c) Oversegmentation metric

(d) Undersegmentation metric

Fig. 2. Segmentation results

Table 1. Segmentation results

Method	# Reg.	Indoor Scene Overseg.			Indoor Scene Underseg.			# Reg.	Outdoor Scene Overseg.			Outdoor Scene Underseg.		
		A	N	F	A	N	F		A	N	F	A	N	F
Tensor-based	151	0.66	0.89	0.44	0.61	0.90	0.38	215	0.53	0.83	0.26	0.83	0.98	0.69
Normals	497	0.33	0.56	0.10	0.69	0.95	0.46	681	0.21	0.38	0.08	0.90	1.00	0.79
Ground truth	52							15						

three regions, which is not a bad result taking into account the distribution of points and noise present in those zones. For undersegmentation, non tensor-based method produced slightly better results driven by the high quantity of generated regions.

4   Concluding Remarks

A graph-based algorithm oriented to 3D image segmentation has been proposed in this paper. Even though the proposed algorithm obtains good results, it is necessary to model the noise in order to improve its performance in regions that are far away from the camera position, taking into account its anisotropic nature. Moreover, it is very important to model the misplacing problem in big discontinuity regions when the depth of every point is estimated from stereo images.

Given the pseudo-pyramid segmentation, future work will consist of deciding the level to stop in a top-down (or bottom-up) exploration at every zone of the space using a multiscale analysis (e.g. scale-space analysis). Furthermore, these results will be compared to other multilevel approaches (e.g [8]).

References

1. Alexa, M., Behr, J., Cohen-Or, D., Fleishman, S., Levin, D., Silva, C.T.: Point set surfaces. IEEE Procedings Visualization, pp. 21–28 (2001)
2. Felzenswalb, P., Huttenlocher, D.: Efficient Graph-Based Image Segmentation. International Journal of Computer Vision vol. 59(2) (2004)
3. Hoover, A., Jean-Baptiste, G., Jiang, X., Flynn, P.J., Bunke, H., Goldgof, D.B., Bowyer, K., Eggert, D.W., Fitzgibbon, A., Fisher, R.B.: An experimental comparison of range image segmentation algorithms. IEEE Transactions on Pattern Analysis and Machine Intelligence 18(7), 673–689 (1996)
4. Levin, D.: Approximation Power of Moving Least-Squares. Mathematics of Computation 67(224), 1517–1531 (1998)
5. Lorensen, W., Cline, H.: Marching Cubes: A High Resolution 3D Surface Construction Algorithm. ACM SIGGRAPH Computer Graphics 21(4), 163–169 (1987)
6. Medioni, G., Lee, M., Tang, C.: A Computational Framework for Feature Extraction and Segmentation (Science). Elsevier, Amsterdam (2000)
7. Point Grey Research Inc. web page: http://www.ptgrey.com
8. Tong, W., Tang, C., Mordohai, P., Medioni, G.: First Order Augmentation To Tensor Voting For Boundary Inference And Multiscale Analysis in 3D. IEEE Trans. on Pattern Analysis and Machine Intelligence 26(5), 594–611 (2004)

Morphological Operators for Flooding, Leveling and Filtering Images Using Graphs

Fernand Meyer and Romain Lerallut

Paris School of Mines

Abstract. We define morphological operators on weighted graphs in order to speed up image transformations such as floodings, levelings and waterfall hierarchies. The image is represented by its region adjacency graph in which the nodes represent the catchment basins of the image and the edges link neighboring regions. The weights of the nodes represent the level of flooding in each catchment basin ; the weights of the edges represent the altitudes of the pass points between adjacent regions.

1 Introduction

It is an inspiring mental exercise to consider an image as a topographic surface or landscape. The image content can then be expressed in terms of peaks, crests, valleys, wells, cliffs, catchment basins, watershed lines. Morphological segmentation for instance is based to a large extent on the extraction of the watershed line of gradient images [2].

Transforming an image is the same as transforming its topography. The landscape is modified during the geological ages by slow erosion of the relief ; much faster and often catastrophic modifications are due to floodings. The flood creates large flat zones or lakes, covering and masking all details underneath. As the flood progresses, it covers more and more structures and only the most salient peaks or crests remain visible above the flood level.

Flooding is an anti-extensive operation : the flood level is above the initial ground level and only valleys and wells are filled. In order to erode peaks, one uses the dual transform of flooding, called razing : one inverts the image $(f \rightarrow -f)$ applies a flooding on $-f$ and inverts again the result.

Combining floodings and razings makes it possible to construct auto-dual filters : they operate in a symmetrical way on the white and the black structures in the images. Floodings are also an important step in morphological segmentation : a flooded topographic surface has far less minima and catchment basins than the unflooded topographic surface ; furthermore each catchment basin of the flooded relief is a union of catchment basins of the unflooded relief. The catchment basins of a series of increasing floodings form a hierarchy, or series of nested partitions [6].

We organize our presentation in two parts, as two companion papers. The first paper is organized as follows. In a first part we present how floodings are used, both for filtering and for segmenting images, In a second part we recall

F. Escolano and M. Vento (Eds.): GbRPR 2007, LNCS 4538, pp. 158–167, 2007.

how floodings are characterized and constructed at the pixel level. In a third part, we show how floodings can be represented on the region adjacency graph. From a local characterization of a valid flooding on a graph we derive an algorithm for constructing it. For this we define two couples of adjunct erosions and dilations on graphs, which combined with edge contractions are the basic tools for construction morphological operators on graphs.

The second paper analyzes the complexity of the flooding construction, both on images and on graphs and discusses in depth under which conditions it will be more interesting to flood an image at the pixel level or at the graph level. Flooding the graph becomes particularly interesting if each node represents a large number of pixels and if the same operation has to be performed a number of times on the same graph.

2   Definition of a Flooding

Definition 1. *A function g is a flooding of a function f if and only if $g \geq f$ and for any couple of neighboring pixels (p, q) : $g_p > g_q \Rightarrow g_p = f_p$*

This definition is equivalent to the following criterion.

Criterion 1. Flood: *A function g is a flooding of a function f if and only if $g = f \vee \varepsilon g$*

The relation $\{g$ is a flooding of $f\}$ is reflexive, antisymmetric and transitive: it is an order relation.

In particular, if f and h are two functions such that $f \leq h$, then the family of floodings (g_i) of f verifying $g_i \leq h$ form a complete lattice for this order relation. The smallest element is f itself and the largest is obtained by repeating the geodesic erosion of h above f : $h^{n+1} = f \vee \varepsilon h^n$ until stability, that is when $h^{n+1} = h^n$. We say that $h^\infty = \mathrm{Fl}(f, h)$ is the flooding of f below h. The dual operator is called razing : $\mathrm{Rz}(f, h) = -\mathrm{Fl}(-f, -h)$.

A flooding g is obtained from a function f, by creating a number of lakes on the topographic surface of f. All connected components where $g > f$ are flat if g is a flooding of f. We will call lake of a flooding g any flat zone of g containing at least a pixel p for which $f_p > g_p$. Let us consider a lake L of a flooding g of a reference function f. If all neighbors of L have a higher altitude, then L is a regional minimum. If L has a lower neighbor, there exists a couple of pixels (p, q), p belonging to L and $g_p > g_q$. This implies that $g_p = f_p$, meaning that the level of the flooding g and the level of the ground f are the same at pixel p : the lake cannot build a wall of water without solid ground to hold the water. This is clearly illustrated in figure 1, where the right figure cannot be a valid flooding, whereas the left figure is a valid one.

Flooding a topographic surface fills lakes, whereas razings suppress peaks. Suppressing peaks and at the same time filling valleys will be obtained by applying both operators in sequence. The resulting operator is called a *leveling* [7] and will be illustrated in the companion paper. The next paragraph presents a simple filter obtained by the composition of a flooding and a razing.

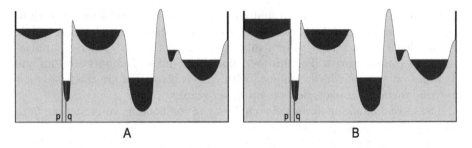

Fig. 1. A: a physically possible flooding ; B : an impossible flooding, where a lake is limited by a wall of water at position p

3   Flooding for Filtering and Segmenting Images

3.1   Flooding for Filtering

A common filter is a flooding followed by its dual razing. A first flooding fills each catchment basin up to its lowest pass point. On the resulting topographic surface a dual razing clips each peak down to its highest pass point. Both operations are parameter-free and can be iterated, leading with each new step to a stronger filtering of the image, creating larger and larger flat zones. figure 2 presents the result of 3 iterations of such a filtering on a painting by Seurat. The flat zones of this simplified image are in much lower number and their large size will simplify the segmentation of the image. Noteworthy is the fact that the image is simplified, but the contours are not blurred nor displaced.

3.2   Flooding and Segmentation

Flooding Associated to Markers. Flooding is also extremely useful in the context of morphological segmentation. For segmenting the image f in figure 3a, first its gradient magnitude is computed (see figure 3b). The watershed line of the gradient shows a severe over-segmentation (see figure 3c). Even the sky, apparently rather homogeneous in the initial image is cut into multiple small pieces. As a matter of fact, the gradient image is extremely sensitive to noise and its minima are extremely numerous, each of them giving birth to a catchment basin.

Fig 4 shows how to reduce this over-segmentation. We would like to segment the image into two particular regions. In figure 4a we have indicated which regions we are interested in. The two areas marked (one of the markers is formed by two connected components) will be set in a marker image to the value of the gradient image, and to the maximal gray value elsewhere. figure 4b presents the highest possible flooding of the gradient entirely below the marker image : it presents 3 minima, corresponding exactly to the 3 chosen markers, this operation is called *swamping*. Finally, the contours which are retained by the flooding of this swamped image are the strongest ones separating the markers, as shown in the segmentation of figure 4c.

Fig. 2. Initial image and filtered image, applying 3 iterations of an extreme flooding followed by an extreme razing

Size Oriented Flooding. Size oriented flooding may be visualized as a process where sources are placed at each minimum of a topographic surface and pour water in such a way that all lakes share some common measure (height, volume or area of the surface). As the flooding proceeds, some lakes eventually become full lakes, as the level of the lowest pass point has been reached. Let L be such a full lake. The source of L stops pouring water and its lake is absorbed by a neighboring catchment basin X, where an active source is still present. Later the lake present in X will reach the same level as L, both lakes merge and continue growing together. Finally only one source remains active until the whole topographical surface is flooded. If we construct the watershed line every time a catchment basin is absorbed by a neighboring basin, we obtain a series of nested partitions, also called *hierarchy* [1]: each region of a coarse partition is the union of a number of regions of a finer partition.

In figure 5, a flooding starts from all minima in such a way that all lakes always have uniform depth, as long as they are not full. The resulting hierarchy is called dynamics in case of depth driven flooding and has first been introduced by M.Grimaud[3]. Deep catchment basins represent objects which are contrasted ; such objects will take long before being absorbed by a neighboring catchment basin. The most contrasted one will absorb all others. This criterion obviously takes only the contrast of the objects into account and not their size. If we control the flooding by the area or the volume of the lakes, the size of the objects also is taken into consideration [8]; in multimedia applications, good results are often obtained by using as measure the volume of the lakes, as if each source would

Fig. 3. a) initial image ; b) morphological gradient ; c) watershed of the gradient image

a) 3 markers placed on the b) result of the swamping of c) resulting segmentation
  cameraman image   the gradient, this image has
                                  three minima

Fig. 4. Using markers as a means to reduce over-segmentation

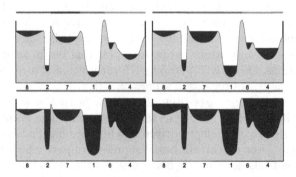

Fig. 5. Example of a height synchronous flooding. Four levels of flooding are illustrated, each of them is topped by a figuration of the corresponding catchment basins.

pour water with a constant flow This is illustrated by the following figures. The topographical surface to be flooded is a color gradient of the initial image (maximum of the morphological gradients computed in each of the R, G and B color channels). Synchronous volumic flooding has been used, and 3 levels of fusions have been represented, corresponding respectively to 15, 35 and 60 regions.

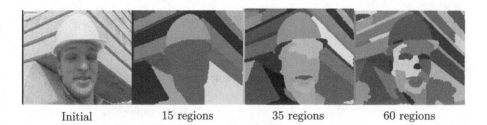

Initial   15 regions   35 regions   60 regions

4 Operators on Graphs

We define a graph $G = [E, N]$, where E is the set of edges and N the set of nodes. (i, j) will be the edge between the neighboring nodes i and j. Two edges are neighbors if they have a common node as extremity. Two nodes are neighbors if they are linked by an edge.

The weights $[e, n]$ of the graph G are represented as two functions e and n, respectively for the edges and the nodes. e_{ij} is the weight of the edge (i, j) and n_i the weight of the node i. The same graph may have various distributions of weights.

4.1 Contraction of Edges

For contracting an edge (i, j) in a graph G, one suppresses this edge and its two extremities are merged into a unique node k (in practice, k may be one of the preexisting nodes i or j). All edges incident to i or to j become edges incident to the new node k and either keep the same weight, or combine their weights in a manner suitable for the application.

4.2 Two Adjunctions on Graphs

Erosions and Dilations

Definition 2. *We define now two couples of an erosion and a dilation:*
- *an erosion* $[\varepsilon_{ne}n]_{ij} = n_i \wedge n_j$ *and its adjunct dilation* $[\delta_{en}e]_i = \bigvee_{(k\ neighbs\ of\ i)} e_{ik}$
- *a dilation* $[\delta_{ne}n]_{ij} = n_i \vee n_j$ *and its adjunct erosion* $[\varepsilon_{en}e]_i = \bigwedge_{(k\ neighbs\ of\ i)} e_{ik}$

Proposition 1. *The operators we defined are pairwise adjunct or dual operators:*
- *(ε_{en}, δ_{ne}) and (ε_{ne} , δ_{en}) are respectively adjunct*
- *(ε_{en}, δ_{en}) and (ε_{ne} , δ_{ne}) are respectively dual*

Proof. Let us prove that δ_{ne} and ε_{en} are adjunct operators

If $G = [e, n]$ and $\overline{G} = [\overline{e}, \overline{n}]$ are two graphs with the same nodes and edges, but with different valuations on the edges and the nodes then:

$$\delta_{ne}n \le \overline{e} \Leftrightarrow \forall i, j : n_i \vee n_j \le \overline{e}_{ij} \Leftrightarrow \forall i, j : n_i \le \overline{e}_{ij}$$

$$\Leftrightarrow \forall i, j : n_i \le \bigwedge_{(j \text{ neighbs of } i)} \overline{e}_{ij} = [\varepsilon_{en}\overline{e}]_i \Leftrightarrow n \le \varepsilon_{en}\overline{e}$$

which establishes that δ_{ne} and ε_{ne} are adjunct operators.
Let us prove that ε_{ne} and δ_{ne} are dual operators:

$$[\varepsilon_{ne}(-n)]_{ij} = -n_i \wedge -n_j = -(n_i \vee n_j) = -[\delta_{ne}n]_{ij}$$

Hence: $\delta_{ne}n = -[\varepsilon_{ne}(-n)]$.

Openings and Closings. As ε_{en} and δ_{ne} are adjunct operators, the operator $\varphi_n = \varepsilon_{en}\delta_{ne}$ is a closing on n and $\gamma_e = \delta_{ne}\varepsilon_{en}$ is an opening on e [4].

Similarly as ε_{ne} and δ_{en} are adjunct operators, the operator $\varphi_e = \varepsilon_{ne}\delta_{en}$ is a closing on n and $\gamma_n = \delta_{en}\varepsilon_{ne}$ is an opening on e.

Also see the works of L. Vincent [9] on erosions and dilations on graph nodes which can be reformulated as the composition of a nodes-to-edges operations, followed by an edges-to-nodes: $\varepsilon_{vincent} = \varepsilon_{nn} = \varepsilon_{en} \circ \varepsilon_{ne}$

5   Floodings on Graphs

A flooding g of f will be perfectly known as soon the level of the lakes in each catchment basin of f is known. For this reason, it is possible to represent a flooding of f as a particular weight distributions on the nodes of its region adjacency graph or the derived MST. However not any distribution of weights on the nodes will represent a valid flooding. We now establish criteria to be respected.

5.1   Criteria for a Weight Distribution to Be a Valid Flooding

Let $T = [e, n]$ and $\overline{T} = [\overline{e}, \overline{n}]$ be respectively the MSTs of a function f and of a flooding g of f . Consider an edge (i, j). Two situations are possible

- the level of both lakes at i and j is lower than the pass-point, then:
  $$\overline{e_{ij}} = e_{ij} > (\varepsilon_{ne}\overline{n})_{ij}$$
- the level of both lakes at i and j is higher or equal than the pass-point e_{ij} ; in this case both lakes and the pass-point $\overline{e_{ij}}$ have the same altitude :
  $$\overline{e_{ij}} = (\varepsilon_{ne}\overline{n})_{ij} > e_{ij}$$

Both situations may be summarized by $\overline{e} = \varepsilon_{ne}\overline{n} \vee e$

5.2   First Algorithm for Constructing a Valid Flooding

We derive from these two relations $\overline{n} = \varepsilon_{en}\overline{e} \wedge \overline{n}$ and $\overline{e} = \varepsilon_{ne}\overline{n} \vee e$ an algorithm for constructing a flooding of a function f, using its MST.

The MST of the neighborhood graph of the reference image f is computed. Two copies of the same graph will be used :

Reference image : The edges of the MST are valuated by the altitude of the pass points between catchment basins.

Flooded image : All edges and nodes are initialized with a valuation ∞, except the nodes containing a regional minimum of the marker function. These nodes are initialized with the altitude of the lowest regional minimum of the marker function.

This process is illustrated in figure 6 for a one-dimensional image.

a) Initialization : $\overline{e}^{(0)} = \infty$,
$\overline{n}^{(0)}$ = lowest regional minimum of the marker function contained in the catchment basin n and $\overline{n}^{(0)} = \infty$ in the catchment basins without a regional minimum of the marker function
b) Repeat until idempotence:
$\overline{e}^{(n+1)} = \varepsilon_{ne}\overline{n}^{(n)} \vee e$
$\overline{n}^{(n+1)} = \varepsilon_{en}\overline{e}^{(n+1)} \wedge \overline{n}^{(n)}$

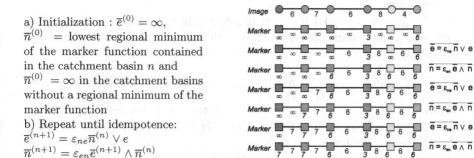

Fig. 6. Construction of the flooding weight distribution $[\overline{e}, \overline{n}]$

5.3   A More Synthetic Flooding Algorithm

The preceding algorithm has the advantage to use the elementary erosions and dilations we defined earlier. Most of the time we are interested by the levels $\overline{n}$ of the lakes in the flooded image but not so much by the levels of the pass points $\overline{e}$. We will derive a unique formula by replacing $\overline{e}_{ij} = (\overline{n}_i \wedge \overline{n}_j) \vee e_{ij}$ by its value in the expression of $\overline{n}_i = \bigwedge_{(j\,\text{neighbs of } i)} \overline{e}_{ij} \wedge \overline{n}_i \wedge \overline{n}_i$. We obtain:

$$\overline{n}_i = (\varepsilon_{en}\,\overline{e})_i \wedge \overline{n}_i = \bigwedge_{(j\,\text{neighbs of } i)} \overline{e}_{ij} \wedge \overline{n}_i = \bigwedge_{(j\,\text{neighbs of } i)} [(\overline{n}_i \wedge \overline{n}_j) \vee e_{ij}] \wedge \overline{n}_i$$
$$= \bigwedge_{(j\,\text{neighbs of } i)} [(\overline{n}_i \vee e_{ij}) \wedge (\overline{n}_j \vee e_{ij})] \wedge \overline{n}_i$$
$$= \bigwedge_{(j\,\text{neighbs of } i)} (\overline{n}_j \vee e_{ij}) \wedge \overline{n}_i \quad \text{since} \quad \overline{n}_i \vee e_{ij} \geq \overline{n}_i$$

Hence we have proved the following proposition:

Proposition 2. *The weight distribution $[\overline{e}, \overline{n}]$ on a tree T is a flooding of the weight distribution of $[e, n]$ if and only if $\overline{n}_i = \bigwedge_{(j\ neighbs\ of\ i)} (\overline{n}_j \vee e_{ij}) \wedge \overline{n}_i$*

From the relation $\bar{n}_i = \bigwedge\limits_{(j\,\text{neighbs of }i)} (\bar{n}_j \vee e_{ij}) \wedge \bar{n}_i$ we derive $\bar{n}_i \leq \bar{n}_j \vee e_{ij}$.

Inversely, if for each neighbor j of i we have $\bar{n}_i \leq \bar{n}_j \vee e_{ij}$ then

$$\bar{n}_i \leq \bigwedge\limits_{(j\,\text{neighbs of }i)} (\bar{n}_j \vee e_{ij}) \text{ and } \bar{n}_i = \bigwedge\limits_{(j\,\text{neighbs of }i)} (\bar{n}_j \vee e_{ij}) \wedge \bar{n}_i$$

The weight distribution $[\bar{e}, \bar{n}]$ on a tree T is a flooding of the weight distribution of $[e, n]$ if and only if for any two neighboring nodes i and j one of the following criteria is verified:

1. $\bar{n}_i \leq \bar{n}_j \vee e_{ij}$, which may be written as $\{\bar{n}_i \leq \bar{n}_j \text{ or } \bar{n}_i \leq e_{ij}\}$. This last formula directly leads to the two following criteria if one recalls the logical equivalence between $\{A \Rightarrow B\}$ and $\{notA \text{ or } B\}$

2. $\{\bar{n}_i > \bar{n}_j \Rightarrow \bar{n}_i \leq e_{ij}\}$: if the level in two neighboring basins is different, then the highest lake has a level lower than or equal to the pass-point separating them

3. $\{\bar{n}_i > e_{ij} \Rightarrow \bar{n}_i = \bar{n}_j\}$ If the level in two neighboring basins is higher than the pass-point separating them, then both adjacent lakes have merged and have the same level.

From this relation $\bar{n}_i = \bigwedge\limits_{(j\,\text{neighbs of }i)} (\bar{n}_j \vee e_{ij}) \wedge \bar{n}_i$ we derive a shorter flooding algorithm, where we compute only the level of the flooded nodes.

This process is illustrated in figure 7. The algorithm may be further speeded up by processing the nodes $\bar{n}$ n increasing order by using a hierarchical queue [5]. Initially the nodes corresponding to the regional minima of g are put in the queue at a level corresponding to their altitude. As soon as a node i goes out of the queue, its neighbors are updated and the edges adjacent to i are cut, producing two sub-trees for each edge which is cut. These sub-trees may then be treated independently from each other.

a) Initialization :
$\bar{n}^{(0)}$ = lowest regional minimum of the marker function contained in the catchment basin n and $\bar{n}^{(0)} = \infty$ in the catchment basins without a regional minimum of the marker function
b) Repeat until idempotence :
if j neighbor of i and $\bar{n}_i > \bar{n}_j \vee e_{ij}$ then $\bar{n}_i = \bar{n}_j \vee e_{ij}$

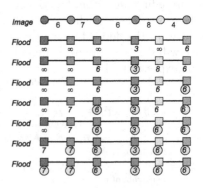

Fig. 7. Progressive construction of $\bar{n}$ using the algorithm if j neighbor of i and $\bar{n}_i > \bar{n}_j \vee e_{ij}$ then $\bar{n}_i = \bar{n}_j \vee e_{ij}$

6   Conclusion

The image-based flooding operations have known three eras, which have guided their implementations. In the beginning, all image operations were performed using dedicated hardware which could only perform erosions, dilations and pixel-wise operations, from this viewpoint comes the expression of the flooding as an erosion and supremum operation. Then came the age of the sequential propagation, in which few passes were required if the image was not too convoluted. The most modern implementations of the floodings now use priority queues in order to optimize away all redundant propagations and thus ensure that the flooding can be performed in a single pass (though it requires random access to the image, and is hard to implement in hardware).

However, no matter the speed of this last implementation, or the regular improvements in computer power, the size of the datasets increases just as fast, if not faster. From the original 2D images, we have come to process 3D, even 4D images, sometimes with vector values (color and hyperspectral images). In the second part of this paper, we will examine the performance of these graph-based methods, compare them to the image-based methods and determine their areas of effectiveness.

References

1. Benzécri, J.P.: Description mathématique des classifications. Revue de Statistique Appliquée 20(3), 23–56 (1972)
2. Beucher, S.: Segmentation d' images et morphologie mathématique. Thèse de doctorat en morphologie mathématique, ENSMP, 1822 (1990)
3. Grimaud, M.: géodésie numérique en morphologie mathématique, La. Thèse de doctorat en morphologie mathématique, ENSMP (December 1990)
4. Heijmans, H.: Morphological Image Operators. Advances in Electronics and Electron Physics, vol. 24. Academic Press, Boston (1994)
5. Meyer, F.: Un algorithme optimal de ligne de partage des eaux. In: Actes 8ème Congrès AFCET Reconnaissance des Formes et Intelligence Artificielle, pp. 847–857, Lyon-Villeurbanne (November 1991)
6. Meyer, F.: Morphological segmentation revisited. In: Bilodeau, M., Meyer, F., Schmitt, M. (eds.) Space, Structure, and Randomness: Contributions in Honor of Georges Matheron in the Field of Geostatistics, Random Sets, and Mathematical Morphology. Springer Lecture Notes in Statistics, vol. 183, pp. 148–315. Springer, Heidelberg (2005)
7. Meyer, F.: The levelings. In: Heijmans, H.J.A.M., Roerdink, J.B.T.M. (eds.) Mathematical Morphology and its Applications to Image and Signal Processing, Proc. ISMM'98, Amsterdam, June 1998, pp. 199–206. Kluwer, Dordrecht (1998)
8. Vachier, C.: Extraction de caractéristiques, segmentation d'images et morphologie mathématique. Thèse de doctorat en morphologie mathématique, ENSMP (1995)
9. Vincent, L.: Mathematical morphology on graphs. Technical Report Proc. Visual Communications and Image Processing'88, Part of SPIE's, Cambridge Symposium, Centre de Morphologie Mathématique / ENSMP, August 1988. 918 (1988)

Graph-Based Multilevel Temporal Segmentation of Scripted Content Videos

Ufuk Sakarya[1,2] and Ziya Telatar[2]

[1] TÜBİTAK-UZAY (The Scientific & Technological Research Council of Turkey – Space Technologies Research Institute), ODTÜ Yerleşkesi, 06531, Ankara, Turkey
[2] Ankara University, Faculty of Engineering, Department of Electronics Engineering, Ankara, Turkey
ufuk.sakarya@uzay.tubitak.gov.tr, Ziya.Telatar@eng.ankara.edu.tr

Abstract. This paper concentrates on a graph-based multilevel temporal segmentation method for scripted content videos. In each level of the segmentation, a similarity matrix of frame strings, which are series of consecutive video frames, is constructed by using temporal and spatial contents of frame strings. A strength factor is estimated for each frame string by using a priori information of a scripted content. According to the similarity matrix reevaluated from a strength function derived by the strength factors, a weighted undirected graph structure is implemented. The graph is partitioned to clusters, which represent segments of a video. The resulting structure defines a hierarchically segmented video tree. Comparative performance results of different types of scripted content videos are demonstrated.

Keywords: Temporal video segmentation, shot clustering, scene detection, video summarization, graph partitioning, normalized cuts.

1 Introduction

In recent years, multimedia data has received continuously increasing interest in humans and this content gets bigger day by day with the advances in technology. Most of this content is related to visual information including video data produced by filmmakers, TV channels, amateur camera users etc. Extracting specific information from such a huge amount of video content creates some difficulties to search the whole media like limitations on time consuming in browsing and retrieving of relevant data. To overcome that application based drawbacks, the current trend is to develop algorithms capable of parsing them by segmenting and then indexing. On the other hand, temporal segmentation of a video is needed to enable an efficient indexing procedure for localizing and accessing the source of relevant information.

A complete video is constructed by shots, which are the collection of consecutive frames that are recorded in one camera record time. Therefore, shots carry out a priori information to be considered as a first step of the temporal video segmentation. Then, shot detection is handled to solve the indexing problem, however, it cannot properly

F. Escolano and M. Vento (Eds.): GbRPR 2007, LNCS 4538, pp. 168–179, 2007.
© Springer-Verlag Berlin Heidelberg 2007

contribute in some application domains and a higher-level segmented video like a DVD chapters is needed. For example, high-level video segments (scenes) in some browsing applications are more important than shots.

Shot clustering is a hot topic in the research community. Many approaches have been presented to contribute the problem until now. One of the earliest proposed methods is a *scene transition graph* (STG) framework [1]. In this approach, firstly shots are clustered. Each shots cluster represents a node on the directed graph and the temporal relationship of shots constructs the edges between nodes. Yeung and Yeo [2] used the time-constrained clustering in STG. Time-adaptive grouping approach was introduced by [3] and a table-of-content (ToC) technique was demonstrated. In this approach, shots are clustered to an intermediate entity called video group and the groups are merged to construct scenes. In [4], shots were clustered into scenes using a strict scene definition. Zhai and Shah [5] used the Markov chain Monte Carlo technique in order to construct scene borders.

Graph-based solution approaches have become very popular for the pattern recognition research community. A novel graph-based method called *normalized cuts* [6] was given as an efficient tool in the image segmentation problem. Graph-based approaches have also been presented in structuring and summarizing of videos. One of the graph-based approaches is the above-mentioned STG [1]. Odobez *et al.* [7] introduced the spectral method for home videos. The method is based on visual and temporal similarities. Rasheed and Shah [8] defined a *shot similarity graph* (SSG) in the similar way. Besides visual and temporal content similarities, it uses the motion content similarity. Ngo *et al.* [9] introduced a graph-based approach worked on two steps. In the first step, shots are clustered using normalized cuts algorithm [6]. In the second step, clustered shots are represented by a temporal directed graph similar to STG. Lu *et al.* [10] presented a novel graph-based dynamic video summarization method. This method also works on two steps. In the first step, shots are grouped into *shot strings* using normalized cuts algorithm [6]. In the second step, shot strings are represented by a spatial-temporal directed graph. Peng and Ngo [11] proposed a clip-based similarity measures based on two bipartite graph-matching algorithms. One is a maximum matching on an unweighted bipartite graph and the other is an optimal matching on a weighted bipartite graph. Gong [12] presented a video summarization method using graph-based representations for audiovisual contents. In [13], a content-adaptive analysis and representation framework using graph-based approaches was proposed for audio event discovery.

In this paper, we propose a graph-based framework for a multilevel temporal segmentation of scripted content videos. Our proposed method is based on a *weighted undirected* graph representation for all segmentation levels. We propose a strength factor approach in order to improve the efficiency of graph-based clustering algorithm on the similarity matrix.

The rest of the paper is organized as follows. The proposed multilevel temporal segmentation approach is introduced in Section 2. In Section 3, comparative performance results are demonstrated. Finally, conclusion remarks are given in section 4.

2 Graph-Based Multilevel Temporal Video Segmentation

In this section, a graph-based framework for a multilevel temporal segmentation of scripted content videos is proposed. In briefly, the proposed method works as follows. In each level of segmentation, a similarity matrix of frame strings is constructed by using temporal and spatial contents of frame strings. Using a priori information about a frame string, a strength factor is estimated for each frame string. The similarity matrix is reevaluated from a strength function derived by these strength factors. Then, a weighted undirected graph is constructed by the similarity matrix. The graph is partitioned by using normalized cuts algorithm [6] with one additional constraint [8]. Each graph cluster represents one segment of a video. Therefore, a hierarchically segmented video tree is constructed.

2.1 Initial Definitions

Basic definitions in the temporal video segmentation are introduced in Figure 1. In addition to the definitions, frame string used in our proposed approach is introduced as follows.

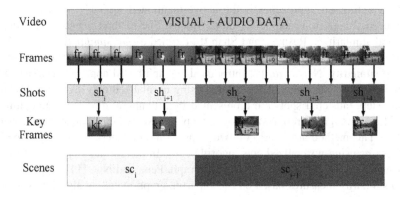

Fig. 1. Temporal video segmentation definitions

Frame: It is the smallest temporal video segment. The ith frame of a video is denoted by fr_i.

Shot: It is the collection of consecutive frames that are recorded in one camera record time. The ith shot of a video is denoted by sh_i. $sh_i = \{ fr_k, fr_{k+1}, ..., fr_{k+n-1}, fr_{k+n} \}$.

Key Frame: It is the best representation of frame/frames in a shot. The jth key frame of the ith shot of a video is denoted by $kf_{i,j}$.

Scene: It is the collection of consecutive shots that are semantically related. The ith scene of a video is denoted by sc_i. $sc_i = \{ sh_k, sh_{k+1}, ..., sh_{k+n-1}, sh_{k+n} \}$.

Frame String: It is the collection of consecutive frames according to one criterion. In the sth segmentation level, i^{th} frame string of a video is denoted by $fs_{s,i}$. It is the more general definition of collection set of frames. A frame string can be equal to a shot, a scene or any video segment. In addition, a frame string can be a collection of frame strings. $fs_{s,i} = \{ fr_k, fr_{k+1}, ..., fr_{k+n-1}, fr_{k+n} \}$.

2.2 The Proposed Method

For each s segmentation level, there is a video segmented by N_s frame strings, which are not intersect. There is a temporal and spatial similarity measurement function Sim_s for each $fs_{s,i}$ and $fs_{s,j}$. For an s segmentation level, there is a similarity matrix $sm_s(i,j)$ for each frame string indexed by i and j.

$$sm_s(i,j) = Sim_s(fs_{s,i}, fs_{s,j}). \tag{1}$$

In order to improve graph-based clustering algorithm depending on the similarity matrix, we propose a strength factor approach. The main idea for this issue can be explained as follows. If high probable correct relations between frame strings are strengthened and high probable false relations between frame strings are weakened, then it is expected that an efficient segmentation is obtained. Using prior information in s segmentation level, frame string strength factor $sf_s(i)$ is estimated for each $fs_{s,i}$. The similarity matrix is reevaluated by a strength factor function SF_s as follows.

$$SM_s(i,j) = sm_s(i,j) . SF_s(sf_s(i), sf_s(j)). \tag{2}$$

A weighted undirected graph $G_s=(V_s,E_s)$ is constructed by using SM_s. Each $fs_{s,i}$ is represented by a graph vertex $v_{s,i}$ and $SM_s(i,j)$ is represented by an undirected edge $e_s(i,j)$ between each vertex $v_{s,i}$ and $v_{s,j}$. The graph is partitioned using normalized cuts algorithm [6] with one additional constraint [8]. The graph is recursively partitioned two disjoint sets A and B, $A \cup B=V_s$, $A \cap B=\varnothing$, according to normalized cuts criteria:

$$Ncut(A, B) = \frac{cut(A,B)}{assoc(A,V_s)} + \frac{cut(A,B)}{assoc(B,V_s)} . \tag{3}$$

Graph cut $cut(A,B)$ and association $assoc(A,V_s)$ are defined as follow equations:

$$cut(A, B) = \sum_{i \in A, j \in B} e_s(i,j) , \tag{4}$$

$$assoc(A,V_s) = \sum_{i \in A, j \in V_s} e_s(i,j) . \tag{5}$$

The graph is partitioned using one additional constraint defined in [8]:

$$(i < j \ or \ i > j) \ for \ all \ v_{s,i} \in A, v_{s,j} \in B . \tag{6}$$

Each graph cluster represents a new segment of a video. Frame strings in the same cluster are merged for a new frame string. Therefore, N_{s+1} frame strings are constructed from N_s frame strings, where $N_{s+1} < N_s$.

In the next two sub-sections, an application of the proposed method to two levels segmentation case is explained. In other words, this application is a framework for scenes construction from shots.

2.3 Level One

In order to start segmentation process, a video is segmented to N_1 frame strings. Shot detection process is applied for this initialization. We assume that N_1 shots are detected and each shot is represented by $fs_{1,i}$. For this level, we modified and extended the scene detection method given in [8] by using our strength factor approach.

In briefly, Rasheed and Shah's method [8] works as follows. For each ith and j^{th} indexed shots, the similarity matrix is calculated as:

$$sm_1(i,j) = w(i,j) \cdot ShotSim(i,j), \qquad (7)$$

where $w(i,j)$ is a temporal distance similarity function and $ShotSim(i,j)$ is a similarity function based on visual and motion contents. $ShotSim(i,j)$ is defined as follows:

$$ShotSim(i,j) = \alpha \cdot VisSim(i,j) + \beta \cdot MotSim(i,j), \qquad (8)$$

where α, β are constants, $VisSim(i,j)$ and $MotSim(i,j)$ are visual and motion similarity functions respectively. Key frame selection, visual and motion content calculation is based on HSV color histogram. A weighted undirected graph is constructed by using the similarity matrix. The graph is partitioned by using Eq. (3) and Eq. (6). Each partitioned graph cluster represents a scene.

We have also adopted the shot goodness property from [8] to our algorithm. The shot that has the biggest shot goodness value in a scene is accepted as the representative shot of a scene. For each of ith shot, a total visual similarity in its scene is defined as:

$$C(i) = \sum_{j \in Scene} VisSim(i,j) \cdot \qquad (9)$$

Then, the shot goodness $F(i)$ is calculated as

$$F(i) = \frac{C^2(i) \times L_i}{log(Mot(i) + \Theta)}, \qquad (10)$$

where L_i, Θ and $Mot(i)$ are the shot length of ith shot, a small positive constant and a function that calculates the motion content value of ith shot respectively.

According to our proposed approach, frame string strength factor $sf_1(i)$ is estimated by using the frame string length and the motion content in a frame string. This factor is increased with the length and decreased with the motion. Eq. (10) is used as a hint in the selection of these features.

$$sf_1(i) = \frac{length(fs_{1,i})}{\mu_{length}} \cdot \frac{\mu_{mot}}{Mot(fs_{1,i})}, \qquad (11)$$

where $length(.)$, μ_{length}, and μ_{mot} are a number of frames measurement function, an average number of frames and an average motion for all frame strings in this level respectively. Strength factor function SF_1 is defined as follows,

$$SF_1(sf_1(i), sf_1(j)) = max(sf_1(i), sf_1(j)). \qquad (12)$$

The similarity matrix is reevaluated by SF_1. Next, the graph-based partitioning is applied for obtaining the next level initialization frame strings.

2.4 Level Two

In this level, N_2 frame strings are obtained from Level One. In this level, a similarity matrix is calculated for each ith and j^{th} indexed frame strings as follows.

$$sm_2(i,j) = ds(i,j) . FSSim(i,j),$$ (13)

where $ds(i,j)$ is a temporal distance similarity function and $FSSim(i,j)$ is a similarity function defined as follows.

$$ds(i, j) = e^{-|i-j|},$$ (14)

$$FSSim(i,j) = c_1 . L2VisSim(i,j) + c_2 . L2MotSim(i,j),$$ (15)

where c_1, and c_2 are constants, and $c_1 + c_2 = 1$. $L2VisSim(i,j)$ is a visual similarity function. It is based on one-shot representation of a frame string. One shot is selected using Eq. (10). Only selected two shots from two frame strings are evaluated by $VisSim(i,j)$ function. $L2MotSim(i,j)$ is the motion content similarity function using an average motion content of all shots in a frame string. Its calculation is structurally similar to $MotSim(i,j)$. It is defined as follows:

$$L2MotSim(i, j) = \frac{2 \times min(Mot2(fs_{2,i}), Mot2(fs_{2,j}))}{Mot2(fs_{2,i}) + Mot2(fs_{2,j})},$$ (16)

where $Mot2(.)$ is an average motion content measurement function for a frame string.

$$Mot2(fs_{2,i}) = \frac{1}{M} \sum_{fs_{1,j} \in fs_{2,i}} Mot(fs_{1,j}),$$ (17)

where M is the total $fs_{1,j}$ number in $fs_{2,i}$.

Frame string strength factor $sf_2(i)$ is estimated using the frame string length, shot number and the motion content in a frame string. The factor is increased with the length and the shot number and decreased with the motion.

$$sf_2(i) = \sqrt{\frac{length(fs_{2,i})}{\mu_{flength}} . \frac{slength(fs_{2,i})}{\mu_{slength}} . \frac{\mu_{mot}}{Mot2(fs_{2,i})}},$$ (18)

where $slength(.)$ is a number of shots measurement function, $\mu_{flength}$ and $\mu_{slength}$ are an average number of frames and an average number of shots for all frame strings in this level respectively. The strength factor function SF_2 is the same as in Level One.

The similarity matrix is reevaluated by SF_2. Then, the graph-based partitioning is applied for obtaining the clustered frame strings. Each of the clusters represents a scene.

3 Experiments and Results

3.1 Data Set and Ground-Truths

In this work, three videos are used in the experiments to test the performance of our proposed method and to compare with the scene detection method presented in [8]. Two movies Hamlet (HA), The Karate Kid (KK) and a television series Hayat Bilgisi (HB) are used in the experiments. Detailed test videos information is given in Table 1. It can be easily seen from Table 1 that all three videos have different shot-scene distribution on the temporal axis. Shots and scenes ground-truths are generated by manually and shot detection is not taken into account in the experiments. Shot detection initialization process is extracted from the human generated ground-truths.

Table 1. Test Videos Information

Video Name	Total Shots	Total Scenes	Duration (Minute)
Hamlet (HA) (Movie)	185	10	17:59
Hayat Bilgisi (HB) (A television series)	195	5	7:30
The Karate Kid (KK) (Movie)	94	6	18:18
Total	474	21	43:47

3.2 Implementation Details

There are three important parameters suggested in the implementation of the method in [8]. Selection values: $\alpha=0.5$ (visual similarity weight), $\beta=0.5$ (motion similarity weight) and d=20 (temporal decay parameter). Key frame selection, visual and motion content calculation is realized by *ColSim(.)* function which is based on HSV color histogram. As our implementation, we used a quantized 3-D HSV histogram. Quantization levels of H, S and V are 18, 3 and 3 respectively. These quantization levels are selected according to the work given in [11]. Moreover, two parameters c_1, and c_2 are selected as 0.5 and 0.5 respectively in the implementation of Level Two method (L2M).

3.3 Performance Evaluation

Two error functions are applied for comparative performance tests. One is F-measure (F) used in [11]. It measures the quality of the detected clusters. It is in the range 0 to 1 and F=1 shows a perfect result. This analysis is performed at the shots level, not at the frames level. *GT* and *DT* are sets of ground-truth clusters and detected clusters respectively. F-measure *F* is calculated as follows.

$$F = \frac{1}{Z} \sum_{c_i \in GT} |C_i| \max_{c_j \in DT} \left\{ f(C_i, C_j) \right\},$$ (19)

where

$$f(C_i, C_j) = \frac{2 \times \text{Re}(C_i, C_j) \times \text{Pr}(C_i, C_j)}{\text{Re}(C_i, C_j) + \text{Pr}(C_i, C_j)},$$ (20)

$$Z = \sum_{c_i \in GT} |C_i|.$$ (21)

Recall *(Re)* and precision *(Pr)* functions are defined as follows:

$$\text{Re}(C_i, C_j) = \frac{|C_i \cap C_j|}{|C_i|},$$ (22)

$$\text{Pr}(C_i, C_j) = \frac{|C_i \cap C_j|}{|C_j|}.$$ (23)

Our Level One method (L1M) and Rasheed & Shah's method presented in [8] (RSM) are the similar except for our strength factor approach extension. In order to measure the effects of it, the shots in errors (SIE) function is also applied. SIE is similar to a function defined with the same name in [7]. In SIE measurement, each detected cluster is assigned to one ground truth cluster according to maximum intersection with it. After the assignment, shots that are not in the assigned ground truth cluster are counted. This count is divided by the total shots in the video. SIE is in the range 0 to 1 and SIE=0 shows no error shot in the video.

There is a variable parameter K that is used to determine the normalized cuts threshold in the experiments. If normalized cuts value is smaller than K value, then the graph is partitioned. Normalized cuts values can be in the range of 0 to 2, and high values generate more clusters. K values are selected in the discrete range between 0.1 and 2 and a step size of 0.1 in the experiments; however, meaningful results are demonstrated in this paper. Moreover, L2M has two normalized cuts threshold parameters K1 for the first level and K2 for the second level. The same above-mentioned procedures are also applied for K1 and K2 in the experiments.

3.4 Results

Table 2 shows F-measure results for L1M and RSM with respect to variable K parameter. The mean values[1] of the three results are given for each K parameter. Maximum mean value of L1M is obtained as 0.75 for K=0.2. RSM gives 0.75 for K=0.1. Therefore, if maximum mean values are taken into account, both methods produce the equal result. If maximum of F-measures criteria is applied for the evaluation, L1M has better results for two test videos and RSM has a better result for the one. Maximum values in L1M are calculated as 0.73, 0.91 and 0.93 for K parameter values of 0.5, 0.2 and 0.5 for videos HA, HB and KK respectively. RSM gives 0.76, 0.88 and 0.89 for K parameter values of 0.7, 0.1 and 0.5. For both methods, a suitable K parameter selection is a critical issue because of the content variation. This drawback is an important point to motivate us to propose the multilevel approach. In that sense, it is needed for a decision of the previous level segmentation of L2M. In order to decide a suitable one, SIE assessments are made. Table 3 shows SIE assessments between L1M and RSM. In the over segmented region of K values, between 0.9 to 1.1, L1M produces less or equal mean of SIE results. Thus, it can be said that L1M is more suitable than RSM for a previous level segmentation choice. In addition to SIE evaluations, next two experiments in the following paragraph also show that L1M is the suitable one.

Tables 4 and 5 demonstrate L2M F-measure analysis using L1M and RSM as a previous level segmentation method respectively. Each of them is constructed by that the highest values of all videos and its mean are cropped from the total results table which has the results for all *K1* and *K2* in the discrete range between 0.1 and 2 with a step size of 0.1. Therefore, *K1* and *K2* values are different for both tables. According to Table 4, the maximum mean value of L2M using L1M is obtained as 0.83 for K1=0.9, K2=1.5. It outperforms both L1M and RSM. If maximum of F-measures criteria is applied for the evaluation, it has better results for all video except for KK. The maximum values in L2M using L1M are calculated as 0.88, 0.94 and 0.88 for videos HA, HB and KK respectively. L2M using L1M outperforms L2M using RSM according to both criterions; maximum of F-measures and maximum mean value.

---

[1] Means of the results are calculated from the exact values not from the quantized values shown in the tables.

Table 2. Comparative F-measure analysis between Level One method and Rasheed & Shah's method [8] with respect to variable K parameter

K	The proposed Level One method				Rasheed & Shah's method [8]			
	HA	HB	KK	Mean	HA	HB	KK	Mean
0.1	0.59	0.89	0.49	0.66	0.69	0.88	0.68	0.75
0.2	0.67	0.91	0.66	0.75	0.65	0.74	0.68	0.69
0.3	0.71	0.80	0.66	0.73	0.72	0.71	0.76	0.73
0.4	0.73	0.72	0.76	0.73	0.72	0.71	0.76	0.73
0.5	0.73	0.55	0.93	0.73	0.72	0.58	0.89	0.73
0.6	0.70	0.54	0.93	0.72	0.70	0.55	0.80	0.68
0.7	0.69	0.54	0.83	0.68	0.76	0.53	0.80	0.69
0.8	0.69	0.46	0.83	0.66	0.72	0.49	0.80	0.67
0.9	0.69	0.44	0.80	0.64	0.70	0.42	0.80	0.64

Table 3. Comparative SIE analysis between Level One method and Rasheed & Shah's method [8] with respect to variable K parameter

K	The proposed Level One method				Rasheed & Shah's method [8]			
	HA	HB	KK	Mean	HA	HB	KK	Mean
0.9	0.05	0.02	0.04	0.04	0.06	0.02	0.10	0.06
1.0	0.05	0.02	0.04	0.04	0.04	0.01	0.10	0.05
1.1	0.04	0.02	0.04	0.03	0.02	0.01	0.06	0.03

Table 4. Level Two method F-measure analysis using Level One method as a previous level segmentation

K1	K2	HA	HB	KK	Mean
0.9	0.9	0.66	0.94	0.53	0.71
0.9	1.0	0.66	0.81	0.66	0.71
0.9	1.1	0.66	0.81	0.66	0.71
0.9	1.2	0.69	0.81	0.66	0.72
0.9	1.3	0.76	0.75	0.76	0.76
0.9	1.4	0.83	0.75	0.76	0.78
0.9	1.5	0.83	0.77	0.88	0.83
0.9	1.6	0.88	0.66	0.88	0.81

Table 5. Level Two method F-measure analysis using Rasheed & Shah's method [8] as a previous level segmentation

K1	K2	HA	HB	KK	Mean
1.1	0.5	0.65	0.93	0.71	0.76
1.1	0.6	0.65	0.85	0.71	0.74
1.1	0.7	0.65	0.90	0.83	0.79
1.1	0.8	0.73	0.90	0.83	0.82
1.1	0.9	0.76	0.77	0.83	0.79
1.1	1.0	0.84	0.77	0.83	0.82
1.1	1.1	0.80	0.72	0.83	0.78
1.1	1.2	0.80	0.67	0.86	0.78

Figure 2 illustrates comparative scenes detection results with respect to ground-truth boundaries for HA. Figure 2(a) and (b) show the scene construction results in a constant cluster number (10 scenes for HA) case and maximum F-measures case respectively.

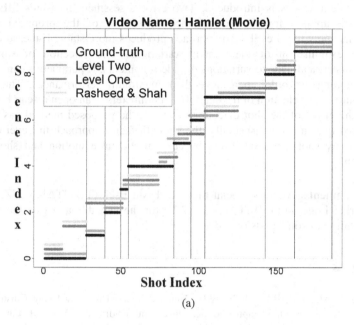

(a)

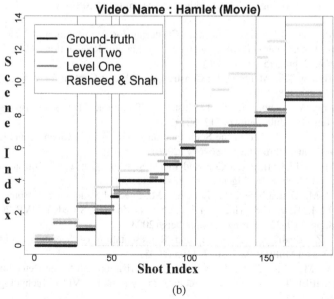

(b)

Fig. 2. Hamlet scene detection results. (a) The constant cluster number case: K=0.5 for RSM, K=0.5 for L1M, K1=0.9 and K2=1.6 for L2M using L1M. (b) Maximum F-measure case: K=0.7 for RSM, K=0.5 for L1M, K1=0.9 and K2=1.6 for L2M using L1M.

4 Conclusions

In this paper, a graph-based framework for a multilevel temporal segmentation of scripted content videos is introduced. Two error assessment measures SIE and F-measure are applied for the performance evaluation of the proposed method. Experiments show that Level Two approach produces more stable F-measure results under the examination of the content variations. The selection of suitable K parameters is critical. Due to variations of videos contents, K parameters can vary in a quite large range in order to achieve their maximum performances. The content adaptive automatic selection of the suitable K parameters is an open research issue. In addition, the effect of the shot detection errors on the proposed method has been an un-answered question yet. Especially, the strength factor approach in Level One can be affected by shot detection errors because of the false motion and shot length information.

Acknowledgments. Authors special thank to Ersin Esen (TÜBİTAK – UZAY) and Kamil Berker Loğoğlu (TÜBİTAK – UZAY) for the valuable assist on the software implementation of some parts of codes.

References

1. Yeung, M.M., Yeo, B.L., Wolf, W.H., Liu, B.: Video Browsing Using Clustering and Scene Transitions on Compressed Sequences. In: Rodriguez, A.A., Maitan, J. (eds.) Multimedia Computing and Networking. Proceedings of SPIE, vol. 2417. pp. 399-413 (1995)
2. Yeung, M., Yeo, B.L.: Time-Constraint Clustering for Segmentation of Video into Story Units. In: Proc. International Conference on Pattern Recognition ICPR 96, Vienna, Austria, vol. C, pp. 375–380 (1996)
3. Rui, Y., Huang, T.S., Mehrotra, S.: Constructing Table of Content for Videos. Multimedia Systems 7, 359–368 (1999)
4. Tavanapong, W., Zhou, J.: Shot Clustering Techniques for Story Browsing. IEEE Transactions On. Multimedia 6, 517–527 (2004)
5. Zhai, Y., Shah, M.: Video Scene Segmentation Using Markov Chain Monte Carlo. IEEE Transactions on Multimedia 8, 686–697 (2006)
6. Shi, J., Malik, J.: Normalized Cuts and Image Segmentation. IEEE Transactions Pattern Analysis and Machine Intelligence 22, 888–905 (2000)
7. Odobez, J.-M., Gatica-Perez, D., Guillemot, M.: Spectral Structuring of Home Videos. In: Bakker, E.M., Lew, M.S., Huang, T.S., Sebe, N., Zhou, X.S. (eds.) CIVR 2003. LNCS, vol. 2728, pp. 310–320. Springer, Heidelberg (2003)
8. Rasheed, Z., Shah, M.: Detection and Representation of Scenes in Videos. IEEE Transactions on Multimedia 7, 1097–1105 (2005)
9. Ngo, C.W., Ma, Y.F., Zhang, H.J.: Video Summarization and Scene Detection by Graph Modeling. IEEE Transactions On Circuits And Systems For Video Technology 15, 296–305 (2005)
10. Lu, S., King, I., Lyu, M.R.: Novel Video Summarization Framework for Document Preparation and Archival Applications. IEEE Conference, pp. 1–10 (March 5-12, 2005)

11. Peng, Y., Ngo, C.W.: Clip-Based Similarity Measure for Query-Dependent Clip Retrieval and Video Summarization. IEEE Transactions on Circuits and Systems for Video Technology 16, 612–627 (2006)
12. Gong, Y.: Summarizing Audiovisual Contents of a Video Program. EURASIP Journal on Applied Signal Processing, pp. 160–169 (2003)
13. Radhakrishnan, R., Divakaran, A., Xiong, Z., Otsuka, I.: A Content-Adaptive Analysis and Representation Framework for Audio Event Discovery from Unscripted Multimedia. EURASIP Journal on Applied Signal Processing (2006), Article ID 89013, 24 pages, doi:10.1155/ASP/2006/89013 (2006)

Deducing Local Influence Neighbourhoods with Application to Edge-Preserving Image Denoising

Ashish Raj, Karl Young, and Kailash Thakur

Radiology, UCSF, San Fransisco, USA
Industrial Research Limited, Wellington, New Zealand
{ashish.raj,karl.young}@ucsf.edu
k.thakur@irl.cri.nz

Abstract. Traditional image models enforce global smoothness, and more recently Markovian Field priors. Unfortunately global models are inadequate to represent the spatially varying nature of most images, which are much better modeled as piecewise smooth. This paper advocates the concept of local influence neighbourhoods (LINs). The influence neighbourhood of a pixel is defined as the set of neighbouring pixels which have a causal influence on it. LINs can therefore be used as a part of the prior model for Bayesian denoising, deblurring and restoration. Using LINs in prior models can be superior to pixel-based statistical models since they provide higher order information about the local image statistics. LINs are also useful as a tool for higher level tasks like image segmentation. We propose a fast graph cut based algorithm for obtaining optimal influence neighbourhoods, and show how to use them for local filtering operations. Then we present a new expectation-maximization algorithm to perform locally optimal Bayesian denoising. Our results compare favourably with existing denoising methods.

Keywords: Influence neighbourhoods, graph cuts, denoising, markov fields, Bayesian estimation.

1 Introduction

Image models have traditionally been stationary, whether deterministic e.g. global smoothness [1], polynomial [2] or spline models [3]. Unfortunately stationarity fails to capture the spatially varying nature of most images, which are much better modeled as piecewise smooth. Spatially varying stochastic methods like adaptive filters [4],[5] and Markov Random Field (MRF) priors [6],[7] are computationally challenging. We propose an image model which allows local processing for speed and adaptivity, but contains higher order statistical information over an extended local neighbourhood.

We introduce the concept of local influence neighbourhoods (LINs) as a bridge between local and global priors. The influence neighbourhood of a pixel is defined as the set of its neighbours having a causal influence on it. Knowledge of LINs allows us to restrict the domain used for estimation of a pixel quantity to

F. Escolano and M. Vento (Eds.): GbRPR 2007, LNCS 4538, pp. 180–190, 2007.

within its appropriate local neighbourhood, without crossing object or texture boundaries. LIN-based priors can provide higher order information about local image statistics than pixel intensity statistics. We focus on image denoising as a canonical application of the proposed approach. Interestingly, we show that full Bayesian denoising under a general non-stationary edge-preserving prior can be well-approximated by an expectation-maximization (EM) type algorithm involving only local filtering using LINs, under mild and reasonable assumptions.

Intuitively, the neighborhood of the point $p = (x_0, y_0)$ is the set of pixels close to p in both space as well as in intensity. A first (but as we shall show, unsatisfactory) attempt at defining a neighborhood around p can be achieved by the box

$$B_{\delta,\epsilon}(p) = \{q \in \mathcal{P} \mid \|p - q\| \le \delta, |I(p) - I(q)| \le \epsilon\}, \tag{1}$$

where $I(p)$ is the intensity of the image at p and parameters δ and ϵ are based on the local image statistics. Unfortunately the connectedness of $B_{\delta,\epsilon}(p)$ is not guaranteed since (1) operates on each pixel independently. It was proposed to use only the central connected component of $B_{\delta,\epsilon}(p)$ [8],[9]; however, connected components are highly susceptible to noise, intensity gradients and fine local features, and can still leave holes. A classic instance where these box methods are liable to fail is near long, thin structures, whose elongated neighbourhoods are hard to reproduce. The idea of influence neighbourhoods is not new [8],[10],[11]. Ad hoc efforts like adaptive morphological structuring elements [8], SUSAN [12] and pixons [11],[13] are all location-intensity "boxes" akin to $B_{\delta,\epsilon}(x_0, y_0)$ above. Pixons are circularly symmetric Gaussian kernels with support on a local disk of variable radius, amounting to boxes of adaptive diameter. A Bayesian belief network method [10] is conceptually similar to ours, but is cumbersome, slow and can only select a neighbourhood from six preset choices. Arbitrarily shaped neighbourhoods are allowed in [9], a variable window method based on maximal connected sets but no geometric constraints are used, and the problem of holes remains. The LIN idea is also related to the work on oversegmentation of images while preserving edges, under the so-called "super-pixel" method [14].

We introduce a MRF prior model for estimating LINs. Our approach does not consider pixels in isolation as above, but considers the *whole* configuration of the neighborhood. Figure 1 shows the neighbourhoods that can be expected from various methods. Noise causes isolated mis-classified pixels in box approaches. In contrast the shape and size of our LINs depicted in (e) are completely adaptive, with no holes or isolated pixels.

Image denoising is a well-studied problem [15],[16],[17],[18], and a comprehensive survey is not attempted here. Our main contribution is to show that an EM algorithm involving local filtering using LINs solves a challenging Bayesian denoising problem.

2 MRF Approach for Local Influence Neighbourhoods

We give now the algorithm to obtain, for each pixel p in the image, the LIN set $\mathcal{B}_p$. Define a sliding rectangular window $\mathcal{W}_p$ of reasonable size around p

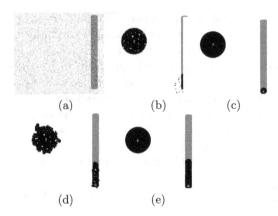

Fig. 1. Possible neighbourhoods obtained by various methods around p and q. (a) Synthetic noisy image, (b) Neighbourhood using location-intensity box [8],[12] and Eq. (1). (c) Adaptive scale [19], [11]. (d) Connected sets [9]. (e) Expected LIN from our approach.

so that $\mathcal{B}_p \subseteq \mathcal{W}_p$. For each $p \in \mathcal{P}$ we set up a small Markov Random Field $\mathbf{F}^p = \{F_q^p | q \in \mathcal{W}_p\}$. Each random variable F_q^p takes on a value f_q^p in the *binary* label set $\mathcal{L} = \{0, 1\}$. Then we define the labeling

$$\mathcal{B}_p = \{q \mid f_q^p = 1, \ q \in \mathcal{W}_p\}. \tag{2}$$

We also define the set of LINs $\mathbf{f} = \{\mathbf{f}^p | p \in \mathcal{P}\}$ which is a realization of the random set $\mathbf{F} = \{\mathbf{F}^p | p \in \mathcal{P}\}$. Define a neighborhood system $\mathcal{N} = \{(p, q) | p, q \in \mathcal{P}, \|p - q\| \le d_0\}$ where d_0 is 1.0 for 4-connected neighbourhoods and 1.5 for 8-connected neighbourhoods. We will abbreviate the joint event $\{\mathbf{F}^p = \mathbf{f}^p | p \in \mathcal{P}\}$ as $\mathbf{F} = \mathbf{f}$. We write $Pr(\mathbf{F} = \mathbf{f})$ as $Pr(\mathbf{f})$, $Pr(\mathbf{F}^p = \mathbf{f}^p)$ as $Pr(\mathbf{f}^p)$ etc. We denote the observed, noisy image as the set $\mathbf{O} = \{O_p, p \in \mathcal{P}\}$, and true image as $\mathbf{I} = \{I_p, p \in \mathcal{P}\}$, with

$$O_p = I_p + n_p,$$

$\{n_p, p \in \mathcal{P}\}$ being i.i.d. Gaussian random variables of variance σ_n^2. Note we model not intensity but local influence neighborhoods as a MRF.

2.1   Formulating a MRF Prior for Local Influence Neighborhoods

To characterize $Pr(\mathbf{f}|\mathbf{I})$ we propose the Generalized Potts potential well:

$$V_{p,q}(f_q^p, f_s^p) = u(q, s) \cdot (1 - \delta(f_q^p - f_s^p)), \tag{3}$$

$$u(q, s) = K - |I_q - I_s| \tag{4}$$

Note $u(q, s)$ specifies the depth of the potential well. The prior probability of the MRF is given by

$$Pr(\mathbf{f}^p|I) = \propto \exp\left(-\lambda \sum_{q,s \in \mathcal{W}_p, \ (q,s) \in \mathcal{N}} u(q, s) \cdot (1 - \delta(f_q^p - f_s^p))\right), \tag{5}$$

Above terms are *separation costs* of assigning binary labels to adjacent pixels. This cost is zero if the adjacent labels are identical, and non-zero if they are different. However the formulation above is incomplete because although it encodes spatial coherence, it does not reflect our expectation that pixels in the same neighbourhood should be close in location as well as intensity. Indeed, (5) is maximized by the trivial solution $\mathcal{B}_p = \{p\}$. One way to incorporate expectations regarding geometric and intensity closeness is via the prior:

$$\bar{P}r(F_q^p = 1) \propto \psi_q^p = \exp(-\frac{\rho_{p,q}^2}{\sigma_\rho^2}) \cdot \exp(-\frac{d^2(p,q)}{\sigma_d^2}), \tag{6}$$

where $\rho_{p,q} = \sqrt{|x_p - x_q|^2 + |y_p - y_q|^2}$ is the Cartesian distance between pixels p and q, $d_{p,q} = |I_p - I_q|$ is the intensity distance, and parameters σ_ρ and σ_d select the appropriate geometric and intensity scale. We note that the above "probability" distribution is in fact a heuristic - a useful model; unless this model is exhaustively verified and valiadated for a vast set of imaging data, Eq (6) will remain a design choice rather than an actual prior. It is possible that other choices of prior might work better, and this area needs some investigation. Maximizing (6) will lead to a "box" similar to (1). Therefore the criterion

$$\hat{f}_q^p = \arg\max_{f_q^p}(f_q^p\psi_q^p + (1 - f_q^p)(1 - \psi_q^p)) \tag{7}$$

clearly constitutes a first-order improvement over box methods [8], [12], [11], [13], [9], [20], [19]. Hence we wish to combine (7) with the MRF model (5):

$$Pr(\mathbf{f}^p|I) = \exp\left(-\lambda \sum_{q,s\in\mathcal{W}_p,\ (q,s)\in\mathcal{N}} u(q,s) \cdot (1 - \delta(f_q^p - f_s^p))\right)$$
$$\cdot \prod_{q\in\mathcal{W}_p} (f_q^p\psi_q^p + (1 - f_q^p)(1 - \psi_q^p)), \tag{8}$$

where $p \in \mathcal{P}$ is a pixel, $\mathcal{W}_p$ its associated window. The new term assigns an independent probability of including any pixel in $\mathcal{W}_p$ within the LIN $\mathcal{B}_p$, while the old terms continue to penalize separation between adjacent pixels. The optimal LIN around pixel p is the field realization which maximizes prior (8), or equivalently, minimizes the energy

$$E_p(\mathbf{f}^p, \mathbf{I}) = \sum_{q,s\in\mathcal{W}_p,\ (q,s)\in\mathcal{N}} \lambda\left(K - |I_q - I_s|\right)(1 - \delta(f_q^p - f_s^p))$$
$$+ \sum_{q\in\mathcal{W}_p} \left(f_q^p \log\psi_q^p + (1 - f_q^p)\log(1 - \psi_q^p)\right). \tag{9}$$

Minimization of $E_p(\mathbf{f}^p, \mathbf{I})$ is over the *binary random field* $\mathbf{F}^p$, given $\mathbf{I}$:

$$\hat{\mathbf{f}}^p|\mathbf{I} = \arg\min_{\mathbf{f}^p}\{E_p(\mathbf{f}^p, \mathbf{I})\}. \tag{10}$$

Implementation of (10) makes it virtually impossible to obtain disjoint object neighborhoods. We now describe a graph cut algorithm to minimize (10).

2.2   A Graph Cut Algorithm to Compute LINs

Equation (10) is a 2-way *mincut* problem which can be efficiently solved in polynomial time by the *maxflow* algorithm; for details see [21]. Let the graph $\mathcal{G} = \langle \mathcal{V}, \mathcal{E} \rangle$ have nodes $\mathcal{V} = \{\mathcal{P}, \alpha, \beta\}$, and the two extra nodes α and β are the source and sink, respectively. Nodes represent pixels and edges represent their neighborhood relationships. Edges $\mathcal{E} = \{\mathcal{T}, \mathcal{N}\}$ comprise two sets - $\mathcal{T}$, the set of so-called t-links, and $\mathcal{N}$ is the set of n-links. Figure 2 depicts the graph construction, the mincut, and the LIN $\mathcal{B}_p$. In the figure, t-link $t_p^\alpha \in \mathcal{T}$ is an edge going from α to the pixel node p in P, t-link $t_p^\beta \in \mathcal{T}$ is an edge going from β to p, and n-link $n_{\{p,q\}} \in \mathcal{N}$ is an edge going from the pixel node p to pixel node q, both in P.

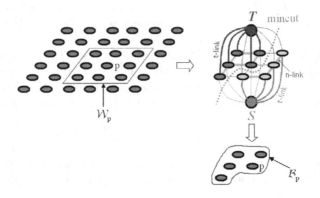

Fig. 2. Constructing the graph over a window $\mathcal{W}_p$ centered at pixel p. Mincut on this graph leads to a binary classification of every pixel in $\mathcal{W}_p$, giving the influence neighbourhood of p.

We wish to perform optimal LIN estimation for each pixel p by minimizing the energy functional $E_p(\mathbf{f}^p, \mathbf{I})$, given $\mathbf{I}$. So we form a graph on $\mathcal{W}_p$, the window around p. Clique potentials in (3) involve only adjacent neighbors, so the set of n-links is simply $\mathcal{N}$.

The graph cut algorithm is described below: For each pixel p in the image

1. Form $\mathcal{G} = \langle \mathcal{V}, \mathcal{E} \rangle$, $\mathcal{V} = \{\mathcal{W}_p, \alpha, \beta\}$, $\mathcal{E} = \{\mathcal{T}, \mathcal{N}\}$; $\mathcal{T} = \{(q, \alpha), (q, \beta) | q \in \mathcal{W}_p\}$.
2. Find Mincut $\mathcal{C}$ on $\mathcal{G}$ using Table 1. Define mapping $\mathbf{f}^p = \{f_q^p, q \in \mathcal{W}_p\}$ as:

$$f_q^p = \begin{cases} 0 & \text{if } t_q^\alpha \in \mathcal{C} \\ 1 & \text{if } t_q^\beta \in \mathcal{C} \end{cases}$$

The mincut on $\mathcal{G}$ minimizes (10), as indicated by Theorem 1, whose proof is suppressed to save space, but follows along the lines of previous results [21]. The edge assignments used here are described in Table 1.

Theorem 1. *Equivalence between Mincut and minimization of $E_p(\mathbf{f}^p, \mathbf{I})$.*

The mapping $\mathbf{f}^p$ produced by the mincut on $\mathcal{W}_p$ with edge capacities as described in Table 1 minimizes $E_p(\mathbf{f}^p, \mathbf{I})$.

Table 1. Edge weight assignment

Edge	Weight	for
t_q^α	$\log \psi_q^p$	$q \in \mathcal{W}_p$
t_q^β	$\log (1 - \psi_q^p)$	$q \in \mathcal{W}_p$
$n_{\{q,s\}}$	$|I_q - I_s|$	$q, s \in \mathcal{W}_p$, $(q,s) \in \mathcal{N}$

3 Expectation-Maximization Algorithm For MAP Image Denoising Using Local Influence Neighbourhoods

Recall the classic stationary image model enforcing global smoothness is given by $Pr(\mathbf{I}) = \exp\left(-\frac{||\mathcal{D}(\mathbf{I})||^2}{2\sigma_d^2}\right)$, where $\mathcal{D}$ is a high-pass operator which may generally be written in terms of a low-pass operator $\mathcal{D} = I - \mathcal{H}$. We modify this by restricting the operation of the low-pass operator $\mathcal{H}$ to follow the domain implied by the LINs, and call it $(\mathcal{H}_\mathbf{f}(\mathbf{I}))$. Hence we propose

$$Pr(\mathbf{I}|\mathbf{f}) = \exp\left(-\frac{||\mathbf{I} - \mathcal{H}_\mathbf{f}(\mathbf{I})||^2}{2\sigma_d^2}\right). \qquad (11)$$

In the presence of i.i.d. Gaussian noise, $Pr(\mathbf{O}|\mathbf{I}) = \exp\left(-\frac{||\mathbf{O}-\mathbf{I}||^2}{2\sigma_n^2}\right)$.

Now $Pr(\mathbf{O}|\mathbf{I}, \mathbf{f}) = Pr(\mathbf{O}|\mathbf{I})$, so joint maximum a posteriori (MAP) estimate of $\mathbf{I}$ and $\mathbf{f}$ maximize

$$Pr(\mathbf{I}, \mathbf{f}|\mathbf{O}) \propto Pr(\mathbf{O}|\mathbf{I}, \mathbf{f}) \cdot Pr(\mathbf{I}, \mathbf{f})$$

Since joint maximization over both $\mathbf{I}$ and $\mathbf{f}$ is challenging, we propose an expectation-maximization (EM) type approach, which involves a two-step optimization. Start with $\hat{\mathbf{I}}^{(0)} = \mathbf{O}$, the noisy image. Then iterate:

1. $\hat{\mathbf{f}}^{(k)} = \arg\max_\mathbf{f} Pr(\mathbf{f}|\hat{\mathbf{I}}^{(k-1)})$
2. $\hat{\mathbf{I}}^{(k)} = \arg\max_\mathbf{I} Pr(\mathbf{I}|\mathbf{O}, \hat{\mathbf{f}}^{(k)}) = \arg\max_\mathbf{I} Pr(\mathbf{O}|\mathbf{I}) \cdot Pr(\mathbf{I}|\hat{\mathbf{f}}^{(k)})$

until $k > k_{max}$ or $||\mathbf{I}^{(k)} - \mathbf{I}^{(k-1)}|| < \epsilon$.

The first step is identical to the method described in §2.1, Eq. (10), which is solved using graph cuts. The second step is easily shown, from (11):

$$\hat{\mathbf{I}}^{(k)} = \arg\min_\mathbf{I} \frac{||\mathbf{O} - \mathbf{I}||^2}{\sigma_n^2} + \frac{||\mathbf{I} - \mathcal{H}_{\hat{\mathbf{f}}^{(k)}}(\mathbf{I})||^2}{\sigma_d^2}.$$

To simplify, we introduce the approximate iteration

$$\hat{\mathbf{I}}^{(k)} = \arg\min_\mathbf{I} \frac{||\mathbf{O} - \mathbf{I}||^2}{\sigma_n^2} + \frac{||\mathbf{I} - \mathcal{H}_{\hat{\mathbf{f}}^{(k)}}(\hat{\mathbf{I}}^{(k-1)})||^2}{\sigma_d^2}.$$

This approximation is good since $\mathcal{H}_{\hat{\mathbf{f}}^{(k)}}$ is a low-pass operator. A closed form solution is obtained by differentiating and equating to zero:

$$\hat{\mathbf{I}}^{(k)} = \frac{\sigma_d^2 \mathbf{O} + \sigma_n^2 \mathcal{H}_{\hat{\mathbf{f}}^{(k)}}(\hat{\mathbf{I}}^{(k-1)})}{\sigma_d^2 + \sigma_n^2}. \qquad (12)$$

This is simply a local LIN-restricted kernel filtering operation.

4   Results

First we show some typical examples of LINs obtained from the proposed technique, in Figure 4. Next we show some typical local influence neighbourhoods obtained on real images using our approach. Figure 3 shows the first six most frequently occurring LINs counted over a large number of test images, including all the images displayed in this paper. These LINs were obtained using parameter choices described below. While parameter choice can influence the relative frequency order of these LINs for a single image, counted over a number of images, the LINs displayed in the figure can be reasonably taken to reflect their true frequency in commercially obtained images. The overwhelming majority of LINs in any image is the circular neighbourhood shown in (a). This is reasonable, since most pixels do not occur near object boundaries, and will get assigned a circular LIN by our algorithm. We note that none of the neighbourhood choices used in [10] figure in the most frequent list. Interestingly, the frequency rank of

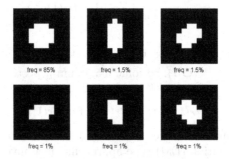

Fig. 3. The 6 most frequently occurring local influence neighbourhoods. The frequency of occurrence is indicated below each figure.

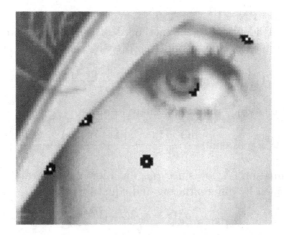

Fig. 4. Some example LINs shown for the "Lena" image

Table 2. Mean Square Error (MSE) and Peak Signal-to-Noise Ratio (PSNR, dB, in brackets) of denoising algorithms

Image	Noisy image	Gaussian	Adaptive-scale	LIN filtering
Lena	35.4	53.0	27.4	23.4
	(32.3)	(30.5)	(33.4)	(34.1)
Lighthouse	72.2	218.7	64.0	63.2
	(29.5)	(24.7)	(30.1)	(30.1)
Mandill	348.1	397.6	264.9	251.8
	(21.9)	(21.3)	(23.0)	(23.3)
Peppers	338.7	66.4	65.9	50.7
	(21.9)	(28.9)	(29.0)	(30.1)
Bike	148.3	311.7	209.6	124.3
	(26.4)	(23.2)	(24.9)	(27.2)

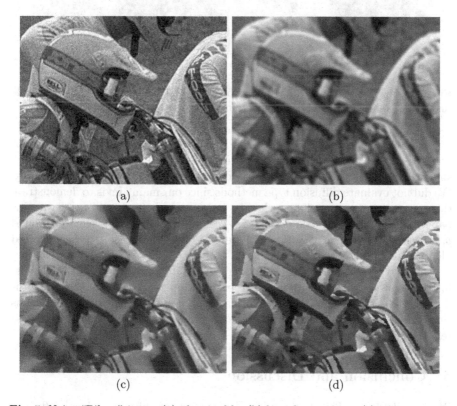

(a)   (b)

(c)   (d)

Fig. 5. Noisy "Bikers" image (a), denoised by (b) kernel smoothing, (c) adaptive-scale smoothing, and (d) LIN-restricted smoothing

the standard 3 x 3 neighbourhood used in many conventional filtering operations was nearly 50, which suggest that using this neighbourhood is inappropriate in almost all cases!

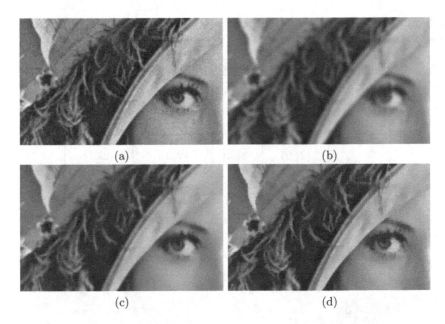

Fig. 6. Noisy "Lena" image (a), denoised by (b) kernel smoothing, (c) adaptive-scale smoothing, and (d) LIN-restricted smoothing

Mean square error (MSE) and peak signal to noise ratio (PSNR) are summarized in Table 2. For comparison we implemented simple Gaussian kernel smoothing and the method of [19], a good example of adaptive kernel smoothing. We did not evaluate diffusion type methods since our main aim is to demonstrate the power of influence neighbourhoods. Parameter choice for all results in this paper are as follows: $\sigma_\rho = 2.5$, $\lambda = 0.5$, $\mathcal{W}_p$ is the 11×11 sliding window around p. Three EM iterations ($k_{max} = 3$) were used in each case. σ_d was selected identical to the method used in [19].

Zoomed regions of denoising results are shown in figures 5, and 6. LIN denoising compares favourably with both kernel smoothing and scale-adaptive smoothing. The adaptive scale method performs poorly in noisy high-frequency regions since the scale chosen by the technique becomes too small for adequate noise removal.

5   Conclusion and Discussion

We described an efficient graph-based algorithm for obtaining local influence neighbourhoods. LINs were shown to provide additional local context information to create an image model which is superior to global smoothness assumptions. We demonstrated the use of LINs for local filtering operations, obtaining LIN-restricted morphological filters as well as kernel filters. We then incorporated LINs within a Bayesian denoising method, and proposed a new expectation-maximization type denoising algorithm. We argued and provided some experimental evidence that using LINs in prior models can be superior to pixel-based statistical models since they provide higher order information about local image statistics. Our results compare favourably with existing denoising methods like scale-adaptive kernel smoothing.

Several parameter and design choices of our algorithm have not been adequately justified experimentally. For instance, we have nto investiagted teh boundary and computational effects of changing the size of $\mathcal{W}_p$, the window size. We have not discussed a propoer determination of noise and signal variances, nor the effect of different smoothing parameters λ. These issues are currently being investigated experimentally.

Apart from the denoising, there are many interesting uses for optimal LINs. LINs can be used as structuring elements for morphological operations, and to improve segmentation and edge detection. A simple method for segmenting images using LINs would be to iteratively merge neighbouring LINs using graph clustering techniques [22]. Similarity metrics between neighbouring LINs can be used as MRF priors. LINs can also be used to improve MRF formulation of images, by restricting cliques so that no pixel pairs occur in violation of their respective LINs. LINs can act as feature vectors for matching, target detection and registration applications. Many workers have used local window or patch-based feature vectors for these tasks, for example using wavelet coefficients of a local window as feature vectors for registration in [23]. We suggest the use of LINs in a similar fashion. LIN-based feature vectors can incorporate higher-order local contour information than is possible by purely intensity based features.

References

1. Keys, R.G.: Cubic convolution interpolation for digital image processing. IEEE Transactions on Acoustics, Speech and Signal Processing 29(6), 1153–1160 (1981)
2. Meijering, E.: A chronology of interpolation. Proceedings of the IEEE 90(3), 319–342 (2002)
3. Lee, S., Paik, J.: Image interpolation using adaptive fast b-spline filtering. IEEE International Conference on Acoustics, Speech and Signal Processing, pp. 177–180 (1993)
4. Erler, K., Jernigan, E.: Adaptive recursive image filtering. In: Proceedings, SPIE, pp. 3017–3021 (1991)
5. Rand, K., Unbehauen, R.: An adaptive recursive 2-d filter for removal of gaussian noise in images. IEEE Transactions on Image Processing 1, 431–436 (1992)
6. Li, S.: Markov Random Field Modeling in Computer Vision. Springer-Verlag, Heidelberg (1995)
7. Schultz, R., Stevenson, R.: A bayesian approach to image expansion for improved definition. IEEE Transactions on Image Processing 3(3), 233–242 (1994)
8. Debayle, J., Pinoli, J.C.: Multiscale image filtering and segmentation by means of adaptive neighbourhood mathematical morphology. IEEE International Conference on Image Processing 3, 537–540 (2005)
9. Boykov, Y., Veksler, O., Zabih, R.: A variable window approach to early vision. IEEE Transactions on Pattern Analysis and Machine Intelligence, vol. 20(12), pp. 1283–1294, An earlier version of this work appeared in CVPR '97 (1998)

10. Smits, P.C., Dellepiane, S.G.: Synthetic aperture radar image segmentation by a detail preserving markov random field approach. IEEE Transactions on Geoscience and Remote Sensing 35(4), 844–857 (1997)
11. Puetter, R.: Pixon-based multiresolution image reconstruction and the quantification of picture information content. International Journal of Image Systems and Technologies 6, 314–331 (1995)
12. Smith, S.M., Brady, J.M.: Susan a new approach to low level image processing. International Journal of Computer Vision 23(1), 45–78 (1997)
13. Descombes, X., Kruggel, F.: A markov pixon information approach for low-level image description. IEEE Transactions on Pattern Analysis and Machine Intelligence 21(6), 482–494 (1999)
14. Ren, X., Malik, J.: Learning a classification model for segmentation. In: International Conference on Computer Vision (2003)
15. Ahuja, N., Davis, L.S., Milgram, D.L., Rosenfeld, A.: Piecewise approximation of pictures using maximal neighbourhoods. IEEE Transactions on Computation 27, 375–379 (1978)
16. Lee, J.S.: Digital image enhancement and noise filtering by use of local statistics. IEEE Transactions on Pattern Analysis and Machine Intelligence 2, 165–168 (1980)
17. Perona, P., Malik, J.: Scale space and edge detection using anisotropic diffusion. IEEE Transactions on Pattern Analysis and Machine Intelligence 12(7), 629–639 (1990)
18. Rudin, L.I., Osher, S., Fatemi, E.: Nonlinear total variation based noise removal algorithms. Physica D 60, 288–305 (1993)
19. Saha, P.K., Udupa, J.K.: Scale-based diffusive image filtering preserving boundary sharpness and fine structures. IEEE Transactions on Medical Imaging 20(11), 1140–1155 (2001)
20. Paranjape, R.B., Rangayyan, R.M.: Adaptive neighbourhood mean and median image filtering. Electronic Imaging 3(4), 360–367 (1994)
21. Boykov, Y., Veksler, O., Zabih, R.: Fast approximate energy minimization via graph cuts. IEEE Transactions on Pattern Analysis and Machine Intelligence 23(11), 1222–1239 (2001)
22. Kostas, H., Serafim, N.: Hybrid image segmentation using watersheds and fast region merging. IEEE Transactions on Image Processing 7(12), 1684–1699 (1998)
23. Zhong, X., Dinggang, S., Davatzikos, C.: Determining correspondence in 3-d mr brain images using attribute vectors as morphological signatures of voxels. IEEE Transactions on Medical Imaging 23(10), 1276–1291 (2003)

Graph Spectral Image Smoothing

Fan Zhang and Edwin R. Hancock

Department of Computer Science, University of York, York, YO10 5DD, UK
{zfan,erh}@cs.york.ac.uk

Abstract. A new method for smoothing both gray-scale and color images is presented that relies on the heat diffusion equation on a graph. We represent the image pixel lattice using a weighted undirected graph. The edge weights of the graph are determined by the Gaussian weighted distances between local neighbouring windows. We then compute the associated Laplacian matrix (the degree matrix minus the adjacency matrix). Anisotropic diffusion across this weighted graph-structure with time is captured by the heat equation, and the solution, i.e. the heat kernel, is found by exponentiating the Laplacian eigen-system with time. Image smoothing is accomplished by convolving the heat kernel with the image, and its numerical implementation is realized by using the Krylov subspace technique. The method has the effect of smoothing within regions, but does not blur region boundaries. We also demonstrate the relationship between our method, standard diffusion-based PDEs, Fourier domain signal processing and spectral clustering. Experiments and comparisons on standard images illustrate the effectiveness of the method.

1 Introduction

Smoothing is one of the most fundamental and widely studied problems in low-level image processing. The main purpose of image smoothing is to reduce undesirable distortions and noise while preserving important features such as discontinuities, edges, and corners. During the last two decades, diffusion-based filters have become a powerful and well-developed tool for image smoothing and multi-scale image analysis [1]. Perona and Malik [2] was the first to formalise anisotropic diffusion scheme for scale-space description and image smoothing. The basic idea of this nonlinear smoothing method was to smooth images with a direction selective diffusion that preserves edges. Catte et al. [3] identified the ill-posedness of the P-M diffusion process and proposed a regularised modification. Weickert [1] formulated anisotropic diffusion in terms of a diffusion tensor. This nonlinear diffusion technique has been subsequentially extensively analysed and developed [1,4,5]. More recently, diffusion-based PDEs has also been developed for smoothing multi-valued images [1,6].

Most diffusion-based PDEs for image smoothing assume that the image is a continuous two dimensional function on R^2 and consider discretization for the purpose of numerical implementation. It is desirable that the implementation be fast, accurate, and numerically stable, but these requirements are sometimes

F. Escolano and M. Vento (Eds.): GbRPR 2007, LNCS 4538, pp. 191–203, 2007.
© Springer-Verlag Berlin Heidelberg 2007

difficult to achieve. Moreover, images, and especially noisy ones, may not be sufficiently smooth to give reliable derivatives. Thus, for filtering noisy images it is more natural to consider the image as a smooth function defined on a discrete sampling structure.

In this paper, we present a discrete framework for anisotropic diffusion which relies on the diffusion process on graphs. We admit the discrete nature of images from the outset, and use graphs to represent the arrangement of image pixels. Here the vertices are pixels. Each edge is assigned a real-valued weight, computed using Gaussian weighted distances between local neighboring windows. This weight corresponds to the diffusivity of the edge. Instead of using diffusion-based PDEs in a continuous domain, our method is based on the heat equation on a graph [7,8]. The advantage of formulating the problem on a graph is that it requires purely combinatorial operators and as a result no discretization is required. We therefore incur no discretization errors. We pose the problem of anisotropic diffusion in a graph-spectral setting using the heat kernel. We exploit the relationship between the graph heat-kernel and the Laplacian eigensystem to develop a new method for edge-preserving image smoothing. This is accomplished by convolving the heat kernel with the image. By varying the diffusion time we control the amount of smoothing resulting from heat diffusion. The resulting algorithm can be implemented in two ways. The exact solution of the algorithm can be efficiently computed without iterations by using the Krylov subspace projection technique [9]. The method is a type of anisotropic diffusion that can be applied to smooth both gray-scale and color images. We have also demonstrated the relationship between our method, standard anisotropic diffusion, Fourier domain signal processing and spectral clustering.

2   A Graph Spectral Approach to Image Smoothing

This section describes the algorithm in three stages. These are a) representing an image as a weighted undirected graph, b) establishing and solving the diffusion equation, and c) practical details of implementation.

2.1   Graph Representation

To commence, we represent a gray-scale or color image using a weighted undirected graph $\mathcal{G} = (V, E)$ with node (vertex) set V and edge set $E \subseteq V \times V$. The nodes V of the graph are the pixels of the image. An edge, denoted by $e_{ij} \in E$, exists if the corresponding pair of pixel sites satisfies the connectivity requirement on the pixel lattice. The weight of an edge, e_{ij}, is denoted by $w(i,j)$. The edge weights play an important role in our graph-based diffusion method, since they control the flow of heat across the graph. If the edge weight $w(i,j)$ is large, then heat can flow easily between nodes v_i and v_j. By contrast, if $w(i,j)$ is small, it is difficult for heat to flow from v_i to v_j, i.e., a weight of zero means that heat may not flow along the edge.

A most popular and simply choice for charac-
terising the relationship between different pixels
is the Gaussian weighting function [2,10]. However, for noisy images, this method is not robust
to image noise. A more reliable approach is to
represent each pixel not only using the intensity
of the pixel itself, but also using the intensities of
the neighbouring pixels. Briefly, we characterise
each pixel by a window of neighbors instead of
using a single pixel alone. Thus, we can measure the similarity between two pixels using the
windows surrounding them. This method of similarity measurement was first used for non-local
mean denoising by Buades et al. [5]. Here, we simply use square windows of fixed size n. Fig. 1 illustrates the square window representation of pixels
and the similarities between them. Let $\mathcal{N}_i$ denote

Fig. 1. Illustration of the window representation of pixels.
w(i,j) is large because their
neighbouring windows are similar. w(i,l) is much smaller.

the window of pixel i, and the intensities within window $\mathcal{N}_i$ are encoded as a
vector $\mathcal{N}_i$. Hence, we can measure the similarity between two pixels i and j using
the Gaussian weighted Euclidean distance between the windows $\mathcal{N}_i$ and $\mathcal{N}_j$, i.e.

$$d_\sigma(i,j) = \|\mathcal{N}_i - \mathcal{N}_j\|_{2,\sigma} = G_\sigma * \|\mathcal{N}_i - \mathcal{N}_j\|_2. \tag{1}$$

The Gaussian filter is used here to improve the stability of the distances to noise.
The measurement in Equation (??) is well suited for removing additive white
noise, and this type of noise alters the distance between windows in a uniform
way [5], i.e. $E(\|\mathcal{N}_i - \mathcal{N}_j\|_{2,\sigma}) = \|\underline{\mathcal{N}_i} - \underline{\mathcal{N}_j}\|_{2,\sigma} + 2\epsilon^2$, where $E(\cdot)$ is the expecta-
tion; $\underline{\mathcal{N}_i}$ is the window at i in the ground truth image; and ϵ^2 is the variance of
the noise. This shows that the Gaussian weighted Euclidean distance preserves
the order of similarity between pixels. So the most similar windows to $\underline{\mathcal{N}_i}$ in the
ground truth image are also expected to be the most similar windows to $\mathcal{N}_i$ in
the noisy one. We thus choose to compute the edge weight using

$$w(i,j) = \begin{cases} \exp(-\frac{\|\mathcal{N}_i - \mathcal{N}_j\|_{2,\sigma}^2}{\kappa^2}) & \text{if } \|X(i) - X(j)\|_2 \leq r, \\ 0 & \text{otherwise.} \end{cases} \tag{2}$$

2.2 Graph Smoothing

Since we wish to adopt a graph-spectral approach, we make use of the weighted
adjacency matrix W for the graph $\mathcal{G}$ where the elements are $W(i,j) = w(i,j)$
if $e_{ij} \in E$, and 0 otherwise. We also construct the diagonal degree matrix T
with entries $T(i,i) = \deg(i) = \sum_{j \in V} w(i,j)$. From the degree matrix and the
weighted adjacency matrix we construct the combinatorial Laplacian matrix
$L = T - W$. The spectral decomposition of the Laplacian is $L = \Phi \Lambda \Phi^T$, where
$\Lambda = \mathrm{diag}(\lambda_1, \lambda_2, ..., \lambda_{|V|})$ is the diagonal matrix with the eigenvalues ordered according to increasing magnitude $(0 = \lambda_1 < \lambda_2 \leq \lambda_3 ...)$ as diagonal elements and

$\Phi = (\phi_1|\phi_2|....|\phi_{|V|})$ is the matrix with the correspondingly ordered eigenvectors as columns. Since L is symmetric and positive semi-definite, the eigenvalues of the Laplacian are all positive.

As noted by Chung [7], we can view the graph Laplacian L as an operator $\mathcal{L}$ over the set of real-valued functions $f : V \mapsto R$ such that, for a pair of nodes, i and $j \in V$, we have $\mathcal{L}f(i) = \sum_{e_{ij} \in E}(f(i) - f(j))W(i,j)$. In matrix form the heat equation on a graph associated with the Laplacian L is [7,8]

$$\frac{\partial H_t}{\partial t} = -LH_t \tag{3}$$

where the $|v| \times |V|$ matrix H_t is the heat kernel and t is time. Recently, the heat kernel has been widely used in machine learning for dimensionality reduction, semi-supervised learning and data clustering [8,11]. The heat kernel satisfies the initial condition $H_0 = I_{|V|}$ where $I_{|V|}$ is the $|V| \times |V|$ identity matrix. The solution to the heat equation is found by exponentiating the Laplacian matrix with time t, i.e.

$$H_t = e^{-tL} = I - tL + \frac{t^2}{2!}L^2 - \frac{t^3}{3!}L^3 + \cdots . \tag{4}$$

If we substitute $L = \Phi \Lambda \Phi^T$ into the second equality of Equation (4), we have $H_t = \Phi e^{-t\Lambda}\Phi^T$. The heat kernel is a $|V| \times |V|$ symmetric matrix. For the nodes i and j of the graph $\mathcal{G}$ the resulting element is $H_t(i,j) = \sum_{i=1}^{|V|} e^{-\lambda_i t}\phi_i(i)\phi_i(j)$ When t tends to zero, then $H_t \simeq I - Lt$, i.e. the heat kernel depends on the local connectivity structure or topology of the graph. If, on the other hand, t is large, then $H_t \simeq e^{-t\lambda_2}\phi_2\phi_2^T$, where λ_2 is the smallest non-zero eigenvalue and ϕ_2 is the associated eigenvector, i.e. the Fiedler vector. Hence, the large time behavior is governed by the global structure of the graph.

In order to use the diffusion process to smooth a gray-scale image, we inject at each node an amount of heat energy equal to the intensity of the associated pixel. The heat initially injected at each node diffuses through the graph edges as time t progresses. The edge weight plays the role of thermal conductivity. According to the edge weights determined from Equation (2), if two pixels belong to the same region, then the associated edge weight is large. As a result heat can flow easily between them. On the other hand, if two pixels belong to different regions, then the associated edge weight is very small, and hence it is difficult for heat to flow from one region to another. This heat diffusion process is again governed by the differential equation in (3), however, the initial conditions are different. Now the initial heat residing at each vertex is determined by the corresponding pixel intensity. If we encode the intensities of the image as a column vector $\mathcal{I}$, then the evolution of the pixel intensities for the image follows the equation

$$\begin{cases} \frac{\partial u_t}{\partial t} = -\mathcal{L}u_t \\ u_0 = \mathcal{I}, \end{cases} \tag{5}$$

where u_t is a real-valued function, i.e. $u(x,t) : V \times R \mapsto [0, 255]$, which means the intensity of the pixel x at time t. The solution of Equation (5) is

$$u_t = e^{-tL}\mathcal{I} = H_t\mathcal{I}. \tag{6}$$

As a result the smoothed image intensity of pixel j at time t is $\boldsymbol{u}_t(j) = \sum_{i=1}^{|V|} \boldsymbol{I}(i) \times H_t(i,j)$. This is a measure of the total intensity to flow from the remaining nodes to node j during the elapsed time t. When t is small, we have $\boldsymbol{u}_t \simeq (I - (D - W)t)\boldsymbol{I}$. Since each row i of the heat kernel H_t satisfies the conditions $0 \le H_t(i,j) \le 1$ for $\forall j$ and $\sum_{j=1}^{|V|} H_t(i,j) = 1$, the total integrated intensity over the set of pixels in the image is identical at all times.

In the case of color, or general vector-valued images, we let each component of the image diffuse separately on the graph constructed from the weighting function in Equation (2) using the neighbouring windows of color values. We thus apply Equation (5) to each of the three independent components (RGB in our case) of the color images, and this forms a system of three coupled heat equations. The coupling results from the fact that the edge weight or diffusivity of the graph depends on all the image channels.

2.3 Numerical Implementation

In practice the number of image pixels is large, e.g. $256 \times 256 = 65536$ pixels, so it is not tractable to calculate smoothed images by first finding the heat kernel. To overcome this problem, here we make use of the Krylov subspace projection technique [9], which is an iterative method for sparse matrix problems. This allows us to compute the action of a matrix exponential operator on an operand vector, i.e. $e^{-tL}\boldsymbol{I}$, without having to compute explicitly the matrix exponential e^{-tL} in isolation. The underlying principal of the Krylov subspace technique is to approximate $\boldsymbol{u}_t = e^{tA}\boldsymbol{I} = e^{-tL}\boldsymbol{I}$ by an element of the Krylov subspace $\mathcal{K}_m \equiv \operatorname{span}\{\boldsymbol{I}, (tA)\boldsymbol{I}, ..., (tA)^{m-1}\boldsymbol{I}\}$ where m is typically small compared to the order of L. The approximation being used is

$$\boldsymbol{u}_t \approx \beta \mathcal{V}_m e^{t\mathcal{H}_m} \tau_1, \tag{7}$$

where τ_1 is the first column of identity matrix I_m; $\mathcal{V}_m$ and $\mathcal{H}_m$ are, respectively, the orthonormal basis of the Krylov subspace $\mathcal{K}_m$ and the upper Hessenberg matrix resulting from the well-known Arnoldi process. Thus, the initial large but sparse e^{tA} problem in Equation (6) is reduced to a much smaller but dense $e^{t\mathcal{H}_m}$ problem in Equation (7) which is computationally more desirable. For our local connected graphs, the Laplacian matrix L is symmetric, positive-definite and very sparse with few circulant nonzero elements in each row. As a result, the above Arnoldi process can be replaced by the Lanczos process which decreases the computational complexity and saves CPU time. The implementation of the Krylov subspace method for solving Equation (6) in this paper relies on the MATLAB subroutines from the Expokit package [9]. Tests on a PC with an Intel P4 2.8GHZ CPU and 1.5GB of memory show that it requires approximately $3 \sim 6$ seconds to solve Equation (6) for the Laplacian matrix of a 4-connected or 8-connected graph with 256×256 nodes.

Continuous vs. Discrete Scale. Although we have presented the exact solution of our diffusion equation in (5), there also exists a discrete approximation of the diffusion equation, as is the case with all diffusion-based PDE methods. If we discretise the time (or scale) t of Equation (5), we obtain the following discrete version of our continuous diffusion process

$$u^{k+1} = (I - \varphi L)u^k, \tag{8}$$

where $\varphi > 0$ is the time step size. The discrete version in Equation (8) converges to the continuous diffusion process in Equation (5) for small φ. It is sometimes difficult to set the step size φ and construct the graph efficiently enough to update graph Laplacian over many time steps. Hence, we choose the exact solution for continuous time in our implementation.

2.4   Algorithm Summary

To summarise, the steps of the algorithm are:

1. Use Equation (2) to generate edge weights of the graph for a gray-scale or color image and encode the image intensities (or each channel of a color image) as a long-vector $\mathcal{I}$.
2. Compute u_t from Equation (6) at time t for a gray-scale image (or each channel of a color image) using the Krylov subspace technique.
3. Unpack the resulting vector u_t to recover the smoothed image.

3   Analysis of the Algorithm

3.1   Relationship to Anisotropic Diffusion

We now turn our attention to the relationship between the continuous nonlinear PDE methods and our framework of graph-spectral smoothing. In our method, the geometry of the image is captured by the weight and connectivity structure of the graph representation. A gray-scale image $\mathcal{I}$ can be regarded as a two-dimensional manifold M embedded in R^3, i.e. $X : (x^1, x^2) \in \Omega \subset \mathrm{R}^2 \rightarrow (x^1, x^2, \mathcal{I}(x^1, x^2)) \in M \subset R^3$. The 2×2 metric tensor J of the manifold M is given by $J = \begin{pmatrix} 1 + \mathcal{I}_{x^1}^2 & \mathcal{I}_{x^1}\mathcal{I}_{x^2} \\ \mathcal{I}_{x^1}\mathcal{I}_{x^2} & 1 + \mathcal{I}_{x^2}^2 \end{pmatrix}$ where $\mathcal{I}_{x^1} = \frac{\partial \mathcal{I}}{\partial x^1}$.

We can thus regard the graph representation of the image as a discrete mesh of the manifold. Based on the idea in [11], if we set the edge weight between two nodes i and j (corresponding to two points $x_i = (x_i^1, x_i^2, \mathcal{I}(x_i^1, x_i^2))$ and $x_j = (x_j^1, x_j^2, \mathcal{I}(x_j^1, x_j^2))$ on M) as $w(i,j) = e^{-\frac{\|x_i - x_j\|^2}{4t}} = e^{-\frac{(x_i^1 - x_j^1)^2 + (x_i^2 - x_j^2)^2 + (\mathcal{I}_i - \mathcal{I}_j)^2}{4t}}$ if $\|x_i - x_j\| \leq r$, and 0 otherwise. then the graph Laplacian converges to the continuous Laplace-Beltrami operator $\triangle_M$ of the manifold M. Thus, our graph-based diffusion process in Equation (5) converges to the continuous heat equation on the manifold M, i.e. $\frac{\partial f}{\partial t} = -\triangle_M f$, where f is a function defined on M, i.e. $f(x,t) : M \times R \rightarrow R$, with initial condition $f(x,0) = \mathcal{I}(x)$. The Laplace-Beltrami operator for M is defined as [12] $\triangle_M f = -\sum_{k=1}^2 \sum_{l=1}^2 \frac{1}{\sqrt{|J|}} \partial_k(\sqrt{|J|} J^{kl} \partial_l f)$, where $|J|$ is the determinant of J; J^{ij} are the components of the inverse of the metric tensor J and $\partial_k = \frac{\partial}{\partial x^k}$. Therefore, we have the following diffusion equation

$$\frac{\partial f}{\partial t} = \frac{1}{\sqrt{1+|\nabla \mathcal{I}|^2}} \mathrm{div}(D\nabla f), \tag{9}$$

where ∇ and div are the gradient and divergence operators defined on R^2, and the diffusion tensor D is given by

$$D = \sqrt{|J|}J^{-1} = \frac{1}{\sqrt{1+|\nabla \mathcal{I}|^2}} \begin{pmatrix} 1+\mathcal{I}_{x^2}^2 & -\mathcal{I}_{x^1}\mathcal{I}_{x^2} \\ -\mathcal{I}_{x^1}\mathcal{I}_{x^2} & 1+\mathcal{I}_{x^1}^2 \end{pmatrix}. \tag{10}$$

Thus, for the above choice of weights, our algorithm has a similar formulation to continuous PDE methods in the literature [1]. However, for a different choice of graph representation, e.g. using the weighting method in Equation (2), it is difficult to formulate our graph diffusion explicitly using a continuous PDE in terms of a diffusion tensor. This is because it may not be easy to analyse the geometry of the underlining graph in this case. Although our method evolves a linear equation on a graph representation of an image, it is a highly non-linear analysis of the image in the original spatial coordinates in R^2. Thus, we recast the problem of finding a sophisticated diffusion-based nonlinear PDE for image smoothing to that of finding a faithful representation of the image using a weighted graph. It is often easier to find a graph representation that preserves image structures than to find a diffusion tensor for an equivalent continuous PDE. This can be regarded as one of the main strengths of our method.

3.2   A Signal Processing View of the Algorithm

The present algorithm can also be understood in terms of Fourier analysis, which is a natural tool for image smoothing. An image (a function defined on R^2) normally contains a mixture of different frequency components. The low frequency components are regarded as the image content, and the high frequency components as the noise content. From the signal processing viewpoint, our approach is an extension of the Fourier analysis to images (signals) defined on graphs. This is based on the observation that the classical Fourier analysis of signals defined in a continuous domain can be seen as the decomposition of the signal into a linear combination of the eigenvectors of the graph Laplacian. The eigenvalues of the Laplacian represent the frequencies of the eigenfunctions. As the frequency (eigenvalue) increases, then so the corresponding eigenvector changes more rapidly from vertex to vertex. This idea has been used for surface mesh smoothing in [13].

The image $\mathcal{I}$ defined on the graph G can be decomposed into a linear combination of the eigenvectors of the graph Laplacian L, i.e. $\mathcal{I} = \sum_{k=1}^{|V|} a_k \phi_k$. To smooth the image using Fourier analysis, the terms associated with the high frequency eigenvectors should be discarded. However, because the Laplacian L is very large even for a small image, it is too computationally expensive to calculate all the terms and the associated eigenvectors for the decomposition. An alternative is to estimate the projection of the image onto the subspace spanned by the low frequency eigenvectors, as is the case with most of the low-pass filters. We wish to pass low frequencies, but attenuate the high frequencies. According to the heat kernel picture in Fig. 2, the function e^{-tx} acts as a transfer function of the filter such that $e^{-tx} \approx 1$ for low frequencies, and $e^{-tx} \approx 0$ for high frequencies. This is illustrated in Fig. 2. As the value of t increases, then the transfer function becomes steeper. Thus, the graph heat kernel can be regarded as a low-pass filter kernel.

3.3 Relationship to Spectral Clustering

Spectral clustering has proved to be a powerful tool for image segmentation [10], and, data analysis and clustering [11]. In its simplest form it uses the second eigenvector (Fiedler eigenvector) of the graph Laplacian matrix constructed from the weighted affinity graph for the sample points to obtain a bi-partition of the samples into two groups [10]. Often, instead of considering only the second eigenvector, one uses the first k eigenvectors (for some small number k) simultaneously to obtain a multi-partition into several sets [11].

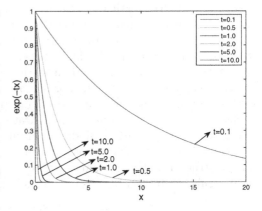

Fig. 2. Graph of the transfer function e^{-tx} with different values of t

In the present algorithm, the heat kernel reduces the effect of the large eigenvalues to zero as time t increases (see Fig. 2). As mentioned in Section 2.2, if t is large the behavior of the heat kernel is governed by the second eigenvector of the graph Laplacian, i.e. $H_t \simeq I - e^{-t\lambda_2}\phi_2\phi_2^T$. To smooth an image, the algorithm then projects the noisy image onto the space spanned by the first few eigenvectors. This process has a similar effect to spectral clustering algorithms [10,11]. Thus, our smoothing method is structure preserving due to the fact that the first few eigenvectors of the Laplacian encode the region-structure of the image.

4 Experiments

In this section, we provide experimental results of applying the graph-spectral smoothing method to a variety of image data. We also give qualitative and quantitative comparisons with several state of the art methods. In our experiments, we simply choose 8-connected graphs to represent all images and set window size 5×5 to compute edge weights from Equation (2).

Gray-scale Case: The top row of Fig. 3 shows the result of applying graph smoothing on the standard *House* image. The algorithm preserves fine structures

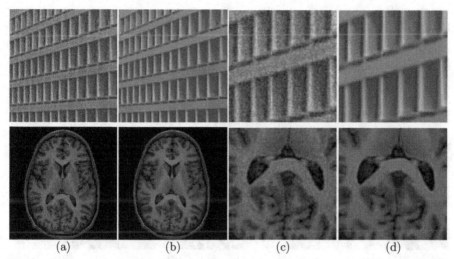

(a)   (b)   (c)   (d)

Fig. 3. Column (a): noisy images. standard *house* (top), Brain MRI (bottom). (b): smoothed results. The parameters for the top is $t = 3, \kappa = 0.15, \sigma = 0.7$, bottom image $t = 3, \kappa = 0.09, \sigma = 0$. (c) and (d): zoomed portions of the noisy and smoothed images.

while removing noise. Because the image has periodic patterns, the performance of the algorithm can hence be improved with a larger connectivity graph (large value of r in Equation (2)). The reason is as follows. For images with repeated texture, the neighbourhoods are very similar [5]. Thus, each pixel not only acquires support from its neighbouring pixels, but also from non-local pixels. The middle row of Fig. 3 gives the result of processing an MRI slice of a head with complex structures. This example demonstrates that the graph smoothing can work well on images with low intensity contrast.

We have qualitatively compared the behavior of our method with state of the art denoisers. Fig. 4 shows results of applying our graph smoothing algorithm (GRAPH), the regularised Perona-Malik (RPM) method [3], non-linear complex ramp-preserving diffusion (NCRD) [4], coherence-enhancing diffusion (CED) [1], total-variation (TV) denoising [14] and wavelet filtering (WAVELET) [15] on the standard *Lenna* image with a large amount of additive zero-mean noise. The figure shows our method gives noticeably better results than the alternative PDE-based methods for both noise elimination and feature preservation. Wavelet denoising may better restore the fine details of the image than the PDE-based filters. This is the case in the region of hair in the *Lenna* image. However, it also introduces some ring-like artifacts in the smooth regions.

In order to better analyse the performance of the graph-based smoothing method, we also compare quantitatively the performance of the proposed algorithm with the aforementioned alternative filters. We test each of these filters with the five images mentioned above, i.e. four standard images widely used in the image processing literature and one MRI image. Table 1 shows the

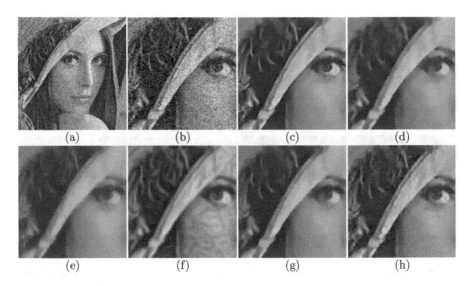

Fig. 4. (a) noisy *Lenna* (b) zoomed portion (c) our graph smoothing with $t = 5, \kappa = 0.1, \sigma = 1.0$. (d) RPM [3] (e) NCRD [4] (f) CED [1] (g) TV [14] (h) WAVELET [15].

root-mean-square (RMS) error of the filtered images. The table shows that our method gives smaller RMS error than the alternative PDE-based methods referred to above. The wavelet filtering [15] gives better results than our method on the standard *Barbara* image and the house image. However, our algorithm gives a smaller RMS error than that of wavelet filtering on the standard *Lenna* image, the *Baboon* image and the MRI image. Moreover, the wavelet filtering [15] requires significantly more CPU time than our method.

Table 1. RMS errors for graph smoothing and state-of-the-art filters

Example	Initial RMS error	Graph smoothing	RPM [3]	NCRD [4]	CED [16]	TV [14]	WAVELET [15]
Standard *Barbara*	7.3240	5.4425	6.6167	6.8175	6.1597	6.4073	5.0114
Standard *Lenna*	9.3648	5.4189	5.9209	7.0591	6.3288	6.0856	5.5158
Standard *Baboon*	9.3519	8.3403	8.7419	9.0323	8.7531	8.5902	8.4752
house	7.3230	3.8535	4.6618	7.0557	3.6956	5.1284	3.4976
MRI brain	5.6200	3.0128	4.1019	4.4023	3.9665	3.8667	3.5866

Note: The standard images of *Barbara* and *Baboon* do not appear in this paper.

We have also evaluated the performance of our method on images with different levels of noise. We take the standard 256×256 gray-scale *Lenna* image as the ground truth. To create noise corrupted images, we scale the intensities of the ground truth image to the range $0 \sim 1.0$ and corrupt it using IID additive zero-mean noise with increasing standard deviation. We then rescale the noisy intensities to the range $0 \sim 255$ to create a sequence of noisy images. Both our method and the aforementioned filters, i.e., RPM [3], NCRD [4], CED [16],

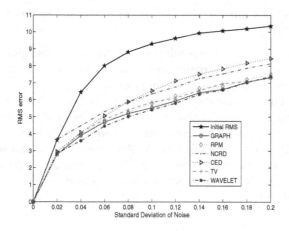

Fig. 5. RMS error comparison of graph smoothing with RPM [3], NCRD [4], CED [16], TV [14] and WAVELET [15] on standard *Lenna* image corrupted by additive zero-mean noise with increasing standard deviation

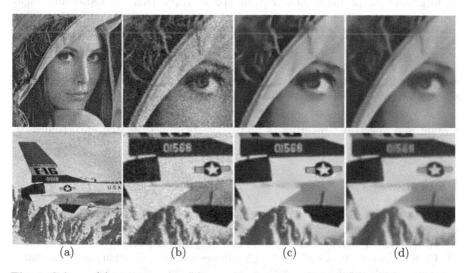

Fig. 6. Column (a): noisy images (b): smoothed results. The parameters for the top is $t = 3, \kappa = 0.15, \sigma = 0.7$, middle $t = 3, \kappa = 0.09, \sigma = 0$, and bottom image $t = 3, \kappa = 0.12, \sigma = 0.7$. (c) and (d): zoomed portions of the noisy and smoothed images.

TV [14] and WAVELET [15], were applied to the sequence of noisy images. For each noisy image, we manually choose the parameters of each method to obtain the best result. To compare these methods, we again computed the RMS error of the reconstructed images as a function of the noise standard deviation for each method. The results are plotted in Fig. 5. The plot shows that our method gives results that are comparable to those of wavelet filtering [15] and also shows that our method outperforms the alternative PDE-based approaches [3,4,16,14].

Color Case: Fig. 6 shows the results of applying our graph smoothing on color images. The results are compared with that of the curvature-preserving PDE regularisation [6]. Fig. 6 gives the results of applying the two methods on the standard *Lenna* image and the standard *Airplane* image. Both images are corrupted by a large amount of additive noise. As the figure shows, our method not only preserves the object boundaries and features, such as the eyes and hair of the face of *Lenna* and the numbers on the *Airplane*, but it is also not sensitive to noise. Although the curvature-preserving PDE [6] works fairly well and is able to eliminate severe noise, but it has a tendency to over smooth and blur the fine details of the objects.

5   Conclusion

In this paper, we have proposed a novel algorithm for gray-scale and color image smoothing. Unlike most continuous PDE-based methods for image smoothing, we formulate the problem using a graph-spectral approach. We represent images using weighted undirected graphs, and the edge weights are computed using distances between neighbouring windows of pixels. We smooth images by allowing the image intensities to diffuse across the graph structure. The solution is found by simply convolving the heat kernel with the original image. The numerical implementation of the algorithm can be efficiently accomplished using the Krylov subspace technique. We theoretically demonstrated the relationships between our graph smoothing, the standard diffusion-based PDE, signal processing and spectral clustering. Experiments demonstrate the effectiveness of the method.

References

1. Weickert, J.: Anisotropic diffusion in image processing. Teubner-Verlag (1998)
2. Perona, P., Malik, J.: Scale-space and edge dectection using anisotropic diffusion. IEEE Trans. Pattern Anal. Mach. Intell. 12(7), 629–639 (1990)
3. Catte, F., Lions, P., Morel, J., Coll, T.: Image selective smoothing and edge detection by nonlinear diffusion. SIAM J. Num. Anal. 29, 182–193 (1992)
4. Gilboa, G., Sochen, N., Zeevi, Y.: Image enhancement and denoising by complex diffusion processes. IEEE PAMI 26(8), 1020–1036 (2004)
5. Buades, A., Coll, B., Morel, J.: A review of image denoising algorithms, with a new one. Multiscale Model. Simul. 4(2), 490–530 (2005)
6. Tschumperle, D.: Fast anisotropic smoothing of multi-valued images using curvature-preserving pde's. IJCV 68(1), 65–82 (2006)
7. Chung, F.: Spectral Graph Theory. American Mathmatical Society (1997)
8. Kondor, R., Lafferty, J.: Diffusion kernels on graphs and other discrete structures. In: Proc. Int. Conf. on Machine Learning, pp. 315–322 (2002)
9. Sidje, R.: Expokit: a software package for computing matrix exponentials. ACM Trans. on Mathematical Software 24(1), 130–156 (1998)

10. Shi, J., Malik, J.: Normalized cuts and image segmentation. IEEE PAMI 22, 888–905 (2000)
11. Belkin, M., Niyogi, P.: Laplacian eigenmaps for dimensionality reduction and data representation. Neural Computation 15(6), 1373–1396 (2003)
12. Rosenberg, S.: The Laplacian on a Riemannian Manifold. Cambridge Press, Cambridge (1997)
13. Taubin, G.: A signal processing approach to fair surface design. In: Proceedings of SIGGRAPH, pp. 351–358 (1995)
14. Rudin, L., Osher, S., Fatemi, E.: Nonlinear total variation based noise removal algorithms. Physica D 60, 259–268 (1992)
15. Portilla, J., Strela, V., Wainwright, M., Simoncelli, E.: Image denoising using scale mixtures of gaussians in the wavelet domain. IEEE TIP 12, 1338–1351 (2003)
16. Weickert, J.: Coherence-enhancing diffusion filtering. International Journal of Computer Vision 31, 111–127 (1999)

Probabilistic Relaxation Labeling by Fokker-Planck Diffusion on a Graph

Hong-Fang Wang and Edwin R. Hancock

Computer Science Department, University of York,
Heslington, York, YO10 5DD, UK
{hongfang,erh}@cs.york.ac.uk

Abstract. In this paper we develop a new formulation of probabilistic relaxation labeling for the task of data classification using the theory of diffusion processes on graphs. The state space of our process as the nodes of a support graph which represent potential object-label assignments. The edge-weights of the support graph encode data-proximity and label consistency information. The state-vector of the diffusion process represents the object-label probabilities. The state vector evolves with time according to the Fokker-Planck equation. We show how the solution state vector can be estimated using the spectrum of the Laplacian matrix for the weighted support graph. Experiments on various data clustering tasks show effectiveness of our new algorithm.

1 Introduction

The problem of data clustering is a pervasive one, and numerous algorithms have been developed to solve it. Recently, spectral clustering methods have attracted widespread interest for this task. These methods use a weighted graph to represent and cluster data in an unsupervised manner using the eigenvectors of a proximity matrix. A comparison of spectral clustering methods can be found in Fischer and Poland's work [6]. Since the steady state random walk on a graph is determined by the Fiedler vector of the Laplacian matrix, spectral methods are closely akin to those that employ random walks on a graph structure. Consequently, various learning methods have exploited random walks for data classification tasks. These are semi-supervised methods that set their labeled data-points as the absorbing states of the walk, and pose the clustering problem in terms of the probability of reaching one of those states. Examples are the approaches described in [19,20,26]. These methods do not address the problem of exploiting knowledge concerning the semantic constraints that exist between different class labels, also they do not use initial confidence in the assignment of class labels. Zhu, Ghahramani and Lafferty [27] also use random walks for clustering tasks. However, they pose the problem in a continuous state space, and address only the binary classification problem. Although of limited generality, binary labelling can be used to tasks such as foreground-background separation in problem domains such as video surveillance and target tracking

F. Escolano and M. Vento (Eds.): GbRPR 2007, LNCS 4538, pp. 204–214, 2007.
© Springer-Verlag Berlin Heidelberg 2007

where interesting objects are usually manually specified at the beginning of a data sequence [1,14].

Relaxation labeling is a widely used contextual labelling method in computer vision and pattern recognition. The pioneering work was Waltz's line-labeling technique [23] which is referred to as discrete relaxation. It uses constraint relations to discard inconsistent object-label assignments. Subsequent work show how a diverse family of relaxation algorithms could be successfully applied to problems including image processing, clustering, pattern recognition [16,7,15,25]. For example, Rosenfeld, Hummel and Zucker's work [16] resulted in a continuous relaxation scheme in which the object-label assignments are represented by a probability vector. The probability vector is updated in a non-linear fashion using a support function to combine evidence for the different object-label assignments. Later work has refined the process in a number of ways. For instance, Faugeras and Berthod [5] have posed relaxation labelling as an optimisation process. Hummel and Zucker [8] have shown how relaxation labelling can be viewed as satisfying a set of variational inequalities. Hancock and Kittler [9] have posed both discrete relaxation and probabilistic relaxation in an evidence combining setting commencing from a Bayesian viewpoint.

Belief propagation [24,18] is also a local evidence combining and propagation process that is widely used in machine learning. Bayesian methods are used for inference and Markov properties are usually assumed. As a result each node in the network depends only on its immediate parent nodes, and is independent of the 'past' nodes given the current immediately connected (parent) nodes. Weiss gives an analysis of the relationship between classical relaxation methods and belief propagation in [24], and demonstrates an advantage of belief propagation over traditional relaxation in terms of the rate of convergence. However, more complex relaxation labeling methods based on product support also converge more rapidly than those based on simple arithmetic support.

In this paper, we aim to develop a new relaxation labeling method using the theory of Fokker-Planck diffusion processes on graphs and apply the method to the problem of data classification. In keeping with the spirit of relaxation labelling, we aim to use prior knowledge of the initial label confidence and the semantic constraints on the class labels. In so doing, we first construct a support graph which combines the structural information of the given object-set and label constraints. Markov diffusion processes are then defined on the graph iteratively on the basis of the graph's neighbourhood system together with its vertex and edge attributes. The object-label probabilities are then the components of the state probability vector of the Fokker-Planck diffusion process on our support graph. Experimental results on various data-sets show encouraging performances of our algorithm.

2   Relaxation Labeling

Relaxation labeling is an iterative process that propagates label probabilities globally via local interactions. Suppose that we are given a set of objects $X = \{x_i\}_{i=1}^n$, and a set of labels $\Omega_i = \{\Omega_{i,j}\}_{j=1}^m$ for each object $x_i \in X$. The task of relaxation labeling is to assign a consistent and unambiguous label $\omega \in \Omega_i$ to each object $x_i \in X$ on the basis of contextual information between the objects, and prior knowledge concerning the compatibility structure of the labeling (i.e., label consistency constraints). For simplicity, we assume that the label sets Ω_i are identical for the different objects, as is the case of most relaxation labeling applications. Denote by $\Omega = \{\omega_i\}_{i=1}^m$, where m is the size of the given label set. The available contextual information concerning label consistency is usually represented using compatibility coefficients or functions in the form of a matrix $R = \{R_{ij}(\omega_i, \omega_j)\}$, which represents the compatibility between the node x_i with a label ω_i and the node x_j with a label ω_j. The matrix R are usually given, but it may also be learned from training data [15]. It is also convenient to assume spatial homogeneity, that is, the entries $R_{ij}(\omega_i, \omega_j)$ are invariant to location [5].

Since the early work by Waltz [23], several different approaches for relaxation labeling have been developed. These approaches can be loosely classified as discrete and continuous (or probabilistic). Here we are concerned with the continuous case where each object maintains a weight (or, probability) vector for each label in the label-set. The labeling process then adjusts the weight for each object-label pair according to current local label confidences and compatibility information. The final labels are chosen as those with the largest weights. The choice of the compatibility functions and the initial label assignment values are crucial to performance. The former represents the constraints between different labels, and the latter concerns the prior confidence in the object-label assignment. The support function also plays an important role as it is responsible for combining evidence for object-label assignments using the current label probability estimates and the label compatibilities. For a successful and efficient relaxation labeling process, a neighbourhood system also needs to be defined in accordance with the structure of the given problem and object arrangement. In [8], Hummel and Zucker use the following arithmetic average support function for assigning label ω_j to object j:

$$S^{(k)}(j \leftarrow \omega_j) = \sum_{i \in \mathcal{N}_j} \sum_{\omega_i \in \Omega} R_{ij}(\omega_i, \omega_j) p^{(k)}(i \leftarrow \omega_i) \tag{1}$$

where $\mathcal{N}_j$ is the set of neighbours of node j, and $p^{(k)}(i \leftarrow \omega_i)$ is the probability that node x_i is assigned label ω_i at the kth iteration. A powerful alternative to the arithmetic average support in Eq.1 is to use the product support function (e.g., [9]) given by:

$$S^{(k)}(j \leftarrow \omega_j) = \prod_{i \in \mathcal{N}_j} \sum_{\omega_i \in \Omega} R_{ij}(\omega_i, \omega_j) p^{(k)}(i \leftarrow \omega_i). \tag{2}$$

With the support function to hand, the label probabilities are revised by using the update equation, e.g., [16]:

$$p^{(k+1)}(j \leftarrow \omega_j) = \frac{p^{(k)}(j \leftarrow \omega_j) S^{(k)}(j \leftarrow \omega_j)}{\sum_{\omega_i \in \Omega} p^{(k)}(j \leftarrow \omega_i) S^{(k)}(j \leftarrow \omega_i)} \tag{3}$$

The process is iterated until a consistent and unambiguous labeling is found.

3 Diffusion Processes on Graphs

Diffusion processes are local probability evolution processes with the property that the behaviour in the future is independent of the past given the current state probabilities. During probability evolution, local state probabilities are updated through time using a neighbourhood system defined on the state space. In this paper, we cast probabilistic relaxation labeling as a diffusion process on a support graph. The first reason for embarking on this study is that graphs are natural representations of the relational information residing in a data-set. Secondly, in common with relaxation labelling, diffusion processes propagate information globally via local evolution. Finally, recently there has been increasing interest in defining diffusion processes on graphs [10,2,3,11,13] and successful applications are reported in [21,22].

3.1 The Graphical Model

Suppose we are given a graph $G = (V, E)$ where V is the node-set, and E is the edge-set. For a pair of nodes $v_i, v_j \in V$ connected by an edge, that is, $(v_i, v_j) \in E$ (denote by '$i \sim j$'), the edge weight matrix W of G has elements $W_{ij} = f_{ij}$ if $i \sim j$, and $W_{ij} = 0$ otherwise. Here f_{ij} is a weight function on vertices v_i and v_j that reflects the strength of the connection. Denote by $D = diag(deg_1, \ldots, deg_n)$ where $deg_i = \sum_j W_{ij}$ is the degree of node i. Then the normalized adjacency matrix of graph G is given by $\mathcal{A} = D^{-1/2}WD^{-1/2}$. For undirected graphs f_{ij} is symmetric; that is, $f_{ij} = f_{ji}$ and thus W and $\mathcal{A}$ are symmetric. Another frequently used representation is the Laplacian matrix. The elements L_{ij} of the Laplacian matrix L are given by:

$$L_{ij} = \begin{cases} -W_{ij} & \text{if } i \sim j; \\ deg_i & \text{otherwise} \end{cases}$$

An alternative is to use the normalized Laplacian, which is given by [3] $\mathcal{L} = D^{-1/2}(D - W)D^{-1/2} = I - \mathcal{A}$. The eigensystems of the above matrices $\mathcal{L}$ and $\mathcal{A}$ are closely related. Suppose that the normalized weighted adjacency matrix $\mathcal{A}$ has the following eigen-decomposition:

$$\mathcal{A} = U\Lambda_\mathcal{A} U^T. \tag{4}$$

The eigen-decomposition of the normalized Laplacian then is $\mathcal{L} = U(I - \Lambda_\mathcal{A})U^T$. That is, their difference in the eigensystems only resides in the eigenvalues with $\Lambda_\mathcal{L} = I - \Lambda_\mathcal{A}$.

The element P_{ij} of the (one-step) transition matrix P is the probability of moving to a node v_j given that the current step is at node v_i. That is, it governs the discrete time random walk on the graph. For weighted graphs, the elements of P are given by $P_{ij} = W_{ij}/\sum_{k,i\sim k} W_{ik}$. The matrix P is, in general, not symmetric. As a result it is not easy to compute its eigensystem. However, P shares the same eigenvalues and has eigenvectors related to those of $\mathcal{A}$. From Eq.4, the right eigenvector of P is $v_i^l = D^{1/2}u_i$ and the left eigenvectors are

$v_j^r = D^{-1/2}u_j$. Denote the left and right eigenvector matrix of P as $V^l = \{v_i^l\}$ and $V^r = \{v_j^r\}$, respectively, we can see that $(V^l)^T \cdot V_r = (V^r)^T \cdot V^l = I$ where I is the identity matrix, and $P = V^l \Lambda_A V^r$.

3.2 Diffusion Processes and Random Walks

Given a state space Γ and a suitable probability measure, a Markov process $\{y(t); t \geq 0\}$ is uniquely defined by an initial state probability vector $\mathbf{p}_0$ and a semi-group transition function $P(t, \mathbf{x}, \Gamma), t \in [0, \infty)$ defined on the state space. The function $P(t, \mathbf{x}, \Gamma)$ evolves according to the diffusion equation:

$$dP(t, \mathbf{x}, \Gamma)/dt = -\mathcal{F}P(t, \mathbf{x}, \Gamma), \qquad \mathcal{F} = \frac{d}{d\mathbf{x}}\left\{a(\mathbf{x})\frac{d}{d\mathbf{x}} + b(\mathbf{x})\right\}, \qquad (5)$$

where $\mathcal{F}$ is the so-called Fokker-Planck operator. That is, $P(t, \mathbf{x}, \Gamma)$ is then the solution of Eq.5, and takes on the matrix exponential form $P(t, \mathbf{x}, \Gamma) = e^{-t\mathcal{F}}$. For simplicity, we write $P(t, \mathbf{x}, \Gamma)$ as $P(t)$. Given the initial state probability distribution vector $\mathbf{p}_0$, the state probability vector at time t is then computed from the formula:

$$\mathbf{p}_t = P(t) \cdot \mathbf{p}_0 = e^{-\mathcal{F}t} \cdot \mathbf{p}_0 \qquad (6)$$

In the discrete approximation, the operator $\mathcal{F}$ is explicitly represented by a matrix. The computation of the matrix exponential $\exp(-t\mathcal{F})$ involved can be naïvely computed by its definition

$$e^{-t\mathcal{F}} = \sum_{k=0}^{\infty} \frac{(-1)^k t^k \mathcal{F}^k}{k!}, \qquad (7)$$

The convergence of the series depends on the values of t and the matrix $\mathcal{F}$, and this process is usually slow. One efficient method of computation is the scaling and squaring method [12]. When $\mathcal{F}$ is real and symmetric, we can perform the eigen-decomposition $\mathcal{F} = U\Lambda U^T$, where $U = (\mathbf{u}_1, \ldots, \mathbf{u}_n)$ is the matrix formed from the right eigenvectors of $\mathcal{F}$, and Λ is the diagonal matrix containing the corresponding eigenvalues. The solution vector of the diffusion equation is thus

$$\mathbf{p}_t = Ue^{-\Lambda t}U^T \cdot \mathbf{p}_0. \qquad (8)$$

From a kernel perspective, the above development can be regarded as defining a kernel function for the graph representation, and seeking a solution from the mapped higher dimensional space. In fact, the construction of an effective kernel function from exponentiation can be found in recent literature. For example, in [10,17] the matrix representation K of a given data-set is given by $K = U\rho(\Lambda)U^T$, where $\Lambda = (\lambda_1, \ldots, \lambda_m)$ are the eigenvalues, and U is the corresponding eigenvector matrix, of the matrix K. The function $\rho(\lambda_h)$ is required to satisfy the condition that $\rho(\lambda_h) \to 0$ as $h \to \infty$. In [17], several functions are introduced from regularization theory, including the regularized Laplacian, $\rho(\lambda_h) = 1 + \sigma^2\lambda_h$, and the diffusion kernel, $\rho(\lambda_h) = \exp(\lambda_h/\sigma^2)$. Of these methods, the diffusion kernel [10] deserves special note. In [10], Kondor and Lafferty derived the diffusion kernel using kernel theory. According to this viewpoint the diffusion kernel is a symmetric and positive semi-definite function which is computed from matrix exponentiation. That is, for a symmetric weight matrix $\mathcal{H}$ whose elements are determined by a weighting function, the kernel matrix is computed as $K_\beta = e^{\beta \mathcal{H}}$. The matrix K_β obtained is thus guaranteed to be symmetric and positive semi-definite, and satisfies the diffusion equation.

4   Relaxation Labeling by Diffusion

In this section, we show how the theory of diffusion processes on graphs can be used to formulate a new method for probabilistic relaxation labelling. We recast relaxation labelling in a graph-setting using a support graph. Denote it by $G_S(V_S, E_S, \mathcal{A}_S)$. The node-set $V_S = X \times \Omega$ of the graph is the Cartesian product of the object-set X and the label-set Ω. That is, each vertex $v_{i\omega_i} = (x_i, \omega_i) \in V_s$ represents the assignment of label $\omega_i \in \Omega$ to object $x_i \in X$. We then define a label probability vector for our relaxation labeling process as $\mathbf{p}_t = [p_t(1 \leftarrow \omega_1), \ldots, p_t(1 \leftarrow \omega_m), \ldots, p_t(n \leftarrow \omega_1), \ldots, p_t(n \leftarrow \omega_m)]$. Each component $p_t(i \leftarrow \omega_i)$ of this vector represents the confidence of assigning a label $\omega_i \in \Omega$ to an object $x_i \in X$. The components of the state-vector can be interpreted as the probability of a random walker residing at the node $v_{i\omega_i}$ at time t.

To satisfy the Markov property for the diffusion process, a neighbourhood system needs to be defined. Here we use the object arrangement topology to define the neighbourhood. We determine the spatial proximity of the objects by thresholding the distance function. We denote the set of neighbours of the object x_i as $\mathcal{N}_i$. Hence, the objects can be represented by an attributed graph $G_X = (X, E_X, W_X)$ with node-set X, edge-set $E_X = \{(x_i, x_j)|x_j \in \mathcal{N}_i\}$ and edge-weight-function W_X. We will give explicit examples of how to compute the edge-weight function W_X when we discus applications of our method in Section 5.

To pose the problem of relaxation labelling as a diffusion process on the support graph we need to compute the infinitesimal generator $\mathcal{F}$ and this requires the weighted edge adjacency matrix $\mathcal{A}$ defined in Eq.9. We assign edge-weights to the support graph so as to incorporate the a priori label compatibilities R and the label probabilities $\mathbf{p}_t$. Following conventional relaxation labeling, the edge weight on the support graph is set equal to the compatible support associated with the object-label assignments for pairs of neighbouring nodes. Accordingly we let

$$\mathcal{A}(v_{i\omega_i}, v_{j\omega_j}) = \begin{cases} p(i \leftarrow \omega_i)p(j \leftarrow \omega_j)W_X(x_i, x_j)R_{ij}(\omega_i, \omega_j) & \text{if } x_j \in \mathcal{N}_i \\ 0 & \text{otherwise} \end{cases} \quad (9)$$

The second step is to define the infinitesimal generator matrix $\mathcal{F}$ for the support graph G_S from the transition matrix $P = D^{-1}\mathcal{A}$ of the support graph:

$$\mathcal{F} = I - D^{-1}\mathcal{A} \quad (10)$$

where D is the degree matrix of $\mathcal{A}$ (as defined in the previous section). To update the state-vector using Eq.6, we can either perform matrix exponentiation directly or compute the exponential from the eigensystem of $\mathcal{F}$. As the matrix P is generally not symmetric, when computing the updated label probability vectors using Eq.8, the relationship between its eigensystem and the symmetric adjacency matrix $\mathcal{A}$ can be used. Suppose that the weighted adjacency matrix $\mathcal{A}$ has the eigen-decomposition $\mathcal{A} = U \Lambda U^T$ and degree matrix D, we have:

$$
\begin{aligned}
\mathbf{p}_t &= e^{-t\mathcal{F}} \cdot \mathbf{p}_0 \\
&= (D^{-1/2}U) \cdot e^{-t\Lambda} \cdot (D^{1/2}U) \cdot \mathbf{p}_0
\end{aligned}
\tag{11}
$$

If viewed from a kernel perspective, the matrix $\mathcal{F}$ must be symmetric in order to be a valid kernel. To symmetrize a matrix P, one frequently used method is to replace P by $P^T P$. In our experiments, this is applied to the adjacency matrix of the object-set since the local average distance between objects is used to scale the weight function. That is, we choose $W_X = \tilde{W}_X^T \tilde{W}_X$.

We also require a stopping criterion to halt iteration of the labeling process. One simple approach is to halt the process after a fixed number of iterations. A more principled approach is to use the asymptotic properties of the label probabilities. To this end we make use of the entropy associated with the label probability distribution:

$$
H_t = -\sum_{i \in X} \sum_{\omega_i \in \Omega} p_t(i \leftarrow \omega_i) \ln p_t(i \leftarrow \omega_i)
\tag{12}
$$

The entropy can be regarded as the amount of disorder of the given system. It reaches its largest value if $p_t(i \leftarrow \omega_i)$ has a uniform distribution, and decreases to zero if $p_t(i \leftarrow \omega_i) \in \{0,1\}$ Thus the iteration of the relaxation process is also halted if the total entropy decreases below a threshold, or if its change between two consecutive iterations is small. The algorithm is summarized as follows:

1. Initialization: Set $\mathbf{p}^{(k-1)} = \mathbf{p}_0$, t, N, $\sigma = \{\sigma_i\}_{i=1}^N$, H;
2. Compute the weight matrix W_X for the object-set X;
3. Compute the adjacency matrix $\mathcal{A}^{(k)}$ in current iteration k using Eq.4 and $\mathbf{p}^{(k-1)}$;
4. Compute the infinitesimal generator matrix $\mathcal{F}^{(k)}$ using current $\mathcal{A}^{(k)}$;
5. For each iteration k, compute the updated label probabilities using Eq.6;
6. Compute the entropy H using Eq.12, and the variation $\Delta H = H^{(n)} - H^{(n-1)}$. If either H or ΔH is below the threshold, go to step 8 otherwise go to step 3;
7. If $k \geq N$, go to step 8; otherwise go to step 3;
8. Assign maximum probability label to each object.

5 Experiments

We first experiment our newly developed algorithm on five synthetic data-sets which are also used in [6]. These are chosen to study the performance of the algorithm with different numbers of clusters, different cluster size, and different

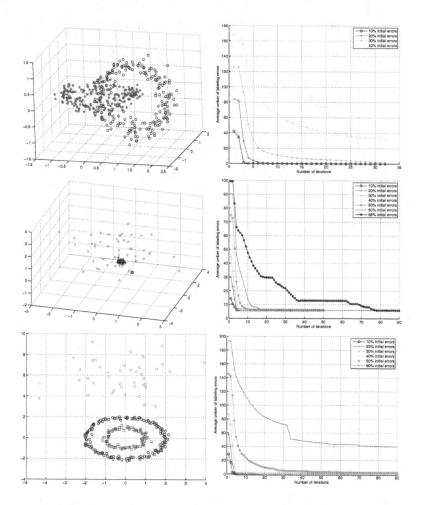

Fig. 1. Labeling results of synthetic data-set. *From top to bottom, left to right: original data-sets R2, G3, & RG and their corresponding labeling results.*

cluster distributions which are considered to be the most common variations in real world data-sets.

One advantage of relaxation labeling is its ability to reduce ambiguities in label assignments. Thus we first assign correct labels to all objects. Next we experiment with our algorithm by randomly selecting a fraction of the assigned object labels and flipping the labels randomly to take on new values. The fraction of labels to be flipped ranges from 10% to $\frac{1}{|\Omega|}$, i.e., the probability of uniform label assignment. We use a Gaussian weight function $W_X(x_i, x_j) = \exp\left(-\frac{d_{ij}}{\sigma_i}\right)$ to compute the weighted adjacency matrix $\mathcal{A}$ in Eq.9, where d_{ij} is the Euclidean distance between objects $x_i, x_j \in X$. The value of σ_i in the weight function are chosen to be a function of the average distance between nearest neighbours.

Table 1. Representations of label compatibilities

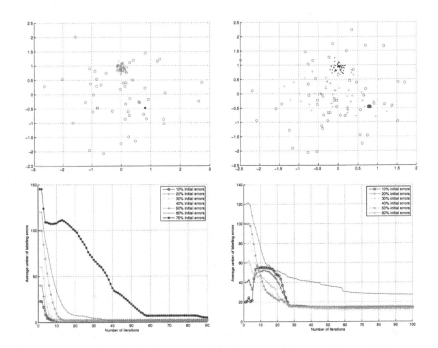

R2	G3	RG	G4_1	G4_2
$\begin{bmatrix} 1 & .3 \\ .3 & 1 \end{bmatrix}$	$\begin{bmatrix} 1 & 0 & .5 \\ 0 & 1 & .5 \\ .5 & .5 & 1 \end{bmatrix}$	$\begin{bmatrix} 1 & 0 & .5 \\ 0 & 1 & .5 \\ .5 & .5 & 1 \end{bmatrix}$	$\begin{bmatrix} 1 & 0 & 0 & .5 \\ 0 & 1 & 0 & .5 \\ 0 & 0 & 1 & .5 \\ .5 & .5 & .5 & 1 \end{bmatrix}$	$\begin{bmatrix} 1 & 0 & .1 & .5 \\ 0 & 1 & .1 & .5 \\ 0 & .1 & 1 & 0 \\ .5 & .5 & 0 & 1 \end{bmatrix}$

Fig. 2. Labeling results of synthetic data-sets. *Top row (left to right): The G4_1 & G4_2 data-set; Bottom row (left to right): Results of G4_1 & G4_2.*

The label compatibility matrices for each data-set are shown in Table 1. The average clustering error on 20 runs for each data-set with different initial label probabilities are shown in the right column of Figures 1 and 2. For each case, as the iteration number increases then so the error falls rapidly.

We next experiment with our algorithm on two real world data-sets, namely Iris and Wine [4]. They have already been studied in a variety of papers, e.g., [6]. Both data-sets contain three clusters. The dimensionality of the Iris data is four, while that of the Wine data is nine. The results are shown in Fig.3.

The algorithms converges quickly to the stationary distribution, where the majority of the labels are assigned correctly. When less than 50% of the labels

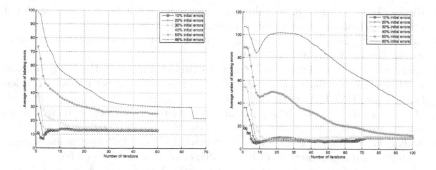

Fig. 3. Labeling results of real world data-sets. *Left: Iris; Right: Wine.*

are initialized in error, the algorithm corrects them in just a few iterations. The numbers of iterations required are significantly smaller than those for traditional relaxation labeling algorithms. The results are comparable with those obtained using alternative data-clustering algorithms.

6 Discussion

In this paper a new probabilistic relaxation labeling method is developed and is applied to the task of data classification. The diffusion process setting of the development ensures the global propagation of local labeling confidence and the reduction in the number of iterations required. The essential component is the kernel function which defines the Markov process. For the task studied in this paper, exponential kernels are used and these are the natural choice for a diffusion process. Experimental results show encouraging performance for both synthetic and real world data-sets.

References

1. Agarwal, A., Triggs, B.: Tracking articulated motion using a mixture of autoregressive models. In: ECCV, pages III, pp. 54–65 (May 2004)
2. Belkin, M., Niyogi, P.: Laplacian eigenmaps and spectral techniques for embedding and clustering. NIPS (2001)
3. Chung, F.R.K.: Spectral Graph Theory. American Mathematical Society (1997)
4. Blake, C.L., Newman, D.J., Hettich, S., Merz, C.J.: UCI repository of machine learning databases (1998)
5. Faugeras, O., Berthod, M.: Improving consistency and reducing ambiguity in stochastic labeling: An optimization approach. IEEE Trans. PAMI, vol. 3(4) (1981)
6. Fischer, I., Poland, J.: New methods for spectral clustering. Technical Report IDSIA-12-04, IDSIA (2004)
7. Hancock, E.R., Kittler, J.: Edge-labeling using dictionary based relaxation. IEEE Trans. PAMI 12, 165–181 (1990)
8. Hummel, R.A., Zucker, S.W.: On the foundations of relaxation labeling processes. IEEE Trans. PAMI, l.PAMI-5, 267 (1983)

9. Kittler, J., Hancock, E.R.: Combining evidence in probabilistic relaxation. Int. J. Pattern Recognition And Artificial Inteligence 3(1), 29–51 (1989)
10. Kondor, R.I., Lafferty, J.: Diffusion kernels on graphs and other discrete input spaces. In: ICML, pp. 315–322 (2002)
11. Lafon, S., Lee, A.B.: Diffusion Maps and Coarse-Graining: A Unified Framework for Dimensionality Reduction, Graph Partitioning and Data Set Parameterization. IEEE Trans. PAMI 28, 1393–1403 (2006)
12. Moler, C., van Loan, C.: Nineteen dubious ways to compute the exponential of a matrix, twenty-five years later. SIAM Review 45(1), 3–49 (2003)
13. Nadler, B., Lafon, S., Coifman, R.R., Kevrekidis, I.G.: Diffusion maps, spectral clustering and eigenfunctions of Fokker-Planck operators. NIPS (2005)
14. Okuma, K., Taleghani, A., de Freitas, N., Little, J., Lowe, D.: A boosted particle filter: Multitarget detection and tracking. In: ECCV, pp. 28–39 (2004)
15. Pelillo, M., Refice, M.: Learning compatibility coefficients for relaxation labeling processes. IEEE Trans. PAMI 16(9), 933–945 (1994)
16. Rosenfeld, A., Hummel, R., Zucker, S.: Scene labeling by relaxation operations. IEEE Trans. Systems. Man and Cybernetics 6, 420–433 (1976)
17. Smola, A.J., Kondor, R.: Kernels and regularization on graphs. In: Warmuth, M., Schökopf, B. (eds) COLT/KW 2003 (2003)
18. Sudderth, E.B., Ihler, A.T., Freeman, W.T., Willsky, A.S.: Nonparametric belief propagation. In: CVPR, pp. 605–612 (2003)
19. Szummer, M., Jaakkola, T.: Partially labeled classification with Markov random walks. In: NIPS 15 (2002)
20. Tishby, N., Slonim, N.: Data clustering by Markovian relaxation and the information bottleneck method. In: NIPS 13 (2000)
21. Tsuda, K., Noble, W.S.: Learning kernels from biological networks by maximizing entropy. Bioinformatics 20, 326–333 (2004)
22. Vert, J.-P., Kanehisa, M.: Graph-driven feature extraction from microarray data using diffusion kernels and kernel CCA. In: NIPS (2002)
23. Waltz, D.L.: Generating semantic descriptions from drawings of scenes with shadows. Technical Report 271, MIT AI Lab (1972)
24. Weiss, Y.: Interpreting images by propagating Baysian beliefs. In: ADVANCES IN NEURAL INFORMATION PROCESSING SYSTEMS, pp. 908–914 (1997)
25. Zheng, Y., Doermann, D.: Robust point matching for two-dimensional nonrigid shapes. In: Proceedings IEEE Conf. on Computer Vision, pp. 1561–1566 (2005)
26. Zhou, D., Schölkopf, B.: Learning from labeled and unlabeled data using random walks. In: Proceedings of the 26th DAGM Symposium, pp. 237–244 (2004)
27. Zhu, X., Ghahramani, Z., Lafferty, J.: Semi-supervised learning using Gaussian fields and harmonic functions. In: ICML, pp. 1561–1566 (2003)

Assessing the Performance of a Graph-Based Clustering Algorithm

P. Foggia[1], G. Percannella[2], C. Sansone[1], and M. Vento[2]

[1] Dipartimento di Informatica e Sistemistica
Università di Napoli Federico II, Via Claudio, 21 I-80125 Napoli (Italy)
{foggiapa,carlosan}@unina.it
[2] Dipartimento di Ingegneria dell'Informazione ed Ingegneria Elettrica,
Università di Salerno, via P.te Don Melillo, I-84084 Fisciano (SA), Italy
{pergen,mvento}@unisa.it

Abstract. Graph-based clustering algorithms are particularly suited for dealing with data that do not come from a Gaussian or a spherical distribution. They can be used for detecting clusters of any size and shape without the need of specifying the actual number of clusters; moreover, they can be profitably used in cluster detection problems.

In this paper, we propose a detailed performance evaluation of four different graph-based clustering approaches. Three of the algorithms selected for comparison have been chosen from the literature. While these algorithms do not require the setting of the number of clusters, they need, however, some parameters to be provided by the user. So, as the fourth algorithm under comparison, we propose in this paper an approach that overcomes this limitation, proving to be an effective solution in real applications where a completely unsupervised method is desirable.

1 Introduction

In Pattern Recognition and Computer Vision there is a significant number of applications that use clustering algorithms [1].

The main drawback of most clustering algorithms is that their performance can be affected by the shape and the size of the clusters to be detected [2]. Some well-known clustering algorithms (e.g. the *k-means* [2] or the *self-organizing maps* [3]), for example, fail if data are distributed in the feature space along a non-smooth manifold [4]. Such algorithms, in fact, are based on the assumption that the data are sampled from a Gaussian or a spherical distribution. Moreover, in order to obtain an adequate clustering result, these algorithms sometimes require some *a priori* knowledge about the actual number of clusters and/or require the setting of a threshold or a parameter.

On the other hand, in some applications there is the need of grouping, in one or more clusters, only a part of the whole dataset. This happens when samples of interest for the application at hand are present together with several *noisy* samples. We can refer to this case as a *cluster detection* problem. It occurs, for example, in the context of image segmentation, as described in [5]. Here, among all the edges coming from an edge detection algorithm, only the interesting ones (*true positives*) have to be grouped together by a cluster detector, in order to use them for achieving a good segmentation result.

F. Escolano and M. Vento (Eds.): GbRPR 2007, LNCS 4538, pp. 215–227, 2007.

In these situations most clustering algorithms yield a not so useful result, as in any case they try to attribute each sample to a cluster. So, noisy samples, that are typically not similar to each other, are grouped together with true positives. Even if it would be theoretically possible to group together in some clusters only noisy samples, it is practically difficult because the number of clusters has to be provided in advance. This cannot be effectively done, since the actual distribution of noisy samples cannot be easily modeled.

A particular family of clustering algorithms that can cope with these problems is the one based on graph theory. The algorithms of this family represent the problem data through an undirected graph. Each node is associated to a sample in the feature space, while to each edge is associated some distance measure between nodes connected under a suitably defined neighborhood relationship. A cluster is thus defined to be a connected sub-graph, obtained according to criteria peculiar of each specific algorithm. Algorithms based on this definition are capable of detecting clusters of various shapes and sizes, at least for the case in which they are well separated [4]. Moreover, isolated samples should form singleton clusters and then can be easily discarded as noise in case of cluster detection problems.

In order to provide some useful suggestions about the convenience of using a specific graph-based clustering approach, in this paper we present a detailed performance evaluation of four different graph-based clustering algorithms. Comparisons have been made by using a set of six validity indices [6,7] that are commonly used for evaluating the quality of a clustering algorithm.

In particular, we performed two kinds of experiments. In both cases we used synthetic data. In the first test, data to be clustered have been generated by a modified version of the model proposed in [8] that we have extended in order to include also a distance between graph nodes. The second test refers to a cluster detection problem. In this case, additional noise has been added to graphs obtained with the previously cited model, in order to simulate samples that do not belong to any cluster.

Three of the algorithms selected for comparison have been chosen from the literature. In particular, we considered the Markov Clustering proposed by van Dongen [9], the Iterative Conductance Cutting proposed by Kannan et al. [10] and the Geometric MST Clustering introduced by Gaertler in [11].

While these algorithms do not require the setting of the number of clusters, as it is usual in case of graph-based clustering algorithms, they need however some parameters to be provided by the user. The fourth algorithm under comparison is an approach developed by the authors that overcomes this limitation. This method, therefore, can be effectively used in real applications where a completely unsupervised method is desirable. Our proposal is based on the algorithm described by Zahn in [12]. The original algorithm constructs the Minimum Spanning Tree (MST) of the graph representing the samples. After that, it identifies inconsistent edges and removes them from the MST. The remained connected components are then the clusters provided by the algorithm. An edge is inconsistent if the distance associated to it is greater than a predefined threshold. The Zahn algorithm does not suggest a criterion for deriving this threshold, leaving it as a manually provided parameter. In order to determine automatically the optimal value of this threshold, in this paper we propose a method based on the use of the *Fuzzy C-Means* algorithm [13].

The organization of the paper is then as follows: in Section 2, the proposed graph-based clustering approach is described. The other algorithms under comparison are presented in Section 3. In Section 4, the database used for benchmarking is described, while, in Section 5, the selected cluster validity indices are reviewed. A comparative analysis of the results obtained by the considered algorithms is reported in Section 6 and some conclusions are finally drawn in Section 7.

2 The Fuzzy C-Means MST Clustering Algorithm (FMC)

The clustering algorithm proposed in this paper is based on graph theoretical cluster analysis. It starts by constructing the complete graph where each node is associated to a sample to be clustered. The weight of each edge accounts for the distance between the connected nodes. Then, the Minimum Spanning Tree (MST) is computed on the graph. By removing all the edges with weights greater than a threshold λ, we arrive at a *forest* containing a certain number of subtrees (clusters). In this way, the method automatically groups nodes into clusters. As demonstrated in [11], the clustering induced by the subtrees is independent of the particular MST. So the clustering results do not depend on the algorithm chosen for deriving the MST: in this paper, we used the Prim's algorithm [14].

It is worth noting that the optimal value of λ typically depends on the specific clustering problem. As a consequence, it is not possible to use a fixed value of λ for every case. Our proposal is then to determine the optimal value of λ by reformulating the problem as the one of partitioning the whole set of edges into two clusters, according to their weights. The cluster of the edges of the MST with small weights will contain edges to be preserved, while the edges belonging to the other cluster will be removed from the MST. In order to solve this problem we employ the *Fuzzy C-Means* clustering algorithm [13].

Fuzzy C-Means is a clustering technique based on the minimization of the following objective function:

$$J_m = \sum_{i=1}^{N} \sum_{j=1}^{C} u_{ij}^m \left(x_i - c_j \right)^2, \qquad 1 \leq m < \infty$$

where m is any real number greater than 1, x_i is the i-th measured data (in our case the weight of the i-th edge of the MST), c_j is the center of the cluster, u_{ij} is the degree of membership of x_i to the cluster j, C is the number of clusters (in our case $C = 2$) and N is the number of objects to be clustered. Fuzzy partitioning is carried out through an iterative optimization of the objective function shown above, with the update of membership u_{ij} and the cluster centers c_j by:

$$u_{ij} = \frac{1}{\sum_{k=1}^{C} \left(\frac{x_i - c_j}{x_i - c_k} \right)^{\frac{2}{m-1}}} \qquad \text{and} \qquad c_j = \frac{\sum_{i=1}^{n} u_{ij}^m \cdot x_i}{\sum_{i=1}^{n} u_{ij}^m}$$

This iteration will stop when:

$$\max_{ij} \left\{ \left| u_{ij}^{(k+1)} - u_{ij}^{(k)} \right| \right\} < \varepsilon$$

where ε is a termination criterion between 0 and 1, whereas k are the iteration steps. This procedure converges to a local minimum or a saddle point of J_m. At the end of the procedure, each edge x_i has been assigned to the cluster r such that:

$$r = \arg\max_j u_{ij}$$

At this point, all the edges of the MST are separated into two clusters. Then, we remove from the MST all the edges belonging to the cluster s whose center exhibits the largest value, i.e.:

$$s = \arg\max_j c_j$$

As said before, the FMC algorithm requires the user to specify the value of ε. Indeed, we verified that a good value for this parameter is substantially independent on the considered application. In particular, the value of ε can be fixed to 0.5 in every case. With this setting, the algorithm can be considered as really unsupervised.

3 Algorithms Selected for the Comparison

In this section, we will provide a brief description of the algorithms that will be used for our experimental comparison, together with the settings used for employing them.

3.1 The Markov Clustering Algorithm

The Markov Clustering algorithm (MCL) was proposed by van Dongen in his PhD thesis [9,15] in 2000. The rationale of the method is based on the observation that if a group of nodes is strongly connected inside and has few connections to the outside (which is the property defining a cluster), a random walk starting from one of the nodes in the group is more likely to remain in the group after a few steps than to go outside. Conceptually, it is possible to define a clustering procedure as follows: each edge is assigned a probability, derived by the edge attribute. Then, a large number of random walks is simulated starting from each node i of the graph and measuring the frequency of the walk arriving at each node j after k steps. Finally, two nodes i, j are considered to be in the same cluster if the probability of the arrival at j starting from i is above a threshold; the transitive closure of this relation determines a partition of the whole graph into clusters.

While this *Monte-Carlo* approach is conceptually sound, it is unacceptably expensive from the computational point of view. So the MCL algorithm proposes a faster procedure to compute the probabilities of arrival. The algorithm has two parameters: an expansion exponent k (a natural number greater than 1) and an inflation exponent r (a positive real number, usually greater than 1). The algorithm alternates between two phases, expansion and inflation, until a fixed point is reached. In the expansion phase, the probability of the random walk is computed by raising the matrix of the edge probabilities to the k-th power. In the inflation phase, the matrix is renormalized after raising each element to r; the resulting matrix is used as input for the subsequent expansion. The goal of the inflation phase is to reduce towards 0 the smaller probabilities and to enhance towards 1 the larger ones. At the ends, the clustering is determined by the resulting probabilities which are sensibly different from 0. Notice that there is no formal proof of convergence of the algorithm, although in practice it has never occurred a case in which a fixed point was not achieved after a few tens of iterations.

3.2 The Iterative Conductance Cutting Algorithm

The Iterative Conductance Cutting algorithm (ICC) was proposed by Kannan et al. in 2000 [10]. This algorithm works in a hierarchical way: it starts with a single cluster comprising the whole graph and at each step it tries to split a cluster into two, as long as a performance measure computed on the two resulting parts is below a threshold α. The iteration stops when there are no more clusters that can be split remaining within the threshold.

The measure used to evaluate the opportunity of the split is *cluster conductance*, defined in the same paper. Basically, this measure compares the sum of the inter-cluster edges with the sum of all the edges incident to the nodes of the clusters. The lower the conductance, the better is the clustering; the maximum value of 1 is attained for degenerate cases such as one-node clusters or whole-graph clusters.

An interesting aspect of this algorithm is the determination of the split to perform among all the possible splits of a given cluster. An exhaustive search of the split minimizing the conductance would require an exponential time complexity with respect to the size of the cluster. The authors propose instead a polynomial approximation based on a spectral technique. In particular, the nodes of the cluster are sorted according to the corresponding component of the second largest eigenvector of the normalized adjacency matrix (whose values are a similarity measure between adjacent nodes). Only the cuts consistent with this ordering (i.e. in which all the nodes of a part are greater than all the nodes in the other) are considered, thus avoiding a combinatorial explosion. The claim of the authors is that this strategy usually gives a good approximation of the optimal split.

3.3 The Geometric MST Clustering Algorithm

The Geometric MST Clustering (GMC) algorithm is an extension of the Minimum Spanning Tree clustering algorithm by Zahn [12]. This method, introduced by Gaertler in his master thesis [11] and in the paper by Brandes et al. [8], solves the problem of finding a suitable threshold for cutting the edges of the minimum spanning tree by computing for each possible threshold a performance measure and choosing the optimal one (note that there are at most $n - 1$ distinct thresholds to be considered, where n is the number of nodes in the graph). For non-attributed graphs, the author propose the use of a *geometric graph embedding* to define a distance between nodes (hence the name of the algorithm); we have not used this part of the method since the edges of our graphs are already attributed with the distance. In the paper by Brandes et al. [8], several performance measures have been used in an experimental comparison (*coverage, performance, conductance*).

3.4 Settings Used for the Above Described Algorithms

As clarified in the previous Section, the FMC algorithm does not require any parameter to be specified in advance.

The MCL algorithm requires a transition probability matrix, together with the two parameters k and r. We have derived the transition probabilities from the distances by assuming an exponential distribution. We have chosen $k = 3$ and $r = 2$, by optimizing the performance over a small subset of the database.

The ICC algorithm requires a similarity matrix that we have defined by taking the inverse of the distance. Also, we have used the value 0.45 for the threshold α, which has been selected by optimizing the performance over a small subset of the database.

The GMC requires the choice of a performance measure; following [8], we have used *conductance* for this purpose. Since the computation of conductance requires a similarity matrix, we have defined one using the same technique adopted for ICC.

4   The Databases Used for Performance Evaluation

We have constructed two databases of synthetic graphs, according to a model derived from [8], that we have extended to include also distance between graph nodes. The original model by Brandes et al. starts by choosing a random partition $P_1,...,P_k$ of the n nodes, such that the average number of nodes $|P_i|$ is equal to a parameter s and the variance is $s/4$. Then, the nodes belonging to the same P_i are randomly linked by assigning to each *internal* edge a probability p given as another parameter. The nodes belonging to different subsets are randomly linked with a probability chosen so as to make the expected number of *external* edges equal to one half of the expected number of internal edges.

In a first database, aimed at modeling pure clustering problems, we have extended this method by adding to each edge a distance value, chosen from an exponential distribution. The mean of the distribution is set to 1 for internal edges and to a parameter $d > 1$ for external ones.

In a second database, aimed at measuring cluster detection performance, we have also added spurious nodes (i.e., nodes that are outside of any cluster). In particular, we take one of the P_i chosen at random and split it into singleton nodes, which are only linked by external edges. Besides this addition, the two databases have been generated according to the same model parameters.

Table 1 resumes all the values we have used for the different model parameters.

Table 1. Values of the parameters used for generating the two graph databases

Parameter	Description	Values
n	number of nodes	100, 200, 300, 400
s	average cluster size	$\sqrt{n}$, $\sqrt{n}/2 + n/6$, $n/3$
p	probability of internal edges	0.4, 0.75
d	average distance of external edges	2, 4, 8

Note that the actual values of the s parameter are dependent on n; we have chosen to have a value giving (on the average) three large clusters ($n/3$), a value giving many small clusters ($\sqrt{n}$) and a value which is intermediate between these two extremes.

For each database, we have generated 20 graphs for each of the 72 possible combinations of the above mentioned parameters, obtaining 1440 "pure clustering" and 1440 "cluster detection" graphs.

5 Cluster Validity Indices

The indices described below have been developed for the case of vector clustering, where a distance measure can be computed between any two vectors, and also the center of a cluster can be determined. In order to apply those indices to graph clustering, we have replaced the notion of the center of a set of vectors with the centroid of a set of nodes, defined as the node which minimizes the largest distance from any other node in the set. Furthermore, we have used as our distance measure the length of the shortest path on the graph between the two considered nodes.

5.1 Davies-Bouldin Index

The Davies-Bouldin index proposed in [16] measures the validity of the cluster as the average ratio between *within-cluster scatter* and *between-cluster separation*. More formally:

$$DB = \frac{1}{C}\sum_{i=1}^{C} \max_{j\neq i} \frac{A_i + A_j}{d_{ij}}$$

where C is the number of clusters, A_i is the average distance of members of cluster i from the centroid of the cluster and d_{ij} is the distance between the centroid of cluster i and the centroid of cluster j. This index should give smaller values for good clusters. Notice, however, that in the undesirable case in which each node belongs to a singleton cluster the index achieves its minimum (0).

5.2 Dunn Index

The index proposed by Dunn in [17] is related to the ratio between the maximum distance within a cluster and the minimum distance between two clusters. More formally,

$$D = \frac{\min_{i,j\neq i} \delta_{ij}}{\max_k \Delta_k}$$

where Δ_k is the maximum distance within cluster k and δ_{ij} is the minimum distance between a node in cluster i and a node in cluster j. The larger the value of this index, the better should be the clustering. However, it has to be noted that, in the undesirable case in which each node belongs to a singleton cluster, the index achieves its maximum ($+\infty$). Another weakness of this index is that it is influenced only by the worst clusters in the partition, instead of providing an integral validity measure.

5.3 Calinski-Harabasz Index

The index by Calinski and Harabasz [18] is based on the so-called *trace* of the between-cluster distances and of the within-cluster distances. Namely, the trace of the between-cluster distances T_B is defined as the weighted sum of the squared distances of the centroids of the clusters from the centroid of the whole node set. Each distance is weighted by the number of nodes in the corresponding cluster. The trace of the within-cluster distance T_W is defined as the sum of the squared distances between each node and the centroid of its cluster. Thus, the Calinski-Harabasz index is defined as:

$$CH = \frac{T_B/(C-1)}{T_W/(n-C)}$$

where n is the number of nodes. This index gives larger values for good clusters. Note that in the undesirable extreme cases where each node belongs to a singleton cluster or all the nodes belong to only one giant cluster, the definition implies an undefined 0/0 ratio.

5.4 Xie-Beni Index

Xie and Beni have defined in [19] a validity index for fuzzy clustering schemes, based on the normalized ratio between the *compactness* of a partition and its *separation*. When applied to crisp clustering, the index can be expressed as:

$$XB = \frac{T_W}{n \min_{ij} d_{ij}^2}$$

This index provides smaller values for good clusters. Note that in the undesirable case in which each node belongs to a singleton cluster, this index achieves its minimum (0); furthermore, in the opposite case in which all the nodes belong to only one giant cluster, the value of the index is not defined.

5.5 C Index

The C index introduced by Hubert and Schultz in [20], is based on the computation of three sets of distances between nodes. The first is the set S_W of all the within-cluster distances. Let m be the cardinality of this set. The second set, S_{min}, is the set of the m smallest distances considering all the pairs of nodes; similarly the third set S_{max} is the set of the m largest distances. Given these sets, the C index is defined as

$$C = \frac{sum(S_W) - sum(S_{min})}{sum(S_{max}) - sum(S_{min})}$$

This index provides small values for good clusters. Notice that in the two extreme undesirable cases in which each node belongs to a singleton cluster, or in which all the nodes belong to only one giant cluster, this index give rise to the undefined ratio 0/0.

5.6 $\mathscr{I}$ index

The $\mathscr{I}$ index defined by Maulik and Bandyopadhyay [7] (sometime referred to as "\cal I", from the LaTeX command required to obtain $\mathscr{I}$), is based on the product of three terms which take into account respectively the number of clusters, the compactness of the clusters and their separation. Like the Xie-Beni index, it is aimed at evaluating fuzzy clustering algorithms. For the special case of crisp clustering, $\mathscr{I}$ is defined as:

$$\mathscr{I} = \left(\frac{1}{C} \frac{E_1}{E_C} \times \max_{i,j \neq i} d_{ij} \right)^2$$

where E_1 is the sum of the distances between each node and the centroid of the whole graph, and E_C is the sum of the distances between each node and the centroid of the corresponding cluster. This index gives larger values for good clusters. Notice that in the undesirable case in which each node belongs to a singleton cluster, this index achieves its maximum ($+\infty$).

6 Experimental Results

In this Section we will present tests aimed at evaluating and comparing performance of the above described algorithms with respect to the following two problems: pure clustering and cluster detection. In order to compare performance of the considered clustering algorithms, we employed the cluster validity indices described in the previous section. It should be noted that the quality of the clustering solution provided by a certain algorithm cannot be straightforwardly derived by considering the absolute values assumed by such indices. However, it is always verified that a better value (where better means higher or lower, depending on the type of index) assumed by an index corresponds to a better solution. Hence, given a certain clustering problem, these indices can be more profitably used for ranking the solutions provided by

Table 2. Performance of the selected algorithms with respect to the pure clustering problem

Cluster validity index	FMC	MCL	ICC	GMC
Davies-Bouldin	395	0	0	1052
Dunn	493	0	0	954
Calinski-Harabasz	92	592	13	707
Xie-Beni	269	4	0	1173
C	265	91	109	941
$\mathscr{I}$	151	6	454	796
Total	**1665**	**693**	**576**	**5623**

Table 3. Performance of the selected algorithms with respect to the cluster detection problem

Cluster validity index	FMC	MCL	ICC	GMC
Davies-Bouldin	1086	0	0	375
Dunn	1192	0	0	269
Calinski-Harabasz	249	24	3	823
Xie-Beni	1095	0	0	366
C	443	3	133	760
$\mathscr{I}$	105	4	193	792
Total	**4170**	**31**	**329**	**3385**

various techniques. For this reason, all the results reported in this Section take into account only the best algorithm for each experimental setting and for each index, without considering the specific value of the index itself.

In the Tables 2 and 3 we report the performance of the four selected algorithms with respect to the pure clustering and the cluster detection problems, respectively. Each cell of the table shows the number of occurrences where a given algorithm obtained the best value of the cluster validity index. From the results presented in Tables 2 and 3, it is evident that the MST based clustering algorithms (FMC and GMC) performs significantly better than the remaining two algorithms with regards to both the considered problems. In particular, GMC performs best in case of pure clustering, while FMC resulted the best cluster detector.

It is worth noting that the sum of the values on each row of the previous tables is not constant (and equal to the total number of graphs, i.e. 1440) as it would be expected; this is ascribable to the following two factors: firstly, in the undesirable cases in which the cluster validity indices are not defined, no winner algorithm is declared; secondly, in the cases in which n different algorithms obtain the same best score, we declare n winners. We have performed an analysis of the statistical significance of the results in Tables 2 and 3 using the Friedman test. Within a significance threshold of 0.1% all the presented results are significant.

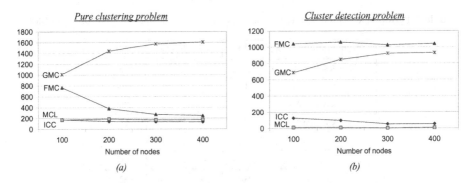

Fig. 1. Performance of the four selected algorithms with respect to the number of nodes in case of *(a)* pure clustering and *(b)* cluster detection

In order to have a deeper insight into the behavior of the selected algorithms with respect to the parameters used for constructing the dataset, we have reported some additional plots. In particular, the dependence of the performance with respect to the number of nodes, to the external/internal edge distance ratio and to the cluster size is shown in the Figures 1, 2 and 3, respectively, where the Y-axis shows the total number of wins over all the indices. We have not explicitly considered the density parameter, since we verified that the performance resulted quite independent of it. Analyzing the Figures 1-3, we can draw the conclusion that GMC performance improves as the number of nodes grows, while the performance of FMC is nearly constant or slightly decreasing; this trend is also confirmed when considering the average cluster size. On the other hand, FMC performance increases when the external/internal edge distance ratio becomes larger; this can be explained considering that a larger distance ratio reinforces the basic assumption underlying FMC that the distances of internal edges and of external edges form two separable modes of the distance distribution.

In conclusion, from the experimental analysis carried out it is possible to observe that the MST based clustering approaches outperform the ICC and the MCL algorithms. The proposed unsupervised approach seems to be better suited for facing the cluster detection problem than the pure clustering. Furthermore, in both cases the FMC can be preferred to the GMC when the distribution of the edge weights is bimodal and the order of magnitude of the problem (overall number of nodes and average cluster size) is low; in all the other cases, the GMC is to be preferred.

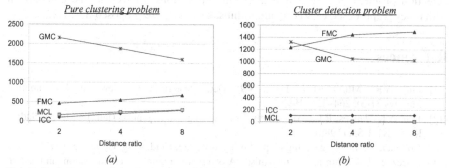

Fig. 2. Performance of the four selected algorithms with respect to the external/internal edge distance ratio in case of *(a)* pure clustering and *(b)* cluster detection

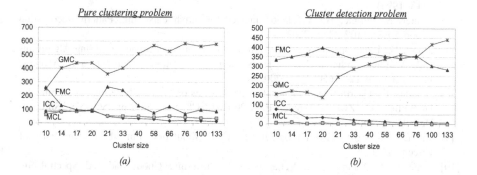

Fig. 3. Performance of the four selected algorithms with respect to the cluster size in case of *(a)* pure clustering and *(b)* cluster detection

7 Conclusions

In this paper we presented a benchmarking activity for assessing the performance of four graph-based clustering algorithms with regards to two different problems: pure clustering and cluster detection. One of the considered algorithms (FMC) has been originally proposed by the authors; it is different from the others in that it does not require the user to set any parameter or threshold. The comparison has been carried out on a database of synthetically generated graphs, using a set of six cluster validity indices. As it could be expected, there is no algorithm that is definitively better than the others. However, in all the experimentations the clustering algorithms based on the Minimum Spanning Tree (MST) proved to be the best. In particular, in case of the pure clustering problem the Geometric MST Clustering (GMC) algorithm provided the highest performance, while the proposed method was able to better cope with the noise of the cluster detection problem. Furthermore, the FCM seems to be better suited to operate when the distribution of the edge weights is bimodal and the order of magnitude of the problem (overall number of nodes and average cluster size) is low.

Future steps of this activity will regard the testing of the algorithms on a cluster detection problem in real computer vision applications, as well as the comparison with other graph-based clustering algorithms, such as those proposed in [21,22].

References

[1] Jain, A.K., Murty, M.N., Flynn, P.J.: Data clustering: a review. ACM Computing Surveys 31(3), 264–323 (1999)

[2] Jain, A.K., Dubes, R.C.: Algorithms for clustering data. Prentice-Hall, Inc, Upper Saddle River, NJ, USA (1988)

[3] Kohonen, T.: Self-organizing maps. Springer-Verlag, Heidelberg, Germany (1995)

[4] Juszczak, P.: Learning to recognise. A study on one-class classification and active learning, PhD thesis, Delft University of Technology, ISBN: 978-90-9020684-4 (2006)

[5] Wu, Z., Leahy, R.: An Optimal Graph Theoretic Approach to Data Clustering: Theory and Its Application to Image Segmentation. IEEE Transactions on PAMI 15(11), 1101–1113 (1993)

[6] Günter, S., Bunke, H.: Validation indices for graph clustering. Pattern Recognition Letters 24(8), 1107–1113 (2003)

[7] Malik, U., Bandyopadhyay, S.: Performance Evaluation of Some Clustering Algorithms and Validity Indices. IEEE Transactions on Pattern Analysis and Machine Intelligence 24(12), 1650–1654 (2002)

[8] Brandes, U., Gaertler, M., Wagner, D.: Experiments on Graph Clustering Algorithms. In: Di Battista, G., Zwick, U. (eds.) ESA 2003. LNCS, vol. 2832, pp. 568–579. Springer, Heidelberg (2003)

[9] van Dongen, S.M.: Graph Clustering by Flow Simulation. PhD thesis, University of Utrecht (2000)

[10] Kannan, R., Vampala, S., Vetta, A.: On Clustering: Good, Bad and Spectral. In: Foundations of Computer Science 2000, pp. 367–378 (2000)

[11] Gaertler, M.: Clustering with spectral methods, Master's thesis, Universitat Konstanz (2002)

[12] Zahn, C.: Graph-theoretical methods for detecting and describing gestalt clusters. IEEE Transactions on Computers C-20, 68–86 (1971)

[13] Bezdek, J.C.: Pattern Recognition with Fuzzy Objective Function Algorithms. Plenum Press, New York (1981)

[14] Horowitz, E., Sahni, S.: Fundamentals of Computer Algorithms, Computer Science Press (1978)

[15] Enright, A.J., van Dongen, S., Ouzounis, C.A.: An efficient algorithm for large-scale detection of protein families. Nucleic Acids Research 30(7), 1575–1584 (2002)

[16] Davies, D.L., Bouldin, D.W.: A Cluster Separation Measure. IEEE Trans. Pattern Analysis and Machine Intelligence 1, 224–227 (1979)

[17] Dunn, C., Fuzzy, A.: A Fuzzy Relative of the ISODATA Process and Its Use in Detecting Compact Well-Separated Clusters. J. Cybernetics 3, 32–57 (1973)

[18] Calinski, R.B., Harabasz, J.: A Dendrite Method for Cluster Analysis. Comm. in Statistics 3, 1–27 (1974)

[19] Xie, X.L., Beni, G., Validity, A.: A Validity Measure for Fuzzy Clustering. IEEE Trans. on Pattern Analysis and Machine Intelligence 13, 841–847 (1991)

[20] Hubert, L., Schultz, J.: Quadratic assignment as a general data-analysis strategy. British Journal of Mathematical and Statistical Psychology 29, 190–241 (1976)

[21] Shi, J., Malik, J.: Normalized Cuts and Image Segmentation. IEEE Transactions on Pattern Analysis and Machine Intelligence 22(8), 888–905 (2000)

[22] Shental, N., Zomet, A., Hertz, T., Weiss, Y.: Pairwise Clustering and Graphical Models. In: Advances in Neural Information Processing Systems, MIT Press, Cambridge, MA (2004)

A New Greedy Algorithm for Improving b-Coloring Clustering

Haytham Elghazel[1], Tetsuya Yoshida[2], Véronique Deslandres[1],
Mohand-Said Hacid[3], and Alain Dussauchoy[1]

[1] LIESP (ex. PRISMa) Laboratory, Claude Bernard University of Lyon I,
43 Bd du 11 novembre 1918, 69622 Villeurbanne cedex, France
{elghazel,deslandres,dussauchoy}@bat710.univ-lyon1.fr
[2] Grad. School of Information Science and Technology, Hokkaido University
N-14 W-9, Sapporo 060-0814, Japan
yoshida@meme.hokudai.ac.jp
[3] LIRIS Laboratory, Claude Bernard University of Lyon I,
43 Bd du 11 novembre 1918, 69622 Villeurbanne cedex, France
mshacid@liris.cnrs.fr

Abstract. This paper proposes a new greedy algorithm to improve the
specified b-coloring partition while satisfying b-coloring property. The b-
coloring based clustering method in [3] enables to build a fine partition of
the data set (classical or symbolic) into clusters even when the number of
clusters is not pre-defined. It has several desirable clustering properties:
utilization of topological relations between objects, robustness to outliers, all types of data can be accommodated, and identification of each
cluster by at least one *dominant object*. However, it does not consider
the *high quality* of the clusters in the construction of a b-coloring graph.
The proposed algorithm in this paper can complement its weakness by
re-coloring the objects to improve the quality of the constructed partition under the property and the dominance constraints. The proposed
algorithm is evaluated against benchmark datasets and its effectiveness
is confirmed.

Keywords: clustering, graph b-coloring, graph re-coloring.

1 Introduction

Clustering, or unsupervised classification, is a fundamental data mining process
that aims to divide a set of data into groups, or clusters, such that the data within
the same group are similar to each other (*intracluster cohesion*) while data from
different groups are dissimilar (*intercluster separation*). Clustering problems are
ubiquitous in pattern recognition. For surveys on the most important clustering
methods used in pattern recognition see, for example, [10].

Clustering of data is generally based on two approaches: *hierarchical* and *partitioning*. *Hierarchical* clustering algorithms build a cluster hierarchy, or a tree of
clusters (*dendrogram*) whose leaves are the data points and whose internal nodes

F. Escolano and M. Vento (Eds.): GbRPR 2007, LNCS 4538, pp. 228–239, 2007.
© Springer-Verlag Berlin Heidelberg 2007

represent nested clusters of various sizes [5]. On the other hand, *partitioning* clustering algorithms give a single partition of the data by fixing some parameters (number of clusters, thresholds, etc.). Each cluster is represented by its centroid [7] or by one of its objects located near its center [12]. When the distances (dissimilarities) among all pairs of data are specified, these can be summarized as a weighed dissimilarity matrix D in which each element $D(v_i, v_j)$ stores the corresponding dissimilarity. Based on D, the data can also be conceived as a graph where each vertex v_i corresponds to a data and each edge corresponds to a pair of vertices (v_i, v_j) with label $d(v_i, v_j)$.

Additional techniques for the grouping operation include graph-theoretic clustering methods. Many graph-theoretic clustering algorithms basically consist of searching for certain combinatorial structures in the similarity graph. In this case, some hierarchical approaches are related to graph-theoretic clustering. The best-known graph-theoretic divisive clustering algorithm (*the single-link algorithm*) is based on construction of the *minimal spanning tree* (MST) of the data [13], and then deleting the MST edges with the largest lengths to generate single-link clusters. The complete-link algorithms are also reduced to a search for a maximal complete subgraph, namely a *clique*[1] [10] which is the strictest definition of a cluster. Some authors have proposed to use the *vertex coloring of graphs* for the hierarchical classification purpose. In [4], the authors propose a divisive classification method based on dissimilarity tables, where the iterative algorithm consists, at each step, in finding a partition by subdividing the cluster with the largest diameter into two clusters in order to exhibit a new partition with the minimal diameter. By mapping each data item to the corresponding vertex, the subdivision is obtained by a *2-coloring* of the vertices of the maximum spanning tree built from the dissimilarity table. The derived classification structure is a hierarchy.

On the other hand, the partitioning methods are also related to graph-theoretic clustering. Hansen and Delattre [6] reduced the partitioning problem of a data set into k clusters with minimal diameter, to the minimal coloring problem of a superior threshold graph. The edges of this graph are the pairs of vertices distanced from more than a given threshold. In such a graph, each color corresponds to one cluster and the number of colors is minimal. Unfortunately, while this method tends to build a partition of the data set with effectively compact clusters, it does not give any importance to the cluster-separation.

Recently, [3] proposed a clustering method based on the notion of b-coloring of a graph [9,2]. A graph b-coloring consists to color the vertices of a graph with the largest number of colors such that (i) two neighbors have different colors (proper coloring) and (ii) for each color there exists at least one dominating vertex which is adjacent to all the other colors.

The b-coloring based clustering method in [3] enables to build a fine partition of the data set (classical or symbolic) in clusters even when the number of clusters is not pre-defined. This approach exhibits more important features (1) it

---

[1] A clique in an undirected graph G is a set of vertices V such that for every two vertices in V, there exists an edge connecting the two.

gives a strong solution to the problem of the single-point representation of clusters by using the topological relations between objects to build clusters: all the objects participate in the building of their clusters, (2) it is robust in the presence of outliers, (3) it can accommodate all types of data as long as a dissimilarity table can be constructed, and (4) it identifies each cluster by at least one dominant object which guarantees the *disparity between the clusters*. The building of the b-coloring of a specified graph is given in two stages: 1) initializing the colors of vertices with maximal colors, and 2) removing, by a *greedy procedure*, the colors without any dominating vertex. However, it does not always consider the *high quality* of the clusters in the construction of a b-coloring of graph (*i.e.* in both two stages).

This paper proposes a new greedy algorithm to improve the specified b-coloring partition while satisfying b-coloring property. Thus, the proposed algorithm can complement the weakness of the method in [3] by improving the constructed partition. The algorithm selects the vertices which do not affect the dominant objects in the partition, and change their colors (assignments to clusters) by monotonically increasing the property of the re-colored partition. The former guarantees both the property of dominance in each cluster and b-coloring of clusters, and the latter guarantees the improvement of the constructed clusters. The proposed algorithm is evaluated against benchmark datasets and its effectiveness is confirmed.

2 A b-Coloring Based Clustering Algorithm

2.1 Overview of a b-Coloring Based Clustering Algorithm

In [3] a new graph b-coloring clustering algorithm is introduced. This algorithm will be briefly reviewed in the present section. Consider the data to be clustered as an undirected complete edge-weighted graph where the vertex set is the set of data and the edge-weights set reflect the dissimilarities pairs of linked vertices. The b-coloring of this complete graph is not interesting for the clustering problem. Indeed, each data is assumed to belong to one and only one cluster (color). We use a bold italic capital letter to denote a set, e.g., V represents a set of vertices. $|V|$ represents the cardinality of V. Our clustering approach requires, then, to construct a superior threshold graph $G(V, E)$, which is a partial graph of the initial graph. In other words, the superior threshold graph $G(V, E)$ is a simple graph where $\forall v_i, v_j \in V$, an edge (v_i, v_j) exists *iff* $d(v_i, v_j) > \theta$ for a specified threshold θ among the dissimilarity matrix.

A Working Example. Suppose a set of data with the weighted dissimilarity matrix D in Table 1 is given. Fig. 1 shows the superior threshold graph $\theta=0.15$ for Table 1. The edges are labeled with the corresponding dissimilarities.

The b-coloring algorithm performed on this graph includes two stages: 1) initializing the colors of vertices with maximal colors (*c.f.* Fig. 2), and 2) removing, by a *greedy procedure*, the colors without any dominating vertex. Therefore, the partition (b-colored graph) in Fig. 3 is constructed. The vertices with the same color (shape) are grouped into the same cluster. Thus, {**A**,**D**}, {**B**}, {**C**,E,G,I}, {**F**} are the clusters, and the nodes with bold letter are dominating vertices.

Table 1. A weighted dissimilarity matrix

v_i	A	B	C	D	E	F	G	H	I
A	0								
B	0.20	0							
C	0.10	0.30	0						
D	0.10	0.20	0.25	0					
E	0.20	0.20	0.15	0.40	0				
F	0.20	0.20	0.20	0.25	0.65	0			
G	0.15	0.10	0.15	0.10	0.10	0.75	0		
H	0.10	0.20	0.10	0.10	0.05	0.05	0.05		
I	0.40	0.075	0.15	0.15	0.15	0.15	0.15	0.15	0

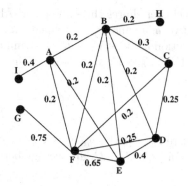

Fig. 1. A threshold graph with $\theta = 0.15$ for the data in Table 1

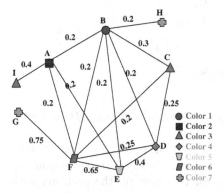

Fig. 2. Initializing the colors of vertices with maximal colors in Fig. 1

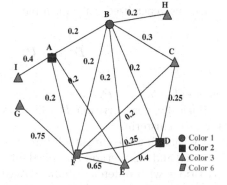

Fig. 3. A b-coloring graph constructed from Fig. 2 by removing colors without any dominating vertex

2.2 Validation Indices for Clustering

The clustering algorithm is an iterative algorithm which performs multiple runs, each of them increasing the value of the dissimilarity threshold θ. The next problem is to validate the partitions of the data set and select the best one as the optimal output clustering, with regards to their separation and tightness values. Many validation indices for clustering have been proposed [1] and adapted to the symbolic framework [11]. Among them, we focus on a validation index called **generalized Dunn's index** $Dunn_G$. $Dunn_G$ is designed to offer a compromise between the *intercluster separation* and the *intracluster cohesion*. So, it is the more appropriated to partition data set in *compact* and *well separated* clusters.

Suppose a set of vertices V is clustered or grouped into a partition $P = \{C_1, C_2, \ldots, C_k\}$ where for $\forall C_i, C_j \in P, C_i \cap C_j = \phi$ for $i \neq j$. We abuse the notation of P to represent both a set of clusters as well as a set of colors, because each cluster $C_i \in P$ corresponds to a color in b-coloring based clustering and no cluster shares the same color. Also, suppose for $\forall v_i, v_j \in V$, the dissimilarity between v_i and v_j is defined and specified as $d(v_i, v_j)$. We assume that the function $d : V \times V \rightarrow R^+$ is symmetric.

For $\forall C_h \in P$, an average within-cluster dissimilarity is defined as

$$S_a(C_h) = \frac{1}{\eta_h(\eta_h - 1)} \sum_{o=1}^{\eta_h} \sum_{o'=1}^{\eta_h} d(v_o, v_{o'}) \tag{1}$$

where $\eta_h = |C_h|^2$, $v_o, v_{o'} \in C_h$.

For $\forall C_i, C_j \in P$, an average between-cluster dissimilarity is defined as

$$d_a(C_i, C_j) = \frac{1}{\eta_i \eta_j} \sum_{p=1}^{\eta_i} \sum_{q=1}^{\eta_j} d(v_p, v_q) \tag{2}$$

where $\eta_i = |C_i|$ and $\eta_j = |C_j|$, $v_p \in C_i$, $v_q \in C_j$.

Dunn's generalized index for a partition P is defined as

$$Dunn_G(P) = \frac{\min_{i,j,i\neq j} d_a(C_i, C_j)}{\max_h S_a(C_h)} \tag{3}$$

where $C_h, C_i, C_j \in P$.

Basically, the partition P which produces the highest $Dunn_G(P)$ indicates the best clustering.

By changing the threshold θ from the table 1, corresponding graphs are constructed and our method is applied for each graph to construct the corresponding partitions with different $Dunn_G$. Actually, for this example, the partition with $\theta=0.15$ has the maximal $Dunn_G$ (1.522) among other ones with different θ.

3   A Greedy Re-coloring Algorithm for b-Coloring Based Clustering

3.1   Motivation

As explained in Section 2.1, for the data in Table 1, our clustering algorithm gives the partition in Fig. 3 as the best one. However, even for the same number of clusters, the graph in Fig. 1 has *different* b-coloring *with larger $Dunn_G$*. An example is shown in Fig. 4. As easily verified, the partition in Fig. 4 is a b-coloring graph, with $Dunn_G = 1.538$, which is larger than 1.522 in Fig. 3.

As illustrated in this example, even when our algorithm described in Section 2 returns a partition P with b-coloring for a graph G, there can be other partitions for the same graph G with *better quality (larger $Dunn_G$)* while satisfying b-coloring. To construct a better partition, it is also important to find a partition with better quality *while satisfying b-coloring*. This problem is formalized as follows.

---

[2] Since C_h contains a set of data (vertices), we use $|C_h|$ to denote its cardinality.

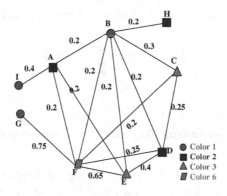

Fig. 4. Another b-coloring with larger $Dunn_G$ for Table 1

Definition 1. Re-Coloring Problem in b-Coloring based Clustering
Find a new partition P' of $G(V, E)$ from a given b-coloring partition P of G such that P' is better than P.

In our current approach, the quality of a partition P is measured by $Dunn_G(P)$. In the following subsection, we describe our algorithm to solve this problem.

3.2 Notations and Definitions

For each $v \in V$, a function $N(v)$ returns neighboring vertices for v, where $\forall v' \in N(v)$, the edge $(v, v') \in E$. $c(v)$ represents the color of v, and a function $N_c(v)$ returns a set of neighboring colors for v (i.e., $N_c(v) = \cup_{v' \in N(v)} c(v')$). A function $C_p(v)$ is defined as $P \setminus N_c(v)$ (here, P as a set of all colors in P). Note that $C_p(v)$ also contains the original color $c(v)$ of v.

Critical Vertices and Non-Critical Vertices. A set of vertices V_d contain the dominating vertices in G. For each $v_d \in V_d$, if $v_s \in N(v_d)$ is the only vertex with the color $c(v_s)$ in $N(v_d)$, v_s is called a supporting vertex of v_d.

V is divided into $V_c \sqcup V_{nc}$ where $V_c \cap V_{nc} = \phi$. Each $v_c \in V_c$ is called a critical vertex, and each $v_{nc} \in V_{nc}$ is called a non-critical vertex. $v_{nc} \in V_{nc}$ can be re-colored in our approach; on the other hand, $v_c \in V_c$ is not re-colored.

V_c is further divided into $V_d \sqcup V_s \sqcup V_f$ where there is no overlap among them. V_f contains a set of finished vertices which are already checked for re-coloring.

When the color $c(v)$ of $v \in V$ is re-colored to c, some $v_{nc} \in V_{nc}$ might become new critical vertices, because some other vertices can become dominating vertices or supporting ones. To reflect the change of colors in G due to re-coloring of v, $V_c^{tmp}(v, c)$ represents the set of vertices which become *new* critical vertices induced from this re-coloring, and $P(v, c)$ represents the partition. In $P(v, c)$, only $c(v)$ is re-colored to c and $c(v')$ is not re-colored for $\forall v' \in V, v' \neq v$.

$\forall v \in \boldsymbol{V}$, $\forall C_i \in \boldsymbol{P}$, an average dissimilarity between v and C_i is defined as

$$d_a(v, C_i) = \frac{1}{\eta_i} \sum_{p=1}^{\eta_i} d(v, v_p) \qquad (4)$$

where $\eta_i = |C_i|$, $v_p \in C_i$.

3.3 A Greedy Re-coloring Algorithm for b-Coloring Based Clustering

A Conservative Approach for Re-coloring of Vertices. We regard that each $v_d \in \boldsymbol{V_d}$ is important in its own and its color should not be changed, because by definition of a dominating vertex, v_d is connected to the vertices with all the other colors. Thus, it is *far away* from the other clusters (at least greater than θ) and contributes to a large between-cluster dissimilarity. Also, we regard that $c(v_s) \in \boldsymbol{V_s}$ should not be changed, because changing $c(v_s)$ to other color can make some $v_d \in \boldsymbol{V_d}$ as non-dominating and thus deteriorates the quality of a partition. Based on these arguments, only $v \in \boldsymbol{V} \setminus \{\boldsymbol{V_c} \sqcup \boldsymbol{V_s}\}$ should be considered for re-coloring. Furthermore, to guarantee the termination of re-coloring, re-coloring of $v \in \boldsymbol{V}$ is tried at most once. To realize this, when re-coloring of $v \in \boldsymbol{V}$ is tried, it is moved into $\boldsymbol{V_f}$.

In summary, we consider re-coloring of only $v \in \boldsymbol{V} \setminus \{\boldsymbol{V_c} \sqcup \boldsymbol{V_s} \sqcup \boldsymbol{V_f}\} = \boldsymbol{V_{nc}}$, and whenever v is checked for re-coloring, it is moved into $\boldsymbol{V_f}$ and its color is fixed. Thus, $\boldsymbol{V_{nc}}$ is monotonically decreased at each re-coloring. Since the color of $v \in \boldsymbol{V_c}$ is fixed once it is inserted into $\boldsymbol{V_c}$, and other possibilities are not explored in later, our algorithm is called a greedy algorithm.

In addition, through re-coloring of $v_{nc} \in \boldsymbol{V_{nc}}$, its color can be changed later. This means that, reflecting $\boldsymbol{V_{nc}}$ to evaluate the quality of a partition $\boldsymbol{P}$ can be an unreliable estimation. Thus, we do not utilize $\boldsymbol{V_{nc}}$ to evaluate the quality of $\boldsymbol{P}$ in the re-coloring process and utilize only $\boldsymbol{V_c}$.

A Vertex Selection Criterion for Re-coloring. Among $v_{nc} \in \boldsymbol{V_{nc}}$, we select $v \in \boldsymbol{V_{nc}}$ with the maximal $d_a(v, c(v))$, i.e., $v^* = \arg\max_{v \in \boldsymbol{V_{nc}}} d_a(v, c(v))$ is selected. Here, $d_a(v, c(v))$ corresponds to the degree of "outlier" for each v, because it represents the average within-cluster dissimilarity of each v. On the contrary, if $v' \neq \arg\max_{v \in \boldsymbol{V_{nc}}} d_a(v, c(v))$ is selected and re-colored before v^*, $|C_p(v)|$ can decrease because some $v'' \in \boldsymbol{V_{nc}}$ can be moved into $\boldsymbol{V_c}$ due to the re-coloring of v'. This amounts to putting more constraints and reducing the possibilities of new color for v^*, $|C_p(v^*)|$, because $N_c(v^*)$ can increase. Thus, if v' is selected before v^*, v^* can be set to the other color, different from the one which can maximally increase $Dunn_G$ by decreasing its numerator. Based on this argument, among $\boldsymbol{V_{nc}}$, we select v^* for re-coloring.

A Color Selection Criterion for Re-coloring. Our objective is to increase the quality of a partition $\boldsymbol{P}$ by re-coloring some vertices while preserving proper coloring. When the vertex v is selected for re-coloring, we check the colors in

$C_p(v)$ and select the one with the maximal $Dunn_G$ in eq.(3). Since re-coloring of v causes to change the values of $S_a(\cdot)$ and $d_a(\cdot, \cdot)$ as well, these should also be updated whenever $c(v)$ is re-colored. We conduct efficient calculation of $S_a(\cdot)$ by utilizing their old values. Furthermore, although naïve calculation of $d_a(\cdot, \cdot)$ takes $O(n^2)$, this can also be reduced to $O(n)$ by utilizing the old values.

For $\exists\, C, C_i \in P$, suppose $v \in V_{nc}$ was initially assigned to C and re-assigned (re-colored) to C_i. Due to this re-coloring, v is moved into $V_f \subset V_c$. For $P = P\backslash\{C, C_i\} \sqcup \{C, C_i\}$, by utilizing the old values, new values of $S_a(\cdot)$ can be efficiently calculated using the following equations:

$$S_a^{new}(C) = S_a^{old}(C) \tag{5}$$

$$S_a^{new}(C_i) = \frac{|C_i|S_a^{old}(C_i) + |C|d_a(v, C_i)}{|C_i| + 1} \tag{6}$$

$$S_a^{new}(C_j) = S_a^{old}(C_j) \quad \forall C_j \in P\backslash\{C, C_i\} \tag{7}$$

Similarly,

$$d_a^{new}(C, C_i) = \frac{|C||C_i|d_a^{old}(C, C_i) + |C|d_a(v, C)}{|C|(|C_i| + 1)} \tag{8}$$

$$d_a^{new}(C_i, C_j) = \frac{|C_i||C_j|d_a^{old}(C_i, C_j) + |C_j|d_a(v, C_j)}{(|C_i| + 1)||C_j|} \tag{9}$$

$$d_a^{new}(C_j, C_h) = d_a^{old}(C_j, C_h) \quad \forall C_j, C_h \in P\backslash\{C, C_i\} \tag{10}$$

A Greedy Re-coloring Algorithm for b-coloring based Clustering. Our greedy update algorithm for b-coloring based clustering is shown in Algorithm 1. When there are multiple candidates with the same value at lines 4 and 10, one of them is selected at random.

The proposed algorithm has the following desirable properties for clustering.

Proposition 1. *Algorithm 1 creates a proper coloring P' of $G(V, E)$ from P.*

Proof. Algorithm 1 re-color $c(v)$ of $v \in V$ only to some $c' \in C_p(v)$. By definition of $C_p(v)$, $c' \notin N_c(v)$ for $\forall c' \in C_p(v)$, which guarantees proper coloring. □

Proposition 2. *Algorithm 1 creates a b-coloring P' of $G(V, E)$ from P.*

Proof. From Proposition 1, P' is proper coloring. We need to show that there is at least one dominating vertex for each color. By definition, this property is satisfied in P. Since Algorithm 1 does *not change* the colors of $v_d \in V_d$, there is at least one dominating vertex for each color in P'. □

Proposition 3. *Algorithm 1 monotonically increase $Dunn_G(C')$ of $G(V, E)$ from P.*

Proof. At lines 10 and 11, the color $c(v^*)$ which maximizes $Dunn_G(P)$ is selected by modifying the original color $c(v)$ (note that it is allowed that $c_{new}(v)$ =c(v), i.e., unchanged). Since this re-coloring is repeated for $\forall v_{nc} \in V_{nc}$, when Algorithm 1 terminates, $Dunn_G(P')$ monotonically increases. □

Algorithm 1. A Greedy Re-coloring Algorithm for b-coloring based Clustering

---

Require: $G(V, E)$; //A graph with a set of vertices and a set of edges.
Require: P; //a partition which is a b-coloring of $G(V, E)$

1: $C' = P$;
2: Divide V into $V_c \sqcup V_{nc}$
3: **while** $V_{nc} \neq \phi$ **do**
4:    $v^* = \arg\max\limits_{v' \in V_{nc}} d_a(v', c(v'))$; //vertex selection
5:    **for each** $c \in C_p(v^*)$ **do**
6:       create $V_c^{tmp}(v^*, c)$ and $P(v^*, c)$ induced from the re-coloring of $c(v^*)$ into c;
7:       For $\forall C_i \in P(v^*, c)$, calculate $d_a(v^*, C_i)$ w.r.t $V_c \sqcup V_c^{tmp}(v^*, c)$;
8:       calculate $S_a^{new}(C_h)$ and $d_a^{new}(C_i, C_j)$ for $\forall C_h, C_i, C_j \in P(v^*, c)$;
9:    **end for**
10:    $c^*(v) = \arg\max\limits_{c \in C_p(v)} Dunn_G(P(v, c))$;
11:    Re-color $c(v*)$ to c^* in C'; //re-coloring of C'
12:    $V_{nc} = V_{nc} \setminus \{v\}$;
13:    $V_f = V_f \cup \{v\}$;
14:    $V_c = V_c \cup V_c^{tmp}(v, c)$ //$v \in V_c^{tmp}(v, c)$ into V_d or V_s or V_f due to its property
15: **end while**
16: **return** C';

---

4   Evaluations

The greedy clustering algorithm was tested by considering two relevant benchmark data sets, viz., *Zoo*, and *Mushroom* from the UCI Machine Learning Repository [8]. To evaluate the quality of the partition discovered by the greedy algorithm (called Improved b-coloring Partition), the results are compared with that of the best partition returned by the *b-coloring clustering algorithm* as the one maximizing the $Dunn_G$ value [3] (called *Original b-coloring Partition*), the *Optimal Hansen's Partition* based on minimal coloring technique [6] and the *Optimal Agglomerative Single-link Partition* [10].

As used in [11], in addition to the value of *Generalized Dunn's index*, our evaluation will be based on a probability matching scheme called *Distinctness* [11]. Such function is very powerful in the cluster validation problem because it works *independently* of the number of clusters and the dissimilarity between objects.

For a partition P with k clusters $\{C_1, C_2, .., C_k\}$, the *Distinctness* is defined as the *intercluster dissimilarity* using a probability match measure, namely the *variance of the distribution match*. The variance of the distribution match between clusters h and l in a given partition is measured as:

$$Var(C_h, C_l) = \frac{1}{m} \sum_i^m \sum_j \left(P\left(a_i = V_{ij} | C_h\right) - P\left(a_i = V_{ij} | C_l\right) \right)^2 \quad (11)$$

where m is the number of attributes a_i characterizing the objects. $P(a_i = V_{ij} | C_h)$ is the *conditional probability* of attribute a_i to take the value V_{ij} in cluster C_h.

Table 2. Evaluation of Zoo Data

Clustering Approach	# Clusters	Distinctness	$Dunn_G$
Re-colored b-coloring Partition	7	0.652	1.120
Original b-coloring Partition	7	0.612	1.071
Optimal Single-link Partition	2	0.506	0.852
Optimal Hansen's Approach	4	0.547	1.028

The above equation assumes that a data object has only one value per attribute (represented by $j \in a_i$). The greater this value, the more dissimilar are the two clusters being compared, and, therefore, the concepts they represent.

The *Distinctness* of the partition P is taken as the average variance between clusters in that partition.

$$Distinctness = \frac{\sum_{h=1}^{k} \sum_{l=1}^{k} Var(C_h, C_l)}{k \times (k-1)} \tag{12}$$

When comparing two partitions, the one that produces the greater distinctness should be the preferred partition since the clusters in this partition represent the more distinct concepts [11].

4.1 Evaluation for Zoo Dataset

The Zoo data [8] uses 100 instances of animals with 17 features and 7 output classes. The name of the animal constitutes the first attribute. There are 15 boolean features corresponding to the presence of hair, feathers, eggs, milk, backbone, fins, tail; and whether airborne, aquatic, predator, toothed, breathes, venomous, domestic, catsize. The numeric attribute corresponds to # legs.

Table 2 provides the clustering results. The Distinctness measure indicates better partitioning for the clusters generated by the *b-coloring clustering approach* (for the original partition as well as for the improved partition). This confirms that our notion of dominating vertex finds more meaningful and well-separated clusters. In the other hand, the improved partition has the larger $Dunn_G$ value. This indicates the pertinence of the *greedy algorithm* to improve the *original b-coloring partition*.

4.2 Evaluation for Mushroom Dataset

The mushroom data set was also obtained from [8]. Each data record contains information that describes the 21 physical properties (e.g., color, odor, size, shape) of a single mushroom. A record also contains a *poisonous* or *edible* label for the mushroom. All attributes are categorical; for instance, the values that the size attribute takes are narrow and broad, while the values of shape can be bell, at, conical or convex, and odor is one of spicy, almond, foul, fishy, pungent

Table 3. Evaluation of Mushroom Data

Clustering Approach	# Clusters	Distinctness	$Dunn_G$
Re-colored b-coloring Partition	17	0.728	0.995
Original b-coloring Partition	17	0.713	0.891
Optimal Single-link Partition	20	0.615	0.866
Optimal Hansen's Approach	19	0.677	0.911

etc. The mushroom database has the largest number of records (that is, 8124) among the benchmark data sets we used in our experiments.

Table 3 provides the results of the clustering obtained, over the mushroom data using the different clustering approaches. According to the Distinctness measure, the proposed *re-colored b-coloring approach* generates the best clustering. The clusters are *compact* and *well-separated*. This confirms the pertinence of the *b-coloring* technique to offer a compromise between the *intercluster separation* and the *intracluster cohesion*. Moreover, the relevance of the new greedy algorithm in the improvement stage of the $Dunn_G$ index is also observed from the table 3.

5 Conclusion and Future Work

This paper has proposed a new greedy algorithm to improve the specified b-coloring partition while satisfying b-coloring property. The b-coloring based clustering method in [3] enables to build a fine partition of the data set (classical or symbolic) into clusters even when the number of clusters is not pre-defined. However, it does not consider the *high quality* of the clusters in the construction of a b-coloring graph. The proposed algorithm in this paper can complement this weakness by re-coloring the objects to improve the quality of the constructed partition under the property and the dominance constraints. We have implemented, performed experiments, and compared our approach to other clustering approaches, and illustrated its efficiency on benchmark datasets (having especially various characteristics). The proposed techniques offers a real compromise between the *intercluster separation* and the *intracluster cohesion*.

There are many interesting issues to pursue: (1) leading more experiments and comparison for our algorithm on a *real medical data set* and *a larger image data set*, and (2) extending the *re-coloring* concept to the *critical vertices* in the sense to better improving the clustering quality, to name a few.

Acknowledgments

The authors are grateful to Prof. Tanaka for his support to conduct this work. This work was partially supported by Core-to-Core Program (No.18001) funded by JSPS in Japan, the grant-in-aid for scientific research (No.18700131) funded by MEXT in Japan.

References

1. Bezdek, J.C., Pal, N.R.: Some new indexes of cluster validity. IEEE Transactions on Systems, Man. and Cybernetics 28(3), 301–315 (1998)
2. Effantin, B., Kheddouci, H.: The b-chromatic number of some power graphs. Discrete Mathematics and Theoretical Computer Science 6(1), 45–54 (2003)
3. Elghazel, H., Deslandres, V., Hacid, M.S., Dussauchoy, A., Kheddouci, H.: A new clustering approach for symbolic data and its validation: Application to the healthcare data. In: Esposito, F., et al. (ed.) ISMIS 2006. LNCS (LNAI), vol. 4208, pp. 473–482. Springer Verlag, Heidelberg (2006)
4. Guénoche, A., Hansen, P., Jaumard, B.: Efficient algorithms for divisive hierarchical clustering with the diameter criterion. Journal of Classification 8, 5–30 (1991)
5. Guha, S., Rastogi, R., Shim, K.: Cure: An efficient clustering algorithm for large databases. In: Proceedings of the ACM SIGMOD Conference, pp. 73–84 (1998)
6. Hansen, P., Delattre, M.: Complete-link cluster analysis by graph coloring. Journal of the American Statistical Association 73, 397–403 (1978)
7. Hartigan, J., Wong, M.: Algorithm as136: A k-means clustering algorithm. Journal of Applied Statistics 28, 100–108 (1979)
8. Hettich, S., Blake, C.L., Merz, C.J.: Uci repository of machine learning databases (1998)
9. Irving, W., Manlov, D.F.: The b-chromatic number of a graph. Discrete Applied Mathematics 91, 127–141 (1999)
10. Jain, A.K., Murty, M.N.: Data clustering: A review. ACM Computing Surveys 31, 264–323 (1999)
11. Kalyani, M., Sushmita, M.: Clustering and its validation in a symbolic framework. Pattern Recognition Letters 24(14), 2367–2376 (2003)
12. Ng, R., Han, J.: Clarans: a method for clustering objects for spatial data mining. IEEE Transactions on Knowledge and Data. Engineering 14(5), 1003–1016 (2002)
13. ZAHN, C.T.: Graph-theoretical methods for detecting and describing gestalt clusters. IEEE Transactions on Computers 20, 68–86 (1971)

Qualitative Spatial Relationships for Image Interpretation by Using Semantic Graph

Yann Hodé[1] and Aline Deruyver[2]

[1] CH Rouffach, 27 rue du 4ième RSM, Rouffach
[2] LSIIT,UMR7005CNRS-ULP Parcd'innovation, Bd Sébastien Brant, BP10413, 67412 ILLKIRCH CEDEX
aline@eavr.u-strasbg.fr

Abstract. In this paper, a new way to express complex spatial relations is proposed in order to integrate them in a Constraint Satisfaction Problem with bilevel constraints. These constraints allow to build semantic graphs, which can describe more precisely the spatial relations between subparts of a composite object that we look for in an image. For example, it allows to express complex spatial relations such as "is surrounded by". This approach can be applied to image interpretation and some examples on real images are presented.

Keywords: Semantic graph, arc-consistency checking, spatial relationship, image interpretation.

1 Introduction

The large expansion of web applications leads to manipulate a huge amount of images. Then, interpreting correctly the content of images is a crucial step to obtain what we want from this huge image database. Image interpretation is also an important issue in medical images, particularly when it is necessary to find automatically anatomical structures such that cerebral structures linked to brain activity. However, a large gap persists between the semantic interpretation of an image and its low-level features. The MPEG-7 standard has been developed to introduce high-level representations (ontology) that capture the semantics of a document [10], [11]. This semantic is sometimes limited to textual descriptors of image components according to a semantic hierarchy. Other semantic descriptions may be more relevant to interpret an image.

Usually, a complex object, like an anatomical structure, is described by the shape of its components and the spatial relationships between these components. Then, a semantic model has to integrate both spatial and morphological constraints. One could think that spatial relations could be simply described by a notion of adjacency as we can find in region adjacency graph (RAG). This formalism has some interesting properties which make it very convenient to describe an image and therefore it has been chosen by many authors [2], [6], [12], [13], [14]. However, the unique notion of adjacency is too poor to describe complex spatial organization of the different parts of an object. Cohn et al. [5] proposed to describe more complex spatial relations between

F. Escolano and M. Vento (Eds.): GbRPR 2007, LNCS 4538, pp. 240–250, 2007.
© Springer-Verlag Berlin Heidelberg 2007

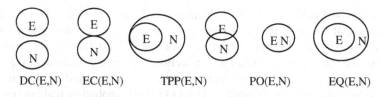

DC(E,N)   EC(E,N)   TPP(E,N)   PO(E,N)   EQ(E,N)

Fig. 1. Illustration of the JEPD relations

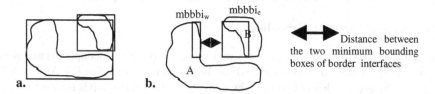

Fig. 2. a. In that case the two regions are not overlapped and the two minimum bounding boxes are overlapped. The analysis of the spatial relation between these two regions is not possible by using minimum bounding boxes. **b.** $mbbbi_w$ is the minimum bounding box of the border interface which is on the left of region A. $mbbbi_e$ is the minimum bounding box of the border interface which is on the right of region B.

regions, in a topological framework, with a set of basic relations and they create for that purpose the RCC8 formalism. RCC8 deals with a set of eight Jointly Exhaustive and Pairwise Disjoint (JEPD) relations called basic relations: DisConnected (DC), Externally Connected (EC), Partial Overlap (PO), Equal (EQ), Tangential Proper Part (TPP), Non Tangential Proper Part (NTPP) and their converses (See Fig.1). However, this formalism does not take into account the shape of the regions and the directional relations. Skiadopoulos and Koubarakis [17] circumvent this drawback by defining formally the Cardinal Direction Relations. These relations exploit the notion of *minimum bounding box* (mbb) and several authors have proposed some ways to combine topological notions with directional relations [18]. This approach has several interesting advantages: it has good properties of computation (computing the minimum bounding box of a region is fast), it is possible to inherit the properties of minimum bounding boxes inside a pyramid of adjacency graphs, it is possible to introduce a notion of absolute or relative metrics and RCC8 relation can be retrieved from it [18]. However, the topological and directional notions should take into account that the notion of distance between two regions is also a very important feature [4]. For example, the difference between different animal faces lies mainly on the difference of distances between each part of the face. The main drawback of working only on minimum bounding boxes is when the two minimum bounding boxes of two regions are overlapped (See Fig. 2.a), it is not possible to compute any useful distance. Moreover, all these works consider that each object is ideally identified. In practice, this is not the case in a segmented image where objects are often arbitrarily over-segmented.

In this paper, we propose new topological and directional relations able to better describe complex spatial relations between two objects made up of several segmented regions. A concrete implementation of these relations is proposed to use it in the context of a constraint satisfaction problem with bilevel constraints. These relations are used as spatial constraints associated with the arcs of a semantic graph. Indeed, this formalism can describe many objects of an image [1], [3], [7], [15], [19]. In section 2, we describe the new spatial relations. In section 3 an implementation of these relations in the context of a CSP with bilevel constraints is proposed. In section 4, we present some experiments with different kinds of models applied on real images.

2 Complex Spatial Relations Between Two Composite Objects

2.1 Cardinal Direction Formalism

In the framework of the cardinal Direction Formalism (CDF), Skiadopoulos and Koubarakis [17] formally defined nine cardinal directions relations. (See Fig. 3). Let A be a region, the greatest lower bound of the projection of A on the x-axis (respectively y-axis) is denoted by infx(A) (respectively infy(A)). The least upper bound of the projection of A on the x-axis (respectively y-axis) is denoted by supx(A) (respectively supy(A)). The minimum bounding box of A, denoted by mbb(A), is the box formed by the rectangle where the coordinates of the left inferior corner are x1=infx(A), y1=infy(A) and the coordinates of the right superior corner are x2=supx(A), y2=supy(A). The single-tile cardinal direction relations can be defined as follows:

A O B iff infx(B)≤infx(A), supx(A)≤supx(B), infy(B)≤infy(A) and supy(A)≤supy(B)
   A S B iff supy(A) ≤ infy(B), infx(B) ≤ infx(A) and supx(A) ≤ supx(B)
   A SW B iff supx(A) ≤infx(B) and supy(A) ≤infy(B)
   A W B iff supx(A) ≤ infx(B), infy(B) ≤ infy(A) and supy(A) ≤ supy(B)
   A NW B iff supx(A) ≤ infx(B) and supy(A) ≤ supy(B)
   A N B iff supy(B) ≤ infy(A), infx(B) ≤ infx(A) and supx(A) ≤ supx(B)
   A NE B iff supx(B) ≤ infx(A) and supy(B) ≤ infy(A)
   A E B iff supx(B) ≤ infx(A), infy(B) ≤ infy(A) and supy(A) ≤ supy(B)
   A SE B iff supx(B) ≤ infx(A) and supy(A) ≤ supy(B)

Each multi-tile cardinal direction relation can be defined as follows:

a R1 : ... : Rk b, 2 ≤ k≤ 9 if there exists regions a1, ..., ak such that a= a1 ∪ ... ∪ak and   a1 R1 b, a2 R2 b, ..., ak Rk b.

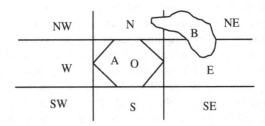

Fig. 3. cardinal direction relation between two regions A and B

2.2 The Connectivity-Direction-Metric Formalism (CDMF)

The minimum bounding boxes of two regions give some information about their spatial relations but this information is sometimes very poor, for example when one box overlapped another (Fig. 2a). In order to describe more complex spatial organization, we use three kinds of basic information:

(1) The notion of connectivity expressed in the topological framework of RCC8 by the primitive dyadic relation $C(x,y)$ read as "x connects with y".

(2) the notion of minimum bounding box introduced in the Cardinal Direction Formalism. Several properties can be deduced from this notion:

- The surface, width and height of a region can be computed.
- The directional relations between two regions. In our context we define four directional relations: N (North), S (South), W (West) and E (East).

$a\ N\ b$ iff $supy(b) \leq infy(a)$, $a\ S\ b$ iff $supy(a) \leq infy(b)$,
$a\ W\ b$ iff $supx(a) \leq infx(b)$, $a\ E\ b$ iff $supx(b) \leq infx(a)$

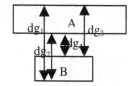 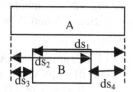

Fig. 4. The 8 metrics between two minimum bounding boxes: distances between A and B and lateral shifts between A and B

- Several metrics between two regions so long as the minimum bounding boxes of the two regions are not overlapped. Eight metrics between two minimum bounding boxes can be defined (see Fig. 4). For the north/south orientation, the definitions are: $dg_1(A,B) = supy(A)-infy(B)$, $dg_2(A,B) =infy(A)-infy(B)$, $dg_3(A,B) = supy(A)-supy(B)$, $dg_4(A,B)= infy(A)-supy(B)$, $ds_1(A,B)= supx(A)-infx(B)$, $ds_2(A,B)= supx(B)-infx(A)$, $ds_3(A,B)= infx(B)-infx(A)$, $ds_4(A,B)=supx(A)-supx(B)$. The definitions for the east/west orientation are similar. The eight relations defined in the CDF can be easily retrieved from our relations with the appropriate metrics.

(3) A new notion of minimum bounding box of border interfaces (mbbbi) between two regions for each main cardinal direction (N, S, E, W). This notion is defined in the following section.

Minimum bounding boxes of border interfaces between two regions. In order to make a more accurate spatial analysis we define the notion of minimum bounding boxes of "border interfaces". We mean by "border interface" the border part of a region which, given a cardinal direction, is in front of another region (See Fig. 2b).

Definition 1. Let R be a region (a set of connected pixels), we note $p(x,y)$ a pixel of R. $E(R)=\{p(x,y) \in R \mid \exists p(x',y')$ one of the 8 connected neighbors of $p(x,y)$, $p(x',y') \notin R\}$. Let be A and B two regions:

-      The border interface Cw(A,B) is defined by {p(x,y) ∈ E(A) such that ∃p(x' ,y) ∈ E(B) and ∀ p(x'',y) such that x< x''<x' p(x'',y) ∉ A and p(x'',y) ∉ B }
-      The border interface Ce(A,B) is defined by {p(x, y) ∈ E(A) such that ∃p(x', y) ∈ E(B) and ∀ p(x'', y) such that x> x''>x' p(x'', y) ∉ A and p(x'', y) ∉ B }
-      The border interface Cn(A,B) is defined by {p(x, y) ∈ E(A) such that ∃p(x ,y') ∈ E(B) and ∀ p(x, y'') such that y< y''<y' p(x, y'') ∉ A and p(x, y'') ∉ B }
-      The border interface Cs(A,B) is defined by {p(x, y) ∈ E(A) such that ∃p(x ,y') ∈ E(B) and ∀ p(x, y'') such that y> y''>y' p(x, y'') ∉ A and p(x, y'') ∉ B }

Definition 2. The minimum bounding box of a border interface in the direction d (mbbbi$_d$) is defined by (infx(Cd(a,b)),infy(Cd(a,b))),(supx(Cd(a,b)),supy(Cd(a,b)))

We can see on the example of Figure 2 that the two mbb of the regions A and B are overlapped. On the contrary the mbbbi$_w$ and the mbbbi$_e$ are not overlapped. Then, it is easy to know on this example that the region A is on the left side of region B.

Additional relations between two regions. Minimum bounding boxes of border interface (mbbbi$_w$, mbbbi$_e$, mbbbi$_n$, mbbbi$_s$) allow to describe additional relations. The four spatial relations between A and B linked to the corresponding mbbbi$_d$ can be defined as follows:

$$A \text{ Ei } B \text{ iff supx } (Cw(B,A)) \leq infx(Ce(A,B)),$$
$$A \text{ Wi } B \text{ iff supx}(Cw(A,B)) \leq infx (Ce(B,A)),$$
$$A \text{ Ni } B \text{ iff supy}(Cn(A,B)) \leq infy(Cs(B,A))$$
$$A \text{ Si } B \text{ iff supy}(Cn(B,A)) \leq infy(Cs(A,B)),$$

All these relations may be associated with the metric d defined as follows: d(A,B)= infz(A)-supz(B) where z = y for Ni or Si relationship, and z = x for Ei or Wi relationship.

Elementary relations in CDMF. CDMF allows to define very complex relationships by a combination of elementary relationships. An elementary relationship is a relation:

- (1) of connectivity or non connectivity
- (2) of directional relationship between mbb with none or one metric relation chosen among the metrics dsi and dgi (i=1…4) defined before (with inferior and superior limits). In that case, we have four directional relationships: N (North), S (South), W (West) and E (East).
- (3) of directional relationship between mbbbi with one metric relation d defined before (with inferior and superior limits). In that case, we have four directional relationships: Ni, Si, Wi and Ei.

Property 1. For each elementary relation ℜe between A and B, A ℜe B ⇒ ∃a ∈ A and ∃b ∈ B, a ℜe b.

Proof: ℜe of type (1). It is straightforward that
A connected to B ⇒ ∃a ∈ A and ∃b ∈ B, a connected to b.
A not connected to B ⇒ ∃a ∈ A and ∃b ∈ B, a not connected to b.

$\Re$e of type (2). Let R be one of the four relations N, S, E, W and d_i be one of the eight metrics defined between two mbb (i=1 ...8). Let min and max be the inferior and superior limits (in number of pixels) of the distance d_i. It is straightforward that A R B and min $\leq d_i$ (mbb(A),mbb(B)) $\leq$ max $\Rightarrow$ $\exists a \in A$ and $\exists b \in B$, a R b min $\leq d_i$ (mbb(a),mbb(b)) $\leq$ max.

$\Re$e of type (3). Let R be one of the four relations Ni, Si, Wi, Ei defined by using the mbbbi in section 2.2. Let d be the metric associated with the two mbbbi. Let min and max be the inferior and superior limits (in number of pixels) of the distance d. It is straightforward that A R B and min $\leq$ d (mbbbi$_R$ (A),mbbbi$_R^{-1}$(B)) $\leq$ max $\Rightarrow$ $\exists a \in A$ and $\exists b \in B$, a R b min $\leq$ d (mbbbi$_R$ (a),mbbbi $_R^{-1}$ (b)) $\leq$ max. R^{-1} is the opposite direction of R.

3 Application of the CDMF Relations to Over-Segmented Objects: Integration of the CDMF in a CSP with Bilevel Constraints

High level interpretation of images consists usually in matching each part of the image with a meaningful representation. The graph formalism is a very natural and convenient way to represent the semantic content of an image and the CDMF may be used to define node and arc constraints. Among several strategies [6], [16], we choose to perform this matching by solving a constraint satisfaction problem (CSP), because it better deals with complex directional spatial relationships. This aspect has been discussed in [9]. To reduce the time complexity of matching a graph with the different subparts of a shape, it is possible to only take into account local constraints. In practice, as problems are usually over-constrained, the arc-consistency checking is enough. Several authors [3], [15], [19] have proposed fast arc-consistency checking algorithms. These algorithms try to associate only one value with one node. This assumption supposes an ideal segmentation (one node of the graph is associated with only one region). In our context, the data are not ideally segmented and usually the objects present in an image are over-segmented in an arbitrary way depending on the grey level distribution in the image. The problem is: assuming A and B as two objects (regions) in an image, such that A $\Re$ B with $\Re$ a combination of $\Re$e of CDMF, how to define the relation $\Re'$ between any subpart a $\in$ A and b $\in$ B such that a $\Re'$ b $\Rightarrow$ A $\Re$ B ?

The elementary relationships of CDMF have an interesting property seen previously. For each elementary relation $\Re$e between A and B, A $\Re$e B $\Rightarrow$ $\exists a \in A$ and $\exists b \in B$, a $\Re$e b. Then the $\Re$e representing arc constraints in the graph formalism are valid to represent constraint on subparts of objects candidate to be matched with a node. However, due to the over-segmentation, a subpart of an object does not always satisfy all the constraints that make classical CSP fail. A solution was described in [7] by introducing two level of constraints in the classical CSP. The first level is the classical constraint between nodes, and the second level called $\mathcal{C}_{mpi}$ is an intra-node constraint. This second level defines how any subpart of an object, which does not satisfy a given inter-node constraint, has to satisfy an intra-node constraint with another region satisfying the inter-node constraint. In the following section, the notion of arc consistency checking with bilevel constraints is defined. Then, an example of its implementation and an example of using CDMF in this context are described. In particular, we will see how to express the complex spatial relationship like "is surrounded by".

3.1 Constraint Satisfaction Problem and Arc Consistency Checking with Bilevel Constraints

We use the following conventions:

- Variables are represented by the natural numbers 1, ... n. Each variable i has an associated domain D_i. We use D to denote the union of all domains and d the size of the largest domain.
- All constraints are binary and relate two distinct variables. A constraint relating two variables i and j is denoted by C_{ij}. $C_{ij}(v,w)$ is the Boolean value obtained when variables i and j are replaced by values v and w respectively. Let $\mathcal{R}$ be the set of these constraining relations.

We defined the Finite-Domain Constraint Satisfaction Problem with Bilevel Constraints (FDCSP$_{BC}$). One level of constraint is between each couple of nodes (spatial relations between objects associated with a node) and the other one level of constraint is between each couple of regions classified inside one node (spatial relations between subparts of the object associated with a node). These constraints are called $\mathcal{C}_{mpi}$ with i=1 ... n. This problem is defined as follows:

Definition 3. Let $\mathcal{C}_{mpi}$ be a compatibility relation, such that $(a,b) \in \mathcal{C}_{mpi}$ iff a and b are compatible. Clearly $\mathcal{C}_{mpi}$ is reflexive. Let C_{ij} be constraint between i and j. Let be a pair S_i, S_j such that $S_i \subset D_i$ and $S_j \subset D_j$, $S_i, S_j \mapsto C_{ij}$ means that (S_i, S_j) satisfies the oriented constraint C_{ij}.

$S_i, S_j \mapsto C_{ij} \Leftrightarrow \forall a_i \in S_i, \exists (a'_i, a_j) \in S_i \times S_j$, such that $(a_i, a'_i) \in \mathcal{C}_{mpi}$ and $(a'_i, a_j) \in C_{ij}$

and $\forall a_j \in S_j, \exists (a'_{j,ai}) \in S_j \times S_i$, such that $(a_j, a'_j) \in \mathcal{C}_{mpj}$ and $(a_i, a'_j) \in C_{ij}$.

Sets $\{S_1 ... S_n\}$ satisfy FDCSP$_{BC}$ iff $\forall C_{ij} \quad S_i, S_j \mapsto C_{ij}$.

A graph G is associated to a constraint satisfaction problem as follows: G has a node i for each variable i. Two directed arcs (i,j) and (j,i) are associated with each constraint C_{ij}. Arc(G) is the set of arcs of G and e is the number of arcs in G. Node(G) is the set of nodes of G and n is the number of nodes in G.

A class of problems called arc-consistency problems with bilevel constraints (AC$_{BC}$), associated with the FDCSP$_{BC}$ is defined as follows:

Definition 4. Let $(i,j) \in$ arc(G). Arc (i,j) is arc consistent with respect to $\mathcal{P}(D_i)$ and $\mathcal{P}(D_j)$ iff $\forall Si \in \mathcal{P}(Di) \exists Sj \in \mathcal{P}(Dj)$ such that $\forall v \in Si \exists t \in Si, \exists w \in Sj, \mathcal{C}_{mpi}(v,t)$ and $C_{ij}(t,w)$.(v and t could be identical)

Definition 5. Let P= $\mathcal{P}(D_1) \times .... \times \mathcal{P}(D_n)$. A graph G is arc-consistent with respect to P iff $\forall (i,j) \in$ arc(G): (i,j) is arc-consistent with respect to $\mathcal{P}(D_i)$ and $\mathcal{P}(D_j)$.

The purpose of an arc-consistency algorithm with bilevel constraints is, given a graph G and a set P, to compute P', the largest arc-consistent domain with bilevel constraints for G in P.

3.2 Implementation of the Arc-Consistency Checking Algorithm with Bilevel Constraints

The AC4 algorithm proposed by Mohr and Henderson [12] has been adapted to solve the AC_{BC} problem. We call this algorithm AC_{4BC} (See [7] for the details of the algorithm). In AC_{4BC}, a node belonging to node(G) is made up of a kernel and a set of interfaces associated with each arc, which comes from another linked node. In addition, an intra-node compatibility relation $\mathcal{C}_{mpi}$ is associated with each node of the graph. It describes the semantic link between different subparts of an object, which could be associated with the node. As in algorithm AC_4, the domains are initialized with values satisfying unary node constraints and there are two main steps: an initialization step and a pruning step. However, whereas in AC_4 a value was removed from a node i if it had no direct support, in AC_{4BC}, a value is removed if it has no direct support and no indirect support obtained by using the compatibility relation $\mathcal{C}_{mpi}$. The indirect supports are found thanks to the notion of interfaces.

3.3 Example of Implementation of the Relation "Is Surrounded" by Introducing the CDMF in the CSP_{BC}

Using the CDM Formalism, it is possible to define the notion "is surrounded by" with over-segmented regions (The graph can be seen in Fig. 5.1). "A is surrounded by B" is defined as follows: $\forall a \in A, \forall R \in \{N, S, W, E\} \exists c \in A$ or $\exists c \in B$, a connected to c and a R c. The possibility to authorized an "or" between the two constraints the consequence of the notion of quasi arc-consistency in AC_{BC} described in [8].

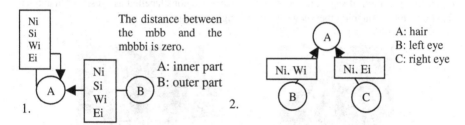

Fig. 5. 1. Graph used to work with the relation "is surrounded by" (for example, centre (A) is surrounded by petals (B) in a flower) 2. Graph used to work with "is partially surrounded by" (for example eyes are partially surrounded by hair)

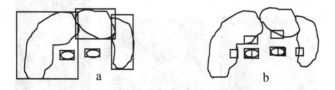

Fig. 6. a) With classical minimum bounding boxes it is not possible to compute the distance between the left eye and the hairs, b) with the mbbbi of the CDMF, it is possible to compute a distance

Another kind of relation is "partially surrounded by" with a given distance. This case can be encountered in the identification of the eyes and hair in a human face (See Fig. 6). In that case, the nodes representing the eyes and the node representing hair have to be related by the three constraints Ni, Ei and Wi. The distance d_{g4} is associated with the three relations (The graph can be seen on Fig. 5.2).

4 Experiments: Application to Check the Semantic Consistency of a Segmentation

Several kinds of test images representing different objects have been chosen. A set of images represents human faces, another set represent cars and finally another set represent flowers. For each kind of objects a semantic graph describing them has been built. The semantic consistency checking has been applied on the segmentation obtained with a pyramidal merging process [12] to stop automatically the merging at the

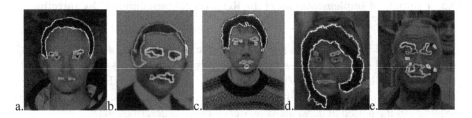

Fig.7. Interpretation of segmentation results of faces. Regions labelled as eyes, mouth and hair are shown by overlapping their edges with the original images.

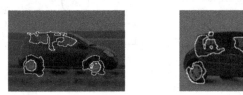

Fig. 8. Interpretation of segmentation results of cars (Labelled regions are tyres and lateral windows)

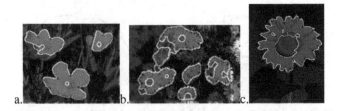

Fig. 9. Interpretation of segmentation results of flowers (labelled regions are centre and petals)

more meaningful pyramidal level (but it could be applied to other methods of segmentation providing a succession of embedded results with respect to the values of their parameters). On Figures 7,8 and 9 the regions with white edges are the obtained segmented regions correctly interpreted by the semantic analysis. In Fig. 7, the use of the quasi-arc consistency checking was necessary [8] to interpret the images because the node "hair" may be empty (see image 'e' of the Fig. 7).

5 Comments and Conclusion

In this article, we have proposed a new way to express complex spatial relation. We have shown that this set of new relations makes possible the expression of cardinal direction relations as well as crucial topological relations such as "is surrounded by ". Thanks to these relations, it is possible to build very precisely semantic graph describing an object made up of several subparts. With the $AC4_{BC}$ algorithm, this semantic graph can be used to retrieve objects inside an image. Some experiments have been made on real images, and we have shown that it is possible to detect very different kind of objects such as faces, cars, and flowers. This approach can be useful in the framework of image indexing to find some categories of images inside very large image databases. This work can be a theoretical foundation and embedding this approach into the MPEG-7 standard or into a realistic system can be a future work.

References

1. Bauckage, C., Braun, E., Sagerer, G.: From image features to symbols and vice versa – Using graphs to loop data- and Model-driven processing in visual assembly recognition. International Journal of Pattern Recognition and Artificial Intelligence 18(3), 497–517 (2004)
2. Bertolino, P., Montanvert, A.: Multiresolution segmentation using the irregular pyramid. In: proceeding IEEE ICIP96, Lausane, pp. 357–360 (1996)
3. Bessière, C.: Arc-consistency ans arc-consistency again. Artificial intelligence 65, 179–190 (1991)
4. Bookstein, F.L.: Morphometric Tools for Landmark data: Geometry and biology. Cambridge University Press, Cambridge (1991)
5. Cohn, A.G., Bennett, B., Gooday, J., Gotts, N.M.: Representing and reasoning with qualitative spatial relations about regions. In: Stock, O. (ed.) Spatial and Temporal reasoning, pp. 97–134. Kluwer, Dordrecht (1997)
6. Conte, D., Foggia, P., Sansone, C., Vento, M.: Thirty years of graph matching in pattern recognition. International Journal Pattern Recognition and Artificial Intelligence 18(3), 265–298 (2004)
7. Deruyver, A., Hodé, Y.: Constraint satisfaction problem with bilevel constraint: application to interpretation of over segmented images. Artificial Intelligence 93, 321–335 (1997)
8. Deruyver, A., Hodé, Y.: Image interpretation with a semantic graph: labeling oversegmented images and detection of unexpected objects. In: proceedings GBR, Ischia 23-25 mai 2001 Italie, Edition Cuen 2001, pp. 137-148 (2001)
9. Deruyver, A., Hodé, Y., Jolion, J.M.: Pyramides adaptatives et graphes sémantiques: segmentation dirigée par la connaissance. In: proceedings of Reconnaissance des Formes et Intelligence Artificielle conference 2006 (CD), Tours, January 2006, France (2006)

10. Hunter, J.: Enhancing the semantic interoperability of multimedia through a core ontology. IEEE Transaction on Circuits and Systems for Video Technology 12(1), 19–58 (2003)
11. Hunter, J.: Adding multimedia to the semantic web: Building an mpeg-7 ontology. In: proceeding of the International Semantic Web Working Symposium, Stanford University California, USA, pp. 261–283 (2001)
12. Jolion, J.M.: Stochastic pyramid revisited. Pattern recognition Letters 24, 1035–1042 (2003)
13. Keselmann, Y., Dickinson, S.: Generic Model Abstraction from Examples. IEEE Transaction on PAMI 27(7), 1141–1156 (2005)
14. Laemmer, E., Deruyver, A., Sowinska, A.: Watershed and adaptive pyramid for determining the apple's maturity state. In: proceeding IEEE ICIP, Rochester USA, pp. 789–792 (2002)
15. Mohr, R., Henderson, T.: Arc and path consistency revisited. Artificial Intelligence 28, 225–233 (1986)
16. Shearer, K., Bunke, H., Venkatesh, S.: Video indexing and similarity retrieval by largest common subgraph detection using decision trees. Pattern Recognition 34, 1075–1091 (2001)
17. Skiadopoulos, S., Koubarakis, M.: Composing cardinal direction relations. Artificial Intelligence 152(2), 143–171 (2004)
18. Hai-Bin, S., Wen-Hui, L.: Qualitative spatial relationships cleaning for spatial data mining. In: procceding of the Fourth International Conference on Machine Learning and Cybernetics Guangzhou, August 18-20, 2005, pp. 1851–1857 (2005)
19. Van Hentenryck, P., Deville, Y., Teng, C.M.: A generic arc-consistency algorithm and its specializations. Artificial Intelligence 57(2), 291–321 (1992)

Separation of the Retinal Vascular Graph in Arteries and Veins

Kai Rothaus, Paul Rhiem, and Xiaoyi Jiang

Department of Mathematics and Computer Science, University of Münster
Einsteinstrasse 62, D-48149 Münster, Germany
{rothaus,rhiem,xjiang}@math.uni-muenster.de

Abstract. The vascular structure of the retina consists of two kinds of vessels: arteries and veins. Together these vessels form the vascular graph. In this paper we present an approach to separating arteries and veins based on a pre-segmentation and a few hand-labelled vessel segments. We use a rule-based method to propagate the vessel labels through the vascular graph. We embed this task as double-layered constrained search problem steered by a heuristical AC-3 algorithm to overcome the NP-hard computational complexity. Results are presented on vascular graphs generated from hand-made as well as on automatical segmentation.

Keywords: Retinal vascular graph, artery, vein, constrained satisfaction problem, constrained propagation.

1 Introduction

The automated analysis of retinal images has been an active research area for a long time. Automated analysis tools together with interactive treatment interfaces help ophthalmologists to interpret the large amount of examination data and draw their diagnostic conclusions. Typical retinal images are shown in Figures 6 to 8. For better visualisation we present the inverted and linearly stretched green channel of the original colour RGB-image. The dark roundish area represents the so-called optical disc, where the optic nerve and all vessels enter the interior of the eye. Starting from this anchor point the vessels spread out in a tree-like structure over the retinal surface and become thinner when they branch until they are invisible as capillaries. At the macula, which is the brightest region of the retina and the region of the highest density of light receptor cells, no vessels are visible anymore.

The most important task in the field of retinal image analysis is the segmentation [2,3,6,9,10] and classification of the vascular structure. Vessels are crucial for a variety of tasks including registration, detection of other features (drusen, microaneurisms, hard exudates or cotton woll spoons), and diagnostic purposes.

Our focus in this work lies on labelling the vessels. Therefore, we assume that the vascular structure is already segmented suitably. More precisely, we claim a segmentation of a retinal image in form of a binary image, where 1

F. Escolano and M. Vento (Eds.): GbRPR 2007, LNCS 4538, pp. 251–262, 2007.
© Springer-Verlag Berlin Heidelberg 2007

represents a vessel pixel (object) and 0 any other (background). There is a strong medical need of the discrimination of arteries and veins. One application is the measurement of the AV-ratio, which is the ratio of the arteries and veins calibre. Another important reason is the selective computation of a feature (e.g. vessel tortuosity) for arteries and veins separately, since some pathological changes effect only one kind of vessels [4].

There are only a few works of automatically discriminating between arteries and veins in retinal images. Simo et al. [8] proposed a Bayesian classifier for pixels to distinguish between arteries, veins, the fovea and the retinal background using image informations.

Furthermore, Akita et al. [1] use a structure-based relaxation scheme to propagate the artery/vein labelling. They model conditional probabilities of segments on basis of the vessel structure. These probabilities influence each other and are updated in each iteration until a stable state is achieved. Thereby, the structure of the vascular graph is kept fixed. Our approach is different to the one of Akita et al. [1] in that we regard the vessel segments to be uniquely an artery or vein. To solve conflicts, we will change the structure of the vascular graph and thus indirectly correct segmentation errors (to some degree). To our best knowledge, there are no other structure-based methods for separation of veins and arteries in retinal images.

Martinez-Perez et al. [5] present an approach to extract separate vascular sub-trees. For the extraction of the vascular graph we use similar methods, but while Martinez-Perez et al. basically are interested in computing geometrical and topological properties of single vessel segments and sub-trees, we focus in this work on the vascular structure itself.

The remainder of this paper is organised as follows. In Section 2 we specify and formalise the problem of separating the vascular graph under anatomical aspects. Subsequently, our algorithm is presented (Sec. 3) and exemplary results are shown (Sec. 4). We conclude this work by discussing the performance of the results and giving an outlook on our future work (Sec. 5).

2   Formal Problem Specification

There are two different kinds of vessels on the retina. The arteries transport oxygenated blood from the heart and the veins discharge the blood back to the heart. We utilise two important anatomical characteristics of these structures:

1. The visible vascular structure is physically cycle-free (although its projection onto the 2D image plane becomes a vascular graph with cycles). One artery enters at the optic nerve head into the interior of the retina and branches without any reconnection (i.e. without anastomosis).
2. At vessel crossings, where one vessel courses over another, only different vessel types are involved. More precisely, an artery could never cross another artery and the same is valid for two veins.

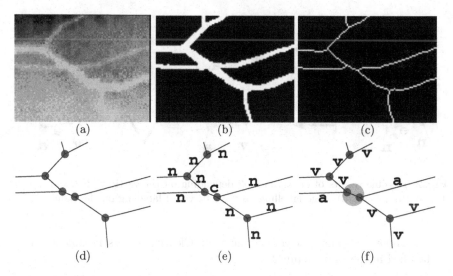

Fig. 1. (a) inverted original; (b) segmentation; (c) skeleton; (d) vascular graph; (e) edge labelling, n=normal edge, c=crossing edge (see Sec. 3) ; (f) vessel labelling

2.1   Graph-Based Representation

We compute the graph representation in a similar way as described by Martinez-Perez et al. [5]. The precondition of the proposed approach is the segmented vascular structure (Fig. 1(b)). Firstly, we apply a sequential skeletonisation procedure that produces an 8-connected skeleton (see Fig. 1(c)). Once the skeleton is calculated the transformation into a graph $\mathcal{G}$ is straightforward. Because of the vessels' anatomical structure the vertices of $\mathcal{G}$ can only have degree 1, 3 or 4. In contrast to Martinez-Perez et al. [5] we do not distinguish the type of vertices on the image analysis level but on anatomical aspects of the vascular structure.

Let us assume that we can represent the vessels as curvilinear segments s_i, which could branch and cross. We construct a planar graph $\mathcal{G} = (\mathcal{E}, \mathcal{V})$, where each edge $e_i \in \mathcal{E}$ $(1 \le i \le m)$ corresponds to a vessel segment s_i in a one-to-one relation. The nodes v_j $(1 \le j \le n)$ of the graph represent the branches or crossings of vessel segments and are of degree three (branches) or four (crossings). Additionally, there are nodes of degree one, where a vessel segment ends. If two vessels cross each other, both are splitted in two vessel segments represented by two edges. Thereby, the opposite segments form one vessel segment pair of the same type in each case.

2.2   SAT-Problem Description

The problem we are faced with is to find a consistent labelling $L\,(s_i) := L_i$ of all vessels segments s_i in arteries $L_i =$ a or veins $L_i =$ v. One can define a rule for a consistent labelling at each vertex v_j depending on its degree: In the case that v_j is a branch, we have three vessel segments with corresponding graph-edges

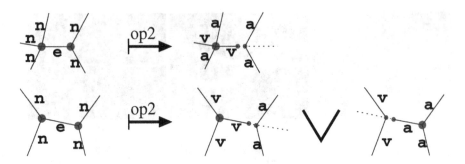

Fig. 2. Possible results of operation op2 depending on the vertices' degrees (top row 4/3, bottom row 3/3), edge labelling on the left, vessel labelling on the right

e_{i_1}, e_{i_2} and e_{i_3} of same type connecting at v_j. Clearly, all vessel segments should be labelled in the same manner:

$$L_{i_1} = \text{a} \quad \Leftrightarrow \quad L_{i_2} = \text{a} \quad \Leftrightarrow \quad L_{i_3} = \text{a} \tag{1}$$

$$\wedge \quad L_{i_1} = \text{v} \quad \Leftrightarrow \quad L_{i_2} = \text{v} \quad \Leftrightarrow \quad L_{i_3} = \text{v} \tag{2}$$

From the second anatomical characteristic, we know that if v_j represents a crossing, then one of the involved vessels is an artery and the other is a vein. Physically, these vessels do not cross, but one is taking course above the other. Since vessels are relatively straight, we can assume that a diagonally opposed vessel segment pair represents the same physical vessel and hence the segments belong to the same vessel type. Let e_{i_1}, e_{i_2}, e_{i_3} and e_{i_4} denote the graph-edges linked at v_j. Since the two pairs (e_{i_1}, e_{i_3}) and (e_{i_2}, e_{i_4}) belong to the same vessel, we can formulate the following two rules:

$$L_{i_1} = \text{a} \quad \Leftrightarrow \quad L_{i_3} = \text{a} \quad \Leftrightarrow \quad L_{i_2} = \text{v} \quad \Leftrightarrow \quad L_{i_4} = \text{v} \tag{3}$$

$$\wedge \quad L_{i_1} = \text{v} \quad \Leftrightarrow \quad L_{i_3} = \text{v} \quad \Leftrightarrow \quad L_{i_2} = \text{a} \quad \Leftrightarrow \quad L_{i_4} = \text{a} \tag{4}$$

In this way, the extracted vascular graph $\mathcal{G}$ leads to a finite set $\mathcal{S}$ of rules. Each rule (1) to (4) is a conjuction of simple logical clauses of the form

$$((L_\mu \neq \alpha) \quad \vee \quad (L_\nu = \beta)) \quad \wedge \quad ((L_\nu \neq \beta) \quad \vee \quad (L_\mu = \alpha)) \tag{5}$$

Note that L_i can be considered as boolean variable, since we can identify the label a with `true` and v with `false`. In other words each node leads to a number of constraints of the form (5). In a natural way we get a satisfiability problem in m variables $L_1, \ldots, L_m$. Obviously, the rules (1) and (2) as well as rules (3) and (4) are redundant so that we keep only the rules (1) and (3) in our rule set $\mathcal{S}$.

The SAT-problem is a special case of constraint satisfaction problems and can thus be solved by standard algorithms like AC-3 (see e.g. [7]) to overcome the NP-hard computation trap. After solving the SAT-problem each edge of the graph $\mathcal{G}$ is labelled by either a or v. Then, the subset of all edges labelled by a and v corresponds to arteries and veins, respectively.

2.3 Unsatisfiable Vessel Labelling

During the segmentation process and the segment identification (e.g. a skeleton algorithm) an erroneous graph representation of the vascular structure may be computed. Typical mistakes are (1) splitting of one crossing into two branches, (2) missing vessel segment on a side of a crossing, (3) falsely detected, i.e. non-existing, vessel segments. The result of these mistakes is usually a globally unsolvable SAT-problem.

For this reason we have to manipulate the graph $\mathcal{G}$ at few selected vertices so that the SAT-problem (Sec. 2.2) becomes solvable. The selection itself is controlled by the algorithm described in Sec. 3. We allow the following operations:

op1 Combining two adjacent branch vertices to one crossing (Fig. 1)
op2 Defining an edge as end segment (connected to only one branch or crossing)
op3 Deleting an edge

Instead of manipulating the graph directly we introduce an auxiliary labelling of vessel segments. To distinguish between the two different types of labelling, we denote the following as *edge labelling* and the discrimination between veins and arteries (Sec. 2.2) as *vessel labelling*. Possible labels for graph-edges are:

c connection between two branches, which should establish a crossing
e artificial end segment, where only one of two vertices is relevant
f falsely detected segment
n normal segment

These edge labelling allow us to adjust the set of applicable clauses at each node in the conflict resolution phase of the algorithm.

The corresponding graph manipulation operation for a c-labelled edge e is op1. To adapt our clause set S (Sec. 2.2) we replace the rules for the two branch vertices, to which e is connected, with the rule (3) of a crossing vertex.

An end segment labelling e (equivalent to op2) is only allowed if at most one of the vertices has degree three and the other has degree three or four (see Fig. 2). The adaption of the rule set S depends on the degree of this second vertex and can be done in a straight-forward way. Note that in the case that both adjacent vertices have degree three, there are two possible interpretations. Furthermore two adjacent edges/segments cannot be labelled as end segments, to avoid an over-fragmentation.

Noisy segments (label f) are simply thrown away by removing the corresponding edge out of $\mathcal{G}$ and deleting the corresponding clauses out of S.

2.4 Optimisation Task

In Section 3 we will introduce plausibility weights for the vertices and edges of $\mathcal{G}$. Based upon these ratings we are interested in that solution, which results in a maximum average plausibility. In other words, we search a labelling of graph-edges (Sec. 2.3) so that the resulting SAT-problem (Sec. 2.2) is solvable and the average plausibility is maximised.

Initially, all vertices v_j are assessed with a plausibility value $w(v_j)$, which should regard the reliability of the assessed rules of v_j. Furthermore each edge e_i is assigned with the plausibility weight $w(e_i) = 1$ or 0 if the corresponding vessel segment is hand-labelled or not. Based upon the order of applying constraints the weights of the new labelled segments are updated by a multiplicative propagation scheme (see Eq. 10 in Sec. 3.2). The optimisation task is then given by minimising

$$\frac{1}{n} \cdot \sum_{i=1}^{n} w(e_i) \tag{6}$$

This average plausibility depends on the number of hand-labelled segments and on the order of solving conflicts. Note that in case of an unsolved conflict, wide parts of the graph are left unprocessed and hold the plausibility weight 0.

3   Graph Separation

The problem we are faced with consists of two layers. The basic layer is the structure of the vascular graph. On this layer graph edges are labelled with either c, e, f or n. This edge labelling defines the constaint set S, which conditions the second, high-level layer. Our approach is to apply on this higher level belief propagation techniques and solve conflicts by adjusting the basic layer. More precisely, if contradictory information are propagated to a vessel segment (competing a/v-labelling), we do not propagate the more likely information, but reorganise the structure. This update results in a different, more realistic graph structure.

We use a two-stage approach for the labelling of the vessel segments. In the first stage (Sec. 3.1), we compute an initial labelling of graph-edges. All edges are labelled by n, except for those connecting two branches. In the latter case the edge is labelled by either n or c (indicating a merge of the two branches to build a crossing according to graph manipulation operation op2). The decision rule for taking one of the two labels n/c is given later in Sec. 3.1.

The second (higher-order) stage performs the labelling of the vessel segments by a variant of AC-3 (Sec. 3.2). This algorithm tries to label all vessel segments as arteries a or veins v. Arising conflicts, which make a consistent labelling impossible (i.e. AC-3 fails), are resolved by a backtracking procedure. The idea is to modify the edge labelling at appropriate edges according to graph manipulation operations op1–op3.

3.1   Initial Edge Labelling

To compute an initial edge labelling, we firstly decide which edges should be labelled with c. All other edges are assumed to be normal edges (label n). It is important to note that a c-label manipulates the graph structure by merging two branch vertices (degree 3) to a crossing vertex (degree 4). Also the rule set S has to be adapted accordingly.

Let s_c denote a vessel segment (respectively a graph-edge) to which we want to assign its initial labelling (c or n). Furthermore $s_1, \ldots, s_4$ are the other involved

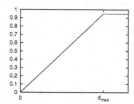 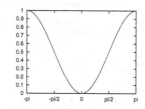 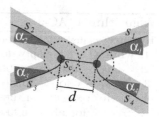

Fig. 3. Plausibility $\mathcal{P}_1$ of adjacency

Fig. 4. Plausibility $\mathcal{P}_2$ of collinearity

Fig. 5. Weighting of adjacent branches

vessel segments (see Fig. 5). Two properties help us decide if the two detected branch vertices of s_c actually belong to one vessel crossing:

P1 The distance d between two branches is relatively small

P2 The two segments s_1 and s_3 are roughly collinear, similarly s_2 and s_4

We are modelling **P1** and **P2** by the following two plausibility functions:

$$\mathcal{P}_1 : (0, \infty) \to (0, 1], \quad \mathcal{P}_1(d) = 0.95 \cdot \min\{d/d_{\max}, 1\}, \quad d_{\max} > 0 \qquad (7)$$
$$\mathcal{P}_2 : [-\pi, \pi] \to [0, 1], \quad \mathcal{P}_2(\beta) = 0.5 \cdot (1 - \cos(\beta)) \qquad (8)$$

The constant factor $d_{\max}$ should depend on the resolution of the retinal image. We compute the argument β of $\mathcal{P}_2$ as

$$\beta = \max_{\alpha_i}\{\alpha_i\} \qquad (9)$$

Here $\alpha_1, \ldots, \alpha_4$ are the inner angles between s_c and $s_1, \ldots, s_4$. If both $\mathcal{P}_1$ as well as $\mathcal{P}_2$ are low, we could assume a crossing instead of two branches. We define thresholds T_1 and T_2 and label the segments with c if $\mathcal{P}_1 < T_1$ and $\mathcal{P}_2 < T_2$.

The thresholds T_1 and T_2 are determined by optimising this classifier. We hypothesise "segment s_c is a normal vessel" and minimise the beta error on a significance level $\alpha = 0\%$. We have examined 11 vascular graphs with 763 "inner" edges. With the conducted thresholds $T_1 = 0.75$ and $T_2 = \mathcal{P}_2(\pi/6)$ all 621 normal edges are correctly labelled with n. On the other hand only 25 of 142 crossing edges are falsely labelled with n. This corresponds to a beta error of about 17.6% and a total error of about 3.3 %.

3.2 Consistent Labelling Search

In the following an extension of the AC-3 algorithm [7] is proposed, which is controlled by a priority queue $\mathcal{Q} \subset \mathcal{V}$. We presume that $\mathcal{Q}$ initially contains some vertices of hand-labelled edges. Since the vascular structure at the optic disc is very compact, which makes the differentiation of the vessels even for human eyes hardly solvable, we define a circle around the optic disc where the algorithm does not proceed the labelling.

Algorithm 1. AC-3*

Require: clause set $\mathcal{S}$, vertices of some hand-labelled edges in queue $\mathcal{Q}$
1: **while** $\mathcal{Q}$ is not empty **do**
2:     $v_i \leftarrow$ REMOVE-HEAD$(\mathcal{Q})$
3:     $s_j \leftarrow$ CHOOSE-RULE$(v_i, \mathcal{S})$
4:     **if** CONSISTENT-LABELLING(v_i, s_j) **then**
5:         **for all** v_k in NEIGHBOURS$[v_i]$ **do**
6:             **if** edge between v_k and v_i was not labelled before **then**
7:                 add v_k to $\mathcal{Q}$
8:             **end if**
9:         **end for**
10:     **else**
11:         BACKTRACKING-SEARCH(v_i)
12:     **end if**
13: **end while**

In each step the best vertex is taken out of $\mathcal{Q}$. Since every vertex in $\mathcal{Q}$ has at least one uniquely labelled edge, we can now propagate the available labels to all other connected edges according to the vertex type.

To arrange the order of processing, we need a heuristic $H(v_i)$ that decides "how good" a vertex v is to be treated in the next processing step. For defining the heuristic H we weight the vertices and edges of $\mathcal{G}$ with values $w \in [0,1]$. The initial weights are defined as follows:

- $w(v) = \mathcal{P}_1(d)$ for a crossing vertex v, where d is the distance to the nearest neighbour of v
- $w(v) = \mathcal{P}_1(d) + \mathcal{P}_2(\beta) - \mathcal{P}_1(d) \cdot \mathcal{P}_2(\beta)$ for a branch vertex v and its nearest neighbour vertex (see Fig. 5)
- $w(e) = 1$ if e is a hand-labelled vessel segment (edge)

All other vertices and edges are initialised with zeros and updated in the labelling process as follows. If the vertex v is processed (i.e. taken out of $\mathcal{Q}$), we take the maximum weighted edge e at v and update the weights for the other edges e', if still unlabelled, by

$$w(e') = w(e) \cdot w(v) \tag{10}$$

Furthermore the other vertex v' of such an e' is added to $\mathcal{Q}$ with heuristic

$$H(v') = w(v') \cdot w(e') \tag{11}$$

Note that in the labelling propagation described above we may encounter a conflict situation, where an edge already has a unique label which is incompatible with the new one. In such a situation a backtracking procedure is started. To solve the conflicts we modify the graph indirectly by using another auxiliary labelling (Sec. 2.3). When AC-3* detects a conflict while processing a vertex v_i the backtracking procedure searches in the neighbourhood of v_i for a suitable edge, which will be labelled differently.

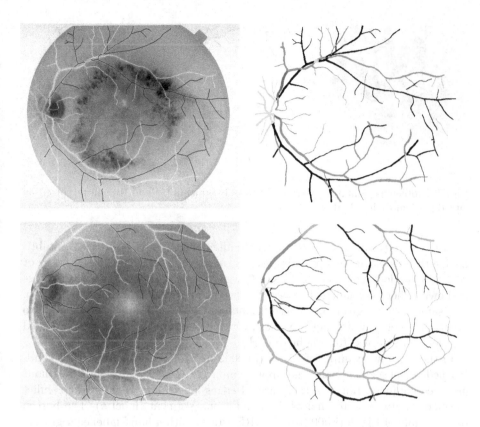

Fig. 6. Vessel classification on hand-segmented retinal images

Firstly, we search either a cycle or a path which ends at two hand-labelled vessels, which caused the conflict. Let $\mathcal{V}'$ denote the vertices on such a chain and $\mathcal{E}' := \{e \in \mathcal{E} | \exists_{v \in \mathcal{V}'} e \text{ incident to } v\}$ the set of incident edges. We inspect all incident edges and relabel the most adequate edge with **c**. Thereby, the most adequate edge is a **n**-labelled edge e, which satisfies the constraints $\mathcal{P}_1 < T_1'$ and $\mathcal{P}_2 < T_2'$ ($T_1' > T_1$ and $T_2' > T_2$) and minimise the plausibility $\mathcal{P}_1 + \mathcal{P}_2 - \mathcal{P}_1 \cdot \mathcal{P}_2$. If no suitable edge is found the algorithm assumes that there is an end segment at v_i and uses the label **e** accordingly.

4   Results

We have tested our vessel classification method on the STARE data set of Hoover et al. [2]. We use their ground-truth segmentation as input for our graph computation procedure, ignore the vessels inside the optical disc and start our AC-3* algorithm with a few hand-classified vessels. This user interaction could be avoided by an automatic detection of the optic disc and an automatic classification of close-by dominant vessels.

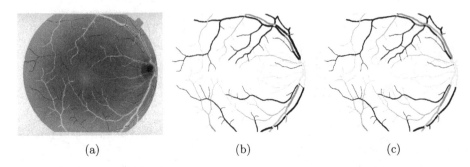

(a)   (b)   (c)

Fig. 7. Problems originated from nearby vessel segments (a) and (b) can be solved by introducing more hand-labelled vessels (c)

In Figure 6 results are depicted, where all conflicts could be treated suitably. Beside the vessels at the optic disc each segmented vessel is labelled as artery or vein. On the left hand side of Figure 6 a gray-scale version of the input image is shown, where the labelled arteries and veins are overlaid in black and white. The same result is given on the right hand side, by shading the segmentation to distinguish between unlabelled vessels, veins, hand-labelled veins, hand-labelled arteries and arteries (from light to dark).

In the top row example of Fig. 6 (im0002 in STARE) we start with 2 hand-labelled vessel segments and end up with an average plausibility of 0.14 without unsolved conflicts. Initially, there are 14 edges labelled with c and the conflict solver recognise 3 additional c-edges and 1 end segment (label e). The botton row example of Fig. 6 (im0081 in STARE) starts with 4 hand-labelled segments and 22 c-edges. The graph separation process solves all conflicts by adding 1 end segment and 3 more c-edges. The average plausibility of the result is 0.2.

The example in Fig. 7 (a) (im0082 in STARE) is more difficult to proceed. If we label 4 vessel segments by hand (Fig. 7 (b)) the program stops with an average plausibility of 0.18. Thereby, only 6 of 10 conflicts are solved and two end segments are introduced. The problem arises mainly from the overlaid crossing and branching in the upper right part of the retina. If such a situation occurs the user can correct the labelling by introducing more hand-labelled vessels at ambiguous regions of the retina. But even if we label 6 vessels by hand (Fig. 7(c)) only 7 of 9 conflicts are solved by adding 4 c-labels and 3 e-labels.

We also conducted some tests on automatic segmentations. The problem hereby is that we need a connected representation of the vasculature. Otherwise, the connectivity is lost and our graph separation algorithm could not proceed the forward labelling. This problem is the main reason why our approach could not handle conflicts suitably. Figure 8 presents two result on an automatic segmented image of the DRIVE data set [10] using the publically available vessel segmentation algorithm of Soares et al. [9]. The difficulty of the first example (top row in Fig. 8, image 16 in DRIVE) is due to the fact that some vessels are non-continuously segmented (main vessels at the upper right region of the retina). Overall we get an average plausibility of 0.08, based on 6 hand-labelled

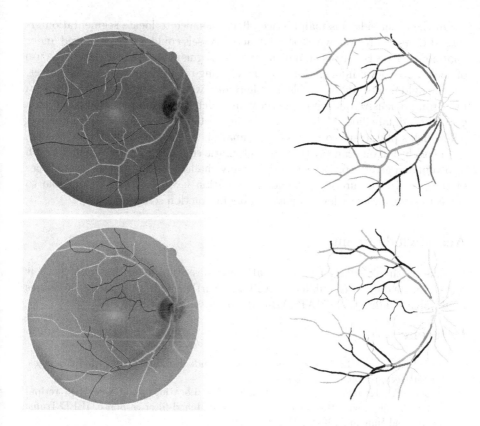

Fig. 8. Vessel classification on automatically segmented retinal image

vessels. All conflicts are solved by introducing 6 new c-label and 0 e-labels. This fact is partly caused by the weak connectivity of the vessel segmentation result, which eases the labelling task since there are fewer constraints compared to a fully connected vascular graph. However, it increases the risk of making erroneous decisions. This effect is shown by the second example (bottom row of Fig. 8, image 10 in DRIVE), where 3 of 6 conflicts are still unsolved, since the connectivity of the vascular structure is not adequate enough to correct falsely labelled edges. The average plausibility of this example is with 0.03 the lowest of the examples, which are presented in this work.

5   Discussion and Conclusions

We have presented an automated graph separation algorithm to distinguish between arteries and veins in retinal images. The performance of our algorithm mainly depends on the quality of the vessel segmentation algorithm. Although we could correct isolated segmentation errors, cumulatively occurred missing or non-continuous vessels are still difficult to handle.

On the other side this malpractice offers a chance to locate segmentation errors. If there is a region, where the extracted vessels could not be arranged under anatomical aspects, this is an indication for a segmentation error. Beside the use of such higher level information, we are working on an extended conflict solver, which could delete edges/vessels. Furthermore, we want to justify our algorithm by a comprehensive quantitative study, in which our results are compared to a ground truth labelling. Moreover, we are interested in an adaptive user interface, which helps physicians to process their analysis in a more efficient way.

In conclusion one can say that the automatic separation of the vascular graph into vein and artery components offers many chances to interpret characteristics of the vessel structure on a higher information level. It could be possible to correct errors on lower levels by a suitable interaction scheme.

Acknowledgement

We like to thank Adam Hoover et al. and Joes Staal et al. for making their retinal image databases publically available. Furthermore, we thank João Soares et al. for providing their MATLAB-code for vessel segmentation.

References

1. Akita, K., Kuga, H.: A computer method of understanding ocular fundus images. Patter Recognition 15(6), 431–443 (1982)
2. Hoover, A., Kouznetsova, V., Goldbaum, M.: Locating blood vessels in retinal images by piece-wise threshold probing of a matched filter response. IEEE Trans. on Medical Imaging 19(3), 203–210 (2000)
3. Jiang, X., Mojon, D.: Adaptive local thresholding by verification-based multi-threshold probing with application to vessel detection in retinal images. IEEE Trans. on PAMI 25(1), 131–135 (2003)
4. Larsen, M., Colmorn, L.B., Bonnelycke, M., Kaaja, R., Immonen, I., Sander, B., Loukovaara, S.: Retinal artery and vein diameters during pregnancy in diabetic women. Investigative Ophthalmology and Visual Science 46, 709–713 (2005)
5. Martinez-Perez, M.E., Hughes, A.D., Stanton, A.V., Thorn, S.A., Chapman, N., Bharath, A.A., Parker, K.H.: Retinal vascular tree morphology: A semi-automatic quantification. IEEE Trans. on Medical Imaging 49(8), 912–917 (2002)
6. Rothaus, K., Jiang, X.: Multi-scale segmentation of the vascular trees in retinal images. In: Proc. of 3rd EMBEC (2005)
7. Russel, S., Norvig, P.: Artificial Intelligence: A Modern Approach, 2nd edn. Prentice Hall International. Prentice-Hall, Englewood Cliffs (2003)
8. Simo, A., de Ves, E.: Segmentation of macular fluorescein angiographies. a statistical approach. Patter Recognition 34(4), 795–809 (2001)
9. Soares, J.V.B., Leandro, J.J.G., Cesar Jr, R.M., Jelinek, H.F., Cree, M.J.: Retinal vessel segmentation using the 2-d morlet wavelet and supervised classification. IEEE Trans. on Medical Imaging 25, 1214–1222 (2005)
10. Staal, J.J., Abramoff, M.D., Niemeijer, M., Viergever, M.A., van Ginneken, B.: Ridge based vessel segmentation in color images of the retina. IEEE Trans. on Medical Imaging 23, 501–509 (2004)

A Fast Construction of the Distance Graph Used for the Classification of Heterogeneous Electron Microscopic Projections

Miroslaw Kalinowski[1], Alain Daurat[2], and Gabor T. Herman[1]

[1] Department of Computer Science
The Graduate Center City University of New York
365 Fifth Avenue New York, NY 10016 USA
mkalinowski@mikconsulting.net, gabortherman@yahoo.com
[2] LSIIT CNRS UMR 7005, Université Louis Pasteur (Strasbourg 1), Pôle API,
Boulevard Sébastien Brant, 67400 Illkirch-Graffenstaden, France
daurat@dpt-info.u-strasbg.fr

Abstract. It has been demonstrated that the difficult problem of classifying heterogeneous projection images, similar to those found in 3D electron microscopy (3D-EM) of macromolecules, can be successfully solved by finding an approximate Max k-Cut of an appropriately constructed weighted graph. Despite of the large size (thousands of nodes) of the graph and the theoretical computational complexity of finding even an approximate Max k-Cut, an algorithm has been proposed that finds a good (from the classification perspective) approximate solution within several minutes (running on a standard PC). However, the task of constructing the complete weighted graph (that represents an instance of the projection image classification problems) is computationally expensive. Due to the large number of edges, the computation of edge weights can take tens of hours for graphs containing several thousand nodes. We propose a method, which utilizes an early termination technique, to significantly reduce the computational cost of constructing such graphs. We compare, on synthetic data sets that resemble projection sets encountered in 3D-EM, the performance of our method with that of a brute-force approach and a method based on nearest neighbor search.

1 Introduction

The motivation for this work comes from three-dimensional electron microscopy (3D-EM) [1]. One of the important challenges encountered in 3D-EM is the problem of partitioning heterogeneous projection sets into their homogeneous components. Due to the unique nature of the projection images, high level of noise and other distortions affecting them, this image classification problem is very difficult. Despite these difficulties, several methods have been recently proposed to solve it [2,3,4]. One of these methods [4] employs graphs to achieve the desired classification. In this method a large complete weighted graph is constructed. Each node of this graph represents a single projection image. The

F. Escolano and M. Vento (Eds.): GbRPR 2007, LNCS 4538, pp. 263–272, 2007.
© Springer-Verlag Berlin Heidelberg 2007

weight of each edge describes the dissimilarity of the images represented by the nodes it connects. The desired image classification is achieved by finding an approximate solution to the Max k-Cut problem [5] (for $k = 2$ this is equivalent to finding an approximate solution of the maximum capacity cut problem [6]).

In general, solving large instances of the Max k-Cut problem is computationally expensive. Even the problem of finding a good approximation to the Max k-Cut is NP-complete [5]. However, an efficient algorithm for finding approximate Max k-Cuts of graphs that originate from (3D-EM) has been developed [4]. Due to unique nature of such graphs, the proposed graph cutting algorithm of [4] is able to find good (from the classification perspective) approximate Max k-Cuts for graphs (containing thousands of nodes) within several minutes of runtime on a standard personal computer.

In comparison to other methods, the graph-based approach to the classification of heterogeneous projection images is quite efficient. However, the cost of constructing the complete weighted graphs, used by this method, increases proportionally to the square of the number of projection images. Consequently, the vast majority of time necessary to classify large sets of projection images is dedicated to graph construction. For example (without optimizations) it takes approximately 24 hours on a single processor (Intel Xeon 1.7GHz) to construct a graph for a data set obtained from 5,000 images. Since the calculations of edge weights between different nodes of the graph are mutually independent, the task of constructing the graph can be easily parallelized. However, significant resources are necessary to construct graphs for large data sets, containing tens of thousands projection images, which are frequently encountered in 3D-EM of biological objects. The objective of the work presented here is to provide an algorithm that reduces the computational cost of constructing such graphs.

The difficulty associated with the classification of projection images comes from the fact that two images belonging to the same class may be less similar (in the traditional sense) than two images belonging to different classes (see Figure 1). In order to overcome this difficulty, a special projection image dissimilarity measure has been proposed [4]. The value of this dissimilarity measure for a pair of 2D images $\bar{x}$, $\bar{y}$ is calculated as

$$S(X, Y) = \min_{x \in X, y \in Y} s(x, y), \text{ with } s(x, y) = \sum_{i=1}^{N}(x_i - y_i)^2, \qquad (1)$$

where X and Y are sets of 1D projections (N-dimensional vectors x and y in $\mathbb{R}^N$, where N is typically in the range between 60 to 150) that are obtained by projecting the images $\bar{x}$ and $\bar{y}$ (respectively) at several hundred (parameter of the measure) evenly distributed projection angles within the plane of each image. This dissimilarity measure is used to calculate the edge weights in the graphs used to classify of 3D-EM projection images. The process of calculating a single weight can be interpreted as finding a squared distance between two sets of points in multi-dimensional Euclidean space.

This distance can be obtained by solving a series of nearest neighbor search (NNS) problems. Since the NNS problem is frequently encountered in database

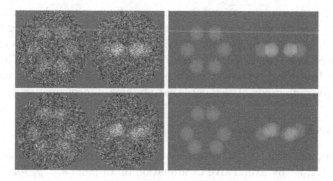

Fig. 1. Left column: sample of images used in experiments; top: two 2D projections of object S6, bottom: two 2D projections of object S6x. Right column: corresponding noiseless projection images

searching, pattern recognition, and data compression [7,8], a number of algorithms to solve it have been proposed. However, due to the high dimensionality of our $\mathbb{R}^N$, not all of these algorithms can be applied to the problem of building graphs for 3D-EM image classification. When the dimensionality of the space is high, the computation time of NNS can be reduced only by the use of metric properties of the distance [9]. Our attempt to apply NNS techniques to the graph building problem is based on the AESA [10] algorithm. AESA was chosen because it is one of the fastest ways of finding the nearest neighbor when the number of points that must be considered is small (which is the case in our graph building problem).

We also developed a new algorithm that utilizes an early termination technique to reduce the cost of constructing graphs for large instances of the image classification problems. This algorithm was experimentally compared with NNS and a brute-force approach, and was found to be computationally more efficient in our application area.

2   Mathematical Background

2.1   Formal Problem Statement

Let N be the dimension of the Euclidean space $\mathbb{R}^N$. The distance $d(x, y)$ between two points $x, y \in \mathbb{R}^N$ is

$$d(x, y) = \sqrt{\sum_{i=1}^{N} (x_i - y_i)^2}. \tag{2}$$

We define the distance $D(X, Y)$ between two nonempty finite sets of points X and Y as

$$D(X, Y) = \min_{x \in X, y \in Y} d(x, y). \tag{3}$$

Note that $d(x, y) = \sqrt{s(x, y)}$ and, hence, $D(X, Y) = \sqrt{S(X, Y)}$. So finding the $S(X, Y)$ of (1) is the same as finding the $D(X, Y)$ of (3) and squaring the result. Our problem can be stated at follows. Given a set Ω, whose elements are subsets of $\mathbb{R}^N$, find $S(X, Y)$ (equivalently, $D(X, Y)$), for all X and Y in Ω. The practical difficulty is that our Ω typically consists of thousands of elements, each one of which contains hundreds of elements that are high-dimensional vectors.

2.2   Brute-Force Approach

Since the computation of the square root is expensive, the brute-force approach is to find $S(X, Y)$ by exhaustive examination of all pairs $s(x, y)$ for $x \in X$ and $y \in Y$.

2.3   Nearest Neighbor Search

One of the ways to compute $D(X, Y)$ is suggested by the formula

$$D(X, Y) = \min_{x \in X} D'(x, Y), \text{ where } D'(x, Y) = \min_{y \in Y} d(x, y). \tag{4}$$

The computation of $D'(x, Y)$ is equivalent to finding the distance between a point x and its the *nearest neighbor* $y^{(0)}$ in the set Y (because $d(x, y^{(0)}) = D'(x, Y)$). In many cases, the computational cost of finding the nearest neighbor $y^{(0)}$ can be significantly reduced by using the triangle inequality

$$d(a, b) \geq |d(a, c) - d(b, c)|, \text{ for } a, b, c \in \mathbb{R}^N. \tag{5}$$

Let us assume that the distances $d(y^{(k)}, y^{(l)})$ are precalculated for all pairs of vectors $y^{(k)}, y^{(l)} \in Y$ and the distances $d(x, y^{(1)})$, $d(x, y^{(2)})$ for $y^{(1)}, y^{(2)} \in Y$ have been already computed. Now, if for some point $y^{(3)} \in Y$ the inequality

$$d(x, y^{(1)}) \leq |d(x, y^{(2)}) - d(y^{(2)}, y^{(3)})| \tag{6}$$

holds, then the computation of distance $d(x, y^{(3)})$ is unnecessary for the purpose of calculating $D'(x, Y)$, since from (6) follows that $d(x, y^{(3)}) \geq d(x, y^{(1)})$.

2.4   Early Termination

Let I such that $I \neq \varnothing$ and $I \subseteq \{1, \ldots, N\}$. We define a partial sum $s_I(x, y)$, for any two points $x, y \in \mathbb{R}^N$, as

$$s_I(x, y) = \sum_{i \in I} (x_i - y_i)^2. \tag{7}$$

Using this definition, $s(x, y)$ can be expressed as

$$s(x, y) = s_I(x, y) + s_{\bar{I}}(x, y), \text{ where } \bar{I} = \{i | 1 \leq i \leq N \text{ and } i \notin I\}. \tag{8}$$

Now let us assume that we have already calculated the value $s(a, b)$ for a pair of points $a \in X$ and $b \in Y$. Then this value constitutes an upper bound for the calculation of $S(X, Y)$ because

$$S(X, Y) \leq s(a, b). \tag{9}$$

If the value of $s_I(x, y)$ in (8), for some pair of points x, y, is greater or equal to $s(a, b)$, then the calculation of $s_{\bar{I}}(x, y)$ for these points is not necessary (the value of $s(x, y)$ for this pair must be greater or equal to $s(a, b)$). If a tight upper bound of $S(X, Y)$ can be found early, then the calculation of $s_{\bar{I}}(x, y)$ may not be necessary for many pairs (x, y), which results in a significant reduction of the computational cost of finding $S(X, Y)$. This technique is called *early termination*, and has been used in many algorithms [11].

3 Algorithms

3.1 AESA-Based Algorithm

AESA [10] is an efficient algorithm for NNS problem in multi-dimensional space. It is based on the technique described in Section 2.3. Our generalization of AESA (Algorithm 2) to compute the distance between two sets (X and Y) of points in $\mathbb{R}^N$ assumes that distance between each pair of points in Y has been precomputed. The algorithm processes one point $x \in X$ at the time and for this point maintains two structures: a set P that contains points $p \in Y$, for which $d(x, p)$ have not been calculated and may be smaller than d (current estimate of $D(X, Y)$) and a vector g that associates with each point $y \in Y$ a lower bound $g(y)$ of $d(x, y)$. Initially P contains all the elements of Y and all $g(y)$ are set to zero. The size of P is gradually reduced by the following process, which is repeated until P is empty, and starts with arbitrarily selected point c.

The point c is removed from P and the value $d(x, c)$ is calculated. If $d(x, c)$ is smaller than d, then d is set to $d(x, c)$. For each $p \in P$ the value of $g(p)$ is updated by setting it to $\max(g(p), |d(x, c) - d(p, c)|)$. All elements of P for which $g(p)$ is larger than d are removed from P. The element of P for which $g(p)$ is smallest becomes the next point c.

When P is empty, the distance $d(x, y)$ has been either calculated or determined to be larger than d for all $y \in Y$. Therefore the algorithm is ready to process next point x. When all $x \in X$ have been processed, $d = D(X, Y)$. The use of the triangle inequality (5) sometimes significantly reduces the number of distances $d(x, y)$ that must be computed, and this lowers the cost of finding the distance $D(X, Y)$. However, for some datasets, only few calculations of $d(x, y)$ can be avoided using this technique. (Unfortunately, this happens to be the case in our application area.) In such cases, the cost of finding the distance $D(X, Y)$ may increase due to the overhead associated with computation and testing of the values $g(p)$.

3.2   Early Termination

Our algorithm (see Algorithm 3) utilizes, with a slight modification, the early termination technique, that is described in Section 2.4. The computation of the value $s(x, y)$ is performed according to the following formula:

$$s(x, y) = s_{I_1}(x, y) + s_{I_2}(x, y) + s_{I_3}(x, y), \tag{10}$$

where I_1, I_2, I_3 form a partition of $\{1, \ldots, N\}$. The algorithm takes these three subsets I_1, I_2, I_3 and an integer l as parameters. The first stage (lines 1-11) of Algorithm 3 computes and saves the values of $s_{I_1}(x, y)$, for all pairs $(x, y) \in X \times Y$, and identifies l pairs (x, y) for which the partial sums $s_{I_1}(x, y)$ are the smallest. In the second stage (lines 14-15) of Algorithm 3, the values of $s(x, y)$ are computed for the l pairs (x, y) identified during the first stage. This is done using (10), in which the value $s_{I_1}(x, y)$ has been already computed in the first stage. The minimum of values $s(x, y)$ computed in this stage is used in the third stage as an upper bound s for value of $S(X, Y)$. In the third stage (lines 16-19) of Algorithm 3, the search for the $S(X, Y)$ is conducted. For each of the pairs $(x, y) \in X \times Y$ the algorithm checks if the value of $s_{I_1}(x, y)$ (calculated in first stage) is greater than s (the current estimate of upper bound of $S(X, Y)$). If the value of $s_{I_1}(x, y)$ for a pair (x, y) is greater or equal to the value of s, then the algorithm examines next pair. If the value of $s_{I_1}(x, y)$ for a pair (x, y) is smaller than the value of s, then the algorithm computes the sum $s_{I_1}(x, y) + s_{I_2}(x, y)$. Since the value of $s_{I_1}(x, y)$ is already known (it was calculated in the first stage), only $s_{I_2}(x, y)$ must be calculated. If the value of sum $s_{I_1}(x, y) + s_{I_2}(x, y)$ is larger than or equal to the value of s, then the algorithm examines next pair. Otherwise the algorithm computes the value of $s(x, y)$ as $s(x, y) = s_{I_1}(x, y) + s_{I_2}(x, y) + s_{I_3}(x, y)$ (the values of the sums $s_{I_1}(x, y)$ and $s_{I_2}(x, y)$ are reused from previous computations). If the computed value of $s(x, y)$ is smaller than s, then the value of s is set to that value. The algorithm terminates after examining all pairs (x, y) in the third stage and returns the square root of s as the distance $D(X, Y)$.

4   Experiments and Results

4.1   Datasets

All data sets used in our experiments were synthetically generated by a process designed to produce sets closely reassembling those found in 3D-EM. Each data set contained $5,000$ images. For each of these images we produced 240 vectors (1D projections) containing 81 real numbers. The computation of the similarity measure between each pair of images required finding a distance between two corresponding sets of points in $\mathbb{R}^{81}$, where the first set contains 240 points representing 1D projections of first image and second one contains 120 points representing 1D projections of second image (since the mirror image of each 1D projection in the set is also in this set, only half of the 1D projections in second set need be considered). In order to construct the graph representing each data set, the similarity measure for 12,497,500 pairs must be calculated.

Algorithm 1. Brute-Force

1: $s \leftarrow +\infty$
2: **for all** $x \in X$, $y \in Y$ **do**
3:     $s \leftarrow \min(s, s(x,y))$
4: **return** $\sqrt{s}$

Algorithm 2. AESA-Based Algorithm

1: $d \leftarrow +\infty$
2: **for all** $x \in X$ **do**
3:     $P \leftarrow Y$
4:     **for all** $p \in P$ **do**
5:        $g(p) \leftarrow 0$
6:     $c \leftarrow$ an arbitrary element of P
7:     **while** $P \neq \emptyset$ **do**
8:        $P \leftarrow P \setminus \{c\}$; $e \leftarrow d(c,x)$
9:        $d \leftarrow \min(d, e)$
10:       $gmin \leftarrow +\infty$
11:       **for all** $p \in P$ **do**
12:          $g(p) \leftarrow \max(g(p), |e - d(p,c)|)$ $\{d(p,c)$ is precomputed, because $p, c \in Y\}$
13:          **if** $g(p) > d$ **then**
14:             $P \leftarrow P \setminus \{p\}$
15:          **else if** $g(p) < gmin$ **then**
16:             $gmin \leftarrow g(p)$; $cmin \leftarrow c$
17:       $c \leftarrow cmin$
18: **return** d

Algorithm 3. Early Termination (I_1, I_2, I_3 and l are parameters).

1: $n \leftarrow 0$, $s \leftarrow \infty$
2: **for all** $(x,y) \in X \times Y$ **do**
3:     $s' \leftarrow s_{I_1}(x,y)$
4:     **if** $s' < s$ **then**
5:        $n \leftarrow \min(n+1, l)$
6:        $k \leftarrow n$
7:        **while** $k > 1$ and $s^{(k-1)} > s'$ **do**
8:          $x^{(k)} \leftarrow x^{(k-1)}$; $y^{(k)} \leftarrow y^{(k-1)}$; $s^{(k)} \leftarrow s^{(k-1)}$; $k \leftarrow k-1$
9:        $x^{(k)} \leftarrow x$; $y^{(k)} \leftarrow y$; $s^{(k)} \leftarrow s'$
10:       **if** $n = l$ **then**
11:          $s \leftarrow s^{(l)}$
12: $\{$Here $(x^{(1)}, y^{(1)}), \ldots, (x^{(l)}, y^{(l)})$ are the l pairs which have the smallest $s_{I_1}(x,y)\}$
13: $s \leftarrow \infty$
14: **for** $i = 1, \ldots, l$ **do**
15:     $s \leftarrow \min(s, s_{I_3}(x^{(i)}, y^{(i)}) + s_{I_2}(x^{(i)}, y^{(i)}) + s_{I_1}(x^{(i)}, y^{(i)}))$
16: **for all** $(x,y) \in X \times Y$ **do**
17:     **if** $s_{I_1}(x,y) < s$ **then**
18:        **if** $s_{I_2}(x,y) + s_{I_1}(x,y) < s$ **then**
19:          $s \leftarrow \min(s, s_{I_3}(x,y) + s_{I_2}(x,y) + s_{I_1}(x,y))$
20: **return** $\sqrt{s}$

4.2 Experiments with AESA-Based Algorithm

We have tested AESA algorithm with 6 datasets. Since the runtimes are quite long, the algorithm was only run with subsets of 100 images taken randomly in each data set. The runtimes of AESA-based algorithm for one such subset on an Athlon 1800+ are between 204.38 and 208.46 seconds. We used the brute-force algorithm, a direct implementation of (1) (see Algorithm 1), as the base line reference to evaluate these results. The runtimes of brute-force algorithm on the same datasets are between 36.56 and 36.77 seconds, which means that the AESA-based algorithm is more than 5 times slower.

4.3 Experiments with Early Termination Algorithm

Parameters Expecting that the majority of the useful information is concentrated in the center of the image we decided to divide the dimensions of $\mathbb{R}^{81}$ among sets I_1, I_2, I_3 in the following way:

$$I_1 = \{41 - n_1, \ldots, 41 + n_1\}$$
$$I_2 = \{41 - n_2, \ldots, 41 - n_1 - 1, 41 + n_1 + 1, \ldots, 41 + n_2\}$$
$$I_3 = \{1, \ldots, 41 - n_2 - 1, 41 + n_2 + 1, \ldots, 81\} = \{1, \ldots, 81\} \setminus (I_2 \cup I_3).$$

In order to determine the optimal values of parameters we tested runtimes of early termination algorithm with different values of n_1, n_2, l on a randomly selected, small subset of the projection images. Figure 2 shows some of the results of our tests. Based on these tests we have chosen $n_1 = 13$, $n_2 = 22$, $l = 20$ for our experiments.

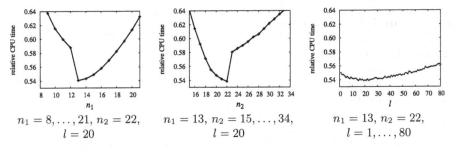

$$n_1 = 8, \ldots, 21, n_2 = 22,\qquad n_1 = 13, n_2 = 15, \ldots, 34,\qquad n_1 = 13, n_2 = 22,$$
$$l = 20\qquad\qquad l = 20\qquad\qquad l = 1, \ldots, 80$$

Fig. 2. CPU-time in function of the three parameters. (The CPU time 1.0 corresponds to the brute-force algorithm).

Results. We have tested Algorithm 3 with 6 different datasets on an Athlon 1800+ computer. The runtimes are between 12.56 and 12.90 hours. As in our experiments with AESA-based algorithm we used runtime of brute-force algorithm to evaluate these results. Since the number of operations performed by brute-force algorithm is fixed (depends only on the size of the data set), we applied it only to one of the data sets. The brute-force algorithm required 23.60 hours of runtime to construct the graph. Based on this test we can conclude that the early termination algorithm is more than 45% faster.

5   Discussion and Conclusion

The AESA-based algorithm attempts to lower the computational cost of finding $D(X, Y)$ by reducing the number of pairs $x \in X$, $y \in Y$ for which the computation of the distance $d(x, y)$ is necessary. This is achieved by calculating the lower bounds of distances $d(x, y)$ for all the pairs x, y and testing them to identify these pairs x, y for which $d(x, y)$ must be larger than $D(X, Y)$. Since the cost of calculating and testing the lower bound of distance $d(x, y)$ is lower than cost of calculating $d(x, y)$, the time is saved on pairs x, y for which the calculation of $d(x, y)$ is avoided. However, the cost of calculating and testing the lower bound of the distance $d(x, y)$ adds to the cost of calculating $d(x, y)$ for those pairs x, y for which the calculation $d(x, y)$ is necessary. The total cost of finding $D(X, Y)$ is reduced only if computation of $d(x, y)$ is avoided for a sufficiently large number of pairs x, y. The poor performance of AESA-based algorithm can be explained by the fact that the computation of $d(x, y)$ was avoided only for less than 20% of pairs x, y. Clearly the spatial distribution of points in sets representing EM images is such that AESA-based approach cannot applied to them.

The good performance of early termination algorithm indicates that the assumption about the concentration of useful information in the center of the image was correct. The calculation of sum $s(x, y)$ (10) was necessary only for approximately 4% of pairs $(x, y) \in X \times Y$. For approximately 65% of pairs only the sum $s_{I_1}(x, y)$ was computed. Remaining pairs required the computation of sums $s_{I_1}(x, y)$ and $s_{I_2}(x, y)$.

Our method significantly reduces the cost of constructing graphs to classify EM projection images by relatively simple means. Future research is necessary to test applicability of other, more advanced methods to this task. Since the problem of searching for nearest neighbor has been intensively studied in many domains, the number of methods which could be considered is quite large. Several nearest neighbor searching algorithms have been developed to quantize image vectors (an early example of such algorithm can be found in [12]). Since these algorithms were designed for searching in spaces with small number of dimensions, their applicability to our problem may be limited. As indicated by Yianilos [9], the methods based on kd-trees or on constructions of computational geometry become inefficient as the number of dimensions increases. However, some methods, such as the one proposed by Lai et. al. [13], are better suited for high-dimensional spaces and are more likely to reduce the computational cost of constructing graphs to classify EM projection images. The results of our initial experiments with the algorithm of [13], adapted to the computation of $D(X, Y)$, suggest that, when applied to the data sets that are discussed above, this algorithm is approximately 13% faster than the brute-force algorithm and 63% slower than the early termination algorithm.

We evaluated the applicability of optimization techniques to the problem of constructing graphs to classify EM projection images. Our results indicate that the cost of constructing such graphs can be significantly reduced by using the algorithm employing early termination. Since even with this reduction, the graph construction remains the most time-consuming part of the EM image classification process, it may be desirable to explore the applicability of additional optimization techniques to the construction of graph used to classify EM projection images.

Acknowledgments

We would like to thank one of the reviewers for pointing out to us the paper by Lai et. al. [13]. This research is supported by the National Institutes of Health through grant HL70472. A large part of the research was done while the second author was visiting the City University of New York.

References

1. Frank, J.: Three-Dimensional Electron Microscopy Of Macromolecular Assemblies: Visualization of Biological Molecules in Their Native State. Oxford University Press, New-York, NY, USA (2006)
2. Scheres, S.H.W., Gao, H., Valle, M., Herman, G.T., Eggermont, P.P.B., Frank, J., Carazo, J.M.: Disentangling conformational states of macromolecules in 3D-EM through likelihood optimization. Nature Methods 4, 27–29 (2007)
3. Fu, J., Gao, H., Frank, J.: Unsupervised classification of single particles by cluster tracking in multi-dimensional space. J. Struct. Biol. 157, 226–239 (2007)
4. Herman, G.T., Kalinowski, M.: Classification of heterogeneous electron microscopic projections into homogeneous subsets (submitted to Ultramicroscopy)
5. Kann, V., Khanna, S., Lagergren, J., Panconesi, A.: On the hardness of approximating Max k-Cut and its dual. Chicago J. Theoret. Comp. Sci (1997) http://cjtcs.cs.uchicago.edu/articles/1997/2/contents.html
6. Schrijver, A.: Combinatorial Optimization: Polyhedra and Efficiency. Springer Verlag, Berlin Heidelberg (2003)
7. Papadopoulos, A.N., Manolopoulos, Y.: Nearest Neighbor Search: A Database Perspective. Springer, New-York, NY, USA (2005)
8. Chávez, E., Navarro, G., Baeza-Yates, R., Marroquín, J.L.: Searching in metric spaces. ACM Comput. Surv. 33(3), 273–321 (2001)
9. Yianilos, P.N.: Data structures and algorithms for nearest neighbor search in general metric spaces. In: Proc. of SODA '93, pp. 311–321 (1993)
10. Vidal, E.: An algorithm for finding nearest neighbours in (approximately) constant time. Pattern Recogn. Lett. 4(3), 145–157 (1986)
11. Bey, C.D., Gray, R.M.: An improvement of the minimum distortion encoding algorithm for vector quantization. IEEE Trans. Commun. 33(10), 1132–1133 (1985)
12. Heckbert, P.: Color image quantization for frame buffer display. In: Proc. of SIGGRAPH '82, pp. 297–307 (1982)
13. Lai, J.Z.C., Liaw, Y.C., Liu, J.: Fast k-nearest-neighbor search based on projection and triangular inequality. Pattern Recogn 40(2), 351–359 (2007)

An Efficient Ontology-Based Expert Peering System

Tansu Alpcan, Christian Bauckhage, and Sachin Agarwal

Deutsche Telekom Laboratories
10587 Berlin, Germany
(tansu.alpcan,christian.bauckhage,sachin.agarwal)@telekom.de

Abstract. This paper proposes a novel expert peering system for information exchange. Our objective is to develop a real-time search engine for an online community where users can query experts, who are simply other participating users knowledgeable in that area, for help on various topics. We consider a graph-based scheme consisting of an ontology tree where each node represents a (sub)topic. Consequently, the fields of expertise or profiles of the participating experts correspond to subtrees of this ontology. Since user queries can also be mapped to similar tree structures, assigning queries to relevant experts becomes a problem of graph matching. A serialization of the ontology tree allows us to use simple dot products on the ontology vector space effectively to address this problem. As a demonstrative example, we conduct extensive experiments with different parameterizations. We observe that our approach is efficient and yields promising results.

1 Introduction

Document retrieval studies the problem of matching user queries to a given set of typically unstructured text records such as webpages or documents. Since user queries may also be unstructured and can range from a few keywords to multi-sentenced descriptions of the desired information, pre-processing steps such as stop word removal, stemming, and keyword spotting usually precede the actual retrieval.

Given similarly purged dictionaries, most systems for document retrieval and text classification rely on the *vector space model* of documents. It represents documents and queries by term-by-document vectors and allows for approaches based on statistical learning. Recent research in this area includes the use of support vector machines [1], probabilistic semantic indexing [2], or spectral clustering [3].

However, despite their dominant role, methods relying on term-by-document vectors suffer from several drawbacks. For instance, they cannot capture relations among terms in a single document and have to assume a static dictionary in order to fix the dimension of the vectors. Graph-based models, in contrast, easily cope with these shortcomings, providing a promising alternative approach to document retrieval.

In an early contribution Miller [4] has considered bipartite matchings between documents and queries which are given in terms of co-occurrence graphs. More recently, Schenker et al. [5,6] have proposed a graph structure for documents and queries that accounts for sequences of words. Matches are computed based on a k-nearest neighbors (kNN) criterion for graphs and it has been shown that this outperforms common vector-based kNN retrieval.

F. Escolano and M. Vento (Eds.): GbRPR 2007, LNCS 4538, pp. 273–282, 2007.
© Springer-Verlag Berlin Heidelberg 2007

In this paper, we assume a different view on graph-based document retrieval. Focusing on development of a peer-to-peer (P2P) communication mechanism for an online community, we describe a retrieval system that exploits semantic structures for text-based classification. The online community users identify themselves as experts for certain domains and fields of knowledge. Users may either address the community with problems they need help on, or –if they are qualified– can respond to other users' questions. Rather than mediating the communication between community members by an online user forum (offline) mechanism, we aim at a solution that automatically proposes appropriate experts given a user query in real time, who can be contacted directly via, for example, instant messaging.

Our approach is based on a comprehensive ontology tree describing relevant fields of knowledge where each node corresponds to a single subject or topic described by a bag of words. Similarly, we associate each query and expert with a bag of words of flexible size and define a similarity measure to compare two such bags. Utilizing an algorithm which will be described in detail in Section 2, this formulation enables us to represent entities such as queries or experts as subtrees of the ontology at hand. Furthermore, serialization of the ontology tree allows for defining an *ontology-space*. Therefore, queries as well as experts can equivalently be represented as vectors in this linear vector space. The problem of peering queries and experts then becomes a problem of tree matching which is addressed using dot product operations between the respective vectors.

1.1   Related Work

Ontology-based document retrieval and search has recently become an active area of research. Especially ontology building from a set of documents and term similarity measures have found increased attention [7,8]. In a recent work more closely related to our scenario, Wu et al. [9] have studied an expert matching problem similar to ours. However, their approach differs from our solution for they apply ontologies to compute path-length-based distances between concepts upon which they base several definitions of similarity measures for documents. Our work, in contrast, exploits hierarchal coarse-to-fine information contained in the ontology and measures document similarities in semantics induced vector spaces. Moreover, while the algorithms in [9] require manual intervention, our scheme is fully automatic. Finally, preliminary experiments we conduct demonstrate that our approach leads to a higher performance in terms of precision and recall than the one in [9].

1.2   Organization

The rest of the paper is structured as follows: in Section 2 we describe our approach and algorithms developed in detail. Section 3 presents a demonstrative experimental study and discusses its results. The paper concludes with a summary and remarks on future research directions in Section 4.

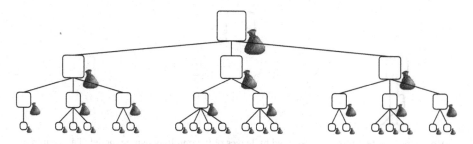

Fig. 1. An example ontology tree where each node is associated with a bag (set) of words. In our implementation, the bag of words of a higher level node contains keywords regarding the corresponding topic as well as the union of all bags of words of its descendants.

2 Model and Approach

We present an ontology-based semantic model and approach to address the query-expert peering problem. Specifically, we describe the structure of the ontology, a simple similarity measure, and a mapping algorithm followed by the expert peering scheme.

2.1 The Ontology

We consider a strictly hierarchical ontology or knowledge tree $T = (\mathcal{N}, \Sigma)$ consisting of a set of nodes or subjects $\mathcal{N} = \{n_1, \dots, n_N\}$ and a set of edges Σ between them such that each subject $n \in \mathcal{N}$ has a unique parent node corresponding to a broader subject (see Fig.1). Other than this assumption the approach we develop in this paper is independent of the nature and contents of the specific ontology tree chosen.

Let us define for notational convenience $\mathcal{C}(n)$ as the set of children and $p(n)$ as the unique parent of node n. We associate each node n with a representative *bag of words* $\mathcal{B}(n) := \{w_1, \dots, w_{B_n}\}$, where w_i denotes the ith word. This set (bag) of words can be for example obtained by processing a collection of related texts from online and encyclopedic resources using well-known natural language processing methods. Subsequently, we optimize all of the bag of words $\mathcal{B}$ in the ontology both vertically and horizontally in order to strengthen the hierarchical structure of the tree and to reduce redundancies, respectively. First, in the vertical direction, we find the union of bag of words of each node n and the ones of its children $\bar{\mathcal{B}}(n) = \mathcal{B}(n) \cup [\cup_{i \in \mathcal{C}(n)} \mathcal{B}(i)]$. Then, we replace $\mathcal{B}(n)$ with $\bar{\mathcal{B}}(n)$ for all $n \in \mathcal{N}$. We repeat this process starting from leaf nodes until the root of the tree is reached. Next, in the horizontal direction, we find the overlapping words among all children of a node n, $\tilde{\mathcal{B}}(n) = \cap_{i \in \mathcal{C}(n)} \mathcal{B}(i)$, subtract these from each $i \in \mathcal{C}(n)$ such that $\mathcal{B}(i) = \mathcal{B}(i) \setminus \tilde{\mathcal{B}}(n)$, and repeat this for all $n \in \mathcal{N}$.

2.2 Mapping to Ontology-Space

The ontology tree can easily be serialized by, for example, ordering its nodes from top to bottom and left to right. Hence, we obtain an associated vector representation of the tree $\mathbf{v}(T) \in \mathbb{R}^N$. An important aspect of our algorithm is the representation of

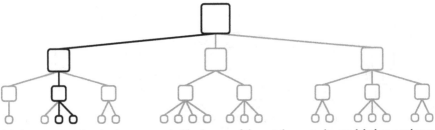

(a) An expert on two topics represented by leaves of the ontology can be modeled as a subtree

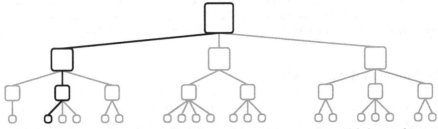

(b) A query on a topic represented by a leave of the ontology can be modeled as a subtree

Fig. 2. Experts and queries can be mapped to subtrees of an ontology, which creates the basis of the peering scheme proposed

entities as subtrees of the ontology (see Fig. 2) and equivalently as vectors on the so called *ontology-space* $S(T) \subset \mathbb{R}^N$, which is a compact subset of $\mathbb{R}^N$. In order to map expert profiles and queries, which are given by arbitrary keyword lists, onto subtrees we use a similarity measure between any entity representable by a bag of words and the ontology tree. In this paper, we choose the subsequently described measure and mapping algorithm. However, a variety of similarity measures can be used towards this end.

Let us consider the following example scenario to further motivate this mapping scheme (see Fig. 2). A query on electromagnetism (a topic represented by a leave node) is, in a wider sense, a query on theoretical physics which, in turn, is a query in the area of physics in general. Therefore, even if there is no expert on electromagnetism is found an expert on quantum mechanics and computational physics might be able to help the user as these are subbranches of theoretical physics.

Let us define, for analysis purposes, a global dictionary set $\mathcal{D} := \cup_{n \in \mathcal{N}} \mathcal{B}(n)$ of cardinality M and an M-dimensional *dictionary-space* $S(\mathcal{D}) \subset \mathbb{R}^M$ by choosing an arbitrary ordering. Thus, each node or item i is associated with an occurrence vector $\mathbf{w}(i)$ in the dictionary space indicating whether or not a word appears in the respective bag of words:

$$\mathbf{w}(i) := [I(w_1), \ldots, I(w_M)], \quad \mathbf{w}(i) \in S(\mathcal{D}), \tag{1}$$

where $I(w_j) = 1$ if $w_j \in B(i)$ and $I(w_j) = 0$ otherwise. Note that the vectors $\mathbf{w}$ are usually sparse as the cardinality of $B(i)$ is usually much smaller than the one of $\mathcal{D}$.

Input: bag of words $B(i)$, ontology tree T, similarity measure r
Output: corresponding subtree describing entity i and its vector representation $\mathbf{v}(i)$

/* *Compute the similarity between the given entity's bag of words $B(i)$ and the ones of the tree nodes iteratively from top to bottom* */

1. consider the highest semantic categories $\{n_{h_1}, \ldots, n_{h_H}\}$, i.e. each node n immediately below the root node, and compute the similarities $r(i, n)$
2. determine the node $n_k \in \{n_{h_1}, \ldots, n_{h_H}\}$ with the highest similarity
3. add n_k to the resulting subtree and set the corresponding entry in the vector to 1
4. consider the child nodes $\{n_{c_1}, \ldots, n_{c_C}\}$ of n_k and for each child node n compute the similarities $r(i, n)$
5. compute the mean μ and the standard deviation σ of the resulting similarities
6. consider all nodes $\{n_k\}$ in the current set of children for which $r(i, n_k) > \mu + \alpha \sigma$ where $\alpha \geq 0$ is a fixed parameter
7. for each node n_k in the set $\{n_k\}$ continue with step 3, until the lowest level of the tree is reached

---

Fig. 3. Algorithm to map an entity i characterized by an arbitrary list of keywords to a subtree of an ontology whose nodes are associated with bags of words

We now define an example similarity measure $r(i, j)$ between two entities i and j (with respective bag of words $B(i)$, $B(j)$ and vectors $\mathbf{w}(i)$, $\mathbf{w}(j)$):

$$r(i, j) := \frac{|B(i) \cap B(j)|}{\sqrt{|B(i)|}\sqrt{|B(j)|}}, \tag{2}$$

where $|\cdot|$ denotes the cardinality of a set. Note that this measure actually corresponds to the the cosine of the angle between occurrence vectors but clearly does not require to assemble a global dictionary $\mathcal{D}$ for its computation.

Given the similarity measure we present an efficient mapping from the dictionary space to the ontology-space $S(\mathcal{D}) \rightarrow S(T)$ through the algorithm in Fig. 3. Using this algorithm, any item i can be represented as a subtree of the ontology or alternatively as a vector $\mathbf{v}(i) \in S(T)$ with one-zero entries on the ontology-space. We note that the algorithm in Fig. 3 is inherently robust due to its top-to-bottom iterative nature and usage of the ontology's hierarchical structure. In other words, it solves a series of classification problems at each level of the tree with increasing difficulty but in a sense of decreasing importance. We will discuss this in the next section in more detail. The optimizations of the ontology tree described in Section 2.1 also add to the robustness of the mapping, especially the aggregation of bag of words from leaves to the root.

2.3 Query-Expert Peering

As the first step of query-expert peering, we convert each query to a bag of words and associate each expert with its own bag. The experts bag of words can be derived, for example, by processing personal documents such as resumes, webpages, blogs, etc.

The algorithm in Fig. 3 enables us then to represent any query or expert as a subtree of the ontology as well as a binary vector on the ontology-space. Thus, the query-expert peering problem becomes one of graph (tree) matching which we in turn address by using the ontology tree to span the corresponding linear space. There are two important advantages of this approach:

1. The ontology (vector) space has a much smaller dimension than the commonly used term-by-document spaces. It also avoids the need for maintaining large, inefficient, and static dictionaries.
2. Each dimension of the ontology-space, which actually corresponds to a node (subject), has inherent semantic relations with other nodes. One such relation is hierarchical and immediately follows from the tree structure of the ontology. However, it is also possible to define other graph theoretic relations, for example, by defining overlay graphs.

We now describe a basic scheme for query-expert peering. Let us denote by q a query with its bag of words $\mathcal{B}(q)$ and by $\mathcal{E} = \{e_1, \ldots, e_E\}$ a set of experts represented by the respective bag of words $\mathcal{B}(e_i)$, $i = 1, \ldots, E$. Our objective is to find the best set of experts given the query. Using the approach in Section 2.2 we map the query and experts to subtrees of the ontology, and hence obtain vectors $\mathbf{v}(q)$, and $\mathbf{v}(e_1), \ldots, \mathbf{v}(e_E)$, respectively, on the ontology space $S(T)$. Then, we define a matching score m between a query and an expert

$$m(q, e) := \mathbf{v}(q) \cdot \mathbf{v}(e), \tag{3}$$

as the dot product of their vectors. Subsequently, those experts with the highest ranking matching scores are assigned to the query.

3   Experiments

We conduct a set of preliminary offline experiments to numerically study the performance of the system developed. We present next the experiment setup followed by the numerical results and their interpretation.

3.1   Experiment Setup

We begin the experimental setup by selecting an ontology and associate each of its nodes with a bag of words as described in Section 2.1. In this paper we choose (rather arbitrarily) a 245 node subset of an ontology[1] prepared by the Higher Education Statistics Agency (HESA), an educational institution in the United Kingdom. The bag of words for each node is obtained via the following procedure:

1. The node's name is used in finding 10 top ranked documents through *Yahoo!* search web services.[2]
2. The obtained HTML documents are converted to text (ASCII) format and concatenated into a single document.

---

[1] http://www.hesa.ac.uk/jacs/completeclassification.htm
[2] http://developer.yahoo.com/search/

3. This resulting document is further processed using the Natural Language Toolkit (NLTK) [10] by (a) tokenizing, (b) stop word removal, and (c) stemming with Porter's stemmer [11], which finally yields the bag of words.

We next randomly generate experts for the purpose of offline experiments. We consider three types of experts: one knowledgeable in a single specific topic (represented by a subtree of ontology ending at a single leaf node), one with two specific topics (branches), and one with three topics. Each randomly generated pool of experts contains equal number of each type.

One can devise a variety of methods for random query generation. However, the procedure for generating queries with a known answer (an ordering of best matching experts) is more involved. We overcome this difficulty by generating a separate "query" bag of words for each node of the ontology following the steps above. We ensure that these bags of words are obtained from documents completely disjoint from the ones used to obtain node-associated bag of words. Thus, we generate queries by randomly choosing a node from the ontology and a certain number of keywords from its "query" bag of words. Since we know which node the query belongs to, we easily find a "ground truth" subtree or vector associated with the query which in turn allows computing the "best" ordering of experts for peering. This yields a basis for comparison with the result obtained from the generated query.

Finally, we use the similarity measure and mapping algorithm described in Section 2 to compute the expert peering, i.e. the set of experts $R(q)$ with highest matching scores given a query q. Then, as described above the "ground truth" vectors are used to calculate the set of "correct" experts $A(q)$. The recall and precision measures are calculated as the average of $N = 1000$ such queries in these experiments:

$$recall = \frac{1}{N} \sum_{i=1}^{N} \frac{\mid A(q_i) \cap R(q_i) \mid}{\mid A(q_i) \mid}, \quad precision = \frac{1}{N} \sum_{i=1}^{N} \frac{\mid A(q_i) \cap R(q_i) \mid}{\mid R(q_i) \mid}.$$

3.2 Numerical Results

We next present and discuss the numerical results. In the experiments we choose the following specific parameter values: the number of query keywords (out of respective "query" bag of words) $\{20, 40, 60, 80, 100\}$, the number of experts $\{50, 100\}$, and the parameter α of the algorithm in Fig. 3 $\{0.0, 1.0\}$.

We first limit the cardinality of A to one, i.e. there is only a single expert in the "correct" peering set. The precision and recall versus the range of parameters in this case is shown in Figures 4(a) and (b), respectively. Aiming to find only the single best matching expert is clearly over restrictive and leads to poor results. In fact, given the uncertainties within the underlying representation mechanisms it is neither very meaningful to expect such degree of accuracy nor required for the application areas considered.

Next, the best matching experts are defined as the ones with the top three highest ranking scores. Notice that this set may contain more than three experts in some cases. The precision and recall improve drastically for all parameter choices as observed in

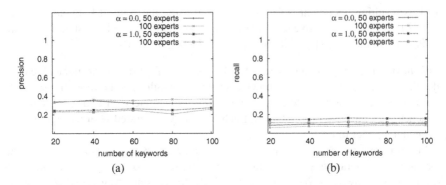

Fig. 4. (a) Precision and (b) recall for a range of parameters when we find only the best matching expert to each query

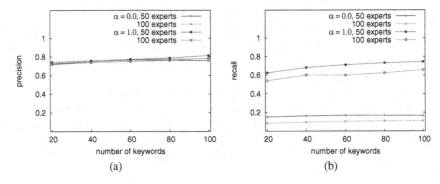

Fig. 5. (a) Precision and (b) recall for a range of parameters when we find the set of experts with the top three rankings to each query

Figures 5(a) and (b), respectively. This result demonstrates the robustness of our expert peering scheme: its performance improves gradually when accuracy restrictions are eased. This is further illustrated by Figures 6(a) and (b), where the performance further increases when the set of best matching experts is defined by the ones belonging to the top six ranks. It is important to note that for each case the set of "correct" experts obtained from the "ground truth" vectors is defined as the set of experts with the single highest ranking value. Our observations on and interpretations of results with respect to the values of other parameters include:

1. Choosing the larger $\alpha = 1$ value for the algorithm in Fig. 3 leads to improved results. Since this parameter affects the branching threshold value when mapping queries to a subtree of ontology we conclude that increasing it restricts unnecessary branching, and hence noise.
2. The precision remains high regardless of the number of experts and α in Figures 5 and 6. We attribute this result to hierarchical structure and robustness of our system.

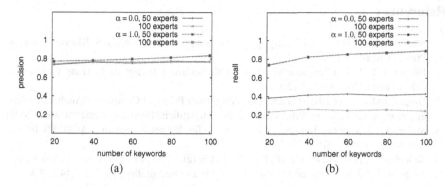

Fig. 6. (a) Precision and (b) recall for a range of parameters when we find the set of experts with the top six rankings to each query

3. With the correct set of parameters we observe in Fig. 5 and especially Fig. 6 that both the precision and recall are relatively insensitive to the number of experts which indicates scalability.
4. Although the precision and recall slightly increase with increasing number of words in the queries these curves are rather flat demonstrating that our system performs well in peering the experts even when given limited information.

4 Conclusion

In this paper we have presented an ontology-based approach for an expert peering and search system. We have studied the underlying principles of a real-time search engine for an online community where users can ask experts, who are simply other participating users knowledgeable in that area, for help on various topics. We have described a graph-based representation scheme consisting of an ontology tree where each node corresponds to a (sub)topic and is associated with a bag of words. This allows us to represent the fields of expertise (profile) of the participating experts as well as incoming queries as subtrees of the ontology. Subsequently, we have addressed the resulting graph matching problem of assigning queries to relevant experts on a vector space, which follows from a serialization of the ontology tree, using simple dot products of respective vectors.

Preliminary experiments utilizing an example ontology demonstrate the efficiency, robustness, and high performance of our algorithm over a range of parameters. These promising results also open the way for future research. One research direction is the refinement of our algorithm toward an adaptive update of the α parameter that controls the branching behavior in subtree generation. Another interesting question is how to make the underlying ontology dynamic by adding, deleting, and merging nodes. Yet another direction is the study of time-varying expert profiles and it's analysis as a dynamic system. We finally note that although the expert peering problem we focus on in this paper has specific properties differing from document retrieval our approach can be applied to that area as well.

References

1. Joachims, T.: Learning to Classify Text Using Support Vector Machines. Kluwer Academic Press, Dordrecht (2002)
2. Hofmann, T.: Latent Semantic Models for Collaborative Filtering. ACM Trans. on Information Systems 22(1), 89–115 (2004)
3. Ding, C.: Document Retrieval and Clustering: from Principal Component Analysis to Self-aggregation Networks. In: Proc. Int. Workshop on Artificial Intelligence and Statistics (2003)
4. Miller, L.: Document Representation Models for Retrieval Systems. ACM SIGIR Forum 14(2), 41–44 (1979)
5. Schenker, A., Last, M., Bunke, H., Kandel, A.: Classification of web documents using a graph mode. In: Proc. Int. Conf. on Document Analysis and Recognition, pp. 240–244 (2003)
6. Schenker, A., Last, M., Bunke, H., Kandel, A.: Classification of Web Documents Using Graph Matching. Int. J. of Patter Recognition and Artificial Intelligence 18(3), 475–496 (2004)
7. Lim, S.Y., Park, S.B., Lee, S.J.: Document retrieval using semantic relation in domain ontology. In: Szczepaniak, P.S., Kacprzyk, J., Niewiadomski, A. (eds.) AWIC 2005. LNCS (LNAI), vol. 3528, pp. 266–271. Springer, Heidelberg (2005)
8. Chung, S., Jun, J., McLeod, D.: A web-based novel term similarity framework for ontology learning. In: ODBASE: Int. Conf. on Ontologies, Databases and Applications of Semantics, Montpellier, France (2006)
9. Wu, J., Yang, G.: An ontology-based method for project and domain expert matching. In: Wang, L., Jiao, L., Shi, G., Li, X., Liu, J. (eds.) FSKD 2006. LNCS (LNAI), vol. 4223, pp. 176–185. Springer, Heidelberg (2005)
10. Bird, S., Klein, E., Loper, E.: The natural language toolkit (NLTK) (2001)
11. Porter, M.: An Algorithm for Suffix Stripping. Program 14(3), 130–137 (1980)

Computing Homology Group Generators of Images Using Irregular Graph Pyramids*

S. Peltier[1], A. Ion[1], Y. Haxhimusa[1], W.G. Kropatsch[1], and G. Damiand[2]

[1] Vienna University of Technology, Faculty of Informatics,
Pattern Recognition and Image Processing Group, Austria
{sam,krw,ion,yll}@prip.tuwien.ac.at
[2] University of Poitiers
SIC, FRE CNRS 2731, France
damiand@sic.univ-poitiers.fr

Abstract. We introduce a method for computing homology groups and their generators of a 2D image, using a hierarchical structure i.e. irregular graph pyramid. Starting from an image, a hierarchy of the image is built, by two operations that preserve homology of each region. Instead of computing homology generators in the base where the number of entities (cells) is large, we first reduce the number of cells by a graph pyramid. Then homology generators are computed efficiently on the top level of the pyramid, since the number of cells is small, and a top down process is then used to deduce homology generators in any level of the pyramid, including the base level i.e. the initial image. We show that the new method produces valid homology generators and present some experimental results.

1 Introduction

Handling 'structured geometric objects' is important for many applications related to geometric modeling, computational geometry, image analysis, etc. One has often to distinguish between different parts of an object, according to properties which are relevant for the application. For image analysis, a region is a (structured) set of pixels or voxels, or more generally a (structured) set of lower-level regions. At the lowest level of abstraction, such an object is a subdivision[1], i.e. a partition of the object into cells of dimension 0, 1, 2, 3 ... (i.e. vertices, edges, faces, volumes ...) [1,2]. In general, combinatorial structures (graphs, combinatorial maps, n-G-maps etc.) are used to describe objects subdivided into cells of different dimensions. The structure of the object is related to the decomposition of the object into sub-objects, and to the relations between these sub-objects: basically, topological information is related to the cells and their adjacency or incidence relations. Further information (embedding information) is associated to these sub-objects, and describes for instance their shapes (e.g. a point, respectively a curve, a part of a surface, is associated with each vertex, respectively

---

* Supported by the Austrian Science Fund under grants P18716-N13 and S9103-N04.
[1] For instance, a Voronoi diagram in the plane defines a subdivision of the plane.

F. Escolano and M. Vento (Eds.): GbRPR 2007, LNCS 4538, pp. 283–294, 2007.
© Springer-Verlag Berlin Heidelberg 2007

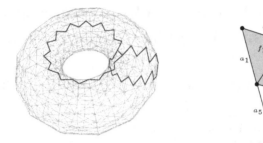

Fig. 1. (a): a triangulation of the torus. (b): a simplicial complex made of 1 connected component and containing one 1−dimensional hole.

each edge, each face), their textures or colors, or other information depending on the application. A common problem is to characterize structural (topological) properties of handled objects. Different topological invariants have been proposed, like Euler characteristics, orientability, homology,... (see [3]).

Homology is a powerful topological invariant, which characterizes an object by its "p−dimensional holes". Intuitively the 0−dimensional holes can be seen as connected components, 1−dimensional holes can be seen as tunnels and 2−dimensional holes as cavities. For example, the torus in Fig.1(a) contains one 0−dimensional hole, two 1−dimensional holes (each of them are an edge cycle) and one 2−dimensional hole (the cavity enclosed by the entire surface of the torus). This notion of p−dimensional hole is defined in any dimension. Another important property of homology is that local calculations induce global properties. In other words, homology is a tool to study spaces, and has been applied in image processing for 2D and 3D image analysis [4]. Although in this paper we use 2D binary images to show the proof of concept, we do not encourage usage of homology groups and generators to find connected components in 2D images, since efficient approaches already exist [5]. However, these 'classical' approaches cannot be easily extended for many problems that exist in higher dimensions, since our visual intuition is inappropriate and topological reasoning becomes important. Computational topology has been used in metallurgy [6] to analyze 3D spatial structure of metals in an alloy and in medical image processing [7] in analyzing blood vessels. In higher dimensional problems (e.g. beating heart represented in 4D) the importance of homology groups and generators becomes clear in analyzing objects in these spaces (number of connected components, tunnels, holes, etc), because of the nice and clean formulation which holds in any dimension. One can think of other applications, as a preprocessing step, to speed up recognition of complex shapes in large image databases, e.g. images are first filtered based on their topological invariants and afterward are matched using shapes, appearances, etc.

The usage of homology groups and generators in image processing is a new topic and is not widely spread. In this paper we introduce a new method for computing homology groups and their generators using a hierarchical structure which is build by using two operations: contraction and removal. These two operations are used also in [8] to incrementally compute homology groups and their generators of 2D closed surfaces, but a hierarchy is not build.

The paper is structured as follows. Basic notions on homology and irregular graph pyramids are recalled in Section 2 and Section 3. The proposed method to compute homology groups and their generators is presented in detail in Section 4. Experimental results on 2D images that show the correctness of the new method are found in Section 5.

2 Homology

In this part, the basic homology notions of chain, cycle, boundary, and homology generator are recalled. Interested readers can find more details in [9].

The homology of a subdivided object X can be defined in an algebraic way by studying incidence relations of its subdivision. Within this context, a cell of dimension p is called a $p-$cell and the notion of $p-chain$ is defined as a sum $\sum_{i=1}^{nb\ p-cells} \alpha_i c_i$, where c_i are $p-$cells of X and α_i are coefficients assigned to each cell in the chain. Homology can be computed using any group $\mathfrak{A}$ for the coefficients α_i. But, the theorem of universal coefficients [9] ensures that all homological information can be obtained by choosing $\mathfrak{A} = \mathbb{Z}$. It is also known [9] that for nD objects embedded in $\mathbb{R}^D$, homology information can be computed by simply considering chains with moduli 2 coefficients ($\mathfrak{A} = \mathbb{Z}/2\mathbb{Z}$). Note that in this case, a cell that appears twice on a chain vanishes, because $c + c = 0$ for any cell c when using moduli 2 coefficients ($i.e.$ if a cell appears even times we discard it otherwise we keep it). In the following, only chains with coefficients over $\mathbb{Z}/2\mathbb{Z}$ will be considered. Note that the notion of chain is purely formal and the cells that compose a chain do not have to satisfy any property. For example, on the simplicial complex illustrated on Fig.1(b) the sums: $a_1 + a_4$, a_3 and $a_2 + a_7 + a_4$ are $1-$chains.

For each dimension $p = 0, \dots, n$, where $n = dim(X)$, the set of $p-$chains forms an abelian group denoted C_p. The $p-$chain groups can be put into a sequence, related by applications ∂_p describing the boundary of $p-$chains as $(p-1)-$chains:

$$C_n \xrightarrow{\partial_n} C_{n-1} \xrightarrow{\partial_{n-1}} \cdots \xrightarrow{\partial_1} C_0 \xrightarrow{\partial_0} 0,$$

which satisfy $\partial_p \partial_{p-1}(c) = 0$ for any $p-$chain c. This sequence of groups is called a *free chain complex*.

The boundary of a $p-$chain reduced to a single cell is defined as the sum of its incident $(p-1)-$cells. The boundary of a general $p-$chain is then defined by linearity as the sum of the boundaries of each cell that appears in the chain e.g. in Fig.1(b), $\partial(f_1 + f_2) = \partial(f_1) + \partial(f_2) = (a_1 + a_2 + a_7) + (a_7 + a_3 + a_6) = a_1 + a_2 + a_3 + a_6$. Note that as mentioned before, chains are considered over $\mathbb{Z}/2\mathbb{Z}$ coefficients i.e. any cell that appears twice vanishes.

For each dimension p, the set of p-chains which have a null boundary are called *p-cycles* and are a subgroup of C_p, denoted Z_p e.g. $a_1 + a_2 + a_7$ and $a_7 + a_5 + a_4 + a_3$ are $1-$cycles. The set of p-chains which bound a $p+1$-chain are

Table 1. Translation of homology notions to graph theory

Homology theory	Graph theory
0-cell, 1-cell, 2-cell	vertex, edge, face
0-chain, 1-chain, 2-chain	set of vertices, set of edges, set of faces
0-cycle, 1-cycle, 2-cycle	set of vertices, closed path of edges, closed path of faces

called *p-boundaries* and they are a subgroup of C_p, denoted B_p e.g. $a_1 + a_2 + a_7 = \partial(f_1)$ and $a_1 + a_6 + a_3 + a_2 = \partial(f_1 + f_2)$ are 1−boundaries.

According to the definition of a free chain complex, the boundary of a boundary is the null chain. Hence, this implies that any boundary is a cycle. Note that according to the definition of a free chain complex, any 0−chain has a null boundary, hence every 0−chain is a cycle.

The pth homology group, denoted H_p, is defined as the quotient group Z_p/B_p. Thus, elements of the homology groups H_p are equivalence classes and two cycles z_1 and z_2 belong to the same equivalence class if their difference is a boundary (*i.e.* $z_1 = z_2 + b$, where b is a boundary). Such two cycles are called *homologous* e.g. let $z_1 = a_5 + a_4 + a_3 + a_7$, $z_2 = a_5 + a_4 + a_6$ and $z_3 = a_1 + a_2 + a_3$; z_1 and z_2 are homologous ($z_1 = z_2 + \partial(f_2)$) but z_1 and z_2 are not homologous to z_3. Let H_p be a homology group generated by q independent equivalence classes $C_1, \cdots, C_q$, any set $\{h_1, \cdots, h_q \mid h_1 \in C_1, \cdots, h_q \in C_q\}$ is called a *set of generators* for H_p. For example, either $\{z_1\}$ or $\{z_2\}$ can be chosen as a generator of H_1 for the object represented in Fig.1(b).

Note that some of the notions mentioned before could be confused with similar notions from graph theory. Tab.1 associates these homology notions with notions classically used in graph theory.

3   Irregular Graph Pyramids

In this part, basic notions of pyramids like receptive field, contraction kernel, and equivalent contraction kernel, are introduced. For more details see [10].

A pyramid (Fig. 2a) describes the contents of an image at multiple levels of resolution. A high resolution input image is at the base level. Successive levels reduce the size of the data by a *reduction factor* $\lambda > 1.0$. The *reduction window* relates one cell at the reduced level with a set of cells in the level directly below. The contents of a lower resolution cell is computed by means of a *reduction function* the input of which are the descriptions of the cells in the reduction window. Higher level descriptions should be related to the original input data in the base of the pyramid. This is done by the *receptive field* (RF) of a given pyramidal cell c_i. The RF(c_i) aggregates all cells (pixels) in the base level of which c_i is the ancestor.

Each level represents a partition of the pixel set into cells, *i.e.* connected sub-sets of pixels. The construction of an irregular pyramid is iteratively local [11]. On the base level (level 0) of an irregular pyramid the cells represent single pixels and the neighborhood of the cells is defined by the 4(8)-connectivity of the pixels. A cell on level $k + 1$ (parent) is a union of neighboring cells on level k (children). This union is controlled by so called *contraction kernels* (CK) [12], a spanning forest which relates two successive levels of a pyramid. Every parent computes its values independently of other cells on the same level. Thus local independent (and parallel) processes propagate information up and down and laterally in the pyramid. Neighborhoods on level $k + 1$ are derived from neighborhoods on level k. Higher level descriptions are related to the original input by the *equivalent contraction kernels* (ECK). A level of the graph pyramid consists of a pair $(G_k, \overline{G_k})$ of plane graphs G_k and its geometric dual $\overline{G_k}$ (Fig. 2b). The vertices of G_k represent the cells on level k and the edges of G_k represent the neighborhood relations of the cells, depicted with square vertices and dashed edges in Fig. 2b. The edges of $\overline{G_k}$ represent the borders of the cells on level k, solid lines in Fig. 2b, including so called pseudo edges needed to represent neighborhood relations to a cell completely enclosed by another cell. Finally, the vertices of $\overline{G_k}$ (circles in Fig. 2b), represent junctions of border segments of $\overline{G_k}$. The sequence $(G_k, \overline{G_k})$, $0 \le k \le h$ is called irregular (dual) graph pyramid. For simplicity of the presentation the dual $\overline{G}$ is omitted afterward.

4   Computing Homology Generators in a Graph Pyramid

There exists a general method for computing homology groups. This method is based on the transformation of incidence matrices [9] (which describe the boundary homomorphisms) into their reduced form called *Smith normal form*. Agoston proposes a general algorithm, based on the use of a slightly modified Smith normal form, for computing a set of generators of these groups [3]. Even if Agoston's algorithm is defined in any dimension, the main drawback of this method is directly linked to the complexity of the reduction of an incidence matrix into its Smith normal form, which is known to consume a huge amount of time and space. Another well known problem is the possible appearance of huge integers during the reduction of the matrix. A more complete discussion about Smith normal algorithm complexity can be found in [13]. Indeed, Agoston's algorithm cannot directly be used for computing homology generators and different kinds of optimisations have been proposed.

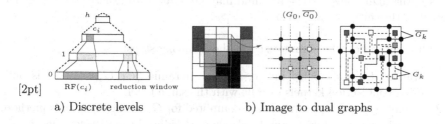

[2pt]   RF(c_i)   reduction window

a) Discrete levels   b) Image to dual graphs

Fig. 2. a) Pyramid concept, and b) representation of the cells and their neighborhood relations by a pair of dual plane graphs at the level 0 and k of the pyramid

Based on the work of [14,15], an optimisation for the computation of homology generators, based on the use of sparse matrices and moduli operations has been proposed [16]. This method avoids the possible appearance of huge integers. The authors also observed an improvement of time complexity dropping from $O(n^2)$ to $O(n^{5/3})$, where n is the number of cells of the subdivision.

An algorithm for computing the rank of homology groups *i.e.* the Betti numbers has been proposed in [17]. The main idea of this algorithm is to reduce the number of cells of the initial object in order to obtain a homologically equivalent object, made out of less cells. In some special cases (orientable objects), Betti numbers can directly be deduced from the resulting object. However, this method cannot directly provide a set of generators. Based on the previously mentioned work, an algorithm for computing a minimal representation of the boundary of a 3D voxel region, from which homology generators can directly be deduced has been defined in [8].

4.1 Description of the New Method

The method we propose in this paper has the same philosophy as the methods of Kaczynski and Damiand [18,19]: reducing the number of cells of an object for computing homology. Moreover, we keep all simplifications that are computed during the reduction process by using a pyramid. In this way, homology generators can be computed in the top level of the pyramid, and can be used to deduce generators of any level of the pyramid. In particular, we show how generators of the higher level can be directly down-projected on the desired level (using equivalent contraction kernels).

Starting from an initial image, we build an irregular graph pyramid. The method we provide here is valid as long as the algorithm used for the construction of the pyramid preserves homology. In particular, we show here that the decimation by contraction kernels, described in Section 3 [12], preserves homology of a subdivided object. Indeed, homology of the initial image can thus be computed in any level of the pyramid, and in particular in the top level where the object is described with the smallest number of cells.

Moreover, we use the notion of receptive field and equivalent contraction kernel, and show that the generators of homology groups of any level of the pyramid can be deduced from those computed on the higher level. Note that in special cases, the higher level of the pyramid may be reduced to exactly a set of generators of the initial image, as shown in [8].

Our method can be summarized in the following steps:

1   Starting from a labeled image, a graph pyramid $\{G_0, G_1, \ldots, G_k\}$ is built using contraction kernels of cells with the same label.
2   Homology groups generators are computed for G_k, using Agoston's method.
3   Homology generators of any level i can be deduced from those of level $i + 1$ using the contraction kernels. In particular, we obtain the homology generators of the initial image.

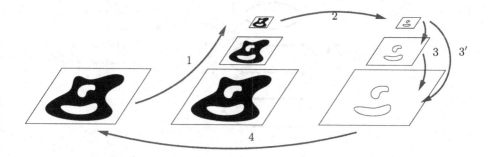

Fig. 3. Computing generators of homology groups using an image pyramid

Note that homology generators of the lowest level can directly be deduced from the highest level using the notion of equivalent contraction kernel (arrow $3'$ in Fig.3). Fig.3 illustrates the general method that we propose for computing homology generators of an image.

4.2   Preserving Homology on Irregular Graph Pyramids

The algorithm described in [17] is based on operations of *interior face reduction* that reduce the number of cells of the subdivision. The main idea is to find a $p-$cell a and a $(p+1)-$cell b, such that a is incident to b. Then a and b are removed and the boundary of the other $p-$cells that were adjacent to a are modified such that the new boundary $\partial(b')$ is defined as its initial boundary added with the boundary of b. Indeed, if a is incident to exactly two $p-$cells b and b', the result of the corresponding interior face reduction can be seen as the removal of a and the merging of b and b'. It is proved in [17] that interior face reduction preserves homology.

Observing the dual graph, the operations of contraction and removal that are used to build each level of the pyramid are interior face reduction: two faces that are merged share a common edge that is removed, and an edge is contracted if one of its endpoints is incident to exactly two different edges. Thus, homology is preserved in every level of the pyramid.

4.3   Delineating Generators

A $1D$ generator in $\overline{G}_k = (\overline{V}_k, \overline{E}_k)$ is a closed path connecting vertices of $\overline{G}_k$ and surrounding at least one hole. Each vertex $\overline{v} \in \overline{G}_k$ is the result of contracting a tree (contraction kernel $\overline{CK}$) of $\overline{G}_{k-1}$. Each edge $(\overline{v}_1, \overline{v}_2) \in \overline{G}_k$ corresponds to a surviving edge $(\overline{w}_1, \overline{w}_2) \in \overline{G}_{k-1}$ with $\overline{w}_1 \in \overline{CK}_{k-1}(\overline{v}_1)$ and $\overline{w}_2 \in \overline{CK}_{k-1}(\overline{v}_2)$ i.e. an edge that has neither been contracted nor removed[2].

Given a generator in $\overline{G}_k$, mapping it to the level below is done by identifying the surviving edges in $\overline{G}_{k-1}$ corresponding to the generator edges in $\overline{G}_k$ and,

---

[2] Not part of any simplification.

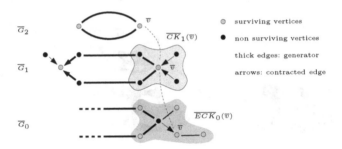

Fig. 4. Top-down delineation of a generator computed in $\overline{G}_2$

where the generator is disconnected, adding paths to fill in the gaps and reconnect. For every two consecutive edges not having a common vertex in $\overline{G}_{k-1}$ but having one in $\overline{G}_k$, the unique path connecting their disconnected endpoints in the contraction kernel $\overline{CK} \subset \overline{G}_{k-1}$ of their shared vertex in $\overline{G}_k$ is added.

Because each path added in $\overline{G}_{k-1}$ is entirely part of a contraction kernel, with contraction being used in the dual only for boundary simplification purposes, never connecting two different boundaries, and because the building process preserves homology (see Sec. 4.2) the obtained generators will be homologous to the ones in $\overline{G}_k$.

Reiterating this process of mapping the generator cycles of $\overline{G}_k$ from k to $k-1, \dots$ to 0, cycles in $\overline{G}_0$ corresponding to the generators of the top level can be identified. By replacing the contraction kernels, with the equivalent contraction kernels, using the same methodology, the generator cycles of $\overline{G}_k$ can be directly mapped to $\overline{G}_0$. For an example, see Fig. 4.

5   Experiments on 2D Images

We present and discuss initial experiments that have been performed on 2D binary shapes. For each shape, we have computed homology generators directly on the initial image, and on the top level of the pyramid.

Tab.2 shows the number of $0D$, $1D$ and $2D$–cells on the initial image, and on the top level of the pyramid for the shape presented on Fig.5 and Fig.6. One can observe that for each shape the total number of cells is considerably reduced on the higher level of the pyramid. Thus, the computation of homology generators can be done on much smaller matrices on the top level instead of the initial image.

Table 2. The number of cells on the initial image and on the top of the pyramid

	Initial image			Top of the pyramid		
	0D-cells	1D-cells	2D-cells	0D-cells	1D-cells	2D-cells
Fig.5.	8153	15785	7630	7	10	1
Fig.6	10352	20148	9793	9	13	1

                    (a)   (b)

Fig. 5. (*a*): the homology generators computed on the initial image. (*b*): the down-projected generator.

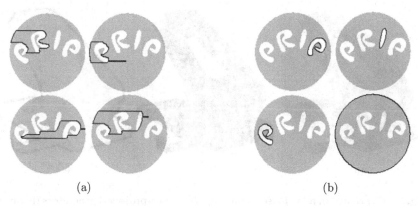

                    (a)   (b)

Fig. 6. (*a*): the homology generators computed on the initial image. (*b*): the down-projected generator.

In Fig.5 and Fig.6, it can be seen that our new method provides a valid set of generators in each case.

Moreover, using the classical method, we cannot have any control of the geometry of the generators computed. More precisely, the aspect of the obtained generators is directly linked to the construction of incidence matrices, which is determined by the scanning of each cell of the initial image. The shape shown on Fig.7 has been obtained from rotating Fig.5. In Fig.7(a), one can observe that the aspect of the generators computed on the initial image "follows" the scanning of the cells (from top to bottom, and left to right). The generators obtained in Fig.7(b) always fit on the boundaries of the image. It is proved in [20] that any generator computed with our new method will always fit on some boundaries of the initial image.

One can note that the sets of cycles obtained in Fig.5(a) and Fig.5(b) do not surround the same (set of) $1D-$holes of the shape S. Indeed, these two sets are two different basis of the same group $H_1(S)$: let a, b and c denote the equivalence class of cycles that surround respectively the left eye, the right eye, and the

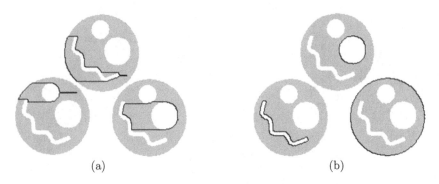

(a)   (b)

Fig. 7. Influence of the scanning (compare with Fig.5)

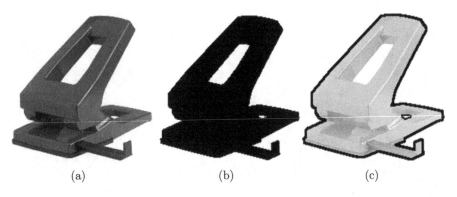

(a)   (b)   (c)

Fig. 8. (*a*): original image. (*b*): segmentation. (*c*): down-projected generators (in black).

mouth. The set of generators in Fig.5(a) describe $H_1(S)$ in the basis $\{a+b, c, a\}$ whereas in Fig.5(b), $H_1(S)$ is described in the basis $\{a, a+b+c, b\}$.

In Fig.8 a real world image is shown. We have first segmented the image (e.g. one can choose the minimum spanning tree based pyramid segmentation [21], and build generators on these segmented images, but for clarity of the presentation we used a binary segmentation). Fig.8(a) shows the original image, Fig.8(b) the used binary segmentation, and Fig.8(c) the brightened image with the obtained generators in black.

6   Conclusion

We have presented a new method for computing homology groups of images and their generators, using irregular graph pyramids. The homology generators are computed efficiently on the top level of the pyramid, since the number of cells is small, and a top down process (down-projection) delineates the homology generators of the initial image. Some preliminary results have been shown for 2D binary images. We have also observed that the generators computed with this new method seem to stay on boundaries.

In a future work, we plan to extend this method to 3D and nD images, using the (already existing) structures of 3D and nD irregular pyramids. We also plan to use the property that down-projected generators always fit on boundaries in order to use homology generators for object matching and object tracking.

References

1. Kovalevsky, V.A.: Finite topology as applied to image analysis. Computer Vision, Graphics, and Image Processing 46, 141–161 (1989)
2. Kovalevsky, V.A.: Digital Geometry Based on the Topology of Abstract Cellular Complexes. In: Chassery, J.M., Francon, J., Montanvert, A., Réveillès, J.P., (eds.): Géometrie Discrète en Imagery, Fondements et Applications, Strasbourg, France, pp. 259–284 (1993)
3. Agoston, M.K.: Algebraic Topology, a first course. Pure and applied mathematics. Marcel Dekker Ed. (1976)
4. Allili, M., Mischaikow, K., Tannenbaum, A.: Cubical homology and the topological classification of 2d and 3d imagery. In: Proceedings of International Conference Image Processing. Vol. 2, pp. 173–176 (2001)
5. Sonka, M., Hlavac, V., Boyle, R.: Image Processing, Analysis and Machine Vision. Brooks/Cole Publishing Company (1999)
6. Kaczynksi, T., Mischaikow, K., Mrozek, M.: Computational Homology. Springer, Heidelberg (2004)
7. Niethammer, M., Stein, A.N., Kalies, W.D., Pilarczyk, P., Mischaikow, K., Tannenbaum, A.: Analysis of blood vessels topology by cubical homology. In: Proceedings of International Conference Image Processing. Vol. 2, 969–972 (2002)
8. Damiand, G., Peltier, S., Fuchs, L.: Computing homology for surfaces with generalized maps: Application to 3d images. In: Bebis, G., Boyle, R., Parvin, B., Koracin, D., Remagnino, P., Nefian, A., Meenakshisundaram, G., Pascucci, V., Zara, J., Molineros, J., Theisel, H., Malzbender, T. (eds.) ISVC 2006. LNCS, vol. 4292, pp. 1151–1160. Springer, Heidelberg (2006)
9. Munkres, J.R.: Elements of algebraic topology. Perseus Books (1984)
10. Jolion, J.M., Rosenfeld, A.: A Pyramid Framework for Early Vision. Kluwer, Dordrecht (1994)
11. Meer, P.: Stochastic image pyramids. Computer Vision, Graphics, and Image Processing 45, 269–294 (1989) Also as UM CS TR-1871, June, 1987
12. Kropatsch, W.G.: Building irregular pyramids by dual graph contraction. IEE-Proc. Vision, Image and Signal Processing 142, 366–374 (1995)
13. Kannan, R., Bachem, A.: Polynomial algorithms for computing the Smith and Hermite normal forms of an integer matrix. SIAM Journal on Computing 8, 499–507 (1979)
14. Dumas, J.G., Heckenbach, F., Saunders, B.D., Welker, V.: Computing simplicial homology based on efficient smith normal form algorithms. In: Algebra, Geometry, and Software Systems, pp. 177–206 (2003)
15. Storjohann, A.: Near optimal algorithms for computing smith normal forms of integer matrices. In: Lakshman, Y.N. (ed.) Proceedings of the 1996 International Symposium on Symbolic and Algebraic Computation, pp. 267–274. ACM Press, New York (1996)

16. Peltier, S., Alayrangues, S., Fuchs, L., Lachaud, J.O.: Computation of homology groups and generators. Computers and graphics 30, 62–69 (2006)
17. Kaczynski, T., Mrozek, M., Slusarek, M.: Homology computation by reduction of chain complexes. Computers & Math. Appl. 34, 59–70 (1998)
18. Kaczynski, T., Mischaikow, K., Mrozek, M.: Computational Homology. Springer, Heidelberg (2004)
19. Damiand, G., Peltier, P., Fuchs, L., Lienhardt, P.: Topological map: An efficient tool to compute incrementally topological features on 3d images. In: Reulke, R., Eckardt, U., Flach, B., Knauer, U., Polthier, K. (eds.) IWCIA 2006. LNCS, vol. 4040, pp. 1–15. Springer, Heidelberg (2006)
20. Peltier, S., Ion, A., Haxhimusa, Y., Kropatsch, W.: Computing homology group generators of images using irregular graph pyramids. Technical Report PRIP-TR-111, Vienna University of Technology, Faculty of Informatics, Institute of Computer Aided Automation, Pattern Recognition and Image Processing Group (2007) http://www.prip.tuwien.ac.at/ftp/pub/publications/trs/
21. Haxhimusa, Y., Kropatsch, W.G.: Hierarchy of partitions with dual graph contraction. In: Michaelis, B., Krell, G. (eds.) Proceedings of German Pattern Recognition Symposium. LNCS, vol. 2781, pp. 338–345. Springer, Heidelberg (2003)

Approximating TSP Solution by MST Based Graph Pyramid[*]

Yll Haxhimusa[1,2], Walter G. Kropatsch[1], Zygmunt Pizlo[2], Adrian Ion[1],
and Andreas Lehrbaum[1]

[1] Vienna University of Technology,
Faculty of Informatics, Institute of Computer Aided Automation,
Pattern Recognition and Image Processing Group, Austria
{yll,krw,ion,lehrbaua}@prip.tuwien.ac.at
[2] University of Purdue,
Department of Psychological Sciences, USA
{yll,pizlo}@psych.purdue.edu

Abstract. The traveling salesperson problem (TSP) is difficult to solve for input instances with large number of cities. Instead of finding the solution of an input with a large number of cities, the problem is approximated into a simpler form containing smaller number of cities, which is then solved optimally. Graph pyramid solution strategies, in a bottom-up manner using Borůvka's minimum spanning tree, convert a 2D Euclidean TSP problem with a large number of cities into successively smaller problems (graphs) with similar layout and solution, until the number of cities is small enough to seek the optimal solution. Expanding this tour solution in a top-down manner to the lower levels of the pyramid approximates the solution. The new model has an adaptive spatial structure and it simulates visual acuity and visual attention. The model solves the TSP problem sequentially, by moving attention from city to city with the same quality as humans. Graph pyramid data structures and processing strategies are a plausible model for finding near-optimal solutions for computationally hard pattern recognition problems.

1 Introduction

Traveling salesperson problem (TSP) is a combinatorial optimization task of finding the shortest tour of n cities given the intercity costs. When the costs between cities are Euclidean distances, the problem is called Euclidean TSP (E-TSP). TSP as well as E-TSP belongs to the class of difficult optimization problems called NP-hard and NP-complete if posed as a decision problem [1]. The straightforward approach by using brute force search would be using all possible permutations for finding the shortest tour. It is impractical for large n since the number of permutations is $\frac{(n-1)!}{2}$. Because of the computational

---

[*] Supported by the Austrian Science Fund under grants P18716-N13 and S9103-N04, and the USA Air Force Office of Scientific Research.

F. Escolano and M. Vento (Eds.): GbRPR 2007, LNCS 4538, pp. 295–306, 2007.

intractability of TSP, researchers concentrated their efforts on finding approximating algorithms. Good approximating algorithms can produce solutions that are only a few percent longer than an optimal solution and the time of solving the problem is a low-order polynomial function of the number of cities [2,3,4]. The last few percent to reach optimality are computationally the most expensive to achieve.

It is by now well established that humans produce close-to-optimal solutions to E-TSP problems in time that is (on average) proportional to the number of cities [5,6,7]. This level of performance can not be reproduced by any of the standard approximating algorithms. Some approximating algorithms produce smaller errors but the time complexity is substantially higher than linear, other algorithms are relatively fast but produce substantially higher errors. It is therefore of interest to identify the computational mechanism used by the human brain.

A simple way to present E-TSP to a subject is to show n cities as points on a computer screen and ask the subject to produce a tour by clicking on the points. In Figure 1a, an E-TSP example of 10 cities is shown and in c the solution given by a human. The tours produced by the subjects are, on average, only a few percent longer than the shortest tours (in Figure 1c and d the cross depicts the starting position and the arrow the orientation used by the subject). The solution time is a linear function of the number of cities [5,6]. Two attempts to emulate human performance by a computational model were undertaken in [5,6]. In [5], authors attempt to formulate a new approximating algorithm for E-TSP motivated by the failure to identify an existing algorithm that could provide a good fit to the subjects' data. The main aspects of the models in [5,7] are its

- (multiresolution) pyramid architecture, and
- a coarse to fine process of successive tour approximations.

They showed that performance of this model (proportion of optimal solutions and average solution error) is statistically equivalent to human performance. Pyramid algorithms have been used extensively in both computer and human vision literature (e.g. [8]), but not in problem solving. The work of [5,9] was the first attempt to use pyramid algorithms to solve the E-TSP. One of the most attractive aspects of pyramid algorithms, which make them suitable for problems

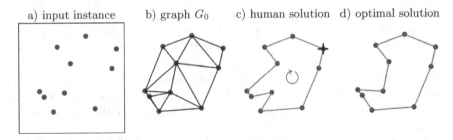

a) input instance   b) graph G_0 c) human solution   d) optimal solution

Fig. 1. E-TSP and solutions given by human and optimal solver

such as early vision or E-TSP, is that they allow to solve (approximately) global optimization tasks without performing a global search. A similar pyramid algorithm for producing approximate E-TSP solutions with emphasis on trade-off between computational complexity (speed) and error in the solution (accuracy) and not on modeling human performance is formulated in [4, Chap.5], and [10].

In this paper we present a computational model for solving E-TSP approximately based on the multiresolution graph pyramid. *The emphasis is on emulating human performance (time and accuracy), and not in finding an algorithm for solving E-TSP as optimally as possible.* The interested reader can consult a large body of the literature in Operations Research for algorithms for E-TSP [4,3] that can produce near to optimal tours. Again, these algorithms have computational complexity that is substantially higher than linear.

Our goal is to show that the results of our model are well fitted to the results of the humans, and the quality and speed are comparable to that of human subjects. The next section presents a short overview of the pyramid representations (Section 2). In Section 3 the solution of the E-TSP using a minimum spanning tree (MST) based graph pyramid is introduced. The bottom-up simplification of the input data is shown in Section 3.1, and in Section 3.2 the top-down approximative solution is described. Psychophysical experiments on E-TSP are presented in Section 4.

2 Irregular Graph Pyramid

In our framework, the TSP input is represented by graphs where cities are represented by vertices, and the intercity neighborhoods by edges (see Figure 1b). Each vertex of the constructed input graph must have at least two edges for the TSP tour to exist. A level (k) of the graph pyramid consists of a graph G_k. Moreover the graph is attributed, $G = (C, N, w_v, w_e)$, where $w_e : N \to \mathbb{R}^+$ is a weighted function defined on edges N. The weights w_e are Euclidean distances in the E-TSP and $w_v : C \to \mathbb{R}^+$ is a weighted function defined on cities C. I.e. each vertex (city) has as a weight its position in the Cartesian coordinate system

Finally, the sequence G_k, $0 \le k \le h$ is called irregular graph pyramid.

In a regular pyramid, the number of vertices at any level k is λ times higher than the number of pixels at the next (reduced) level $k + 1$. The so called reduction factor λ is greater than one and it is the same for all levels k. The number of levels on top of G amounts to $\log_\lambda(|G|)$. This implies that a pyramid is build in $\mathcal{O}[\log(diameter(G))]$ parallel steps [8]. Regular image pyramids are confined to globally defined sampling grids and lack shift invariance [11]. In [12,13] it is shown how these drawbacks can be avoided by adaptive irregular pyramids.

In Graham's model [5], clusters are not explicitly represented. Instead, the centers of the clusters were used in the E-TSP solution process. The centers were modes (peaks) of the intensity distribution produced by blurring the image. To make clusters explicit, Pizlo et. al [14] used an adaptive model in which adaptive top-down partitioning of the plane along the axis of Cartesian system was used. The hierarchy was represented by a binary tree. This top-down clustering had the advantage that the entire E-TSP did not have to be represented at once in the memory. The disadvantage was that although this algorithm was invariant to translation, it was not invariant to rotation. Our new model uses graphs as representation, which are invariant to both translation and rotation of the input city constellation. However, the clustering is performed in bottom-up fashion.

3   Solving E-TSP by a Graph Pyramid

Let $G_0 = (C, N, w_v, w_e)$ be the input graph, with weights on edges given as distances in L_2 space. The goal of the TSP is to find an nonempty ordered sequence of vertices and edges $(v_0, e_1, v_1, ..., v_{k-1}, e_k, v_k, ..., v_0)$ over all vertices of G_0 such that all the edges and vertices are distinct, except the start and the end vertex v_0. This tour is called the optimal tour τ_{opt} and the sum of edge weights in this tour is minimal, i.e.

$$\tau_{opt} = \sum_{e \in \tau} w_e \rightarrow \min,$$

where w_e is the weight of edge e.

We use local to global and global to local processes in the graph pyramid to find a good solution τ^*, approximating the E-TSP. The main idea is to use:

- bottom-up processes to reduce the size of the input, and
- top-down refinement to find an (approximate) solution.

The size of the input (number of vertices in the graph) is reduced such that an optimal (trivial) solution can be found by the combinatorial search, e.g. for a 3 city instance (not all cities are co-linear) there is only one solution, not needing any search, and this is the optimal one. For a 4 city input (not all co-linear) there are three solutions from which two are non-optimal since they cross edges. A pyramid is used to reduce the size of the input in the bottom-up process. The (trivial) solution is then found at the top of the pyramid and refined in a process emulating fovea by humans using lower levels of this pyramid, i.e. the vertical neighborhoods (parent-children relations) are used in this process to refine the tour. The final, in general non-optimal, solution is found when all the cities at the base level of the pyramid are in the tour. The steps needed to find the E-TSP solution are shown in Algorithm 1. Partitioning of the input space is treated in Section 2. Sections 3.1 and 3.2 discuss steps 2 and 4 of Algorithm 1 in more detail.

3.1   Bottom-Up Simplification Using an MST Pyramid

The main idea is that cities being close neighbors are put into a cluster and considered as a single city at reduced resolution. By doing this recursively one produces a pyramid representation of the problem. It is well known that the human visual system represent images on multiple level of scales and resolution [15,16].

Algorithm 1. Approximating E-TSP Solution by an MST Graph Pyramid

---

Input: Attributed graph $G_0 = (C, N, w_v, w_e)$, and parameters r and s

1: partition the input space by preserving approximate location:
   create graph G_0
2: reduce number of cities bottom-up until the graph contains s vertices:
   build graph pyramid $G_k, \forall k = 0, ..., h$, where $s = |G_h|$
3: find the optimal tour τ_a for the graph G_h
4: refine solution top-down until all vertices at the base level are processed:
   refine τ_a until level 0 is reached

Output: Approximate TSP solution τ^*.

There are many different algorithms to make hierarchical clustering of cities [17]. We choose for this purpose the MST principle, especially Borůvka's algorithm [18] since it hierarchically clusters neighboring vertices. The time complexity of Borůvka's algorithm is $\mathcal{O}(|E| \log |V|)$. It can be shown that MST can be used as the natural lower bound and for the case of the TSP with the triangle inequality, which is the case for the E-TSP, it can be used to prove the upper bound as well [19]. The first step in Christofides' heuristics [2] is finding an MST as an approximation of TSP. Christofides shows that it is possible to achieve at least $\frac{3}{2}$ times of the optimal solution of TSP i.e. Christofides heuristics solution of TSP is at most 50% longer than the optimal solution.

For a given graph $G_0 = (C, N, w_v, w_e)$ the vertices are hierarchically grouped into trees (clustered) as given in Algorithm 2. The idea of Borůvka is to do greedy steps like in Prim's algorithm [20], in parallel over the graph at the same time. The size of trees (clusters) are not allowed to contain more than $r \in \mathbb{N}^+$ cities. These trees must contain at least 2 cities, due to the fact that the pyramid must have a logarithmic height [21], since the reduction factor λ is $2 \leq \lambda \leq r$. This parameter can be related also to the number of 'concepts' that humans can have in their 'memory buffer', and is usually not larger than 10.

The number $s \in \mathbb{N}^+$ of vertices in the top level of the pyramid is chosen such that an optimal tour can be found easily (usually $s = 3$, or $s = 4$). Note that

---

Algorithm 2. Reduction of the E-TSP Input by an MST Graph Pyramid

---

Input: Attributed graph $G_0 = (C, N, w_v, w_e)$, and parameters r and s

1: $k \leftarrow 0$
2: **repeat**
3:   $\forall v_k \in G_k$ find the edge $e' \in G_k$ with minimum w_e incident into this vertex
4:   using e' create trees T with no more than r vertices
5:   contract trees T into parent vertices v_{k+1}
6:   create graph G_{k+1} with vertices v_{k+1} and edges $e_k \in G_k \setminus T$
7:   attribute vertices in G_{k+1}
8:   $k \leftarrow k + 1$
9: **until** there are s vertices in the graph G_{k+1}.

Output: Graph pyramid – $G_k, 0 \leq k \leq h$.

---

larger s means a shallow pyramid and larger graph at the top, which also means higher time complexity to find the optimal tour at the top level. Thus r and s are used to control the trade off between speed and quality of solution.

An example of how Algorithm 2 builds the graph pyramid (only the last two levels) is shown in Figure 2. Each vertex (black in G_{h-1}) finds the edge with the minimal weight (solid lines in G_{h-1}). These edges create trees of no more than r ($= 4$) cities. These trees are then contracted to the parent vertices (enclosed black vertices in G_{h-1} are contracted into white vertices in G_h). The parent vertices together with edges not touched by the contraction are used to create the graph of the next level (parallel edges and self loops can be removed, since they are not needed for the clustering of vertices). The dotted lines between vertices in different levels represent the parent-child relations. The new parent vertex attribute can be the gravitational center of its child vertices, or by using the position of the vertex near this gravitational center. The algorithm iterates until there are s vertices at the top of the pyramid, and since s is small a full search can be employed to find the optimal tour τ_a at the top quickly.

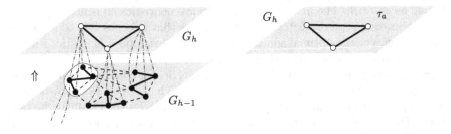

Fig. 2. Building the graph pyramid and finding the first TSP tour approximation

In our current software implementation we use the fully connected graph to represent the input instance, as expected the bottom-up simplification algorithm has at least $\mathcal{O}(|E|^2)$ time complexity [22]. This time complexity can be reduced easily to $\mathcal{O}(|E| \log |V|)$ if instead of the fully connected graph one uses a planar graph e.g. Delaunay triangulation.

3.2 Top-Down Approximation of the Solution

The tour τ_a found at level h of the graph pyramid is used as the first approximation of the TSP tour τ^*. This tour is then refined using the pyramid structure already built. Similar to Pizlo et. al. [14] we have chosen to use the most simple refinement, the one-path refinement. The one-path refinement process starts by choosing (randomly) a vertex v in the tour τ_a. Using the parent-child relationship, this vertex is expanded into the subgraph $G'_{h-1} \subset G_{h-1}$ from which it was created i.e. its receptive field in the next lower level. In this subgraph a path between vertices (children) is found that makes the overall path τ'_a the shortest

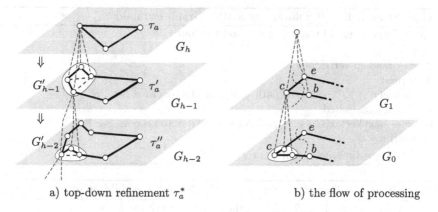

a) top-down refinement τ_a^* b) the flow of processing

Fig. 3. Refining the E-TSP tour by a graph pyramid

one (see Figure 3a). Since the number of vertices (children) in G_h' cannot be larger than r, a complete search is a plausible approach to find the path with the smallest contribution in the overall length of the tour τ_a'. Note that edges in the τ_a' are not necessarily the contracted edges during bottom-up construction.

The refinement process then choses one of the already expanded vertices in G_{h-1}', say v' and expands it into its child at the next lower level G_{h-2}', and the tour τ_a'' is computed. The process of tour refinement proceeds recursively until there are no more parent-children relationships (graph G_0, Figure 3b vertices of the receptive field of c, $RF(c)$, i.e. vertices at the base of the pyramid are reached. E.g. in Figure 3b, the tour is refined as the shortest path between the start vertex b and end vertex e and all the vertices (children of c) of the $RF(c)$. After arriving at the finest resolution, the process of refinement continues by taking a vertex in the next upper level in the same cluster (Figure 3 vertex b or e), and expanding it to its children and computing the tour. Note that the process of vertex expansion toward the base level emulates the movement of fovea (attention) in the process of solving the problem by a human observer. The tour is refined to the finest resolution in one part whereas other parts are left in their coarse resolution. The process converges when all vertices in the pyramid have been 'visited'[3]. More formally the steps are depicted in Algorithm 3, and Procedure 1, and 2.

Other refinement approaches can be chosen as well. One can use different approaches of refinement for e.g. one can think of using many vertices and expanding them in parallel (multi-path refinement), or use the one-path refinement until a particular level of the pyramid and continue with the multi-path refinement afterward. In these cases one needs to change Procedure 1. Note that there is a randomness in choosing which of the vertices to refine, which is may correspond to individual differences on how humans choose from which vertex to start the tour. In this case one needs to change Procedure 2.

---

[3] A demo is given in http://www.prip.tuwien.ac.at/Research/twist/results.php.

Algorithm 3. E-TSP Solution by a MST Graph Pyramid

Input: Graph pyramid G_k, $0 \le k \le h$ and the tour τ_a

1: $\tau^* \leftarrow \tau_a$
2: $v \leftarrow$ random vertex of τ^*
3: **repeat**
4:    refine(τ^*, v) /* refine the path using the children of v. See Prc. 1 */
5:    mark v as visited
6:    $v \leftarrow$ nextVertex(G_k, v, τ^*) /* get next vertex to process. See Prc. 2 */
7: **until** $v = \emptyset$

Output: Approximation E-TSP tour τ^*.

Procedure 1. refine(τ^*, v): refine a path τ^* using the children of v

Input: Graph pyramid G_k, $0 \le k \le h$, the tour τ^*, and the vertex v.

1: $(c_1, \ldots, c_n) \leftarrow$ children of v /* vertices that have been contracted to v */
2: **if** $n > 0$ /* v is not a vertex from the bottom level */ **then**
3:    $v_p, v_s \leftarrow$ neighbours of v in τ^* /* predecessor and successor of v */
4:    $p_1, \ldots, p_n \leftarrow \text{argmin}\{$length of path $\{v_p, c_{p_1}, \ldots, c_{p_n}, v_s\}\}$ such that $p_1, \ldots, p_n$ is a permutation of $1, \ldots, n$ /* optimal order of new vertices in the tour */
5:    replace path $\{v_p, v, v_s\}$ in τ^* with path $\{v_p, c_{p_1}, \ldots, c_{p_n}, v_s\}$

Output: refined TSP tour τ^*.

Procedure 2. nextVertex(G_k, v, τ^*): get next vertex to process

Input: Graph pyramid G_k, $0 \le k \le h$, the vertex v, and the tour τ^*

1: **repeat**
2:    **if** v has unvisited children **then**
3:        $v \leftarrow$ first unvisited child of v in τ^* /* given an orientation */
4:    **else if** v has unvisited siblings **then**
5:        $v \leftarrow$ first unvisited sibling of v in τ^* /* given an orientation */
6:    **else if** v has a parent i.e. v is not a vertex of the top level **then**
7:        $v \leftarrow$ parent of v
8:    **else**
9:        $v \leftarrow \emptyset$
10: **until** (v not visited) $\bigvee$ ($v = \emptyset$)

Output: new vertex to process v.

4   Psychophysical Evaluation of Solutions

Four subjects (including one author) were tested. Each subject solved the same 100 E-TSP problems in a different order. There were 4 different sizes 6, 10, 20, and 50 cities, with 25 instances per problem size. The cites in each problem were generated randomly on a 256×256 square grid [7]. Examples of 10 city tours produced by the subject and by the model are presented in Figure 4. The crosses depict the starting point chosen by the subjects and the model. BSL, OSK, and ZP chose the clock-wise tour, whereas ZL the counter-clock-wise tour. The MST based pyramid model choses randomly the orientation of the tour. To test how well the model fits the subject data, the algorithm is run 15 times with different parameters r $(2 \leq r \leq 7)$. The results of the best model fitting (as well as the standard deviation) to the subject data are shown in Figure 5. It can be seen that fit are quite good. The worst fit is for the case of 50-city problems (especially for OSK). Specifically, the model's performance is not as good as that of the subjects. To improve the models's performance, higher values of r would have to be used. This is how the simulation were performed in [14].

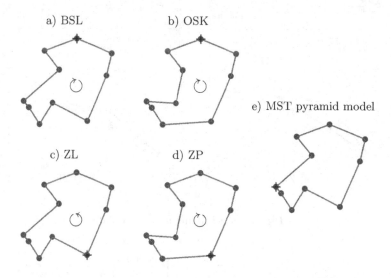

a) BSL   b) OSK

e) MST pyramid model

c) ZL   d) ZP

Fig. 4. E-TSP solutions by humans subjects and the MST pyramid model

For larger instances (> 100 cities) data with human subjects are difficult to obtain. Therefore we tested the results of the Algorithm 1 with the state-of-the-art Concorde TSP solver[4] with respect to time and with *adaptive pyramid* [14] with respect to the solution error. The test is done with respect to the quality of results, and the time needed to solve input examples with 200, 400, 600, 800, and 1000 cities. The error values are shown in Figure 6a and the time performance in Figure 6b. The time plot is normalized to the time needed for methods to solve the 200 city instance in one second. We have fixed the values of the parameter $r = 7$ and $s = 3$ for these experiments. Note that the Concorde algorithm solves the problem optimally, i.e. no error. We show that the results of the MST-based model are comparable to humans in quality and speed, and scale well with large input instances. This solution strategy emulates human fovea by moving attention from city to city.

---

[4] http://www.tsp.gatech.edu/concorde/index.html

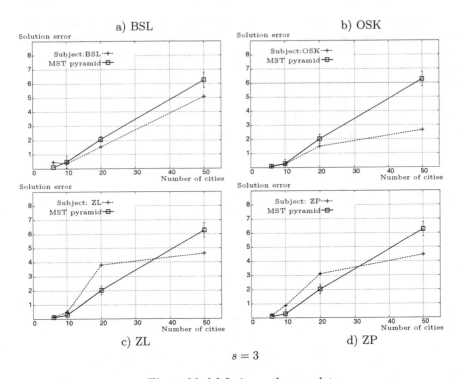

a) BSL   b) OSK

c) ZL   d) ZP

$$s = 3$$

Fig. 5. Model fitting on human data

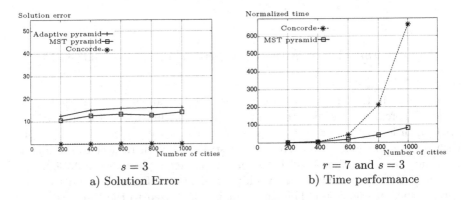

$s = 3$

a) Solution Error

$r = 7$ and $s = 3$

b) Time performance

Fig. 6. The solution error and the time performance

5   Conclusion

Pyramid strategies convert in a bottom-up process a 2D Euclidean TSP problem
with a large number of cities into successively smaller problems with similar
layout and solution until the number of cities is small enough to seek the optimal
solution. Expanding this solution in a top-down manner to the lower levels of

the pyramid approximates the solution. The introduced method uses a version of Borůvka's MST construction to reduce the number of cities. A top-down process is then employed to approximate the E-TSP solution of the same quality and at the same speed as humans do. The new model has an adaptive spatial structure and it simulates visual acuity and visual attention. Specifically, the model solves the E-TSP problem sequentially, by moving attention from city to city, the same way human subjects do. We showed that the new model fits the human data. Pyramid data structures and processing strategies are a plausible model for finding near-optimal solutions for NP-hard pattern recognition problems, e.g. matching.

Acknowledgment. The authors would like to thank anonymous reviewers for their valuable comments.

References

1. Johnson, D.S., McGeoch, L.A.: Local Search in Combinatorial Optimization. In: Aarts, E.H.L., Lenstra, J.K. (eds.) The Traveling Salesman Problem: A Case Study in Local Optimization, pp. 215–310. John Wiley and Sons, Chichester (1997)
2. Christofides, N.: Graph Theory - An Algorithmic Approach, New York, London. Academic Press, San Francisco (1975)
3. Lawler, E.L., Lenstra, J.K., Rinnooy Kan, A.H.G., Shmoys, D.B.: The Traveling Salesman Problem. Wiley, New York (1985)
4. Gutin, G., Punnen, A.P.: The traveling salesman problem and its variations. Kluwer, Dordrecht (2002)
5. Graham, S.M., Joshi, A., Pizlo, Z.: The travelling salesman problem: A hierarchical model. Memory and Cognition 28, 1191–1204 (2000)
6. MacGregor, J.N., Ormerod, T.C., Chronicle, E.P.: A model of human performance on the traveling salesperson problem. Memory and Cognition 28, 1183–1190 (2000)
7. Pizlo, Z., Li, Z.: Pyramid algorithms as models of human cognition. In: Proceedings of SPIE-IS&T Electronic Imaging, Computational Imaging, SPIE, pp. 1–12 (2003)
8. Jolion, J.M., Rosenfeld, A.: A Pyramid Framework for Early Vision. Kluwer, Dordrecht (1994)
9. Pizlo, Z., Joshi, A., Graham, S.M.: Problem solving in human beings and computers. Technical Report CSD TR 94-075, Department of Computer Sciences, Purdue University (1994)
10. Arora, S.: Polynomial-time approximation schemes for euclidean tsp and other geometric problems. Journal of the Association for Computing Machinery 45, 753–782 (1998)
11. Bister, M., Cornelis, J., Rosenfeld, A.: A critical view of pyramid segmentation algorithms. Pattern Recognition Letters 11, 605–617 (1990)
12. Montanvert, A., Meer, P., Rosenfeld, A.: Hierarchical image analysis using irregular tesselations. IEEE Transactions on Pattern Analysis and Machine Intelligence 13, 307–316 (1991)
13. Jolion, J.M., Montanvert, A.: The adaptive pyramid, a framework for 2D image analysis. Computer Vision, Graphics, and Image Processing: Image Understanding 55, 339–348 (1992)

14. Pizlo, Z., Stefanov, E., Saalweachter, J., Li, Z., Haxhimusa, Y., Kropatsch, W.G.: Traveling salesman problem: a foveating model. Journal of Problem Solving 1, 83–101 (2006)
15. Watt, R.J.: Scanning from coarse to fine spatial scales in the human visual system after the onset of a stimulus. Journal of the Optical Society of America 4, 2006–2021 (1987)
16. Pizlo, Z., Rosenfeld, A., Epelboim, J.: An exponential pyramid model of the time-course of size processing. Vision Research 35, 1089–1107 (1995)
17. Duda, R.O., Hart, P.E., Stork, D.G.: Pattern Classification. John Wiley & Sons, Chichester (2001)
18. Neštřil, J., Miklovà, E., Neštřilova, H.: Otakar Borôvka on minimal spanning tree problem translation of both the 1926 papers, comments, history. Discrete Mathematics 233, 3–36 (2001)
19. Atallah, M.J. (ed.): Algorithms and Theory of Computational Handbook. CRC Press, Boca Raton, FL (1999)
20. Prim, R.C.: Shortest connection networks and some generalizations. The. Bell. System Technical Journal 36, 1389–1401 (1957)
21. Kropatsch, W.G., Haxhimusa, Y., Pizlo, Z., Langs, G.: Vision pyramids that do not grow too high. Pattern Recognition Letters 26, 319–337 (2005)
22. Papadimitiou, C.H., Steiglitz, K.: Combinatorial Optimization: Algorithms and Complexity. Dover Publication, Mineola, NY (1998)

The Construction of Bounded Irregular Pyramids with a Union-Find Decimation Process

R. Marfil, L. Molina-Tanco, A. Bandera, and F. Sandoval

Grupo ISIS, Dpto. Tecnología Electrónica, E.T.S.I. Telecomunicación,
Universidad de Málaga
Campus de Teatinos s/n 29071-Málaga, Spain
rebeca@uma.es

Abstract. The Bounded Irregular Pyramid (BIP) is a mixture of regular and irregular pyramids whose goal is to combine their advantages. Thus, its data structure combines a regular decimation process with a union-find strategy to build the successive levels of the structure. The irregular part of the BIP allows to solve the main problems of regular structures: their inability to preserve connectivity or to represent elongated objects. On the other hand, the BIP is computationally efficient because its height is constrained by its regular part. In this paper the features of the Bounded Irregular Pyramid are discussed, presenting a comparison with the main pyramids present in the literature when applied to a colour segmentation task.

1   Introduction

The structure of a pyramid can be described as a graph hierarchy in which each level l is represented by a graph $G_l = (N_l, E_l)$ consisting of a set of nodes, N_l, linked by a set of arcs or edges E_l, named *intra-level edges*. In this hierarchy, each graph G_{l+1} is built from G_l by computing the nodes of N_{l+1} from the nodes of N_l. Each node of N_{l+1} is linked to the set of nodes of N_l which generate it by a set of arcs or edges, named *inter-level edges* $E_{l,l+1}$.

The efficiency of a pyramid to represent the information is strongly influenced by the *data structure* used within the pyramid and the *decimation scheme* used to build one graph from the graph below [1]. Depending on these two features, pyramids have been classified as regular and irregular ones. Regular pyramids [2,3] have a rigid structure where the decimation process is fixed. In these pyramids, the inter-level edges are the only relationships that can be changed to adapt the structure to the image layout. Thanks to this rigid structure, regular pyramids can be efficiently represented as a hierarchy of bidimensional arrays. Each of these arrays is an image where two nodes are neighbours if they are placed in adjacent positions of the array. This is the main advantage of this kind of pyramids because they can be built and traversed with a low computational cost. But of course the simplicity of the rigid structure of regular pyramids comes with a cost [4]: non-connectivity of the obtained receptive fields and incapability to represent elongated objects. In contrast to regular pyramids, irregular ones

F. Escolano and M. Vento (Eds.): GbRPR 2007, LNCS 4538, pp. 307–318, 2007.

have variable data structures and decimation processes which dynamically adapt to the image layout. Thus, the reduction factor between adjacent levels is not fixed and the size of each level and the height of the structure are unknown. Due to that, the classical irregular structures [5] are not computationally efficient. This efficiency has been recently improved using novel strategies [6,1,7,8]. These new approaches are more computationally efficient than the classical ones, but they still require an execution time which make impossible to use them in real time applications, as it is showed in the results section of this paper. A detailed revision of the main regular and irregular structures may be found in [4].

In this paper, the original structure of the Bounded Irregular Pyramid [9] has been modified to reduce its computational cost and to correctly deal with elongated objects. The BIP is a mixture of regular and irregular pyramids, whose goal is to combine their advantages: low computational cost and accurate results. The irregular part of the BIP allows to solve the main weaknesses of regular structures, preservation of connectivity and representation of elongated objects. On the other hand, the BIP is computationally efficient as its height is constrained by its regular part. A first version of this new structure was proposed in [9]. Although computationally efficient, this first version had problems to combine the regular and irregular parts of the pyramid. These problems has now been solved. In this paper the features of the Bounded Irregular Pyramid are discussed, presenting a comparison with the main regular and irregular pyramids present in the literature. In order to do the comparisons, the height of the pyramids, their processing time and the number of obtained regions have been studied. Besides, a quantitative quality measurement has been employed: the Q function.

This paper is organized as follows: the data structure and the decimation process used in the BIP are explained in Section 2. Section 3 presents the segmentation procedure. The obtained experimental results and the comparisons with other pyramidal segmentation algorithms are showed in Section 4. Finally, Section 5 summarizes the extracted conclusions.

2   Data Structure and Decimation Process

The data structure of the Bounded Irregular Pyramid is a combination of regular and irregular data structures: a 2x2/4 regular structure and a simple graph. The regular structure is used in the homogeneous regions of the input image and the irregular structure in the non-homogeneous ones. The mixture of both regular and irregular structures generates an irregular configuration which is described as a graph hierarchy in which each level $G_l = (N_l, E_l)$ consists of a set of nodes, N_l, linked by a set of intra-level edges E_l. There are two types of nodes: nodes belonging to the 2x2/4 structure, named regular nodes, and virtual nodes or nodes belonging to the irregular structure. Two nodes $n_i \in N_l$ and $n_j \in N_l$ which are neighbours at level l are linked by an intra-level edge $e_{ij} \in E_l$.

Each graph G_{l+1} is built from G_l by computing the nodes of N_{l+1} from the nodes of N_l and establishing the inter-level edges $E_{l,l+1}$. Therefore, each node n_i of G_{l+1} has associated a set of nodes of G_l, which is called the *reduction window* of n_i. This includes all nodes linked to n_i by an inter-level edge. The node n_i is called *parent* of the nodes in its reduction window, which are called *sons*. Two nodes n_i and n_j of N_l are said to be adjacent or *neighbours* at level l, if their corresponding reduction windows w_i and w_j are neighbours at level $l - 1$. Two reduction windows $w_i \in N_{l-1}$ and $w_j \in N_{l-1}$ are neighbours if there are at least two nodes $n_r \in w_i$ and $n_s \in w_j$ which are connected by an intra-level edge $e_{r,s} \in E_{l-1}$. The set of nodes in N_l which are neighbours of a node $n_i \in N_l$ is called the *neighbourhood* of n_i. An *intra-level path* is a sequence of ordered nodes linked by intra-level edges. Two nodes $n_i \in N_l$ and $n_j \in N_l$ are said to be *connected* if there exists an intra-level path that includes them both. Equivalently, an *inter-level path* is a sequence of ordered nodes linked by inter-level edges. Two nodes $n_i \in N_p$ and $n_j \in N_q$ are said to be *connected* if there exits an inter-level path that includes them both. The *receptive field* r_i of a node $n_i \in N_l$ is the set of nodes at level 0 which are connected to it by an inter-level path.

2.1 Regular Data Structure Building

Although regular pyramids can be explained as a graph hierarchy, it is more usual to represent them as a hierarchy of image arrays due to their rigid structure. Therefore, in the regular part of the BIP, each regular node is represented by (i, j, l), where l represents the level and (i, j) are the x- and y-coordinate within the level. In each of these arrays two nodes are neighbours if they are placed in adjacent positions of the array in an 8-neighbourhood.

The first step to build the 2x2/4 structure is a 4 to 1 decimation procedure. In order to perform this decimation, each regular node has associated two parameters: homogeneity $Hom(i, j, l)$ and parent link $Parent(i, j, l)$. Regular nodes have $Hom(i, j, l) = 0$ or $Hom(i, j, l) = 1$. $Hom(i, j, l)$ of a regular node is set to 1 if the four nodes immediately underneath are similar according to some criteria and their homogeneity values are equal to 1. Otherwise, it is set to 0. If the node (i, j, l) is a node of the regular structure with $Hom(i, j, l) = 1$, then the parent link of the four cells immediately underneath (sons) is set to (i, j). It indicates the position of the parent of a regular node in its upper level. A regular node without parent has its parent link set to a NULL value. Parent links represent the inter-level edges of the regular part of the BIP.

All the regular nodes presenting a homogeneity value equal to 1 form the regular structure. Regular nodes with an homogeneity value equal to 0 are removed from the structure. Fig. 1.a) shows the regular part of the BIP data structure after being built. White nodes are the non-homogeneous ones. In this example the used similarity criteria is the colour distance. Two nodes are similar if they have similar colour. The base level of the structure is formed by the pixels of the 8x8 original image. The 4 to 1 decimation procedure generates a 4x4 level and a subsequent 2x2 level.

Once the regular structure is generated using the 4 to 1 decimation procedure, there are some regular orphan nodes (regular nodes without parent). From each

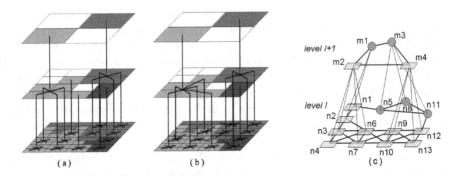

Fig. 1. a) Regular nodes of the BIP and their inter-level edges after the generation step, b) regular nodes of the BIP and their inter-level edges after the parent search step, and c) two levels of the graph hierarchy

of these nodes, a search is made for a node (i_1, j_1, l) in its 8-neighbourhood $\xi_{(i,j,l)}$ which satisfies the following conditions:

$$
\begin{aligned}
&- \; Hom(i_1, j_1, l) = 1 \\
&- \; Parent(i_1, j_1, l) = (i_p, j_p, l+1) \\
&- \; d((i, j, l), (i_1, j_1, l)) < T \\
&- \; d((i, j, l), (i_1, j_1, l)) \le d((i, j, l), (i_k, j_k, l)) \;\; \forall (i_k, j_k, l) \in \xi_{(i,j,l)}
\end{aligned}
\tag{1}
$$

being $d(n_i, n_j)$ a similarity measurement between the nodes n_i and n_j and T a similarity threshold. (i, j, l) is linked to $(i_p, j_p, l+1)$ (*parent search step*). For example, in Fig. 1.b), there are four orphan nodes at level 1, but only for two of them a suitable parent node is found that satisfies (1).

2.2 Irregular Data Structure and Decimation Process

The irregular decimation process used to build the BIP together with the parent search step previously explained are an implementation of a union-find strategy that has been also used by Brun and Kropatsch [1]. The union-find algorithm was proposed by Tarjan [10] as a general method for keeping track of disjoint sets. Basically, it allows performing of set-union operations on sets which are in some way equivalent, while ensuring that the end product of such a union is disjoint from any other set. Union-find algorithms use tree structures to represent sets. Each non-root node in the tree points to its parent, while the root is flagged in some way. A find operation looks for the parent of a node. If two nodes of different sets or trees are similar a union operation is performed by setting one of the roots to be the parent of the other root.

In order to implement this union-find strategy and build the irregular part of the BIP, each virtual node n_i has associated a parent link parameter $Parent(n_i)$ and a neighbourhood vector ξ_{n_i} which stores all the nodes n_j in its neighbourhood. The union-find process to compute the irregular part of G_{l+1} from G_l has three stages:

- Intra-level twining: This stage links two orphan neighbour nodes of the regular structure if they are similar. To do that, from each regular orphan node, (i, j, l), a search is made for a regular neighbour node (i_1, j_1, l) which satisfies the following conditions:

  - $Hom(i_1, j_1, l) = 1$
  - $Parent(i_1, j_1, l) = NULL$ $\qquad\qquad\qquad$ (2)
  - $d((i, j, l), (i_1, j_1, l)) < T$
  - $d((i, j, l), (i_1, j_1, l)) \leq d((i, j, l), (i_k, j_k, l)) \ \forall (i_k, j_k, l) \in \xi_{(i,j,l)}$

  (i, j, l) is linked to (i_1, j_1, l), generating a virtual node at level $l + 1$. In Fig. 1.c) the two regular nodes n_1 and n_2 are linked, generating the virtual node m_1.

- Virtual node linking: this process links two virtual orphan nodes of the level l if they are similar, generating a virtual node at level $l + 1$. From each virtual orphan node, n_i, a search is made for a virtual node n_j:

  - $Parent(n_j) = NULL$
  - $d(n_i, n_j) < T$ $\qquad\qquad\qquad\qquad\qquad$ (3)
  - $d(n_i, n_j) \leq d(n_i, n_k) \ \forall n_k \in \xi_{n_i}$

  In Fig. 1.c) the two virtual nodes n_5 and n_8 are linked, generating the virtual node m_3.

- Virtual parent search: Each virtual orphan node n_i searches for a virtual node n_j in ξ_{n_i}:

  - $Parent(n_j) = n_{j_p}$
  - $d(n_i, n_j) < T$ $\qquad\qquad\qquad\qquad\qquad$ (4)
  - $d(n_i, n_j) \leq d(n_i, n_k) \ \forall n_k \in \xi_{n_i}$

  n_i is linked to n_{j_p}. An example of this is showed in Fig. 1.c) where the virtual node n_{11} is linked with m_3.

When all virtual nodes at level $l + 1$ have been generated, the intra-level edges of G_{l+1} are computed by taking into account the neighbourhood of the reduction windows of the nodes N_{l+1} in G_l.

The regular data structure building and the intra-level twinning step were the only stages to build the first version of the BIP [9]. Specifically, in the first version of the BIP the virtual nodes did not belong to the graph structure. They were used only to store information about the linking of regular nodes and they are not used when a level is built from the level below. Therefore, the first version of the BIP could be seen as an incomplete regular structure where some nodes are fused (using virtual nodes) in order to increase its accuracy. On the contrary, in the proposed BIP structure each level is formed by regular and virtual nodes. Thus, when a level is built from a level below, not only regular nodes are used but also virtual nodes which are linked and generated level by level. In this way, the irregularity of the BIP has been increased improving the adaptation of the structure to the image layout.

3   Colour Image Segmentation Using the BIP

In order to introduce colour information within the BIP, all the nodes of the structure have associated 3 parameters: chromatic phasor $S_{\angle H}(n)$, luminosity

$V(n)$ and area $A(n)$, where S, H and V are the saturation, hue and value of the HSV colour space. The chromatic phasor and the luminosity of a node n are equal to the average of the chromatic phasors and luminosity values of the nodes in its reduction window. The area of a node is equal to the sum of the areas of the nodes in its reduction window, i.e. the cardinality of its receptive field.

The employed similarity measurement between two nodes is the HSV colour distance. Thus, two nodes are similar or have a similar colour if the distance between their HSV values is less than a similarity threshold T. This threshold is not fixed for all levels. Its mathematical expression is the following:

$$T(l) = T_{max} * \alpha(l) \tag{5}$$

being

$$\alpha(l) = \begin{cases} 1 - \frac{l}{L_{reg}} * 0.7 & \text{if } l \leq L_{reg} \\ 0.3 & \text{if } l > L_{reg} \end{cases} \tag{6}$$

L_{reg} is the highest level of the regular part of the BIP. This threshold takes into account that usually the receptive field of a vertex in a high level is bigger than the receptive field of a vertex in a low level. Therefore, the linking of two vertices of a high level implies the merging of two larger regions at the base. This threshold makes more difficult the linking process at upper levels and then, the merging of large regions at the base.

The graph $G_0 = (N_0, E_0)$ is a 8-connected graph where the nodes are the pixels of the original image. All the nodes of $G_0 = (N_0, E_0)$ are initialized with $Hom(i, j, 0) = 1$ and $A(i, j, 0) = 1$. The chromatic phasors and the luminosity values of the nodes in $G_0 = (N_0, E_0)$ are equal to the chromatic phasors and luminosity values of their corresponding image pixels.

The process to build the graph $G_{l+1} = (N_{l+1}, E_{l+1})$ from $G_l = (N_l, E_l)$ is the following:

1. Regular decimation process (Section 2.1).
2. Union-find process:
   (a) Parent search and intra-level twining. In this step, the parent search and the intra-level twining processes are simultaneously performed. Thus, for each regular orphan node the parent search step is carried out (Section 2.1). The found parent can be a regular or a virtual node of G_{l+1}. If for the studied node a parent is not found, then the intra-level twinning step is performed (Section 2.2).
   (b) Virtual parent search and virtual node linking. For each virtual orphan node the virtual parent search step is performed (Section 2.2). If for the studied node a parent is not found the virtual node linking step is performed in order to generate a virtual node in G_{l+1} (Section 2.2).
3. Intra-level edge generation in G_{l+1}. The intra-level edges of G_{l+1} are computed by taking into account the neighbourhood of their reduction windows in G_l.

The hierarchy stops to grow when it is no longer possible to link together any more nodes because they are not similar.

The segmentation process can be roughly divided in two stages: a regular stage which includes two steps (the regular decimation process and the parent search and intra-level twinning) and an irregular stage which includes the virtual parent search and virtual node linking step. The order in which these two stages are performed does not modify the segmentation results. On the other hand, the regular decimation process must be always carried out before the parent search and the intra-level twinning steps because it is the only way to generate regular nodes in the structure. If the parent search and intra-level twinning is performed before, then any regular node will be generated. Fig. 2 shows an example of the construction of the BIP. It can be appreciated as the regular decimation process is the responsible of the generation of all regular nodes of the structure (Fig. 2.a). The parent search process links regular nodes with nodes of the level above (e.g., the node r_3 with r_4) and the intra-level twining process generates new virtual nodes by merging regular nodes of the level below (e.g., the virtual node v_1 is generated by merging nodes r_1 and r_2). The goal of these processes is to find a parent for the regular nodes which do not have a parent after the regular decimation. Besides, the intra-level twining process generates all the virtual nodes at level 1 because all nodes at base level are considered as regular nodes. In subsequent levels, the generation of new virtual nodes can be achieved due to the intra-level twining process or to the virtual node linking process. Thus, Fig. 2.d shows the generation of a new virtual node at level 2 (v_3) due to the merging of two virtual nodes (v_1 and v_2). Finally, the virtual parent search process is necessary to maintain the connectivity of irregular shaped objects.

In order to perform the segmentation, the orphan nodes are used as roots. The receptive field of each of these nodes is a region of the segmented image.

The combination of a regular and an irregular data structures and decimation processes within the BIP allows to solve the two segmentation problems of regular structures with a low computational cost. Therefore, the capability of the BIP to always obtain connected regions and to represent elongated objects are proven below.

Connectivity is preserved when the receptive field associated to every node of the structure is connected. The receptive field r_i of a node $n_i \in N_l$ is connected if for every pair of nodes $(i_p, j_p, 0), (i_q, j_q, 0) \in r_i$ there is at least an intra-level path with all its nodes included in r_i, which connects $(i_p, j_p, 0)$ and $(i_q, j_q, 0)$.

In the process to build the BIP, new nodes of level $l + 1$ are generated either by the regular decimation, intra-level twinning or virtual node linking. In these processes, new nodes in $l + 1$ are generated from neighbour nodes of level l. The new receptive fields of the nodes of level $l + 1$ are generated by grouping the receptive fields of the nodes of level l which generate them. Besides, in the generation of the BIP, there are two parent search processes in which regular or virtual nodes without parent are linked to the parent of a neighbour node. In these cases, the receptive field of the parent node is extended with the receptive field of its new son. Therefore, in the process to build the BIP, a receptive field is always generated or extended by the linking of neighbour nodes to the same parent, following a bottom-up procedure that starts at the base level. It must be

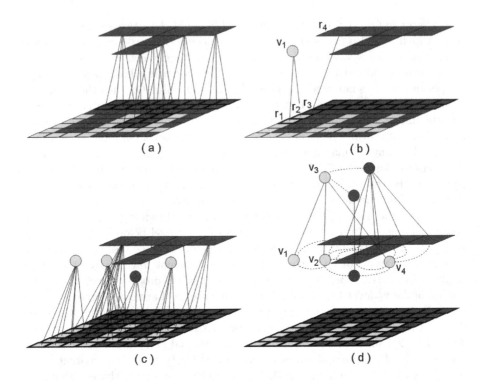

Fig. 2. a) Regular nodes of the BIP and their inter-level edges after the regular decimation process, b) individual examples of a parent search and a intra-level twinning
processes, c) regular and virtual nodes of the BIP and their inter-level edges after the
parent search and intra-level twining processes, and d) individual examples of a virtual
parent search and virtual node linking processes (see text for details)

noted that the receptive field of a node in the base level is formed by itself (and
is therefore connected). Thus, in order to demonstrate that the BIP preserves
connectivity, it suffices to proof that the linking of two neighbour nodes $n_i \in N_l$
and $n_j \in N_l$ to the same parent generates a new receptive field $r_t = r_i \cup r_j$ that
is always connected.

Theorem 1. *If two nodes $n_i \in N_l$ and $n_j \in N_l$ are neighbours at level l, their
receptive fields r_i and r_j are neighbours at the base level.*

Proof: *If $n_i \in N_l$ and $n_j \in N_l$ are neighbours at level l, then their reduction
windows w_i and w_j are by definition neighbours at level $l-1$. If w_i and w_j are
neighbours, then there are at least two nodes at $l-1$, $n_k \in w_i$ and $n_m \in w_j$, which
are neighbours. Therefore, the reduction windows of n_k and n_m are neighbours
at $l - 2$. This can be extended until $l = 0$ where the studied reduction windows
will be part of the receptive fields r_i and r_j. Then, if these reduction windows are
neighbours, the studied receptive fields are neighbours.*

According to theorem 1, if two nodes $n_i, n_j \in N_l$ are neighbours in G_l, their
receptive fields r_i and r_j are neighbours. This means that there is at least a pair

of nodes $(i_1, j_1, 0) \in r_i$ and $(i_2, j_2, 0) \in r_j$ which are neighbours at the base level. Thus, if the receptive fields r_i and r_j are connected, then for every pair of nodes $(i_p, j_p, 0) \in r_i$ and $(i_q, j_q, 0) \in r_j$ there will be at least an intra-level path with all its nodes included in $r_i \cup r_j$ that connects them through the edge between $(i_1, j_1, 0)$ and $(i_2, j_2, 0)$. Therefore, the receptive field $r_t = r_i \cup r_j$ formed by the linking of n_i and n_j to the same parent is connected.

The representation of elongated objects with homogenous colour is a problem in regular pyramidal structures where the reduction windows have always a predefined shape. Therefore, the regular part of the BIP cannot represent these objects. However, irregular pyramids can represent elongated objects thanks to the irregular shape of the generated reduction windows. In the BIP, elongated objets are represented by its irregular part. The most difficult elongated object to represent is the one-pixel width. An elongated object with homogeneous colour of one pixel width is composed by image pixels with similar colour that only have one or two pixels belonging to the same object in their neighbourhoods. The representation of this type of elongated object using the BIP is proven below.

Theorem 2. *If all the nodes of an elongated object of one-pixel width have similar colour at the base level of the BIP, then there exists a virtual node whose receptive field is formed by them.*

Proof: *The nodes of the homogeneous coloured elongated object of one-pixel width conform an intra-level path at the base level formed by similar nodes. That is, they form a sequence of ordered similar nodes linked by intra-level edges. These nodes will be grouped in different subsets of nodes (reduction windows) in the parent search and virtual nodes linking steps, generating a set of parents in the upper level. These parents are connected because their reduction windows are connected and they also form an intra-level path of similar nodes. Therefore, they generate new nodes in the next level which form an intra-level path in that level. This process stops when a level is reached where only one node is generated. The receptive field of this node is the elongated object.*

4   Evaluation of Segmentation Results

In this work, the performance of the BIP has been evaluated using a quantitative measurement, the Q *function* [11]. This function takes into account the following goodness indicators: i) regions must be uniform and homogeneous according with the similarity criterium employed to perform the segmentation, i.e. colour; ii) the interior of the regions must be simple, without too many small holes; iii) adjacent regions must present significantly different values for uniform characteristics; and iv) the existence of small regions is penalized.

$$Q(I) = \frac{1}{1000(N \cdot M)\sqrt{R} \sum_{i=1}^{R} [\frac{e_i^2}{1+log A_i} + (\frac{R(A_i)}{A_i})^2]} \tag{7}$$

being NxM the image size and R the number of segmented regions. A_i and e_i are the area of the region i and its average colour error, respectively. $R(A_i)$ is the number of segmented regions with area equal to A_i.

Table 1. Q value, processing time, height of the hierarchy employed by the segmentation algorithm and number of obtained regions. Average values have been obtained from 50 different images.

	Q_{min}	Q_{ave}	Q_{max}	t_{min} t_{ave}	t_{max}	h_{min}	h_{ave}	h_{max}	NR_{min}	NR_{ave}	NR_{max}
		Q		Processing times (sec)		Hierarchy height			Number of regions		
LRP	1052.1	1570.3	2210.3	0.96 1.35	1.84	9	9	9	17	81.4	208
WRP	1133.7	1503.5	2080.8	0.31 0.42	0.59	9	9	9	18	79.6	149
ClIP	481.7	1132.8	1586.9	2.51 3.95	7.68	16	36.6	72	9	84.3	212
LIP	489.4	1011.5	1334.8	1.70 2.75	6.16	8	25.5	52	12	73.7	210
MIP	355.6	818.5	1301.1	2.41 3.42	4.49	11	32.9	64	45	107.6	201
HIP	460.5	955.1	1530.7	4.07 4.23	4.91	9	11.4	19	23	76.1	150
CoIP	430.7	870.2	1283.7	1.31 2.85	12.9	9	74.2	202	24	91.2	238
BIP	343.2	1090.9	1911.3	0.13 0.16	0.39	8	8.7	15	8	83.6	230

In order to compare the BIP with the main pyramids present in the literature, two segmentation algorithms based on regular pyramids have been implemented: the linked pyramid [2] (LRP), and the weighted linked pyramid with possibilistic linking (WRP) [3]. Comparisons with five segmentation algorithms based on irregular pyramids are also presented: the classical RAG hierarchy [5] (ClIP); the localized pyramid [8] (LIP); the segmentation algorithm proposed by Lallich et al. [7] (MIP); the hierarchy of image partitions by dual graph contraction [6] (HIP) and the hierarchical segmentation algorithm based on combinatorial pyramids [1] (CoIP). In order to quantitatively evaluate the efficiency of the different segmentation algorithms, 50 colour images from Waterloo and Coil 100 databases have been chosen. All these images have been resized to 256x256 pixels. A 3GHz Pentium IV PC has been employed. Two of the images and the obtained results are shown in Fig. 3. It should be mentioned that the selection of the parameters of all algorithms has been conducted to obtain the best results according to the Q function. In the case of the proposed BIP structure the only used parameter is the maximum value of the colour threshold $Tmax$.

Table 1 presents the comparison measurements among methods. This table shows that all irregular pyramids obtain better Q values than regular ones. It can be also noted that the MIP and the CoIP present the best global results. The behaviour of the MIP is excellent, although it is the method that provides the highest number of obtained segmentation regions. In contrast, the BIP and the LIP obtain the lowest number of regions. The results obtained by the BIP are very similar to the ones obtained by the ClIP or the LIP. Finally, it can be noted that although the BIP presents a Q value less than regular structures, this value is slightly higher than in the rest of irregular approaches due to the regular processing used in the BIP. However, the Q value obtained with the BIP is very similar than the obtained one with the rest of irregular approaches.

Table 1 also shows the processing times, the hierarchy height and the number of obtained regions. It can be appreciated that the two regular representations and the BIP and HIP irregular pyramids present the minimum heights. On the contrary, the CoIP and the MIP present the maximum height values. The BIP

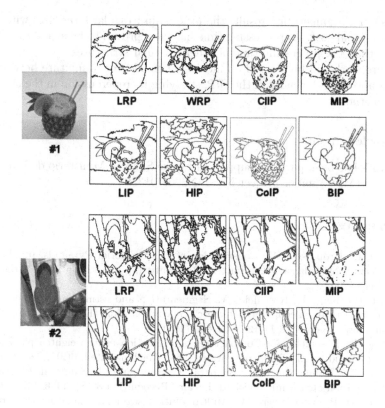

Fig. 3. Segmentation results

is the irregular pyramid with minimum height as it is constrained by its regular part. The number of segmentation regions is very similar in all the algorithms. The fastest algorithms are the BIP and the algorithms based on regular pyramids. The main advantage of the BIP is that it is at least ten times faster than the other irregular structures when run in a sequential PC. The BIP is faster than irregular approaches because a large part of the image is processed following a classical regular pyramid approach. Besides, it is faster than regular algorithms because it does not have a relinking process. The inter-level edges are computed in only one pass.

5   Conclusions and Future Work

This paper has presented a new implementation of the Bounded Irregular Pyramid which has been tested in a colour segmentation task. The main contribution of this pyramid is that it combines the advantages of regular and irregular approaches by mixing a regular data structure and decimation process with an irregular scheme. The irregular part of the BIP consists in a simple graph data structure combined with a union-find decimation strategy. This new pyramid obtains similar segmentation results than the main irregular structures with lower computational cost. Its processing time has been proven to be at least ten times smaller than the processing time of other irregular structures.

Future work will be focused on increasing the degree of mixture between the regular and irregular parts of the BIP, studying its repercussion in the efficiency of the method.

Acknowledgments

This work has been partially supported by the Spanish Ministerio de Educación y Ciencia (MEC) under project No. TIN2005-01359.

References

1. Brun, L., Kropatsch, W.: Construction of Combinatorial Pyramids. In: Hancock, E., Vento, M. (eds.) GbR in Pattern Recognition, pp. 1–12. Springer Verlag, Heidelberg (2003)
2. Burt, P., Hong, T., Rosenfeld, A.: Segmentation and estimation of image region properties through cooperative hierarchical computation. IEEE Trans. on Systems, Man And Cybernetics 11(12), 802–809 (1981)
3. Hong, T., Rosenfeld, A.: Compact region extraction using weighted pixel linking in a pyramid. IEEE Trans. on Pattern Anal. Machine Intell. 6(2), 222–229 (1984)
4. Marfil, R., Molina-Tanco, L., Bandera, A., Rodriguez, J., Sandoval, F.: Pyramid segmentation algorithms revisited. Pattern Recognition 39(8), 1430–1451 (2006)
5. Bertolino, P., Montanvert, A.: Multiresolution segmentation using the irregular pyramid. Int. Conf. On Image Processing 1, 257–260 (1996)
6. Haxhimusa, Y., Kropatsch, W.: Segmentation graph hierarchies. In: Proceedings of the Joint IAPR International Workshop on Syntactical and Structural Pattern Recognition and Statistical Pattern Recognition, pp. 343–351 (2004)
7. Lallich, S., Muhlenbach, F., Jolion, J.: A test to control a region growing process within a hierarchical graph. Pattern Recognition 36, 2201–2211 (2003)
8. Huart, J., Bertolino, P.: Similarity-based and perception-based image segmentation. IEEE. Int. Conf. on Image Processing - ICIP 2005 3, 1148–1151 (2005)
9. Marfil, R., Rodriguez, J., Bandera, A., Sandoval, F.: Bounded irregular pyramid: a new structure for colour image segmentation. Pattern Recognition 37(3), 623–626 (2004)
10. Tarjan, R.: Efficiency of a good but not linear set union algorithm. Journal of the ACM 22, 215–225 (1975)
11. Borsotti, M., Campadelli, P., Schettini, R.: Quantitative evaluation of color image segmentation results. Pattern Recognition Letters 19, 741–747 (1998)

A New Contour Filling Algorithm Based on 2D Topological Map

Guillaume Damiand* and Denis Arrivault

SIC - bât. SP2MI, Bvd M. et P. Curie
BP 30179, 86962 Futuroscope Chasseneuil Cedex - France
{damiand,arrivault}@sic.univ-poitiers.fr

Abstract. In this paper, we present a topological algorithm which allows to fill contours images. The filling problem has been widely treated and it recently appeared that it can always be split into two different process : a generic topological process and a dedicated geometrical post-processing which depends on the application. Our algorithm, based on a 2D topological map description of the image, addresses the first step of processing. It is fast, generic and robust. Moreover, the complete topological description allows to easily integrate geometrical constrains and makes our approach an interesting basis for every filling process.

Keywords: topological maps, filling process, character reconstruction.

1 Introduction

Filling algorithms are used in many applications but especially in character image generation [1]. The motivation of our work comes from the handwritten characters description. In order to use structural methods for describing a character (for example fuzzy hierarchical graphs as in [2]) one needs to extract a clean skeleton. The filling algorithm was developed for the characters reconstruction phase of this process. Actually, a skeleton-based graph is built from character image through a contour approximation and a character reconstruction (Fig. 1).

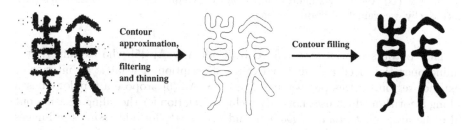

Fig. 1. Processing of a chinese handwritten characters

* Partially supported by the ANR Fundation under grants ANR-06-MDCA-008-05/FOGRIMMI.

F. Escolano and M. Vento (Eds.): GbRPR 2007, LNCS 4538, pp. 319–329, 2007.

The problem of filling the contour of a region has been widely treated during the last three decades. Depending on the application, this can be solved by using an a priori knowledge on the contour topology (with contour approximations for example) or directly with the raster graphics. Pavlidis [3] separated also the "polygon based" techniques from the "pixel based" ones. The purpose of this article is to present a "pixel based" algorithm using a topological description without a priori knowledge.

There are many "pixel based" filling algorithms in the literature. They can be divided into two broad categories [4]: parity check filling algorithms (also called scan-line or edge filling) and seeds growing (also referred to as connectivity filling or region growing). Filling by parity check is fast and requires less or no additional working memory comparing with seeds growing filling. Line by line, the background pixels are associated with a depth number according to the number of black pixels previously encountered during the line scan. Then the filling is done by using the depth number parity. The major difficulty facing such a scheme is that different arcs of the contour may be mapped on the same pixel (Fig. 2 (A)). That is the reason why it often fails to correctly handle complex objects while seeds growing methods are theoretically more robust. Starting from interior starting points (the seeds), seed growing algorithms are propagation procedures that color the regions of interest. Nevertheless, the seed choice is a non trivial issue and can not be automated without an a priory knowledge of the contours relation.

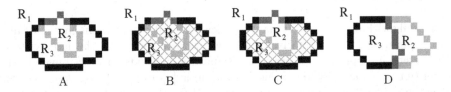

Fig. 2. The problem which occurred when two boundaries are shared. (A) Regions R_2 and R_3 share a part of a boundary (drawn in dark grey in the figure). (B) What we will obtain by our algorithm: both regions are filled. (C) What we want to obtain intuitively. (D) Another configuration, topologically equivant to (A), but in this case, intuitively, we want to fill both regions.

The contours relation are defined by Codrea & Al. [5] who explain that a filling operation need a formal or explicit description of what is an interior or exterior region. In this perspective, Martin & Al. [6] propose a topology-based filling algorithm which uses not only inclusion relation for the filling decision but also the ideas of *dominant subobject* and *exteriority* for addressing the images with ambiguities. The dominant subobject is defined as the region which comprises most of the external perimeter of an object and exteriority allows to fill subobjects that are outside enough of a dominant subobject even if the sharing boundary is small. This algorithm is powerful but requires complex definitions and thresholds. Moreover it does not allow to deal with non closed contours.

Nevertheless, Martin & Al. outline a fundamental aspect of the filling problem. The contour images has to be processed at two levels : a topological process first that is quite generic for all applications and a geometrical process that is dedicated to the application.

The topological process given by Martin & Al. is based on the inclusion relation. This approach is quite poor and does not allow to propose efficient geometrical constraints. The 2D topological maps used in our approach provide a complete topological description that can be efficiently adapt to every application. The filling algorithm proposed in this article is a topological one, generic, complete and fast. We are not addressing the geometrical constraints at the moment but we will explain why this approach is an interesting basis for every filling problems.

In the next section, recalls on 2D-topological maps will be given. Then we will present our algorithm which allows to fill contours image by using this structure. Finally we will provide some examples and try to highlight the advantages and the geometrical extensions of such an approach. The article will end with a conclusion and some perspectives.

In this article, the background pixels of the images are the white ones and contours are drawn in black. Furthermore contours are composed of digital curves (closed or not) which connectivity is 8 with no redundancy. These properties are guaranteed by our processing chain which applies a thinning algorithm.

2 Recalls

Topological maps are an extension of combinatorial maps [7,8,9,10] in order to represent in a unique and minimal way a labeled image. We present here briefly the main notions of combinatorial maps and of topological maps (see [11,12] for more details).

2.1 Combinatorial Maps

Intuitively, a 2D combinatorial map (called also a *2-map*) is an extension of a planar graph that keeps the orientation of edges around each vertex. Each edge of the graph is divided in two parts. Basic elements obtained are called *darts* and are the unique basics of the combinatorial map definition. A 2D combinatorial map can represent the topology of a 2D subdivision of orientable spaces without boundary. This model has been extended to represent any type of subdivision, orientable or not, and with or without boundaries (see [11] and Fig. 3).

More precisely, a subdivision of a 2D topological space is a partition of the space into 3 subsets whose elements are 0D, 1D and 2D *cells* (respectively called vertices, edges and faces, and noted i-cell, $i = 0\ldots2$). Border relations are defined between these cells, where the border of an i-cell is a set of $(j<i)$-cells. Two cells are *incident* when one belongs to the border of the second, and two i-cells are *adjacent* if they are both incident to a common $(j<i)$-cell.

A combinatorial map is an algebra composed by a set of darts that represents the elements of the subdivision, and 2 mappings (called β_1 and β_2) defined on these darts that represent adjacency relations (this can be easily extended in nD, with n mappings). β_1 puts in relation a dart and the next dart of the same face, and β_2 puts in relation both darts incident to a same edge. These β_i have to verify some particular properties in order to ensure the validity of the represented subdivision (β_1 is a permutation and β_2 is an involution, see for example [11] for the formal definition).

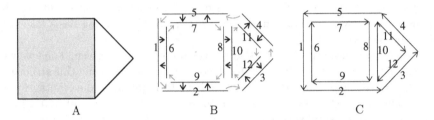

A   B   C

Fig. 3. Usual representation of a 2D combinatorial map. (A) A 2D object. (B) Explicit representation where each dart and each one to one mapping are drawn. Darts are represented by black segments, β_1 by grey arrows and β_2 by black arrows. (C) Implicit representation, where β_i applications are not explicitly drawn but can be deduced from the shape of the objects. Two darts in relation by β_1 are drawn consecutively, and the arrow on darts shows the orientation of β_1. Two darts in relation by β_2 are drawn near, parallel and in reverse orientation.

We can see in Fig. 3B the combinatorial map representing the object shown in Fig. 3A. In this example, each dart and each one to one mapping are drawn. In general, we do not use this representation but we prefer the one shown in Fig. 3C where β_i are not explicitly drawn but can be (generally) deduced from the shape of objects.

Within the combinatorial map framework, all cells are implicitly represented through the notion of *orbit*. Intuitively, an orbit $< \beta_{i_1}, \ldots, \beta_{i_j} > (d)$ is the set of darts that can be reached with a breadth-first search algorithm, starting with d, and using all combinations of all β_{i_k} or $\beta_{i_k}^{-1}$ permutations $\forall k, 1 \leq k \leq j$. With this notion, each cell is defined as a particular orbit. Based on the cells definition, we can retrieve the classical *cell degree* notion. The degree of an i-cell c is the number of distinct $(i+1)$-cells incident to c. Note that in a n-dimensional space, the degree is not defined for n-cells, since $(n+1)$-cells do not exist in such a space.

2.2   Topological Maps

Topological maps are an extension of combinatorial maps in order to represent in a unique and minimal way a labeled image.

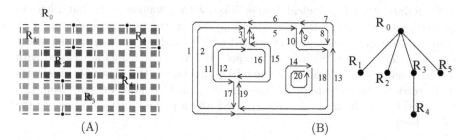

Fig. 4. (A) A 2D labeled image drawn with its interpixel boundaries. (B) The corresponding topological map with its inclusion tree.

We can see in Fig. 4 a 2D labeled image and the corresponding topological map. A topological map is a combinatorial map that represents a labeled image and that verifies particular properties. Indeed, this map is minimal, complete and unique. These properties lead to another characteristic of the topological map: each edge represents exactly an interpixel boundary between two regions of the image (this can be seen in Fig. 4). An interpixel boundary between two regions R_i and R_j, is the set of maximal interpixel curves such that each linel of these curves is incident to exactly one pixel of R_i and one pixel of R_j (see [12] for proofs concerning topological map properties).

When a region is included into another region (as region R_4 in Fig. 4 which is included into region R_3), the corresponding topological map is composed of several connected components. There is no information in the map that allows to place relatively the different connected components, and thus we have lost the topological information concerning the inclusion. To solve this problem, we add an inclusion tree to the topological map. This tree contains each region of the image, rooted by R_0[1], and a region R_i is son of a region R_j when R_i is included into R_j. With this tree, we are now able to retrieve each inclusion relation. Moreover, each region of this tree R is linked with a dart of the topological map that belongs to its external boundary (called *representative dart* of R), and each dart of the map is linked with its belonging region. With these two links, we can efficiently run through all the boundaries of a given region.

Combinatorial map represents the topological part of our model: all the cells of the space subdivision and all the adjacency and incidence relations. But it is also necessary to represent the geometry of the image. We speak about *embedding* to design this geometrical model. There are many different possibilities to represent the geometry and the choice of one of them depends on the application. In this work, we have chosen to use an interpixel matrix.

This matrix contains all the interpixel elements that belong to interpixel boundaries of the corresponding image. We can see in Fig. 4(A) the corresponding embedding of the topological map shown in Fig. 4(B). A linel is present in the matrix if it is between both pixels that belong to two different regions. A pointel is present in the matrix if it is incident to more than two linels.

---

[1] R_0 is the region which surrounds all the image, called the infinite region.

Each dart of the topological map is linked with a doublet (p, l) that allows to retrieve, given a dart d, the corresponding cells in the interpixel matrix. With p we can retrieve the pointel associated with d, and so the coordinates of the corresponding vertex. With (p, l) we can retrieve the first linel associated with d. This linel is oriented and gives the initial direction of the edge associated with d. To retrieve the embedding of the edge incident to dart d, we start from this linel, and follow the path of linels until we find a pointel, or we go back to the initial linel.

3   Using Topological Map for Filling Contours

Given a contour image obtained from a character image after contour approximation, filtering and thinning (Fig. 1), we want to fill each region which corresponds to the interior of a character. The main idea of our solution is to use topological map, and more precisely the inclusion tree associated with topological map in order to retrieve one pixel for each region to fill, and then to use a classical flood-fill algorithm starting from these germs.

Due to the type of our images, each region to fill R can be characterized by two specific properties:

1. the color of R is white in the image;
2. the depth of R in the inclusion tree is even.

The first property can easily be deduced since black pixels in the images belong to a boundary of a character. The second property is deduced from the type of boundaries present in the image as we can see in Fig. 5. Indeed, in our images, two types of boundaries are alternated: external boundaries and internal boundaries. Indeed, each region is always composed by one external boundary. When it has some holes, each one is represented by one internal boundary. This

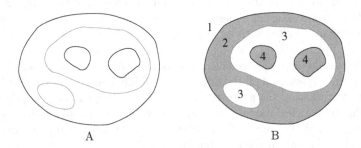

Fig. 5. An example of images we have to process. (A) Initial image. We have drawn external borders in black and internal borders in grey, but this is only for the explanations and all the borders are represented by black pixels in the real image. (B) Image we want to obtain after interiors of characters are filled. The numbers correspond to the depth of each region in the inclusion tree.

process is repeated if there is another region included in one hole, this region is represented by an external boundary and so on.

With these considerations, we can characterized each boundary by its depth in the inclusion tree. By considering that a character is never incident to the border of the image[2], we are sure that the region R_e associated to the external boundary is included in the region that corresponds to the background of the image R_b. The depth of R_b region is 1 since it is directly included in the infinite region, and thus the depth or R_e is 2 since R_e is directly included in R_b.

It is important to note that the regions represented in topological map are 4-connected regions and that each external boundary is a 8-connected path. Consequently, if an external boundary is not a single straight line, it is always associated with more than one region. Those regions are all at a depth 2 in the tree since each one is directly included in R_b. Furthermore, an external boundary can not include any region. Then, the white region R_f delimited by an external boundary is also to a depth 2 in the tree. Indeed, R_f is adjacent to regions representing the external boundary, and by definition of inclusion tree, two adjacent regions have the same depth. Thus we can conclude that each white region with a depth 2 in the tree is a region delimited by an external boundary and thus need to be filled.

Now if we consider an internal boundary, and the region R_i associated with this boundary (we can do the same remark than for external boundary, it is possible to have several regions associated with an internal boundary, but each one has the same depth in the tree). The depth of region R_i is 3 since this region is included into R_f. Indeed, otherwise, the boundary is not an internal boundary since it is adjacent to an external boundary. The white region delimited by this boundary has the same depth in the tree, but this region must not be filled since this is a hole in the surrounding region.

Now, we can do exactly the same remarks for the next external boundary, associated to a region with depth 4, and so on, to conclude that we need to fill each white region with an even depth in the inclusion tree.

Thanks to these properties, we can deduce Algorithm 1 which, given a topological map, computes the list of each germ that belongs to a region to fill.

Given a dart that belongs to a region to fill, we need to find a pixel inside the region. For that, we first recover the doublet (p, l) associated to the dart. Then, depending on the linel, we can compute a pixel inside the region. Indeed, edges of the map are counter-clockwise oriented, and thus we know that given an edge of an external border, the interior of the region is always on the right of the oriented edge. As we can see in Fig. 6, there are only four possible configurations, and depending on the configuration we can directly retrieve the coordinates of the pixel, given the coordinates of the pointel.

The complexity of Algorithm 1 is linear in number of regions of the image. Indeed, this algorithm runs through all regions of the image and for each region to fill, just computes the coordinates of one pixel inside the region by atomic oper-

---

[2] Even if this property is not always true, we can easily modify the image by adding white pixels all around it in order to verify the property.

Algorithm 1. Computation of the list of germs that belong to all regions to fill.

Input: A topological map M
Output: The list of pixels that are all the germs belonging to regions to fill.

$res \leftarrow \emptyset$;
foreach *region r of the inclusion tree* **do**
  if *the depth of r is even* **and** *the color of r is white* **then**
      $d \leftarrow$ representative dart of r;
      $(p, l) \leftarrow$ doublet associated with d;
      add the pixel associated with (p, l) in res;

return res;

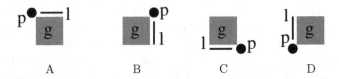

A   B   C   D

Fig. 6. The four possible configurations of a doublet (p, l) and the associated pixel g for each case. We note (p_x, p_y) the coordinates of the pointel and (g_x, g_y) the coordinates of the pixel. (A) $(g_x, g_y) = (p_x, p_y)$. (B) $(g_x, g_y) = (p_x - 1, p_y)$. (C) $(g_x, g_y) = (p_x - 1, p_y - 1)$. (D) $(g_x, g_y) = (p_x, p_y - 1)$.

ations. This algorithm is thus very efficient since we do not need to run through all the pixels of the image. Note that this algorithm needs a topological map, but the computation of topological map can be considered as a pre-processing operation. Moreover this computation can be achieved very quickly by using optimal extraction algorithm [12] with a complexity linear in number of pixels of the image, but also with a single image scan and with only the minimal number of operations to applied for each pixel.

Note that this algorithm does not work when both regions share a common boundary (as the example presented Fig. 2). In such a case, both regions have the same depth in the inclusion tree and thus they will be both filled or both kept empty. This result is not intuitive but the only way to take the good decision is to consider geometrical properties of the common boundaries. This is part of a geometrical post-processing.

4   Experiments and Results

As mentioned in the introduction and at the end of the previous section, our algorithm only addresses the topological part of the filling process. Fig. 7 provides some filling examples with no contour overlapping. Note that in (A), the non closed border over the "t" does not introduce any mistake in the filling process as it does with simple scan line algorithms for example.

Fig. 7. Example of filling results, contour images are on the left when results are on the right

The 2d-topological map is built with only one image scan and the seed growing is calculated locally in linear time. The table 1 presents some computation times obtained with a Athlon 3200+ CPU. We can notice that, comparing to the map construction, the filling process is much quicker. It demonstrates the efficiency of our approach.

Table 1. Computation times in milisecond

Image	Dimensions	Map construction time	Filling time
Chinese (Fig. 1)	368×423	180	8
Seventeen (Fig. 7,(A))	238×99	28	1
Arab (Fig. 7,(B))	458×100	49	2
Symbol (Fig. 7,(C))	514×514	290	5

We can see in Fig. 8 one example where the problem of common boundary occur. In this example, we can see that three problems occur due to the problem of common boundary: the "o" of "Lloyd", the "o" of "done" and the "s" of "this". In these three cases, inner contour and outer contour are merged and thus considered as a unique contour. Note that in these three cases, the common boundary

---

[1] Andrew Senior's Handwriting Database:
  http://www.andrewsenior.com/papers/thesis.html
[2] IFN/ENIT Database: http://www.ifnenit.com/
[3] GREC Symbols Database.

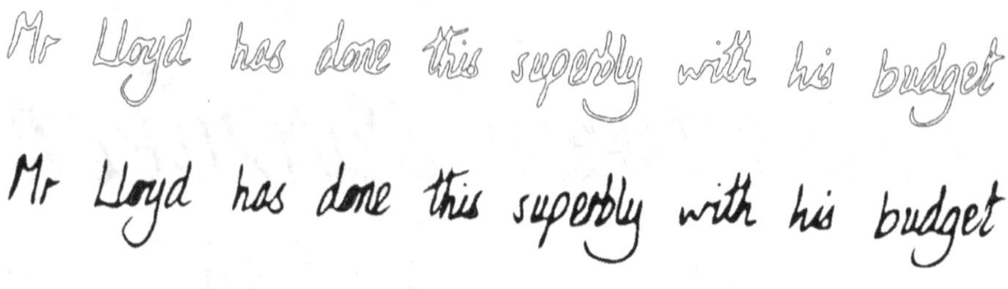

Fig. 8. One example where the problem of common boundary occur. The first line shows the contour image obtained from a character image after contour approxima-tion, filtering and thinning. The second line shows the image obtained after our filling algorithm. We can see the three problems: the "o" of "Lloyd", the "o" of "done" and the "s" of "this".

part is very small (one or two pixels) and can thus be detected easily by adding a geometrical criteria.

5   Conclusion

In this work we present a complete topological filling process based on 2d-topological maps. Starting from a concrete problem of filling character contours images, we follow the idea of Martin & Al. [6] and propose a robust, efficient and simple algorithm for the topological part of the filling problem. Our algo-rithm computes every regions to fill and uses a simple topological property (even depth inclusion) for choosing the seeds which are used for the filling process. We demonstrate on some examples the efficiency of our algorithm which computa-tion time is in $O(n_R)$ with n_R the number or regions. That is to say that we do not need to run through all the pixels of the image.

Our topological process of filling, in this first version, does not integrate ge-ometrical properties which would allow to fill images with contour overlapping (Fig. 2). Nevertheless, those informations are now easily integrable as they can be computed locally. If we consider the Martin & Al. [6] approach, dominant sub-object and exteriority concepts can be simply addressed by a pseudo-inclusion relation based on the common border length over the total border length ratio. In Fig. 2 (C), R_2 will be considered as pseudo-included in R_3, and colored in white, even if both regions are topologically identical. On the contrary, in Fig. 2 (D), R_2 will considered as pseudo-included in R_1 and colored in black. Such a pseudo-inclusion tree can be calculated directly with the borders properties and then computed locally with the 2d-topological map description.

The geometrical properties which allow to compute the pseudo-inclusion and to address the contour overlapping problem, are still not integrated in the al-gorithm but can be easily addressed in the next version since topological maps allow to retrieve efficiently the geometry of region boundaries. Furthermore we will extend this work for every kind of application and any type of boundaries not only 8-connected.

References

1. Lejun, S., Zhou, H.: A new contour fill algorithm for outlined character image generation. Computers & Graphics 19, 551–556 (1995)
2. Arrivault, D., Richard, N., Bouyer, P.: A fuzzy hierarchical attributed graph approach for handwritten hieroglyphs description. In: 11th International Conference on Computer Analysis of Images and Patterns, Versailles, France, 748 (2005)
3. Pavlidis, T.: Contour filling in raster graphics. In: SIGGRAPH '81: Proceedings of the 8th annual conference on Computer graphics and interactive techniques, pp. 29–36. ACM Press, New York, NY, USA (1981)
4. Ren, M., Yang, W., Yang, J.: A new and fast contour-filling algorithm. Pattern Recognition 38, 2564–2577 (2005)
5. Codrea, M., Nevalainen, O.: Note: An algorithm for contour-based region filling. Computers & Graphics 29, 441–450 (2005)
6. Martin, M., Alberola-López, C., Ruiz-Alzola, J.: A topology-based filling algorithm. Computers & Graphics 25, 493–509 (2001)
7. Edmonds, J.: A combinatorial representation for polyhedral surfaces. Notices of the American Mathematical Society vol. 7 (1960)
8. Tutte, W.: A census of planar maps. Canad. J. Math. 15, 249–271 (1963)
9. Jacques, A.: Constellations et graphes topologiques. Combinatorial Theory and Applications 2, 657–673 (1970)
10. Cori, R.: Un code pour les graphes planaires et ses applications. In: Astérisque. Soc. Math. de France, Paris, France, vol. 27 (1975)
11. Lienhardt, P.: Topological models for boundary representation: a comparison with n-dimensional generalized maps. Computer-Aided Design, vol. 23 (1991)
12. Damiand, G., Bertrand, Y., Fiorio, C.: Topological model for two-dimensional image representation: definition and optimal extraction algorithm. Computer Vision and Image Understanding 93, 111–154 (2004)

Extending the Notion of AT-Model for Integer Homology Computation[*]

Rocio Gonzalez-Diaz, María José Jiménez, Belén Medrano[**], and Pedro Real

Applied Math Department, University of Sevilla, Seville, Spain
{rogodi,majiro,belenmg,real}@us.es
http://alojamientos.us.es/gtocoma

Abstract. When the ground ring is a field, the notion of algebraic topological model (AT-model) is a useful tool for computing (co)homology, representative (co)cycles of (co)homology generators and the cup product on cohomology of nD digital images as well as for controlling topological information when the image suffers local changes [6,7,9]. In this paper, we formalize the notion of λ-AT-model (λ being an integer) which extends the one of AT-model and allows the computation of homological information in the integer domain without computing the Smith Normal Form of the boundary matrices. We present an algorithm for computing such a model, obtaining Betti numbers, the prime numbers p involved in the invariant factors (corresponding to the torsion subgroup of the homology), the amount of invariant factors that are a power of p and a set of representative cycles of the generators of homology mod p, for such p.

1 Introduction

There are many tasks in Vision and Image Processing that involve computing certain topological characteristics of objects in a given image such as, for example, connectivity and the number of holes and cavities. We focus here on homology groups (connectivity and the number of holes and cavities can be obtained from them), which are known to be computable in finite dimensions. The classical algorithm for computing integer homology is based on performing row and column operations on the boundary matrices in order to reduce them to the Smith Normal Form (SNF). The integer homology groups can be then determined from this canonical form (see, for example, [13]). However, explicit examples can be given for which this algorithm has a worst-case computational complexity which grows exponentially in both space and time [4].

Our aim is the computation of integer homology information avoiding the computation of the SNF in the integer domain. In fact, our approach allows the computation of Betti numbers, the prime numbers p involved in the invariant

---

[*] Partially supported by Junta de Andalucía (FQM-296 and TIC-02268) and Spanish Ministry for Science and Education (MTM-2006-03722).

[**] Fellow associated to University of Seville under a Junta de Andalucia research grant.

F. Escolano and M. Vento (Eds.): GbRPR 2007, LNCS 4538, pp. 330–339, 2007.
© Springer-Verlag Berlin Heidelberg 2007

factors (corresponding to the torsion subgroup of the homology), the amount of invariant factors that are a power of p and a set of "moduli" representative cycles of the generators of homology in polynomial time. Moreover, our method is not only valid for simplicial complexes but also for other combinatorial objects such as cubical complexes or simploidal complexes since we deal with the group structures.

In the first part of the paper, we recall classical definitions from Algebraic Topology. We also present previous tools for computing topological information: AT-models and AM-models, and we recall the main properties of these structures. Furthermore, we define the notion of λ-AT-model, study its properties, give an algorithm for computing it and study its complexity. Finally, we describe how to obtain homology information in the integer domain from a λ-AT-model. The last section is devoted to conclusions and future works.

2   Definitions and Prior Work

This section introduces the background needed throughout the paper which is essentially extracted from Munkres' book [13]. We also recall briefly the concepts of AT-model and AM-model and their properties.

A *chain complex* $\mathcal{C}$ is a sequence $\{C_q, d_q\}$ of abelian groups C_q and homomorphisms $d_q : C_{q+1} \to C_q$,

$$\cdots \xrightarrow{d_3} C_2 \xrightarrow{d_2} C_1 \xrightarrow{d_1} C_0 \xrightarrow{d_0} 0,$$

such that, for all q, $d_q d_{q+1} = 0$. The set of all the homomorphisms d_q $(q \geq 0)$ is called the *differential* of $\mathcal{C}$. The chain complex $\mathcal{C}$ is *free* if C_q is a free abelian group for each q. It is *finite* if there exists an integer $n > 0$ such that $C_q = 0$ for $q > n$ and each abelian group C_q is finitely generated. In this case, if $C_n \neq 0$, we say that dim of $\mathcal{C}$ is n, and then, $\mathcal{C}$ can be encoded as a pair (C, d), where $C = \bigcup_{q=0}^{n} B_q$, being B_q a basis of C_q and d the matrix corresponding to the differential of $\mathcal{C}$ with respect to the basis C. Suppose that $B_q = \{a_1, \ldots, a_{m_q}\}$. A *q-chain* $a \in \mathcal{C}$ is a formal sum of elements of B_q, $a = \sum_{i=0}^{m_q} \lambda_i a_i$, where $\lambda_i \in \mathbf{Z}$ and $a_i \in C_q$. In this case, dim $a = q$ and $c_a(a_i)$ denotes the coefficient λ_i.

Since our goal is the computation of homology information of "finite" objects (for example, objects explicitly represented within a computer), all chain complexes are finite and free.

Example 1. Shapes are classically modelled with a cellular subdivision. Several combinatorial structures may represent such subdivision. Simplicial complexes have proven to be a useful tool to model a geometric object. Roughly speaking, they are collections of simplices (convex hulls of a set of affinely independent points) that fit together in a natural way to form the object. For every simplicial complex K, one can define a chain complex $C(K)$ canonically associated to it. The homology of K is then defined as the homology of $C(K)$ [13]. Another way to extract combinatorial information from a geometric structure arising naturally, for example, from tomography, numerical computations and graphics, is by means of cubical grids, which subdivide the space into cubes with vertices in an integer lattice. This approach, that can be generalized to an arbitrary dimension, is a cubical complex. The homology of a given cubical complex is the homology of the cubical chain complex associated to it [11]. Finally, simploidal sets [1] include simplicial complexes and cubical complexes as particular cases. They can be used for representing "hybrid" grids coming from finite element methods. In [15], a free chain complex is associated to a simploidal set and the homology of the simploidal set is defined as the homology of the associated chain complex.

We base all formulas and algorithms in this paper on an ordered basis of the chain complex $\mathcal{C}$ where each prefix of the ordering contains the basis of a subcomplex. We call such an ordering a *filter*. In other words, given a chain complex $\mathcal{C}$, $C = \{a_1, \ldots, a_m\}$ is a filter if it is a basis of the chain complex $\mathcal{C}$ and for each j (where $1 \leq j \leq m$), $C = \{a_1, \ldots, a_j\}$ is a basis of a subcomplex of $\mathcal{C}$. For instance, given a chain complex $\mathcal{D} = (D, d)$, a reordering $D' = \{c'_1, \ldots, c'_m\}$ of D such that $\dim c'_i \leq \dim c'_j$ when $i < j$, is always a filter of $\mathcal{D}$.

Example 2. Consider the simplicial complex S derived from the triangulation of the Klein bottle given in Figure 1 and the chain complex $\mathcal{C}(S)$ associated to S. Then,

$$C(S) = \{ \, a, d, ad, f, af, df, adf, b, ab, bf, abf, c, bc, cf, bcf, g, cg, fg, cfg,$$
$$ac, ag, acg, e, ae, eg, aeg, de, ef, def, h, fh, eh, efh, gh, fgh, i,$$
$$gi, hi, ghi, ei, egi, di, dei, ah, aeh, bh, abh, bi, bhi, ci, bci, ai, aci, adi \, \},$$

where $v_0 \cdots v_n$ denotes the simplex spanned by the vertices $v_0, \ldots, v_n$, is a filter of $\mathcal{C}(S)$.

The chain a is a *q-cycle* if $a \in \operatorname{Ker} d_q$. If $a \in \operatorname{Im} d_{q+1}$ then a is called a *q-boundary*. Denote the groups of q–cycles and q–boundaries by Z_q and B_q respectively. Define the integer *qth homology group* to be the quotient group Z_q / B_q, denoted

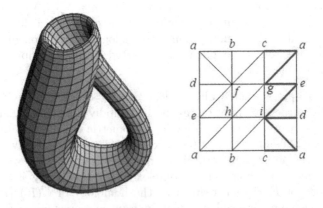

Fig. 1. The Klein bottle and a triangulation of it

by $H_q(\mathcal{C}; \mathbf{Z})$. We say that a is a *representative q–cycle* of the homology generator $a + B_q$ (denoted by $[a]$). For each q, the integer qth homology group $H_q(\mathcal{C}; \mathbf{Z})$ is a finitely generated abelian group. Then $H_q(\mathcal{C}; \mathbf{Z})$ is isomorphic to $F_q \oplus T_q$ where

$$F_q = \mathbf{Z} \oplus \cdots \oplus \mathbf{Z} \text{ and } T_q = (\mathbf{Z}/\alpha_{(q,1)}) \oplus \cdots \oplus (\mathbf{Z}/\alpha_{(q,s)})$$

are the *free subgroup* and the *torsion subgroup* of $H_q(\mathcal{C}; \mathbf{Z})$, respectively. The rank of F_q, denoted by β_q, is called the qth *Betti number* of $\mathcal{C}$. Each $\alpha_{(q,i)}$ is a power of a prime, $\alpha_{(q,i)} = p_i^{t_{(q,p_i)}}$. They are called the *invariant factors* of $H_q(\mathcal{C}; \mathbf{Z})$. The numbers β_q and $\alpha_{(q,i)}$ are uniquely determined by $H_q(\mathcal{C}; \mathbf{Z})$ (up to a rearrangement). Therefore, this representation is in some sense a "canonical form" for $H_q(\mathcal{C}; \mathbf{Z})$.

The qth homology group of $\mathcal{C}$ with coefficients in $\mathbf{Z}/p$ for p a prime, denoted by $H_q(\mathcal{C}; \mathbf{Z}/p)$, is a vector space. Its rank, denoted by $\beta_{(q,p)}$, depends on the prime p. Universal Coefficient Theorem for Homology [13, p. 332] implies that for each prime p,

$$T_{(0,p)} = \beta_{(0,p)} - \beta_0 \text{ and } T_{(q,p)} = \beta_{(q,p)} - \beta_q - T_{(q-1,p)} \text{ for } q > 0;$$

where $T_{(i,p)}$ is the number of invariant factors of $H_i(\mathcal{C}; \mathbf{Z})$ that are a power of p.

Let $\mathcal{C} = \{C_q, d_q\}$ and $\mathcal{C}' = \{C'_q, d'_q\}$ be two chain complexes. A *chain map* $f : \mathcal{C} \to \mathcal{C}'$ is a family of homomorphisms $\{f_q : C_q \to C'_q\}$ such that $d'_q f_q = f_{q-1} d_q$ for all $q \geq 0$. A chain map $f : \mathcal{C} \to \mathcal{C}'$ induces a homomorphism $f_* : H(\mathcal{C}; \mathbf{Z}) \to H(\mathcal{C}'; \mathbf{Z})$ where $f_*[a] = [f(a)]$ for $[a] \in H(\mathcal{C}; \mathbf{Z})$. If $f, g : \mathcal{C} \to \mathcal{C}'$ are chain maps, then *a chain homotopy* $\phi : \mathcal{C} \to \mathcal{C}'$ of f to g is a family of homomorphisms $\{\phi_q : C_q \to C'_{q+1}\}$ such that $f_q - g_q = d'_{q+1} \phi_q + \phi_{q-1} d_q$.

A *chain contraction* of a chain complex $\mathcal{C}$ to another chain complex $\mathcal{C}'$ is a set of three homomorphisms (f, g, ϕ) such that: $f : \mathcal{C} \to \mathcal{C}'$ and $g : \mathcal{C}' \to \mathcal{C}$ are chain maps; fg is the identity map of $\mathcal{C}'$ and $\phi : \mathcal{C} \to \mathcal{C}$ is a chain homotopy of the identity map of $\mathcal{C}$ to gf, that is, $\phi d + d\phi = id_c - gf$. Important properties of chain contractions are: $\mathcal{C}'$ has fewer or the same number of generators than $\mathcal{C}$; and $\mathcal{C}$ and $\mathcal{C}'$ have isomorphic homology groups.

An *AT-model* [6,7,9] for a chain complex $\mathcal{C} = (C, d)$ is a chain contraction of $\mathcal{C}$ to a chain complex $\mathcal{H}$ with null differential. An AT-model can be stored as a set $((C, d), H, f, g, \phi)$, where C and H are basis of $\mathcal{C}$ and $\mathcal{H}$, and f, g and ϕ are the matrices corresponding to the homomorphisms that defines the chain contraction of $\mathcal{C}$ to $\mathcal{H}$. Observe that the homology of $\mathcal{C}$ is isomorphic to $\mathcal{H}$. If the ground ring is $\mathbf{Z}/p$, being p a prime, the following algorithm computes an AT-model for a given chain complex. This algorithm is a straightforward modification of that in [9].

Algorithm 1. *Computing an AT-model for a chain complex $\mathcal{C}$ over $\mathbf{Z}/p$.*

INPUT: a filter $C = \{a_0, \ldots, a_m\}$ of the chain complex $\mathcal{C}$, and the matrix of the differential d for the basis C.

$H := \{\}, \; f := 0, \; g := 0, \; \phi := 0.$

For $i = 0$ to m do
    If $fd(a_i) = 0$, then
        $H := H \cup \{a_i\}$, $f(a_i) := a_i$, $\phi(a_i) := 0$, $g(a_i) := a_i - \phi d(a_i)$.
    If $fd(a_i) \neq 0$, then:
        $\mu_i := \min \{c_{fd(a_i)}(a_j), j = 0, .., i - 1\}$,
        $k := \max \{j \text{ such that } c_{fd(a_i)}(a_j) = \mu_i, j = 0, ..., i - 1\}$,
        $H := H \backslash \{a_k\}$, $f(a_i) := 0$, $\phi(a_i) := 0$.
        For $j = 0$ to $i - 1$ do,
            $\lambda_{a_j} := c_{f(a_j)}(a_k)$,
            $f(a_j) := f(a_j) - \mu_i^{-1} \lambda_{a_j} fd(a_i)$,
            $\phi(a_j) := \phi(a_j) + \mu_i^{-1} \lambda_{a_j}(a_i - \phi d(a_i))$,
OUTPUT: the set $((C, d), H, f, g, \phi)$.

The key idea of this algorithm is the same as in [3]: in the ith step, the element a_i of the filter C is added and then, a homology class is created or destroyed. The algorithm runs in time at most $\mathcal{O}(m^3)$, where m is the number of elements of C. Recall that the notion of AT-model is an useful tool for computing (co)homology, representative cycles of (co)homology generators and the cup product on cohomology of nD digital images as well as for controlling topological information when the image suffers local changes [6,7,9]. The main problem of the computation of AT-models over $\mathbf{Z}/p$ is that if the object under study contains torsion, then the Betti numbers $\beta_{(q,p)}$ can change when p varies.

Example 3. The Betti numbers of the simplicial complex S (see Figure 1) computed over the field $\mathbf{Z}/p$, for $p = 2, 3, 29$.

	$\beta_{(0,p)}$	$\beta_{(1,p)}$	$\beta_{(2,p)}$
$\mathbf{Z}/2$	1	2	2
$\mathbf{Z}/3$	1	1	0
$\mathbf{Z}/29$	1	1	0

An AM-model [5] for a chain complex $\mathcal{C} = (C, d)$ is a chain contraction (f, g, ϕ) of $\mathcal{C}$ to $\mathcal{M} = (M, d')$ such that the matrix A of the differential d' coincides with its Smith normal form and satisfies that any non-null entry of A is greater than 1. Working with coefficients in the integer domain, an AM-model for $\mathcal{C}$ can always be computed. Moreover, the integer (co)homology of $\mathcal{C}$ and representative cycles of (co)homology generators can be directly obtained from $\mathcal{M}$ [8].

The algorithm for computing AM-models given in [8] needs to reduce the matrix of the differential to its Smith Normal Form (SNF). Explicit examples can be given for which the computation of SNF has a worst-case computational complexity which grows exponentially in both space and time [4]. Many algorithms have been devised to improve this complexity bound [10,17,2,14].

Our aim in this paper is the computation of integer homology information avoiding the computation of the SNF in the integer domain.

3   Extending the Notion of Algebraic Topological Model

In this section we will consider that the ground ring is $\mathbf{Z}$. We first define the notion of λ-AT-model which is a generalization of the one of AT-model. We study its properties, give an algorithm for computing it and study its complexity.

Definition 1. *Let $\mathcal{C} = (C, d)$ be a chain complex, λ a non-null integer and $\mathcal{H} = (H, d')$ a chain complex with null differential (that is, $d' = 0$). Let $f : C \to H$, $g : H \to C$ and $\phi : C \to C$ be three homomorphisms. Then the set $((C, d), H, f, g, \phi, \lambda)$ is a λ-AT-model if f and g are chain maps, $fg = \lambda \cdot id_{\mathcal{H}}$ and ϕ is a chain homotopy of $\lambda \cdot id_c$ to gf, that is, $\lambda \cdot id_c - gf = \phi d + d\phi$.*

Proposition 1. *Given a λ-AT-model, a rational AT-model (i.e., an AT-model over $\mathbf{Q}$) can directly be obtained as well as rational (co)homology and representative cycles of (co)homology generators. Concretely, if $((C, d), H, f, g, \phi, \lambda)$ is a λ-AT-model for a chain complex $\mathcal{C}$, then $((C, d), H, \frac{1}{\lambda}f, g, \frac{1}{\lambda}\phi)$ is an AT-model for $\mathcal{C}$ over $\mathbf{Q}$ and $\{g(h) : h \in H\}$ is a set of representative cycles of the generators of $H(\mathcal{C}; \mathbf{Q})$.*

Corollary 1. *Let $((C, d), H, f, g, \phi, \lambda)$ be a λ-AT-model. Then $\mathcal{H}$ (the chain complex generated by H with null differential) is isomorphic to the free subgroup of $H(\mathcal{C}; \mathbf{Z})$. Moreover, the set $\{g(h) : h \in H\}$ is a set of independent non-boundary cycles of $\mathcal{C}$ over $\mathbf{Z}$.*

Proposition 2. *Given a λ-AT model $((C, d), H, f, g, \phi, \lambda)$ and a prime p such that p does not divide λ, then $((C, d_p), H, f_p, g_p, \phi_p)$, where $d_p = d \bmod p$, $f_p = \lambda^{-1}f \bmod p$, $g_p = g \bmod p$ and $\phi_p = \lambda^{-1}\phi \bmod p$, is an AT-model for $\mathcal{C}$ over $\mathbf{Z}/p$ and $\{g_p(h) : h \in H\}$ is a set of representative cycles of the generators of $H(\mathcal{C}; \mathbf{Z}/p)$.*

Proposition 3. *Let $((C, d), H, f, g, \phi, \lambda)$ be a λ-AT-model. Let $a \in C$ such that $d(a) = 0$. If there exists $b \in C$ such that $d(b) = \alpha a$ where $\alpha \in \mathbf{Z}$ and $\alpha \neq 0$, then $f(a) = 0$.*

Proof. Suppose that $b \in C$ such that $d(b) = \alpha a$ where $\alpha \in \mathbf{Z}$ and $\alpha \neq 0$, and $f(a) \neq 0$. Then $\alpha f(a) \neq 0$ (since the ground ring is $\mathbf{Z}$). On the other hand, $\alpha f(a) = f(\alpha a) = f(d(b)) = 0$, a contradiction. □

Corollary 2. *Let $((C, d), H, f, g, \phi, \lambda)$ be a λ-AT-model. Let $a \in C$ such that $d(a) = 0$. If there exists $b \in C$ such that $d(b) = \alpha a$ where $\alpha \in \mathbf{Z}$, $\alpha \neq 0$, and for each β, where $0 < \beta < \alpha$, $\beta a \notin Im\, d$. then α divides λ.*

Proof. By Proposition 3, we have $f(a) = 0$. Since $\lambda a - gf(a) = \phi d(a) + d\phi(a)$, then $\lambda a = d\phi(a)$. Suppose that α does not divide λ. There exists $c, r \in \mathbf{Z}$, such that $0 < r < \alpha$ and $\lambda = c\alpha + r$. On one hand, $ra \notin Im\, d$; on the other hand, $ra = (\lambda - c\beta)a = d(\phi(a) - cb) \in Im\, d$, a contradiction. We conclude that $alpha$ divides λ. □

Corollary 3. *Let* $((C,d), H, f, g, \phi, \lambda)$ *be a* λ-*AT-model. If* $\alpha = p^{t(q,p)}$ *is an invariant factor of* $H_q(C; \mathbf{Z})$, *then* p *divides* λ.

Algorithm 2. *Algorithm for computing a* λ-*AT-model for a chain complex* C.

INPUT: a filter $C = \{a_0, \ldots, a_m\}$ of the chain complex $\mathcal{C}$,
        and the matrix of the differential d for the basis C.
$H := \{\,\}$, $\lambda := 1$, $f := 0$, $g := 0$, $\phi := 0$.
For $i = 0$ to m do
    If $fd(a_i) = 0$, then
        $H := H \cup \{a_i\}$, $f(a_i) := a_i$, $\phi(a_i) := 0$, $g(a_i) := \lambda a_i - \phi d(a_i)$.
    If $fd(a_i) \neq 0$, then
        $\mu_i := \min\ \{|c_{fd(a_i)}(a_j)|, \ j = 0, .., i-1\}$,
        $k := \max\ \{j \text{ such that } |c_{fd(a_i)}(a_j)| = \mu_i, \ j = 0, \ldots, i-1\}$,
        $\lambda_k := c_{fd(a_i)}(a_k)$,
        $H := H \backslash \{a_k\}$, $f(a_i) := 0$, $\phi(a_i) := 0$.
        For $j = 0$ to $i-1$,
            $\lambda_{a_j} := c_{f(a_j)}(a_k)$,
            $f(a_j) := \lambda_k f(a_j) - \lambda_{a_j} fd(a_i)$,
            $\phi(a_j) := \lambda_k \phi(a_j) + \lambda_{a_j}(\lambda a_i - \phi d(a_i))$,
            $\lambda := \lambda \lambda_k$.
OUTPUT: The set $((C,d), H, f, g, \phi, \lambda)$.

Theorem 3. *The set* $((C,d), H, f, g, \phi, \lambda)$ *obtained applying Algorithm 2 defines a* λ-*AT-model for the chain complex* $\mathcal{C} = (C,d)$.

Proof. Assume that $((C_{i-1}, d), H_{i-1}, f_{i-1}, g_{i-1}, \phi_{i-1}, \lambda_{i-1})$ is the λ_{i-1}-AT-model obtained using the algorithm above for the filter $C_{i-1} = \{a_0, \ldots, a_{i-1}\}$. Assume that the annihilation properties $f_{i-1}\phi_{i-1} = 0$, $\phi_{i-1}g_{i-1} = 0$ and $\phi_{i-1}\phi_{i-1} = 0$ hold. We will prove that the set $(C_i, d), f_i, g_i, \phi_i, \lambda_i)$ obtained after adding a_i to the filter C_{i-1} is a λ_i-AT-model. More concretely, we will prove that $f_i d = 0$, $dg_i = 0$, $f_i g_i = \lambda_i \cdot id$, $\lambda_i \cdot id - g_i f_i = \phi_i d + d\phi_i$, $f_i \phi_i = 0$, $\phi_i g_i = 0$ and $\phi_i \phi_i = 0$. We deal only with the case $f_{i-1}d(a_i) \neq 0$; the other case is left to the reader. First, $f_i d(a_i) = \lambda_k f_{i-1} d(a_i) - \lambda_k f_{i-1} d(a_i) = 0$. Second, $\phi_i d(a_i) + d\phi_i(a_i) = \lambda_k \phi_{i-1} d(a_i) + \lambda_k(\lambda_{i-1} a_i - \phi_{i-1} d(a_i)) = \lambda_i a_i = \lambda_i a_i - g_i f_i(a_i)$. Finally, it is clear that $f_i \phi_i(a_i) = 0$ and $\phi_i \phi_i(a_i) = 0$. Now, let $a_j \in C_{i-1}$, then $f_i d(a_j) = \lambda_k f_{i-1} d(a_j)$ which is null by induction; $\phi_i d(a_i) + d\phi_i(a_i) = \lambda_k \phi_{i-1} d(a_j) + \lambda_{d(a_j)}(\lambda_{i-1} a_i - \phi_{i-1} d(a_i)) + \lambda_k d\phi_{i-1}(a_j) + \lambda_{a_j}(\lambda_{i-1} d(a_i) - d\phi_{i-1} d(a_i)) = \lambda_k(\lambda_{i-1} a_j - g_{i-1} f_{i-1}(a_j)) + \lambda_{a_j} g_{i-1} f_{i-1} d(a_i) = \lambda_i a_j - g_i f_i(a_j)$. Moreover, $f_i \phi_i(a_j) = f_i(\lambda_k \phi_{i-1}(a_j) + \lambda_{a_j}(\lambda_{i-1} a_i - \phi_{i-1} d(a_i))) = 0$; $\phi_i \phi_i(a_j) = \phi_i(\lambda_k \phi_{i-1}(a_j) + \lambda_{a_j}(\lambda_{i-1} a_i - \phi_{i-1} d(a_i))) = 0$. If $a_j \in H_i$, then $dg_i(a_j) = dg_{i-1}(a_j) = 0$, by induction; $f_i g_i(a_j) = \lambda_k f_{i-1} g_{i-1}(a_j) - \lambda_{g_{i-1}(a_j)} f_{i-1} d(a_j) = \lambda_k \lambda_{i-1} a_j = \lambda_i a_j$. Finally, it is easy to see that $\phi_i g_i(a_j) = 0$. $\qquad\square$

To study the complexity, fix the dimension of the complex, n, and count the number of elementary operations involved in the algorithm. In the ith step, we have to evaluate $f_{i-1}d(a_i)$. The numbers of elements of C involved in $d(a_i)$

and $f_{i-1}(a_j)$ for $1 \leq j < i$ is at most n and m, respectively. Therefore, the evaluation of $f_{i-1}d(a_i)$ costs $O(nm) = O(m)$. If $f_{i-1}d(a_i) \neq 0$, we have to update $f_{i-1}(a_j)$ and $\phi_{i-1}(a_j)$ for $1 \leq j < i$. The total cost of these operations is $O(m^2)$. Therefore, the total algorithm runs in time at most $O(m^3)$.

The following proposition shows that AT-models over $\mathbf{Z}/p$, p being any prime, can also be computed using Algorithm 2.

Proposition 4. *If the output of Algorithm 2, working with coefficients in $\mathbf{Z}/p$, p being any prime, is $(C, d), H, f, g, \phi, \lambda)$, then $((C, d), H, \lambda^{-1}f, g, \lambda^{-1}\phi)$ is an AT-model over $\mathbf{Z}/p$. Furthermore, $\{g(h) : h \in H\}$ is a set of representative cycles of the generators of $H(C; \mathbf{Z}/p)$.*

Example 4. Consider the simplicial complex S derived from the triangulation of the Klein bottle given in Figure 1 and the chain complex $C(S)$ associated to S. Let $C(S)$ be the filter of $C(S)$ given in Example 1. Running the algorithm above, we obtain a 2-AT-model of $C(S)$, $(C(S), H_s, f_s, g_s, \phi_s, 2)$, where $H_s = \{a, ac\}$. The value of f_s on each vertex of S is $2a$. The value of f_s on each edge marked in red in Figure 1, is $2ac$. The value of f_s on the rest of the simplices of S is zero. For the map g_s, we obtain that $g_s(a) = a$ and $g_s(ac) = ac - bc - ab$. On each vertex of S, ϕ_s gives a path connecting this vertex with a, multiplied by 2; for example, $\phi_s(g) = 2(ab + bc + cg)$. On the edges of S, the key idea is the same, that is, on each edge of S, ϕ_s gives a "path" connecting this edge with ac, multiplied by 2; for example, $\phi_s(gh) = 2(fgh - cfg + bcf + abf)$. On each triangle of S, the value of ϕ_s is zero.

Summing up, given a filter C of a chain complex C, it is possible to compute a λ-AT-model, $\lambda AT = (C, d), H, f, g, \phi, \lambda)$, in $O(m^3)$ if C has m elements. The Betti numbers and a set of independent non-boundary cycles of C over $\mathbf{Z}$ can directly be obtained from λAT. Moreover, the integer λ provides the prime numbers involved in the invariant factors of the torsion subgroup of $H(C; \mathbf{Z})$. This last information will be essential in the next section for designing an algorithm for computing "moduli" representative cycles of the generators of the free and the torsion subgroups of $H(C; \mathbf{Z})$.

4   Computing Integer Homology Information

As we have said before, a λ-AT-model for a given chain complex C provides information of the free subgroup of $H(C; \mathbf{Z})$ as well as the prime numbers involved in the invariant factors of $H(C; \mathbf{Z})$. For obtaining "moduli" representative cycles of the generators of the free and the torsion subgroups of $H(C; \mathbf{Z})$ we only have to compute AT-models for C over $\mathbf{Z}/p$, for each prime p dividing λ. Observe that for this last task, since we work with coefficients in $\mathbf{Z}/p$, we can use either Algorithm 1 or Algorithm 2.

Algorithm 4. *Computing integer homology information and "moduli" representative cycles of homology generators of a chain complex C.*

INPUT: a filter $C = \{a_0, \ldots, a_m\}$ of the chain complex $\mathcal{C}$ of dim n,
    and the matrix of the differential d for the basis C.
Apply Algorithm 2 with coefficients in $\mathbf{Z}$ for computing a
    λ-AT-model for (C, d), $((C, d), H, f, g, \phi, \lambda)$;
    $\beta_q :=$ number of elements of H of dim q, for $q = 0$ to n;
    $G := \{g(h) : h \in H\}$.
For each prime p dividing λ do
    Apply Algorithm 2 with coefficients in $\mathbf{Z}/p$ for computing an
    AT-model for C over $\mathbf{Z}/p$, $((C, d_p), H_p, f_p, g_p, \phi_p)$;
    $T_{(0,p)} = \beta_{(0,p)} - \beta_0$;
    $T_{(q,p)} = \beta_{(q,p)} - \beta_q - T_{(q-1,p)}$, for $q = 1$ to n;
    $G_p = \{g_p(h_p) : h_p \in H_p\}$.
OUTPUT: The sets G, $\{G_p : p$ being a prime dividing $\lambda\}$, $\{\beta_1, \ldots, \beta_n\}$,
    and $\{T_{(q,p)} : 0 \leq q \leq n$ and p being a prime dividing $\lambda\}$.

Summing up, after computing a λ-AT-model and an AT-model over $\mathbf{Z}/p$, for each p dividing λ, for a given chain complex $\mathcal{C}$, we obtain:

- the Betti numbers β_q for $0 \leq q \leq n$, and a set G of independent non-boundary cycles of $\mathcal{C}$ over $\mathbf{Z}$ (in fact, G is also a set of generators of $H(\mathcal{C}; \mathbf{Q})$);
- the prime numbers p involved in the invariant factors corresponding to the torsion subgroup of $H(\mathcal{C}, \mathbf{Z})$, the amount of invariant factors in each dimension q that are a power of p, $T_{(q,p)}$, and a set G_p of representative cycles of the generators of $H(\mathcal{C}; \mathbf{Z}/p)$ for each prime p dividing λ.

Example 5. In Example 1, we applied the Algorithm 2 and computed a 2-AT-model for $\mathcal{C}(S)$ and the Betti numbers of $\mathcal{C}$; $\beta_0 = 1$, $\beta_1 = 1$ and $\beta_2 = 0$. Now, we apply Algorithm 2 for compute an AT-model for (C, d) with coefficients in $\mathbf{Z}/2$ to obtain $((C, d)), H_2, f_2, g_2, \phi_2)$ where $H_2 = \{a, ac, de, adi\}$. Then, $\beta_{(0,2)} = 1$, $\beta_{(1,2)} = 2$, $\beta_{(2,2)} = 1$ and $G_2 = \{a,\ ac + bc + ab,\ ad + de + ae,\ adf + abf + bcf + cfg + acg + aeg + def + efh + fgh + ghi + egi + dei + aeh + abh + bhi + bci + aci + adi\}$. Therefore, $t_{(0,2)} = 0$, $t_{(1,2)} = 1$ and $t_{(2,2)} = 0$. We conclude that $H_0(S) = \mathbf{Z}$ and $H_1(S) = \mathbf{Z} \oplus \mathbf{Z}/2$.

5   Conclusions and Future Work

A λ-AT-model for a chain complex $\mathcal{C}$ can be computed in cubic time. It provides information of the free subgroup of $H(\mathcal{C}; \mathbf{Z})$ and also the primes p that are candidates to be involved in an invariant factor of $H(\mathcal{C}; \mathbf{Z})$. For obtaining the amount of invariant factors that are a power of p and "moduli" representative cycles of homology generators, we compute an AT-model with coefficients in $\mathbf{Z}/p$ for such primes p.

A future work is to study if it is possible to obtain generators with integer coefficients of the torsion subgroup of $H(\mathcal{C}; \mathbf{Z})$. Another task is to study if cohomology features can be computed over $\mathbf{Z}$ from a λ-AT-model.

Concerning to the complexity, Algorithm 4 runs in time $O(m^3 \psi(\lambda))$ in the worst case, ψ being the Euler function. Therefore, one important question is to bound the coefficient λ. In order to improve the complexity we might first compute a chain contraction to obtain a smaller chain complex with same homology in the integer domain and apply Algorithm 4 to a "thinned" complex.

References

1. Dahmen, W., Micchelli, C.A.: On the Linear Independence of Multivariate b-splines I. Triangulation of simploids. SIAM J. Numer. Anal., vol. 19 (1982)
2. Dumas, J.G., Saunders, B., Villard, G.: On Efficient Sparse Integer Matrix Smith Normal Form Computations. J. of Symbolic Computation (2001)
3. Edelsbrunner, H., Letscher, D., Zomorodian, A.: Topological Persistence and Simplification. In: Proc. 41st Symp. on Foundations of Computer Science, pp. 454–463 (2000)
4. Gui, X., Havas, G.: On the Worst-case Complexity of Integer Gaussian Elimination. In: Proc. of ISSAC 1997, pp. 28–31 (1997)
5. González-Díaz, R., Real, P.: Computation of Cohomology Operations on Finite Simplicial Complexes. Homology, Homotopy and Applications 5(2), 83–93 (2003)
6. González-Díaz, R., Real, P.: Towards Digital Cohomology. In: Nyström, I., Sanniti di Baja, G., Svensson, S. (eds.) DGCI 2003. LNCS, vol. 2886, pp. 92–101. Springer, Heidelberg (2003)
7. González-Díaz, R., Real, P.: On the Cohomology of $3D$ Digital Images. Discrete Applied Math. 147, 245–263 (2005)
8. Gonzalez-Diaz, R., Medrano, B., Real, P., Sánchez-Peláez, J.: Reusing Integer Homology Information of Binary Digital Images. In: Kuba, A., Nyúl, L.G., Palágyi, K. (eds.) DGCI 2006. LNCS, vol. 4245, pp. 199–210. Springer, Heidelberg (2006)
9. Gonzalez-Diaz, R., Medrano, B., Real, P., Sánchez-Peláez, J.: Simplicial Perturbation Technique and Effective Homology. In: Ganzha, V.G., Mayr, E.W., Vorozhtsov, E.V. (eds.) CASC 2006. LNCS, vol. 4194, pp. 166–177. Springer, Heidelberg (2006)
10. Iliopoulus, O.S.: Worst-case Complexity Bounds on Algorithms for Computing the Canonical Structure of Finite Abelian Groups and the Hermite and Smith Normal Forms of an Integer Matrix. SIAM J. Comput. 18, 658–669 (1989)
11. Massey, W.M.: A Basic Course in Algebraic Topology. New York (1991)
12. Kaczynski, T., Mischaikow, K., Mrozek, M.: Computational Homology. Applied Mathematical Sciences, vol. 157. Springer-Verlag, Heidelberg (2004)
13. Munkres, J.R.: Elements of Algebraic Topology. Addison–Wesley Co., London (1984)
14. Peltier, S., Alayrangues, S., Fuchs, L., Lachaud, J.: Computation of Homology Groups and Generators. In: Andrès, É., Damiand, G., Lienhardt, P. (eds.) DGCI 2005. LNCS, vol. 3429, pp. 195–205. Springer, Heidelberg (2005)
15. Peltier, S., Fuchs, L., Lienhardt, P.: Homology of Simploidal Set. In: Kuba, A., Nyúl, L.G., Palágyi, K. (eds.) DGCI 2006. LNCS, vol. 4245, pp. 235–246. Springer, Heidelberg (2006)
16. Sergeraert, F.: Homologie effective. I, II. C. R. Acad. Sci. Paris Sér. I Math. 304, vol. 11, pp. 279–282, vol. 12, pp. 319–321 (1987)
17. Storjohann, A.: Near Optimal Algorithms for Computing Smith Normal Forms of Integral Matrices. Proc. of ISSAC 1996, pp. 267–274 (1996)
18. Zomorodian, A., Carlsson, G.: Computing Persistent Homology. In: Proc. of the 20th annual Symposium on Computational Geometry, pp. 347–356 (2004)

Constellations and the Unsupervised Learning of Graphs

B. Bonev, F. Escolano, M.A. Lozano, P. Suau, M.A. Cazorla[1], and W. Aguilar[2]

[1] Robot Vision Group, Departamento de Ciencia de la Computación e IA
Universidad de Alicante, Spain
[2] IIMAS: Instituto de Investigaciones en Matemáticas Aplicadas y Sistemas
Univesidad Nacional Autónoma de México UNAM, México

Abstract. In this paper, we propose a novel method for the unsupervised clustering of graphs in the context of the constellation approach to object recognition. Such method is an EM central clustering algorithm which builds prototypical graphs on the basis of fast matching with graph transformations. Our experiments, both with random graphs and in realistic situations (visual localization), show that our prototypes improve the set median graphs and also the prototypes derived from our previous incremental method. We also discuss how the method scales with a growing number of images.

1 Introduction

Structural criteria, graph matching, and even graph learning, have been considered as fundamental elements in the set up of the constellation (part/features-based) approach to object recognition [12]. Most of research in such direction has been concentrated in exploiting feature (local) statistics, whereas structural (global) statistics have been typically confined to the joint Gaussian of feature locations [5]. However, there has been a recent interest in modelling and learning structural relationships. This is the case of the *tree-structured models* [6][10] and the *k-fans* graph model [3]. However, models with higher relational power are often needed for solving realistic situations. In this regard, a key question is to find an adequate trade-off between the complexity of the model and the computational cost of learning and using it.

In this paper, we present a novel method for the unsupervised learning of general graph models under the constellation approach. Here, we follow central graph clustering [2][14][7], and the core element is *prototype building* or *graphs fusion*. In [16] we proposed an incremental method which depends on the order in which the graphs are fused. In this paper, we present an alternative method which overcomes such problem. It is based on the information provided by the diffusion kernels [4][11] in order to decide which matches are preferable to be considered in order to fuse the nodes of the graphs in the set. Our algorithm works both with continuous graph-matching methods like Softassign, or

F. Escolano and M. Vento (Eds.): GbRPR 2007, LNCS 4538, pp. 340–350, 2007.
© Springer-Verlag Berlin Heidelberg 2007

our kernelized version [15], and with faster alternative discrete matching methods. From this point of view, as in the constellation approach node attributes coming from describing salient features play a key role, here we also propose *graph-transformation matching* [1] a novel fast and reliable method, emerging from putative matches between feature sets, which yields a *consensus graph*, provided that such subgraph exists.

Our graph-learning method for the constellation approach is tested in a *visual localization* (scene recognition) context. The early approach is coarse-to-fine: (i) Given an input image, an appearance-based classifier, trained with the optimal (minimal) number of features finds the most probable submap; (ii) The statistics of the sub-map are exploited to speed-up the extraction of invariant salient features [9]; (iii) Given proper feature descriptors [13] graph-transformation matching finds common subgraphs with images in the same submap; (iv) The image with the highest number of nodes in the subgraph is chosen as output and the viewing coordinates are reported. Here, we compare this early design with the one resulting from replacing (iii) by finding the closest structural prototype in the submap and then match the input image only to the images in such cluster.

The rest of the paper is organized as follows. The core of our proposal, the graph-fusion method, is presented in Section 2. In Section 3 we describe the graph-transformation matching and its implications in the EM clustering algorithm, together with graph-fusion. Experimental results are presented in Section 4, and, finally, in Section 5 we present our conclusions and future works.

2   Mapping Graphs to Prototypes Via Diffusion Kernels

2.1   Building the Super-Graph

Given a set of graphs S, with $N = |S|$, to be clustered, each graph $\mathbf{G}_i \in S$ is a 3-tuple $\mathbf{G}_i = (V_i, E_i, \beta_i)$ where: V_i is the set of nodes, $E_i \subseteq V \times V$ is the set of edges, $\beta_i : V_i \longrightarrow \mathbb{R}^n$ are the node attributes (descriptors of salient points). In order to obtain the prototype, firstly it is necessary to obtain $O(N^2)$ pairwise matching matrices M^{ij} between all pairs $< \mathbf{G}_i, \mathbf{G}_j > \in S \times S$ with $i \neq j$. With respect to the incremental method [14][16], pairwise matchings will be computed only once, which is critical for the efficiency of the EM-clustering (more precisely to the E-steps).

Super-Graph. The latter matching matrices will be used in order to build a super-graph $\mathbf{G}_M$ which encodes the possible matchings among the graphs in S. This super-graph is a 5-tuple $\mathbf{G}_M = (V_M, E_M, \theta, \nu, \xi)$, where

- $V_M = \cup_{i=1}^{|S|} V_i$ is the union of the nodes from all the graphs in S,
- $\theta : V_M \longrightarrow S$, is a function assigning each node in the super-graph with its corresponding graph in the original set,
- $\nu : V_M \longrightarrow \cup_{i=1}^{|S|} V_i$, is a function assigning each node in the super-graph with its corresponding node on the graphs of the original set,

- $E_M = \{<i,j>, i,j \in V_M : M_{\nu_i\nu_j}^{\theta_i\theta_j} = 1\}$, that is, two nodes will be connected if, and only if, their corresponding nodes in the graphs in the set S are matched, and
- $\xi : E_M \longrightarrow \mathbb{R}^+$ is a weighting function for the edges.

Graph Partitions. Discrete matchings $M_{\nu_i\nu_j}^{\theta_i\theta_j}$ (when applying Softassign-like methods continuous variables before cleanup are even more useful) induce disjoint partitions $P_\alpha = \{i : i \in V_M\}$. In an ideal case, each partition would have at the most one node coming from each graph in S:

$$\forall i \in P_\alpha, \nexists j \in P_\alpha : \theta_i = \theta_j, j \neq i, \forall P_\alpha \subset V_M \qquad (1)$$

In this case, the fusion is easy. Each partition corresponds to a node in the prototype graph (see Fig. 2-top-left). However, in a real case, due to the matching ambiguity and errors, a partition could have some nodes from the same graph (see Fig. 2-bottom-left). We must then decide which matches are going to be taken into account in order to build the prototype, and which ones will be discarded. Matches with a higher value in the matching matrix will be preferred, because the higher is this value, the lower is the ambiguity of this match. However, there will be many nodes with the same value in the matching matrix. In order to decide which of them is preferred their kernel values will be used. Therefore, each edge $<i,j>$ from the super-graph will be weighted by a function ξ that is defined as

$$\xi(<i,j>) \longleftarrow M_{\nu_i\nu_j}^{\theta_i\theta_j} + \alpha\Phi_{\nu_i\nu_j}^{\theta_i\theta_j}, \forall <i,j> \in E_M \qquad (2)$$

where α is a small value (i.e. $\alpha \sim 0.01$) and Φ is an affinity measure between matched vertices ν_i and ν_j. In this case, we define $\Phi_{\nu_i\nu_j}^{\theta_i\theta_j} = \exp\{-(K_{\nu_i}^{\theta_i} - K_{\nu_j}^{\theta_j})^2\}$ being K the diffusion kernel associated to the Laplacian of the graph $\theta \in S$ containing vertex ν_i (respectively ν_j), that is, $K = \exp\{-(\beta/m)L\}$ being $L = D - A$ where m is the number of vertices of θ, D is the diagonal matrix registering the degree of each vertex and A is the adjacency matrix. Consequently, we have that $K_{\nu_i}^{\theta_i} = K_{\nu_i\nu_i}^{\theta_i}$ is the ν_i-th element of the diagonal (similarly $K_{\nu_j}^{\theta_j} = K_{\nu_j\nu_j}^{\theta_j}$). As it is well known , the values in the diagonal of a diffusion kernel encode the probability that a lazy random walk remains at such vertex, and such probability encodes how the graph structure *is seen* from a given vertex.

The latter weights $\Phi_{\nu_i\nu_j}^{\theta_i\theta_j}$, which encode structural compatibility, will be used to insert all the edges in E_M into a sorted list $\mathcal{L}_e$. The elements with higher weights will be taken first. These edges will be used in order to build the partitions of the graph, taking into account the constraints in 1. For each edge $<i,j>$, there are 4 possible cases:

- *Neither i nor j are assigned to any partition.* In this case a new partition is created, and both i and j are assigned to it.
- *i is assigned but j is not.* Add j to the partition of i if doing this the constraints are satisfied. If not, add j to a new partition.
- *j is assigned but i is not.* Add i to the partition of j if doing this the constraints are satisfied. If not, add i to a new partition.

```
Algorithm KERNELIZED FUSION {                Switch
    G_M ← BuildSuperGraph(S)                     Case L_i = ∅, L_j = ∅
    P ← ObtainPartitions(G_M)                        l ← new label
    G_P ← BuildPrototype(S, P)                       L_i, L_j ← l
    Return G_P                                       P_l ← {i, j}
}                                                    O_l ← [0]_{1×|S|}
                                                     O_{lθ_i}, O_{lθ_j} ← 1
                                                 Case L_i = ∅, L_j ≠ ∅
                                                     If O_{L_j θ_i} ≠ 1
Algorithm BUILD SUPER-GRAPH {                            L_i ← L_j
    α ← 0.01                                            P_{L_i} ← P_{L_i} ∪ {i}
    For each pair of graphs < G_A, G_B >,               O_{L_i θ_i} ← 1
            G_A, G_B ∈ S                             End If
        M^{AB} ← GraphMatching(G_A, G_B)         Case L_i ≠ ∅, L_j = ∅
        K^A ← Kernel(G_A)                            If O_{L_i θ_j} ≠ 1
    End For                                              L_j ← L_i
    G_M = (V_M, E_M, θ, ν, ξ)                            P_{L_j} ← P_{L_j} ∪ {j}
    For each i ∈ V_M                                     O_{L_j θ_j} ← 1
        For each j ∈ V_M                             End If
            If M^{θ_i θ_j}_{ν_i ν_j} > 0          Case L_i ≠ ∅, L_j ≠ ∅
                E_M ← E_M ∪ < i, j >                 If L_i ≠ L_j
                ξ(< i, j >) ← M^{θ_i θ_j}_{ν_i ν_j} + αΦ^{θ_i θ_j}_{ν_i ν_j}
                                                         If (O_{L_i} O_{L_j}^T = 0)
            End If                                           O_{L_i} ← O_{L_i} + O_{L_j}
        End For                                          For each k : L_k = L_j
    End For                                                  L_k ← L_i
    Return G_M                                           End For
}                                                        P_{L_i} ← P_{L_i} ∪ P_{L_j}
                                                         Remove partition P_j
                                                     End If
Algorithm OBTAIN PARTITIONS {                    End If
    L_e ← Sort_w({< i, j, w >:< i, j >∈ E_M,     End If
            w = ξ(< i, j >)})                Return P
    While L_e ≠ ∅                            }
        < i, j, w >← Remove first L_e
```

Fig. 1. Kernelized graph-fusion algorithm

- *Both i and j are assigned to a partition. If both i and j are assigned to the same partition, there is nothing to do. In other case, fuse the partitions of i and j if it satisfies the constraints.*

The algorithm for obtaining the partitions is detailed in Fig. 1. In this algorithm a set of auxiliary variables L_i and $O_{\alpha\theta}$ are introduced. Each variable L_i maps a vertex $i \in V_M$ to its corresponding partition α. The boolean variables $O_{\alpha\theta}$ indicate whether the partition α contains a node from the input graph θ.

These variables are used to ensure that in each partition there is no more than one node from each graph from the input set.

2.2 Building the Prototypes

After the process described above, a set of partitions P_S will be obtained, satisfying $\bigcup_{P_i \in P_S} P_i = V_M$. Each partition $P_i \in P_S$ corresponds to a node in the fusion graph (prototype). Such graph is an approximation of the median graph[8] and it is defined by the 6-tuple $\bar{\mathbf{G}} = (\bar{V}, \bar{E}, \bar{\beta}, \gamma, \lambda, \mathcal{M})$, where:

- $\bar{V} = \{P_i \in P_S\}$ and $\bar{E} = \{< i,j >: \exists k \in P_i, l \in P_j \; |< k,l >\in E^{ij}\}$ where $E^{ij} = \{< k,l >: k \in P_i, l \in P_j, \theta_k = \theta_l, < \nu_k, \nu_l >\in E^{\theta_k} \equiv E^{\theta_l}\}$.
- $\bar{\beta} : \bar{V} \longrightarrow \mathbb{R}^n$ are the averaged attributes defined as $\bar{\gamma}_{P_i} = \sum_{k \in P_i} \pi_{\theta_k} \beta_{\nu_k}$, where $\pi_{\theta_k} : S \longrightarrow [0,1]$ indicate the probability that graph θ_k belongs to the class defined by prototype.
- $\gamma : \bar{V} \longrightarrow [0,1]$ is the probability density of node P_i in the prototype, and it is defined as $\gamma_{P_i} = \sum_{k \in P_i} \pi_{\theta_k}$. Such probabilities will be properly normalized so that the sum of probabilities of all nodes is unitary.
- $\lambda : \bar{E} \longrightarrow [0,1]$ are the edge weights defined as $\lambda(< i,j >) = \sum_{<k,l>\in E^{ij}} \pi_{\theta_k}$. Thus, such weights are defined by integrating the weights of the graphs to which the nodes implied in the connections belong.
- $\mathcal{M} : \bar{V} \times S \longrightarrow \cup_{i=1}^{|S|} V_i$ defines the correspondence of a vertex in the prototype and a graph with the matched vertex in the latter graph, that is $\mathcal{M}^{P_i A} = \nu_k, k \in P_i : \theta_k = A$. Having such matches we bypass the solving of a graph matching problem between each graph in S and each prototype.

As stated above, the probabilities π_{θ_k} that a graph belongs to a given prototype are here considered as external information comming from the EM algorithm (see next section) and we define the prototype as the mixture

$$\bar{\mathbf{G}} = \sum_{k=1}^{N} \pi_k \mathbf{G}_{\theta_k} = \pi_1 \mathbf{G}_{\theta_1} + \ldots + \pi_N \mathbf{G}_{\theta_N} \tag{3}$$

where $\pi_k \mathbf{G}_{\theta_k}$ denotes the weighting of each graph by its probability.

3 Graph-Transformation Matching and EM Clustering

3.1 One-to-One Matching

Given two images I_i and I_j, to be clustered, let $\mathcal{L}_i = \{\mathbf{s}_k\}$ and $\mathcal{L}_j = \{\mathbf{p}_l\}$ be their respective sets of salient points. Such salient points are obtained through a Bayesian optimization of the entropy-based Kadir and Brady dectector [17]. However, for matching purposes we consider their SIFT 128−length descriptors $\mathbf{D}$ and for each $\mathbf{s}_k$ we match it with $\mathbf{p}_l$ when $\mathbf{D}_{kl} = \arg\min_{\mathbf{p}_l \in \mathcal{L}_j} \{\|\mathbf{D}_k - \mathbf{D}_l\|\}$ and $\mathbf{D}_{kl}/\mathbf{D}_{kl(2)} \leq \tau$ being $\mathbf{D}_{kl(2)}$ the Euclidean distance to $\mathbf{s}_{l(2)}$ the second best match for $\mathbf{s}_k$, and $\tau \in [0,1]$ a distinctivity threshold usually set as $\tau = 0.8$. Consequently, we obtain a set of, say M matchings $\mathcal{M} = \{(k,l)\}$, and we denote by $\hat{\mathcal{L}}_i$ and $\hat{\mathcal{L}}_j$ the sets resulting from filtering, in the original ones, features without a matching in the $\mathcal{M}$ set.

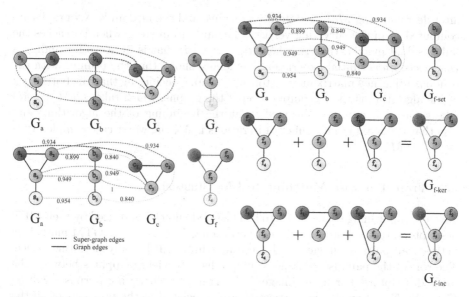

Fig. 2. Illustrating kernelized fusion. Left: Prototype building in an ideal case (top), and a real one where kernels are needed (bottom). Rigth: Step-by-step fusion showing partitions wrt median graph (left) and the difference between kernelized (middle) and incremental (bottom) fusion.

3.2 Iterative Filtering and Consensus Graph

Considering the two sets of M points $\hat{\mathcal{L}}_i \ni \mathbf{s}_k$ and $\in \hat{\mathcal{L}}_j \ni \mathbf{p}_l$, where $\mathbf{s}_k$ matches $\mathbf{s}_l$ we build their associated *median K-NN graphs* as follows. Graph $\mathbf{G}_i = (V_i, E_i)$ is given by vertices V_i associated to the positions of the M points. A non-directed edge $< k, a >$ exists in E_i when $\mathbf{s}_a$ is one of the $K = 4$ closest neighbors of $\mathbf{s}_k$ and also $\|\mathbf{s}_k - \mathbf{s}_a\| \leq \eta$, being $\eta = med_{<r,t>\in V_i \times V_i}\|\mathbf{s}_r - \mathbf{s}_t\|$ the median of all distances be tween pairs of vertices, which filters structural deformations due to outlying points. If there are not K vertices that support the structure of $\mathbf{s}_k$ then this vertex is disconnected completely. The graph $\mathbf{G}_i$, which is not necessarily connected, has the $M \times M$ adjacency matrix A_{ka} where $A_{ka} = 1$ when $< k, a >\in E_i$ and $A_{ka} = 0$ otherwise. Similarly, the graph $\mathbf{G}_j = (V_j, E_j)$ for points $\mathbf{p}_l$ has adjacency matrix B_{lb}, also of dimension $M \times M$ because of the one-to-one initial matching $\mathcal{M}$.

  Graph Transformational Matching (GTM) relies on the hypothesis that outlying matchings in $\mathcal{M}$ may be iteratively removed: (i) Select an outlying matching; (ii) Remove matched features corresponding to the outlying matching, as well as this matching itself; (iii) Recompute both *median K-NN graphs*. Structural disparity is approximated by computing the residual adjacency matrix $R_{ij} = |A_{ka} - B_{lb}|$ and selecting $j^{out} = \arg\max_{j=1...M}\sum_{i=1}^{M} R_{ij}$, that is, the one yielding the maximal number of different edges in both graphs. The selected structural outliers are the features forming the pair $(\mathbf{s}_{j^{out}}, \mathbf{p}_{j^{out}})$, that is, we remove matching (k, j^{out}) from $\mathcal{M}$, $\mathbf{s}_k$ from $\hat{\mathcal{L}}_i$, and $\mathbf{p}_{j^{out}}$ from $\hat{\mathcal{L}}_j$. Then, after decrementing M, a new iteration begins, and the median K-NN graphs are computed from the surviving vertices. The algorithm stops when it reaches the null residual matrix, that is, when $R_{ij} = 0, \forall i, j$, that is, it seeks for finding a *consensus graph* (initial experimental evidence shows that the pruning with the residual adjacency matrix may be too agressive). Considering that the bottleneck of the algorithm is the re-computation of the graphs, which takes $O(M^2 \log M)$ (the same as computing the median at the beginning of the algorithm) and also that the maximum number of iterations is M, the worst case complexity is $O(M^3 \log M)$.

3.3   From Pairwise Matching to EM Clustering

Given N input images $I_1, \ldots, I_N$ to be clustered and characterized by their SIFT descriptors, the first step consists of performing $N \times (N-1)/2$ GTM matchings between all pairs of images, and these matching will be only performed once. The role of the pairwise consensus subgraphs is to yield mappings between the SIFT descriptors. For input image I_i, its graph for clustering purposes will be $\mathbf{G}_i = (V_i, E_i, \beta_i)$ where the vertices V_i are associated to the positions of all the salient points in the image, the edges in E_i are derived from the median K-NN graph considering all salient points, and $\beta_i = \mathbf{D}_i$.

Given N input graphs $\mathbf{G}_i = (V_i, E_i, \beta_i)$, the goal of the Asymmetric Clustering Model (ACM) for graphs [15][16] is to find K (also unknown) graph prototypes $\bar{\mathbf{G}}_\alpha = (\bar{V}_\alpha, \bar{E}_\alpha, \bar{\beta}_\alpha, \gamma_\alpha, \lambda_\alpha, \mathcal{M}_\alpha)$ and the class-membership variables $I_{i\alpha} \in \{0, 1\}$ maximizing the cost function

$$L(\bar{\mathbf{G}}, I) = -\sum_{i=1}^{N} \sum_{\alpha=1}^{K} I_{i\alpha} F_{i\alpha}, \quad F_{i\alpha} = \sum_{k \in \bar{V}_\alpha} \|\bar{\beta}_{\alpha k} - \beta_{i \mathcal{M}_\alpha^{ki}}\| \tag{4}$$

Alternatively, $F_{i\alpha}$ may be defined in terms of the number of matchings, that is the number of vertices $k \in \bar{V}_\alpha$ satisfying $\|\bar{\beta}_{\alpha k} - \beta_{i \mathcal{M}_\alpha^{ki}}\| \leq \tau$ after GTM (the dimension of the consensus graph).

Initialization. For a fixed K, after a greedy process yielding initial prototypes and membership variables, the supergraph $\mathbf{G}_M = (V_M, E_M, \pi, \theta, \nu, \xi)$ (in which all graphs are mapped) is built. As stated above, this step, which implies a quadratic number of GTM processes, will be done only once.

E-step. Membership variables are updated following a deterministic annealing process (with temperature T) and depending on the disparities $F_{i\alpha}$ with respect to the prototypes ($N \times K$ evaluations *without* performing graph matching):

$$I_{i\alpha}^{t+1} = \frac{\rho_\alpha^t e^{-\frac{F_{i\alpha}}{T}}}{\sum_{\delta=1}^{K} \rho_\delta^t e^{-\frac{F_{i\delta}}{T}}}, \text{ being } \rho_\alpha^t = \frac{1}{N} \sum_{i=1}^{N} I_{i\alpha}^t, \tag{5}$$

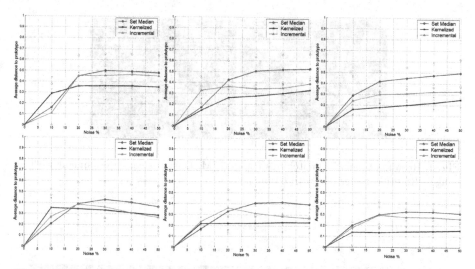

Fig. 3. Robustness with respect to increasing noise levels. Top: Edge noise. Bottom: Node noise. Left: 10% edge density. Center: 30% edge density. Right: 50% edge density.

M-step. After the E-step we have the new membership variables $I_{i\alpha}^{t+1}$ and it is time to update the K prototypes on the basis of graph mixtures whose weights rely on the current membership variables:

$$\bar{G}_\alpha^{t+1} = \sum_{i=1}^N \pi_{i\alpha} G_i \text{ , where } \pi_{i\alpha} = \frac{I_{i\alpha}^{t+1}}{\sum_{k=1}^N = I_{k\alpha}^{t+1}} , \tag{6}$$

Modifying weights $\pi_{i\alpha}$ implies changing the configuration (recompute partitions) of the associated fusion graph $G_{M\alpha} = (V_{M\alpha}, E_{M\alpha}, \pi_\alpha, \theta_\alpha, \nu_\alpha, \xi_\alpha)$ and hence changing the prototypes (but not the supergraph), and hence their attributes $\bar{\beta}_\alpha$. After such recomputation, we proceed to prune the prototypes by discarding vertices (edges) with $\gamma_\alpha < 0.5$ ($\lambda_\alpha < 0.5$) and also their attributes.

Fusion-step. For a variable K, the complete process is started with K_{max} classes and at the end of each EM epoch a statistical test determines whether the two closest prototypes may be fused or not. Then, we compute a fused prototype

$$\bar{G}_\gamma = \sum_{i=1}^N \pi_{i\gamma} G_i \text{ when } h_\gamma < (h_\alpha + h_\beta)\mu \tag{7}$$

being $h_\alpha = \sum_{i=1}^N F_{i\alpha} \pi_{\alpha i}$ the heterogeneity of a class, and $\mu \in [0, 1]$ a merge factor usually set to 0.6.

4   Experimental Results and Discussion

Experiment 1. We have performed two kind of experiments: random graphs, and realistic visual localization. In the first case (see Fig. 3) we have evaluated

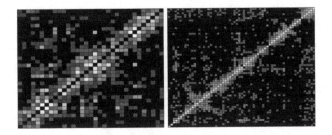

Fig. 4. Results of pairwise matchings using GTM. Left: A small environment with 34 images. Right: A larger one with 64 images.

Fig. 5. Graph prototypes and classified images. Each column shows the prototype and sample images of the corresponding class.

how representative is a prototype by measuring the average distance of the graphs in the class to that prototype in different situations. Compared to the set median graph and the results of our previous incremental method, the new method yields more representative (informative) prototypes: it yields a slower rate of increase of distances with the prototype as the noise level increases.

Experiment 2. Realistic visual localization experiments where performed by considering two types of indoor environments: a small one ($N = 34$ images), and a larger one ($N = 64$ images). Initial pairwise matchings (confusion matrices) are showed in Fig. 4. The obtained prototypes for the first environment ($K = 6$ classes) are showed in Fig. 5, and the likelihood (expressed as the number of matched nodes with the prototype) of each image with respect to each prototype

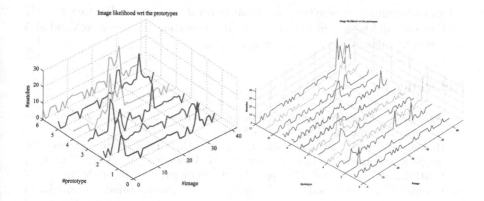

Fig. 6. Likelihoods for both environments. Left: Small environment. Right: Larger (medium-size) one.

are plotted in Fig. 6-left. Some bimodalities (due to geometric ambiguities) arise, but in general it is possible to find, in this case, a simple threshold (above ≈ 5 matches) to report membership. In addition, we observe that the overlap between classes is minimal. We have also estimated the localization error both for the early version, which does not use graph clustering, and the new one proposed in this paper. We have found that the percentage of error with respect to the ideal localization is 86.4% in the early version, but 58.13% in the new one. This indicates an improvement of the localization quality besides the computational savings derived from comparing only with images in the same cluster for fine localization.

Experiment 3. The good results outlined above encouraged us to find the limit of scalability of the approach when the number of images to cluster increases significantly. In this experiment we have tested the method in a larger environment ($N = 64$ images) where our algorithm has unsupervisedly found $K = 12$ classes. Analyzing the pairwise matching matrix (see Fig. 5-right) it has a consistent diagonal with medium-size clusters. On the other hand, the analysis of the likelihoods (Fig. 6-right) reveals few multi-modal classes, and none of them has a unique member. With respect to the localization error, clustering yields a 12.22% less than our early version.

5   Conclusions and Future Works

We have presented a novel method for unsupervised central graph clustering and we have successfully tested it in the context of scene recognition (visual localization). We have found a good generalization conditioning which in turn yields useful structural indexing in coarse-to-fine visual localization provided that the number of ambiguous images does not grow significantly, specially in indoor environments where there are many natural symmetries. We are currently working on building a wearable device for incorporating these elements and also in testing the algorithm in other environments.

References

1. Aguilar, W.: Object recognition based on the structural correspondence of local features. MsThesis, UNAM, México (2006)
2. Bunke, H., Foggia, P., Guiobaldi, C., Vento, M.: Graph clustering using the weighted minimum common supergraph. In: Hancock, E.R., Vento, M. (eds.) GbRPR 2003. LNCS, vol. 2726, pp. 235–246. Springer, Heidelberg (2003)
3. Crandall, D., Felzenszwalb, P., Huttenlocher, D.: Spatial priors for part-based recognition using statistical models. In: Proc. Intl. Conf. on Computer Vision and Pattern Recognition, San Diego, CA, pp. 10–17 (2005)
4. Chung, F.R.K.: Spectral graph theory. In: Conference Board of Mathematical Science CBMS, American Matematical Society, Providence, RI, vol. 92 (1997)
5. Fei-Fei, L., Fergus, R., Perona, P.: One-shot learning of object categories. IEEE Trans. Pattern Anal. Mach. Intell. 28(4), 594–611 (2006)
6. Felzenszwalb, P., Huttenlocher, D.: Pictorial structures for object recognition. International Journal on Computer Vision 61(1), 57–59 (2005)
7. Jain, B., Wysotzki, F.: Central clustering for attributed graphs. Machine Learning 56, 169–207 (2004)
8. Jiang, X., Münger, A., Bunke, H.: On median graphs: properties, algorithms, and applications. IEEE Trans. Pattern Anal. Mach. Intell. 23(10), 1144–1151 (2001)
9. Kadir, T., Brady, M.: Saliency, scale and image description. International Journal of Computer Vision 45(2), 83–105 (2001)
10. Kokkinos, I., Maragos, P., Yuille, A.L.: Bottom-up & top-down object detection using primal sketch features and graphical models. In: Proc. Intl. Conf. on Computer Vision and Pattern Recognition, New York, NY, pp. 1893–1900 (2006)
11. Kondor, R., Lafferty, J.: Diffusion kernels on graphs and other discrete input spaces. In: Proc. Intl. Conf. on Machine Learning, Los Altos CA, pp. 315–322 (2002)
12. Leung, T.K., Burl, M.C., Perona, P.: Finding faces in cluttered scenes using random labeled graph matching. In: Proc. IEEE Intl. Conf. on Computer Vision, Cambridge MA, pp. 637–644 (1995)
13. Lowe, D.G.: Distinctive image features from scale-invariant keypoints. International Journal of Computer Vision 60(2), 91–110 (2004)
14. Lozano, M.A., Escolano, F.: ACM attributed graph clustering for learning classes of images. In: Hancock, E.R., Vento, M. (eds.) GbRPR 2003. LNCS, vol. 2726, pp. 247–258. Springer, Heidelberg (2003)
15. Lozano, M.A., Escolano, F.: A significant improvement of softassign with diffusion kernels. In: Fred, A., Caelli, T.M., Duin, R.P.W., Campilho, A., de Ridder, D. (eds.) Structural, Syntactic, and Statistical Pattern Recognition. LNCS, vol. 3138, pp. 76–84. Springer, Heidelberg (2004)
16. Lozano, M.A., Escolano, F.: Protein classification by matching and clustering surface graphs. Pattern Recognition 39(4), 539–551 (2006)
17. Suau, P., Escolano, F.: Bayesian optimization of the Kadir saliency filter. Submitted to Image and Vision Computing (2006)

On the Relation Between the Median and the Maximum Common Subgraph of a Set of Graphs

Miquel Ferrer[1], Francesc Serratosa[2], and Ernest Valveny[1]

[1] Computer Vision Center, Dep. Ciències de la Computació
Universitat Autònoma de Barcelona, Bellaterra, Spain
{mferrer,ernest}@cvc.uab.es
[2] Departament d'Enginyeria Informàtica i Matemàtiques
Universitat Rovira i Virgili, Tarragona, Spain
francesc.serratosa@urv.cat

Abstract. Given a set of elements, the median can be a useful concept to get a representative that captures the global information of the set. In the domain of structural pattern recognition, the median of a set of graphs has also been defined and some properties have been derived. In addition, the maximum common subgraph of a set of graphs is a well known concept that has various applications in pattern recognition. The computation of both the median and the maximum common subgraph are highly complex tasks. Therefore, for practical reasons, some strategies are used to reduce the search space and obtain approximate solutions for the median graph. The bounds on the sum of distances of the median graph to all the graphs in the set turns out to be useful in the definition of such strategies. In this paper, we reduce the upper bound of the sum of distances of the median graph and we relate it to the maximum common subgraph.

1 Introduction

A fundamental problem in pattern recognition is the selection of suitable representations for objects and classes. In the structural approach to pattern recognition, an object can be represented using graphs. Nevertheless, the main drawback of representing the data and prototypes by graphs is the computational complexity of comparing two graphs. The time required by any of the optimal algorithms may in the worst case become exponential in the size of the graphs. The approximate algorithms, on the other hand, have only polynomial time complexity, but do not guarantee to find the optimal solution. Moreover, in some applications, the classes of objects are represented explicitly by a set of prototypes, which means that a huge amount of model graphs must be matched with the input graph and so the conventional error-tolerant graph matching algorithms must be applied to each model-input pair sequentially. As a consequence, the total computational cost is linearly dependent on the size of the database of model

---

[1] This work was sponsored research Fellowship number 401-027 (UAB) / Cicyt TIN2006-15694-C02-02 (Ministerio Ciencia y Tecnología).

F. Escolano and M. Vento (Eds.): GbRPR 2007, LNCS 4538, pp. 351–360, 2007.

graphs and exponential (or polynomial in subgraph methods) with the size of the graphs. For applications dealing with large databases, this may be prohibitive.

To alleviate these problems, some attempts have been made to try to reduce the computational time of matching the unknown input patterns to the whole set of models from the database. Assuming that the graphs that represent a set or class are not completely dissimilar in the database, only one structural model is used to represent the set, and thus, only one comparison is needed for each class. While in the domain of statistical pattern recognition it is easy to define the representative of a set of objects, it is not so clear how to define the representative of a set of graphs in the structural domain. Thus, there are some different methodologies to represent the set in the literature. In the probabilistic methods, the clusters are described in the most general case through a joint probability space of random variables ranging over graph vertices and arcs. They represent the graphs in the cluster, according to some synthesis process, together with its associated probability distribution [1,2,3]. In the non-probabilistic methods, sets are usually represented by attributed graphs. The set might be represented by a network of graphs [4] or by only one graph. In this last case, a common choice is the median graph. Given a set of graphs, the median is defined as the graph that has the smallest sum of the distances to all graphs in the set [5].

The computation of the median graph is exponential in the number and size of the input graphs. As a consequence, in order to make the practical use of the median-graph concept possible, we have to resort to approximate solutions. In [5], a genetic algorithm is used to synthesize good approximations of the median graph. Nevertheless, it was crucial to deduct the bounds of the sum of distances between the median graph and the graphs of the set to achieve a good solution. To that aim, they show that the sum of the distances between the median graph and the graphs of the set was lower or equal than the sum of the number of nodes of all the graphs.

In this paper, using a particular cost function and the relation between the edit distance and the maximum common subgraph, both introduced in [6], we reduce the upper bound of the sum of distances of the median graph and we relate it to the maximum common subgraph of a set of graphs. This reduction may lead to an increase in the efficiency of both exact and approximate algorithms for the computation of the median graph.

The rest of the paper will be as follows. In section 2 we introduce the basic terminology used in the paper. In section 3, we present the previous results that are the basis of our work. Section 4 contains the main contribution of this paper and a practical example to verify the theoretical results. Finally, some discussions conclude the paper.

2   Definitions

2.1   Basic Definitions

Let L be a finite alphabet of labels for nodes and edges.

Definition 1. *A graph is a triple* $g = (V, \alpha, \beta)$ *where,* V *is the finite set of nodes,* α *is the node labeling function (* $\alpha : V \longrightarrow L$ *), and* β *is the edge labeling function (* $\beta : V \times V \longrightarrow L$ *).*

We assume that our graphs are fully connected, i.e., $E = V \times V$. Consequently, the set of *edges* is implicitly given. Such assumption is only for notational convenience, and it doesn't impose any restriction in the generality of our results. In the case where no edge exists between two given nodes, we can include the special label *null* in the set of labels L. The number of nodes of a graph g is denoted by $|g|$.

Definition 2. *Given two graphs* $g = (V, \alpha, \beta)$*, and* $g' = (V', \alpha', \beta')$*,* g' *is a subgraph of* g*, denoted by* $g' \subseteq g$ *if,*

- $V' \subseteq V$
- $\alpha'(x) = \alpha(x)$ *for all* $x \in V'$
- $\beta'((x,y)) = \beta((x,y))$ *for all* $(x,y) \in V' \times V'$

From definition 2 it follows that, given a graph $g = (V, \alpha, \beta)$, a subset $V' \subseteq V$ of its vertices uniquely defines a subgraph. Such subgraph is called the subgraph *induced* by V'.

Definition 3. *Given two graphs* $g_1 = (V_1, \alpha_1, \beta_1)$*, and* $g_2 = (V_2, \alpha_2, \beta_2)$*, a graph isomorphism between* g_1 *and* g_2 *is a bijective mapping* $f : V_1 \longrightarrow V_2$ *such that,*

- $\alpha_1(x) = \alpha_2(f(x))$ *for all* $x \in V_1$
- $\beta_1((x,y)) = \beta_2((f(x), f(y)))$ *for all* $(x,y) \in V_1 \times V_1$

In the real world, when encoding objects into graph-based representations some degree of distortion may be introduced due to multiple reasons. Hence, graph representations of two identical objects may not have an exact match. Therefore, it is necessary to introduce some degree of error tolerance into the matching process. Hence, we need an algorithm for error-correcting graph matching [7] or equivalently, a method to compute a similarity measure between two given graphs.

Definition 4. *Let* $g_1 = (V_1, \alpha_1, \beta_1)$ *and* $g_2 = (V_2, \alpha_2, \beta_2)$ *be two graphs. An* error-correcting *graph matching (ecgm) from* g_1 *to* g_2 *is a bijective function* $f : \hat{V}_1 \longrightarrow \hat{V}_2$*, where* $\hat{V}_1 \subseteq V_1$ *and* $\hat{V}_2 \subseteq V_2$*.*

We say that node $x \in \hat{V}_1$ is substituted by node $y \in \hat{V}_2$ if $f(x) = y$. If $\alpha_1(x) = \alpha_2(f(x))$ then the substitution is called identical. Otherwise it is called non-identical. In addition, any node from $V_1 - \hat{V}_1$ is deleted from g_1 and any node from $V_2 - \hat{V}_2$ is inserted in g_2 under f.

As described above, the mapping f directly implies an *edit operation* on each node in g_1 and g_2, i.e. nodes are substituted, inserted or deleted. Indirectly, the mapping f implies the same edit operations on the edges of g_1 and g_2. Thus,

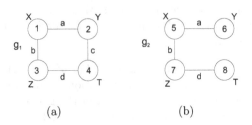

Fig. 1. Two graphs g_1(a) and g_2 (b)

if $f(x_1) = y_1$ and $f(x_2) = y_2$, then the edge (x_1, x_2) is substituted by edge $(f(x_1), f(x_2)) = (y_1, y_2)$. In addition, if a node x is deleted in g_1 then any edge incident to x is also deleted. Similarly, if a node x' is inserted in g_2, then any edge incident to x' is also inserted. In this way, any *ecgm* f can be seen as a sequence of edit operations namely, substitution, insertion and deletion of both nodes and edges, that transform a given graph g_1 into another given graph g_2.

A possible *ecgm* from g_1 to g_2 of figure 1 is $f : 1 \longrightarrow 5, 2 \longrightarrow 6, 3 \longrightarrow 7$. Under this *ecgm* nodes 1, 2 and $3 \in V_1$ are substituted by nodes 5, 6 and 7 $\in V_2$ respectively. In consequence, edges $(1, 2)$ and $(1, 3) \in E_1$ are substituted by edges $(5, 6)$ and $(5, 7) \in E_2$. Note that in all of these substitutions no changes in node or edge labels are involved. In addition, under f, node 4 and edges $(2, 4)$ and $(3, 4)$ are deleted, while node 8 and edge $(7, 8)$ is inserted. There are, of course other *ecgm* from g_1 to g_2.

Definition 5. *The cost of an ecgm* $f : \hat{V}_1 \longrightarrow \hat{V}_2$ *from a graph* $g_1 = (V_1, \alpha_1, \beta_1)$ *to a graph* $g_2 = (V_2, \alpha_2, \beta_2)$, *denoted by* $c(f)$, *is the sum of the costs of insertion, deletion and substitution of both nodes and edges. These costs are represented by* $c_{ni}(x), c_{nd}(x), c_{ns}(x), c_{ei}(e), c_{ed}(e), c_{es}(e)$ *respectively.*

All costs are real non-negative numbers and are used to model the probability of errors and distortions that may change the original model. Usually, the higher the probability of a distortion is to occur, the lower is its cost. Normally, it is assumed that the cost of an identical node/edge substitution is zero, while the cost of any other edit operation is greater than zero. The set of all costs is the *cost function* γ and is usually written in a tuple form, i.e. $\gamma = \{c_{ni}, c_{nd}, c_{ns}, c_{ei}, c_{ed}, c_{es}\}$. If the cost function γ is explicitly given the notation $c_\gamma(f)$ for the *ecgm* is used instead of $c(f)$.

Definition 6. *Given a cost function* γ, *and an ecgm* f *from* g_1 *to* g_2, f *is called an optimal ecgm under* γ *if there is no other ecgm* f' *from* g_1 *to* g_2 *such that* $c_\gamma(f') < c_\gamma(f)$. *The cost of an optimal ecgm,* $c_\gamma(f)$ *is also called the* edit distance *between* g_1 *and* g_2 *denoted by* $d(g_1, g_2)$, *and it can be seen as the sequence of graph edit operations that transforms* g_1 *into* g_2 *with the minimum cost.*

$$d(g_1, g_2) = min(c_\gamma(f)) \tag{1}$$

Notice that for a given cost function γ there are usually more than one optimal ecgm from a graph g_1 to another graph g_2.

2.2   Maximum Common Subgraph

Definition 7. *Let $g_1 = (V_1, \alpha_1, \beta_1)$ and $g_2 = (V_2, \alpha_2, \beta_2)$ be two graphs, and $g'_1 \subseteq g_1$, $g'_2 \subseteq g_2$. If there exists a graph isomorphism between g'_1 and g'_2 then, both g'_1 and g'_2 are called a* common subgraph *of g_1 and g_2.*

Definition 8. *Let $g_1 = (V_1, \alpha_1, \beta_1)$ and $g_2 = (V_2, \alpha_2, \beta_2)$ be two graphs. A graph g_M is called a* maximum common subgraph *(MCS) of g_1 and g_2 if g_M is a common subgraph of g_1 and g_2 and there is no other common subgraph of both g_1 and g_2 having more nodes than g_M.*

2.3   Generalized Median Graph

Definition 9. *Let U be the set of graphs that can be constructed using labels from L. Given $S = \{g_1, g_2, ..., g_n\} \subseteq U$, the* generalized median graph *$\bar{g}$ of S is defined as follows:*

$$\bar{g} = arg \left(\min_{g \in U} \sum_{g_i \in S} d(g, g_i) \right) \tag{2}$$

In other words, the generalized median graph is a graph $g \in U$ which minimizes the sum of distances (SOD) from g to all the graphs in S.

3   Interesting Results Based on the Previous Definitions

In this section we introduce three properties derived from the definitions given above which will be the basis to develop our hypothesis. We first show a particular cost function introduced in [6]. Then, we remind an interesting relation between the edit distance and the maximum common subgraph also given in [6] derived from the introduced cost function. Finally, the bounds for the median graph based on the sum of distances given in [5] are shown.

3.1   A Particular Cost Function

From this point to the rest of this paper we will use a particular cost function given in [6] where the cost of node deletion and insertion ($c_{nd}(x)$ and $c_{ni}(x)$) is always 1, the cost of edge deletion and insertion ($c_{ed}(e)$ and $c_{ei}(e)$) is always 0 and the cost of node and edge substitution ($c_{ns}(x)$ and $c_{es}(e)$) takes the values 0 or ∞ depending on whether the substitution is identical or not, respectively.

3.2 Relation Between Edit Distance and MCS

In [6] it has been proven that, using the previous cost function, the edit distance between two graphs is related to their MCS, g_M, by means of this expression:

$$d(g_1, g_2) = |g_1| + |g_2| - 2 |g_M| \tag{3}$$

3.3 Bounds for the Median Graph

Let U be the set of graphs that can be constructed using labels from L and $S = \{g_1, g_2, ..., g_n\} \in U$. In [5] it is shown that the empty graph $\bar{g}_e$ and the union graph $\bar{g}_u$ are meaningful candidates for the median graph. In this situation, for the true median graph $\bar{g}$, $SOD(\bar{g}) \leq min\{SOD(\bar{g}_e), SOD(\bar{g}_u)\}$ holds. In addition, for any partition $\wp = ((g_{l1}, g_{l2}) \cdots (g_{ln-1}, g_{ln})) \in S$ and for the sum of distances between its elements, $SOD(\wp) = d(g_{l1}, g_{l2}) + d(g_{l3}, g_{l4}) + \cdots + d(g_{ln-1}, g_{ln})$, the following inequality holds, $max\{SOD(\wp)\} \leq SOD(\bar{g})$. Thus, the bounds for the true median graph related to the sum of distances are,

$$max\{SOD(\wp)\} \leq SOD(\bar{g}) \leq min\{SOD(\bar{g}_e), SOD(\bar{g}_u)\} \tag{4}$$

4 Reducing the Upper Bound for the Median Graph

In this section we prove that it is possible to reduce the upper bound for the median graph given in expression (4), using the concept of *maximum common subgraph of a set of graphs*, the cost function introduced in the previous section and expression (3).

Definition 10. *Let* $S = \{g_1, g_2, ..., g_n\}$ *be a set of graphs. A graph* g_{M_S} *is called a* maximum common subgraph *of* S *if* g_{M_S} *is a common subgraph of* $\{g_1, g_2, \cdots, g_n\}$ *and there is no other common subgraph of* $\{g_1, g_2, \cdots, g_n\}$ *having more nodes than* g_{M_S}.

As an example, if we take the set of graphs $S = \{g_1, g_2, g_3, g_4\}$ of figure 3(a), then a possible common subgraph of S is shown in figure 2.

Fig. 2. g_{M_S} corresponding to the set $S = \{g_1, g_2, g_3, g_4\}$ in the figure 3(a)

The following theorem relates the upper bound of $SOD(\bar{g})$ to g_{M_S}:

Theorem 1. *Given the cost function presented in section 3, the* $SOD(\bar{g})$ *falls in the limits*

$$max\{SOD(\wp)\} \leq SOD(\bar{g}) \leq SOD(g_{M_S}) \leq min\{SOD(\bar{g}_e), SOD(\bar{g}_u)\} \tag{5}$$

Proof. First, we start by computing the term $min\{SOD(\bar{g}_e), SOD(\bar{g}_u)\}$. Using the definition of distance given in expression (3):

$$SOD(\bar{g}_e) = \sum_{i=1}^{n} d(g_i, \bar{g}_e) = \sum_{i=1}^{n} |g_i| + |\bar{g}_e| - 2|\bar{g}_e| = \sum_{i=1}^{n} |g_i|$$

Notice that, in this expression $\bar{g}_e$ is the empty graph. Then, the MCS between any graph g_i and $\bar{g}_e$ in expression (3) is $\bar{g}_e$, and $|\bar{g}_e| = 0$. A similar reasoning can be done for $SOD(\bar{g}_u)$. In this case, the MCS between any graph g_i and $\bar{g}_u$ is g_i, and $|\bar{g}_u| = \sum_{i=1}^{n} |g_i|$. Therefore,

$$SOD(\bar{g}_u) = \sum_{i=1}^{n} d(g_i, \bar{g}_u) = \sum_{i=1}^{n} |g_i| + |\bar{g}_u| - 2|g_i| = (n-1)\sum_{i=1}^{n} |g_i|$$

Thus, for $n \geq 2$

$$min\{SOD(\bar{g}_e), SOD(\bar{g}_u)\} = min\left\{\sum_{i=1}^{n} |g_i|, (n-1)\sum_{i=1}^{n} |g_i|\right\} = \sum_{i=1}^{n} |g_i|$$

Now we derive an expression for the term $SOD(g_{Ms})$. If g_{Ms} is the maximum common subgraph of S, then any g_i will have precisely g_{Ms} as a maximum common subgraph between itself and g_{Ms}. Therefore,

$$SOD(g_{Ms}) = \sum_{i=1}^{n} d(g_i, g_{Ms}) = \sum_{i=1}^{n} |g_i| + |g_{Ms}| - 2|g_{Ms}| = \sum_{i=1}^{n} |g_i| - n|g_{Ms}|$$

$$(6)$$

Thus, we have that $SOD(g_{Ms}) \leq min\{SOD(\bar{g}_e), SOD(\bar{g}_u)\} = \sum_{i=1}^{n} |g_i|$. In addition, by the definition of median graph, the inequality $SOD(\bar{g}) \leq SOD(g_{Ms})$ holds. Consequently, equation 5 holds. □

This result shows that the upper bound for the term $SOD(\bar{g})$ can be reduced from $\sum_{i=1}^{n} |g_i|$ to $\sum_{i=1}^{n} |g_i| - n|g_{Ms}|$. This reduction can be used to introduce some improvements in the existing algorithms to compute the true median graph. For instance, in the exact algorithm presented in [8], the configurations where the SOD is greater than the $SOD(g_{Ms})$ do not have to be explored, reducing in this way the search space and the computational complexity. In [5], the same authors present an approximate algorithm for the median graph based on genetic search. In this case, the reduction in the upper bound for the median graph could be used to introduce some improvements in the initialization and mutation of the population of chromosomes, discarding such chromosomes with SOD greater than some factor of $SOD(g_{Ms})$.

4.1   Practical Example

In this section we present a detailed example in order to show, in a more intuitive way, the implications of theorem 1. Consider the situation where $S =$

$\{g_1, g_2, ..., g_n\}$. In this framework, basically 4 situations can appear regarding to the maximum common subgraph of S:

1. At least one maximum common subgraph of all graphs in S exists: $g_{Ms} \neq \varnothing$
   (a) All pairs of graphs in any possible partition share only this maximum common subgraph: $g_{Mij} = g_{Ms} \; \forall$ pair $(g_i, g_j) \; i, j = 1..n; i \neq j$.
   (b) Some pairs of graphs in any partition share more than this maximum common subgraph: $\exists$ a pair $(g_i, g_j) \; i, j = 1..n; i \neq j$ such that $g_{Mij} > g_{Ms}$.
2. No maximum common subgraph between all graphs in S exists: $g_{Ms} = \varnothing$
   (a) All pairs in any partition are disjoint: $g_{Mij} = \varnothing \; \forall$ pair $(g_i, g_j) \; i, j = 1..n; i \neq j$.
   (b) Some pairs in any partition share some subgraph of them: $\exists$ a pair (g_i, g_j) $i, j = 1..n; i \neq j$ such that $g_{Mij} \neq \varnothing$.

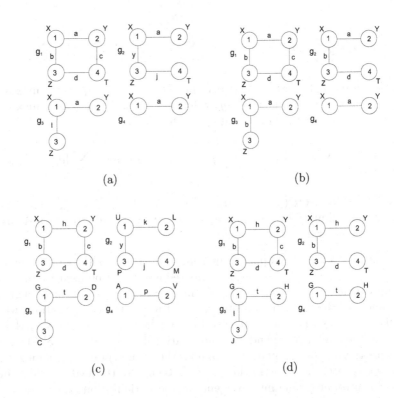

Fig. 3. A set $S = \{g_1, g_2, g_3, g_4\}$ of graphs and the 4 possible situations 1a (a), 1b (b), 2a (c) and 2b (d) in terms of the maximum common subgraph of S

Figure 3 shows an example of such situations for $S = \{g_1, g_2, g_3, g_4\}$. Figures 3(a) and 3(b) correspond to situations, 1.a and 1.b respectively, i.e. a *maximum common subgraph* of S exists. While in figure 3(a) only the maximum common

Table 1. Results for $\bar{g}$, g_{M_S}, $SOD(\bar{g})$, $SOD(g_{M_S})$ and $\sum_{i=1}^{n}|g_i|$ for the situations of figure 3

Situation	$\bar{g}$	g_{M_S}	$SOD(\bar{g})$	$SOD(g_{M_S})$	$\sum_{i=1}^{n}|g_i|$
1a	X(1) —a— Y(2)	X(1) —a— Y(2)	5	5	13
1b	X(1) —a— Y(2), b, Z(3)	X(1) —a— Y(2)	3	5	13
2a	∅	∅	13	13	13
2b	X(1), b, Z(3) —d— T(4), G(1) —t— H(2)	∅	13	13	13

subgraph equal to g_{M_S} exists between any pair of graphs, in figure 3(b) some graphs have a maximum common subgraph greater than g_{M_S}. Figures 3(c) and 3(d) correspond to situations 2.a and 2.b respectively. In these situations, no common subgraph exists among all the graphs in S. While in figure 3(c) there is no maximum common subgraph between any pair of graphs in S, in figure 3(d) some pairs of graphs have a maximum common subgraph.

For each situation in figure 3, the true median $\bar{g}$ and the maximum common subgraph g_{M_S} of S were manually obtained. Then the sum of distances for $\bar{g}$ and g_{M_S}, $SOD(\bar{g})$ and $SOD(g_{M_S})$, respectively, were calculated. Finally, the term $\sum_{i=1}^{n}|g_i|$ was also computed. The results for these 5 features are summarized in table 1. Each row corresponds to one of the four situations in figure 3.

As can be seen in table 1, the equation 5 holds in all cases. In particular, in the situations 1.a and 1.b, i.e. when the maximum common subgraph among all graphs in S exists, the upper bound fixed by the term $SOD(g_{M_S})$ is less than the upper bound fixed by the expression $\sum_{i=1}^{n}|g_i|$. Another interesting fact from table 1 is the relation between the terms $SOD(\bar{g})$ and $SOD(g_{M_S})$. Concretely, in the situations 1.a and 2.a, i.e. when any pair of graphs have the same maximum common subgraph, the true median and the maximum common subgraph among all graphs in S are the same, and consequently, the terms $SOD(\bar{g})$ and $SOD(g_{M_S})$ coincide. However, in the situations 1.b and 2.b, i.e. when some pair of graphs have a maximum common subgraph greater than the maximum common subgraph among all graphs in S, the true median is (or may be in general) greater or equal than the maximum common subgraph of all graphs and then the term $SOD(\bar{g})$ is lower or equal than $SOD(g_{M_S})$.

5   Conclusions

The median graph concept as an alternative to represent prototypes of a set of graphs has been turned out very useful, but the computation of both exact and approximate solutions has been shown very hard.

In this paper we show that under a certain cost function, the upper bound for the median graph related to the sum of distances can be reduced using the concept of maximum common subgraph. This result is not only interesting from the theoretical point of view. In order to prove the usefulness of this result, a detailed example has been presented. The results show that, in some cases, when the maximum common subgraph of all graphs in the set exists, the term $SOD(g_{M_S})$ is less than $\sum_{i=1}^{n} |g_i|$, and then some reduction in the space where the median is searched for can be introduced. As a consequence the time spent for the computation of the median graph could be reduced. For example, this fact could be used to introduce some improvements in the computation of both exact [8] and approximate solutions [5] for the median graph. Finally, under more restrictive conditions about the MCS of any pair of graphs, the maximum common subgraph of all graphs in the set is equal to the median graph. In this sense, it should be investigated in further detail if the maximum common subgraph can be a good approximation of the median graph in more general situations.

References

1. Wong, A., You, M.: Entropy and distance of random graphs with application to structural pattern recognition. T-PAMI 7, 599–609 (1985)
2. Serratosa, F., Alquézar, R., Sanfeliu, A.: Function-described graphs for modelling objects represented by sets of attributed graphs. Pattern Recognition 36(3), 781–798 (2003)
3. Sanfeliu, A., Serratosa, F., Alquézar, R.: Second-order random graphs for modeling sets of attributed graphs and their application to object learning and recognition. IJPRAI 18(3), 375–396 (2004)
4. Messmer, B.T., Bunke, H.: A new algorithm for error-tolerant subgraph isomorphism detection. IEEE Trans. Pattern Anal. Mach. Intell. 20(5), 493–504 (1998)
5. Jiang, X., Münger, A., Bunke, H.: On median graphs: Properties, algorithms, and applications. IEEE Trans. Pattern Anal. Mach. Intell. 23(10), 1144–1151 (2001)
6. Bunke, H.: On a relation between graph edit distance and maximum common subgraph. Pattern Recognition Letters 18(8), 689–694 (1997)
7. Bunke, H., Allerman, G.: Inexact graph matching for structural pattern recognition. Pattern Recognition Letters 1(4), 245–253 (1983)
8. Bunke, H., Münger, A., Jiang, X.: Combinatorial search versus genetic algorithms: A case study based on the generalized median graph problem. Pattern Recognition Letters 20(11-13), 1271–1277 (1999)

A Graph Classification Approach Using a Multi-objective Genetic Algorithm
Application to Symbol Recognition

Romain Raveaux[1], Barbu Eugen[2], Hervé Locteau[2], Sébastien Adam[2], Pierre Héroux[2], and Eric Trupin[2]

[1] L3I Laboratory – University of La Rochelle, France
[2] LITIS Labs – University of Rouen, France
Romain.Raveaux01@univ-lr.fr
{Barbu.Eugen,Herve.Locteau,Sebastien.Adam,Pierre.Heroux,
Eric.Trupin}@univ-rouen.fr

Abstract. In this paper, a graph classification approach based on a multi-objective genetic algorithm is presented. The method consists in the learning of sets composed of synthetic graph prototypes which are used for a classification step. These learning graphs are generated by simultaneously maximizing the recognition rate while minimizing the confusion rate. Using such an approach the algorithm provides a range of solutions, the couples (confusion, recognition) which suit to the needs of the system. Experiments are performed on real data sets, representing 10 symbols. These tests demonstrate the interest to produce prototypes instead of finding representatives which simply belong to the data set.

Keywords: graph classification, multi-objective optimization, machine learning, graph dissimilarity measure.

1 Introduction

Graphs are powerful tools to represents structured objects and they have been applied in many fields of computer science. Graphs unify in a single formalism, web pages [1], molecules [2] and graphic symbols [3] since their vertices represent object components while edges represent relations between components. Symbols can be naturally described in a graph model using primitives (vectors, arcs, connected components, loops...) and geometric relations between these primitives (neighborhood, connection, parallelism...). In this context, the pre-segmented symbol recognition question turns into a graph classification problem which involves comparing graphs, i.e., matching graphs to identify their common features [4]. Only error tolerant matching methods can be efficient due to the noise and the shape variability present in graphic documents. The identification phase is to assign a graph describing an unknown symbol to its class using a learning database.

In this paper, a system able to classify graphs representing symbols is described. It uses a learning database to take into account the variability which can occur in

F. Escolano and M. Vento (Eds.): GbRPR 2007, LNCS 4538, pp. 361–370, 2007.

symbol image representation. Our algorithm aims at learning sets of graph prototypes to consider the possible distortions. Then, these prototypes are used in a classification step in order to determine the class of an unknown and noisy symbol in a recognition system. Our approach can be decomposed into 3 steps. First, a corpus of noisy symbol images [5], representing N symbol classes with M distorted symbol images per class, is used to extract a set of M graphs per class. Then, from this learning set, a graph based Genetic Algorithm (GA) is applied. Its aim is to generate sets of K graph prototypes for each class. The values to be optimized by the multi-objective GA are the recognition rate and the confusion rate which are obtained in the simulation of a classification algorithm using the selected prototypes as learning samples and a test database. Both steps (prototypes learning and classification) use a dissimilarity measure called graph probing in order to evaluate the similarity between graphs [20]. This measure has been chosen after a comparative study between different approaches. Finally, in a validation step, a classification algorithm is applied using the selected prototype set as learning elements, a validation database and the same dissimilarity measure. The paper is organized as follows: In the second section, the graph probing concept is introduced. Then, the third section presents the genetic algorithm in use, and particularly the specific genetic operators involved. The fourth section presents the application to the symbol recognition problem, the comparative study between the tested dissimilarity measures and the obtained classification results. Finally, a conclusion is given and future works are brought in section 5.

2 Dissimilarity Measures

Measures of dissimilarity between complex objects which have a structure (sets, lists, strings, ...) are based on the quantity of shared terms. The simplest similarity measure between two objects is the matching coefficient, which is based on the number of common terms. Using this idea as a starting point, dissimilarity measures which take into account the maximal common subgraph (MCS) of two graphs were proposed in [6]. Another method which proposes a metric distance in the universal set of graphs is the edit distance. It represents the minimum-cost sequence of basic editing operations (e.g. insertion or deletion of vertices and edges with associated costs). The graph edit distance and MCS computation are equivalent to each other under a certain cost function associated to edit operations [7]. These distances between graphs have worst case exponential running times. In our application, we use a genetic algorithm which employs intensively computations of dissimilarities between graphs.

Hence, we have to find faster algorithms which compute dissimilarities, eventually approximations. In such a context, the graph topology can be approximated considering independently the set of vertices and arcs, for instance, edge matching distance or vertex matching distance [8]. Edge matching distance proposes a cost function for the matching of edges and then derives a minimal weight maximal matching between the edge sets of two graphs. This matching has a worst case complexity of $O(n^3)$, where n is the number of edges of the largest graph.

Another possibility to define a similarity measure is to count the number of occurrences of a set of sub graphs (named fingerprints or probes in different contexts) from each graph and to describe the objects to be compared as vectors [9]. In this setting (named graph probing), the similarity between graphs is the similarity between the two associated vectors. These methods are fast since they can be run in linear time, however, when the distance between two graphs is 0, it does not imply that the two graphs are isomorphic. However, a lower bound relation within a factor of four exists between the graph probing and the edit distance [9]. An experimental comparison between graph probing and other approaches is presented in section 4.

3   The Genetic Algorithm in Use

3.1   Genetic Operators Dedicated to Graphs

Genetic Algorithms (GAs) are adaptive heuristic optimization algorithms based on the evolutionary ideas of natural selection and genetics. The basic concept of GAs is designed to simulate natural processes, necessary for evolution of artificial systems. They represent an intelligent exploitation of a random search within a defined search space to solve a problem. As can be seen on fig 1, after a random initialization of a population of possible solutions, GA's are based on a sequential ordering of four main operators: selection, replication, crossover and mutation. In order to apply genetic algorithms to a given problem, three main stages are necessary: the coding of the problem solutions, the definition of the objective function which attributes a fitness to each individual, and the definition of the genetic operators which promote the exchange of genetic material between individuals. In most existing GA applications, a linear representation of individuals is used. Problem parameters are encoded through a binary or a real string. Crossover is then applied through a single-point or two-point based exchange of genes. Regarding mutation, it is applied through a random modification of a small number of genes chosen randomly.

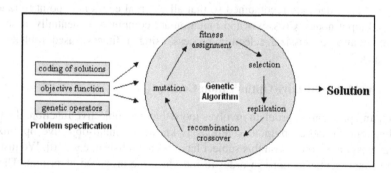

Fig. 1. Overview of a genetic algorithm

In our context of pattern recognition using graph, each individual has to encode a set of graphs (the KxN prototypes). Consequently, the evolution of the individuals through GA implies to revisit classical operators since they have to modify graphs. Concerning mutation, our operator is based on the six unary edit operations which can be applied to a graph: add or remove a node or an edge and modify a node or an edge

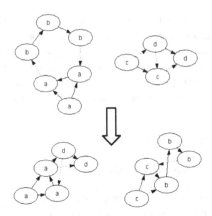

Fig. 2. The crossover operator

label. For each mutation operation, we first decide to apply or not the operator according to a pre-defined rate. If mutation has to be applied, one of the six possibilities is chosen randomly, as well as the new label if the operation is a label modification.

To perform crossover between individuals (see fig. 2), we first randomly partition the set of nodes of each graph in two subsets (see the label of nodes on figure 2). We call *internal edges*, the edges of the initial graph the nodes of which are in the same subset (continuous lines). At the opposite, edges the nodes of which are in different subsets are called external edges (dotted lines). Then, a node is said to be an output node if it is a source of external edge, and an input node if it is the destination of an external edge. Finally, according to the nature of nodes and edges, fragments are swapped and edges are recombined so that all external edges now point to randomly selected input nodes. Crossover and mutation are combined sequentially as shown in figure 1, after a classical selection process using a fitness based roulette wheel approach.

3.2 The Multi-objective Optimization Concept

When an optimization problem involves more than one objective function, the task of finding one or more optimum solutions is known as multi-objective optimization. Some classical textbooks on this subject have been published, e.g. [10]. We just recall here some essential notions in order to introduce the proposed algorithm. The main difference between single and multi-optimization tasks lies in the requirement of compromises between the various objectives in the multi-optimization case. Even with only two objectives, if they are conflicting, the improvement of one of them leads to a deterioration of the other one. For example, in the context of graph classification, the decrease of the reject rate generally leads to an increase of the confusion rate. Two main approaches are used to overcome this problem in the literature.

The first one consists in the combination of the different objectives into a single one (the simpler way being to use a linear combination of the various objectives), and then to use one of the well-known techniques of single objective optimization (like gradient based methods, simulated annealing or classical genetic algorithm). In such a case, the compromise between the objectives is a priori determined through the choice of the combination rule. The main critic addressed to this approach is the difficulty to choose a priori the compromise. It seems a better idea to postpone this choice after having several candidate solutions at hand. This is the goal of Pareto based method using the notion of dominance between candidate solutions. A solution dominates another one if it is better for all the objectives. This dominance concept is illustrated on figure 3. Two criteria J1 and J2 have to be minimized. The set of non-dominated points that constitutes the Pareto-Front appears as 'O' on the figure, while dominated solutions are drawn as 'X'. Using such a dominance concept, the objective of the optimization algorithm becomes to determine the Pareto front, that is to say the set of non-dominated points. Among the optimization methods that can be used for such a task, genetic algorithms are well-suited because they work on a population of candidate solutions. They have been extensively used in such a context. The most common algorithms are VEGA – Vector Evaluated Genetic Algorithm – [11], MOGA – Multi-Objective Genetic Algorithm –approach [12], SGA – Non-Dominated Sorting Genetic Algorithm – [13], NSGA II [14], PAES – Pareto Archived EvolutionStrategy – [15] and SPEA – Strength Pareto Evolutionary Algorithm – [16]. A good review can be found in [17].

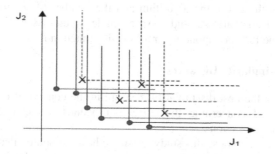

Fig. 3. Illustration of the Pareto Front concept

The proposed genetic algorithm is elitist and steadystate. This means that (i) it manages two populations and (ii) the replacement strategy of individuals in the populations is not made as a whole, but individual per individual. The two populations are a classical population, composed of evolving individuals and an "archive" population composed of the current Pareto Front elements. These two populations are mixed during the iterations of the genetic algorithm. The first population guarantees space exploration while the archive guarantees the exploitation of acquired knowledge and the convergence of the algorithm. This algorithm has been designed in order to be applied to various problems. In the current implementation, the replacement strategy is defined in such a way that the candidate has to be inserted within the archive if no archive element dominates it. In the same time, archive elements dominated by the candidate are eliminated from the archive.

4   Application

In this section, the graph construction step is explained. Then, a comparative study concerning dissimilarity measure is described. It justifies our decision to use graph probing in our context. Finally, the symbol recognition application is presented, results are measured up to another approach and a two objective optimization is performed taking into account the notion of reject.

4.1   Graph Data Set Construction

Our data are made of graphs corresponding to a corpus of 180 noisy symbol images, generated from 10 ideal models proposed in a symbol recognition contest (GREC workshop). In a first step, considering the symbol binary image, we extract both black and white connected components. These connected components are automatically labeled with a partitional clustering algorithm [18] applied on a set of features called Zernike moments [4]. Using these labeled items, a graph is built. Each connected component represents an attributed vertex in this graph. Then, edges are built using the following rule: two vertices are linked with an undirected and unlabeled edge if one of the nodes is one of the h nearest neighbors of the other node in the corresponding image. The two values h and c, concerning respectively, the cluster number found by the clustering algorithm and the number of significant neighbors, are issued from a comparative study. An example of the association between two symbol images and the corresponding graphs is illustrated in fig 4.

4.2   Test on Dissimilarity Distances

In order to choose the best dissimilarity measure in the context of our application, a study has been led concerning the correlation values between the dissimilarity measures proposed in section 2.

Two experiments compose this study. First, we have computed Pearson correlation coefficients (cor) between the different dissimilarity measures. Results are presented on the first line of table 1. The second experiment concerns the correlation between a userdefined ground truth order (or partial order) and the order calculated using the distance between representations. Such a correlation has to be as high as possible since our objective is the classification of graphs. This correlation can be measured using the Kendall rank correlation coefficient (tau) [19]. Using these values, we can select a graph representation and a dissimilarity measure which satisfies both running time constraints and high correlation with the groundtruth of our application. The obtained values, associated with the corresponding run time complexity, point out the trade-off to be made between the quality (agreement with the ground truth) of a similarity measure and its run time complexity. Since our application is quite demanding of dissimilarity measures, graph probing seems more suitable, showing a better trade-off: meaningful and operating in linear time.

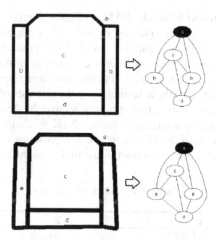

Fig. 4. From symbols to Graphs

Table 1. Correlation between the edit distance: (ED), the edge matching distance(EMD), graph probing(GP) and Ground Truth (GT)

	ED	GP	EMD
ED using cor	1	0.58	0.63
ED using tau	1	0.53	0.63
GT using tau	0.699	0.622	0.657

4.3 Classification Experiments: Mono-Objective

In a first step, with an aim of comparison, we focus on a mono-objective problem which will be extended in the next par to multi-objective. Under such conditions, the learning algorithm consists in the generation of K graph prototypes per symbol class for a group of N classes.

These prototypes are produced by a graph based GA, the aim of which is to find the near optimal solution of the recognition problem using the selected prototypes. In such a context, each individual in our GA is a vector containing K graphs per class, that is to say K feasible solutions (prototypes) for a given class. Hence, an individual is composed of KxN graphs. For the initialization of the population, each graph of each individual is selected randomly from the initial graph corpus. The fitness (the suitability) of each individual is quantified thanks to the classification rate obtained using the corresponding prototypes and a test database. The classification is processed by a 1-NN classifier using the graph probing distance. Then, using the operators described in section 3, the GA iterates, in order to optimize the classification rate. The stopping criterion is the generation number. At the end of the GA, a classification step is applied on a validation database in order to evaluate the quality of the selected prototypes. The obtained results are compared with an approach which also finds K representatives in a set of objects, described only by their reciprocal matrix distance. This approach which minimizes (unsquared) distances from objects to representatives is called Partition Around Medoids (PAM) [18]. PAM gives us its K best prototypes for a given class. Using these prototypes and the graph probing distance, we can compute the recognition rate. Table 2 gives the comparative results for K=1,2,3 prototypes per class, and a group of 10 classes. One can also note that using only the ideal models as learning set, a 1NN classification using graph probing provides a 88,28% recognition rate. All these results show the interests of the prototype selection using GA combined with the graph probing approach. According to us, the main reason is that the learning application creates NxK synthetic elements thanks to the genetic operators in order to obtain the best representation of a particular class. Hence, our range of possibilities is not limited to the graphs constituting the class.

Table 2. Global classification rate

K	1	2	3
PAM	91,42%	94,28%	96,66%
GA	95,29%	96,47%	98,23%

4.4 Multi-objective Experiments

In another step, we add one more objective to the problem: the confusion rate minimization. Therefore, from now, the classifier has distance rejection capability, the capacity not to take decision in case of ambiguity. Consequently, the relation between the recognition rate and the confusion rate is defined as follow:

Confusion rate = 1 – (Recognition rate + Reject rate).

Hence, the problem becomes to find all dominant solutions, the L couples(confusion, recognition), where one couple represents a set of KxN graphs, K prototypes per class

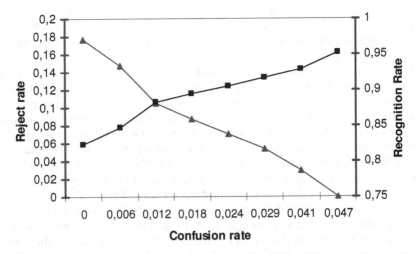

Fig. 5. Example for K=1, Pareto Front of two criteria: [Confusion, Recognition]. The left axis corresponds to the reject rate curve (triangles) and the axis on the right goes with recognition rate (squares).

for N classes. We perform our multi-objective algorithm on the learning data in order to discover the Pareto Front and finally the found solutions are evaluated on the test database (Fig 5).

The Figure 5 illustrates clearly the tradeoff to be made between Confusion rate and Recognition rate. Among the range of possibilities found by the method, we spot two specific cases: (i)A confusion rate equal to zero has been found but in against part, the classifier has rejected a lot. (ii)On the other hand, a high recognition rate has been discovered too but this solution implies some classification errors. This choice depends on what are the system needs. The advantage of providing a range of solutions corresponds completely with an adaptive system which will see its constraints changed dynamically, and at any time, the most suitable prototype set could be picked up among the heap of dominant solutions. In such a case, no need to relearn new graph prototypes, since each solution represents an adaptation of the learning. The way of reaching many objectives at once, gives to our method a wider field of action and even if we have performed only a two objective optimization during our experiments, we precise that the approach can be generalized to other criteria.

5 Conclusion

In this paper, a graph classification algorithm has been proposed with an application to symbol recognition. The approach is based on the learning of graph prototypes using multi-objective genetic algorithm and a fast dissimilarity measure called graph probing. This measure has been judged more efficient from the computation speed point of view. The obtained results, compared with a classification using PAM to select prototype, have shown the interest to generate synthetic prototypes through the use of genetic operators rather than finding them among the elements defining the classes.

In addition, the reject is integrated to the method in terms of multi-optimization without including a priori knowledge. A wide range of solutions is provided to fulfill the system needs. Our further works concern different points. The first of them consists in enriching the symbol description as graph, for example through the use of contour vectorization results. A second one consists in testing the approach on a more important database. Another one consists in comparing the approach with the use of graph kernel SVM. And finally, we are investigating some studies to increase the dissimilarity measure relevance. In this direction, we improve the information extracted from graphs, in taking into account in each probe the neighborhood notion. The idea is to give a more significant interpretation of graph topologies in increasing the sub-graph sizes used as probes. In such a context, each probe is a vision of graphs for a certain level of neighborhood.

References

1. Schenker, A., Last, M., Bunke, H., Kandel, A.: Classification of web documents using a graph model. In: Proceedings of the 7th International Conference on Document Analysis and Recognition (ICDAR), pp. 240–244 (2003)
2. King, R.D., Sternberg, M.J.E., Srinivasan, Muggleton, S.H.: Knowledge discovery in a database mutagenetic chemicals. In: proceedings of the workshop Statistics, machine leaning, discovery in databases at the ECML-95 (1995)

3. Cordela, L.P., Vento, M.: Symbol recognition in documents: a collection of techniques? International Journal on Document Analysis and Recognition 3(2), 73–88 (2000)

4. Khotazad, A., Hong, Y.H.: Invariant image recognition by Zernike Moments. PAMI 12(5), 489–497 (1990)

5. Valveny, E., Dosch, P.: Symbol Recognition Contest: A Synthesis. In: Lladós, J., Kwon, Y.-B. (eds.) GREC 2003. Lecture Notes in Computer Science N°, vol. 3088, pp. 368–385. Springer, Heidelberg (2004)

6. Bunke, H., Shearer, K.: A graph distance metric based on the maximal common subgraph. Pattern Recogn. Lett. 19, 255–259 (1998)

7. Bunke, H.: On a relation between graph edit distance and maximum common subgraph. Pattern Recogn. Lett. 18, 689–694 (1997)

8. Kriegel, H.P., Schönauer, S.: Similarity Search in Structured Data. In: Kambayashi, Y., Mohania, M.K., Wöß, W. (eds.) Data Warehousing and Knowledge Discovery. Lecture Notes in Computer Science, N°, vol. 2737, pp. 309–319. Springer, Heidelberg (2003)

9. Lopresti, D.P., Wilfong, G.T.: A fast technique for comparing graph representations with applications to performance evaluation. International Journal on Document Analysis and Recognition 6, 219–229 (2003)

10. Deb, K.: Multi-Objective optimization using Evolutionary algorithms. Wiley, London (2001)

11. Schaffer, J.D., Grefenstette, J.J.: Multiobjective learning via genetic algorithms. In: Proceedings of the 9th international joint conference on artificial intelligence, Los Angeles, California, pp. 593-595 (1985)

12. Fonseca, C.M., Fleming, P.J.: Genetic algorithm for multi-objective optimization: formulation, discussion and generalization. In: Stephanie editor, Proceedings of the fifth international conference on genetic algorithm, San Mateo, California, pp. 416–423 (1993)

13. Srinivas, N., Deb, K.: Multiobjective optimization using nondominated sorting in genetic algorithm. Evolutionary Computation 2, 221–248 (1994)

14. Deb, K., Agrawal, S., Pratab, A., Meyarivan, T.: A fast and elitist multi-objective genetic algorithm: NSGA-II. IEEE Transactions on Evolutionary Computation 6, 182–197 (2000)

15. Knowles, J.D., Corne, D.W.: Approximating the nondominated front using the Pareto archived evolution strategy. Evolutionary computation 8, 149–172 (2000)

16. Zitzler, E., Thiele, L.: Multiobjective evolutionary algorithms: a comparative study and the strength pareto approach. IEEE Transactions on Evolutionary Computation 3, 257–271 (1999)

17. Coello, C.A.: Coello: Coello, "A short tutorial on Evolutionary Multiobjective Optimisation", In Eckart Zitzler, Kalyanmoy Deb, Lothar Thiele, Carlos A. In: Zitzler, E., Deb, K., Thiele, L., Coello Coello, C.A., Corne, D.W. (eds.) EMO 2001. Lecture Notes in Computer Science n°, vol. 1993, pp. 21–40. Springer, Heidelberg (2001)

18. Kaufman, L., Rousseeuw, P.J.: Finding groups in data. John Wiley & Sons, Inc, New York (1990)

19. Kendall, M.G.: Rank Correlation Methods. Hafner Publishing Co, New York (1955)

20. Sorlin, S., Solnon, C.: Reactive Tabu Search for Measuring Graph Similarity. In: Brun, L., Vento, M. (eds.) GbRPR 2005. LNCS, vol. 3434, pp. 172–182. Springer, Heidelberg (2005)

Graph Embedding Using Quantum Commute Times

David Emms, Richard C. Wilson, and Edwin Hancock

Department of Computer Science
University of York
YO10 5DD, UK

Abstract. In this paper, we explore analytically and experimentally the commute time of the continuous-time quantum walk. For the classical random walk, the commute time has been shown to be robust to errors in edge weight structure and to lead to spectral clustering algorithms with improved performance. Our analysis shows that the commute time of the continuous-time quantum walk can be determined via integrals of the Laplacian spectrum, calculated using Gauss-Laguerre quadrature. We analyse the quantum commute times with reference to their classical counterpart. Experimentally, we show that the quantum commute times can be used to emphasise cluster-structure.

1 Introduction

Random walks on graphs have been used to develop a powerful battery of pattern analysis algorithms. The steady state random walk on a graph is determined by the leading eigenvector of the weighted adjacency matrix or equivalently the Fiedler vector of the Laplacian matrix [6]. Random walks are therefore intimately related to graph-spectra. For example Melia and Shi [8] have used random walks to learn image segmentation. Zhu, Ghahramani and Lafferty [13] have performed semisupervised learning using random walks on a labelled graph structure. Robles-Kelly and Hancock [12] have developed a graph-spectral method inspired by random walks to seriate graphs, i.e. to place the nodes in string order. Borgwardt et al have developed a kernel that preserves the path length distribution of a random walk on a graph, and have used this to analyse protein data [1]. This kernel has been used by Neuhaus and Bunke [10] to kernelise the computation of graph edit distance, and measure the similarity of graphs. Finally, Qiu and Hancock [11] have shown how the commute times of random walks can be used to render graph-spectral clustering algorithms robust to edge weight errors, and have explored the application of the method to image segmentation, multibody motion tracking and graph-matching. The commute time allows the nodes of a graph to be embedded in a low dimensional space, and the geometry of this embedding allows the nodes to be clustered into disjoint subsets.

One of the problems of spectral approaches to the analysis of graphs using random walks is that of cospectrality. That is graphs of different structure can

F. Escolano and M. Vento (Eds.): GbRPR 2007, LNCS 4538, pp. 371–382, 2007.

give the same the pattern of eigenvalues. Emms et al [4] have recently shown how coined quantum walks can be used to lift the problem of co-spectrality, particularly for strongly regular graphs. Quantum walks [5] differ from their classical counterparts in that the real-valued vector of probabilities is described indirectly via a complex-valued state vector. The evolution of the walk depends on the richer representation given by this state vector, allowing effects such as quantum interference to take place.

Quantum walks clearly offer powerful tools for the analysis of graphs, and the aim in this paper is to take their study one step further. Results for hitting times on certain graphs with high levels of symmetry have been studied [2]. However, to date there has been little effort devoted to the study of properties such as hitting time or commute time of walks on more general graphs. Specifically, our aim is to explore whether the commute time associated with the continuous-time quantum walk can be used to embed the nodes of a graph in a low dimensional vector space and to explore the properties of the embedding. When contrasted with the classical walk there are a number of interesting conclusions that can be drawn. First, while the embedding associated with the classical commute time is close to unidimensional, that associated with the quantum walk needs more dimensions to capture its behaviour. This is closely linked to the problem of cospectrality. Second, although mean commute times of the quantum and classical walks are both correlated with path length, in any particular instance they are not correlated with each other: both measure different properties of the graph. Thirdly, the embeddings obtained can distinguish clusters more clearly than those obtained using the classical commute times.

2   Random Walks

Let $G = (V, E, W)$ be a weighted graph with vertex set, V, and edge set, $E = \{\{u, v\} | u, v \in V,$ u adjacent to $v\}$. The graph has a weighted adjacency matrix, W, which is symmetric, $W(u, v) = W(v, u)$, and the entry $W(u, v)$ gives the weight on the edge $\{u, v\}$. Let $n = |V|$ be the total number of vertices in the graph. We define the degree matrix to be the matrix $D = diag(d(1), d(2), \ldots, d(n))$ where $d(u) = \sum_v W(u, v)$ is the degree of vertex u. An unweighted graph corresponds to the particular case where $W(u, v) = 1$ if $\{u, v\} \in E$ and 0 otherwise. In this case, the degree of a vertex is the number of vertices adjacent to it.

The Laplacian matrix, $L = W - D$, for the graph is the graphical equivalent of the Laplacian operator in Euclidean space and has elements

$$L_{uv} = \begin{cases} W(u, v) & \text{if } \{u, v\} \in E; \\ -d(u) & \text{if } u = v; \\ 0 & \text{otherwise.} \end{cases}$$

The Laplacian matrix is used to define the time evolution of the quantum walk on the graph. Before describing the quantum walk, however, we give a brief summary of the classical random walk.

2.1   Classical Random Walk

The continuous-time quantum walk is motivated by (classical) continuous-time Markov chains. We introduce the quantum walk by way of the classical random walk. The state space for the classical random walk is the set of vertices, V. The state of the walk at time t is given by a vector, $\mathbf{p}(t) \in \mathbf{R}^{|V|}$, whose u^{th} entry, $p_u(t)$, is the probability that the walk is at the vertex u at time t. The axioms of probability give that $p_u(t) \in [0, 1]$ for all $u \in V$, $t \in \mathbf{R}^+$ and $\sum_{u \in V} p_u(t) = 1$ for all $t \in \mathbf{R}^+$. The connectivity structure of the graph is respected by requiring that transitions are only allowed between adjacent vertices. Additionally, a transition between a vertex and one of the vertices adjacent to it is proportional to the weight of the edge connecting them. Thus, the state of the walk at time $t + 1$ is given by $\mathbf{p}(t + 1) = T\mathbf{p}(t)$ where T is the transition matrix, given by $T = D^{-1}W$. The classical continuous-time random walk is obtained by introducing a transition rate, μ, that gives the probability of a transition between any pair of neighbouring vertices per unit time. The state vector the walk satisfies the differential equation

$$\frac{d}{dt}\mathbf{p}(t) = \mu L \mathbf{p}(t).$$

2.2   Quantum Random Walk

The state space for a the continuous-time quantum random walk is again the set of vertices, V. However, the state is described by a complex state vector which we write (using Dirac's notation) as $|\psi_t\rangle \in \mathbf{C}^{|V|}$. This can be written componentwise as

$$|\psi_t\rangle = \sum_{u \in V} a_u |u\rangle$$

where $|u\rangle$ is the vector corresponding to the walk being at vertex u, in the following sense. Let X^t be the random variable giving the state of the walk at time t. The probability of the walk being at a particular vertex $u \in V$ is given, indirectly, by the complex state vector according to the rule $P(X^t = u) = a_u a_u^*$ where a_u^* is the complex conjugate of a_u. (We can rewrite this as $P(u) = |\langle u|\psi_t\rangle|^2$ where, for any vector $|\phi_a\rangle = \sum_u a_u |u\rangle$ in the Hilbert space $\mathcal{H} \equiv \mathbf{C}^{|V|}$, $\langle \phi_a|$ is the linear functional that maps every vector $|\phi_b\rangle = \sum_u b_u |u\rangle \in \mathcal{H}$ to the standard inner product $\langle u|\phi\rangle = (|u\rangle, |\phi\rangle) = \sum_u a_u b_u^*$.) For the quantum walk, the axioms of probability give that $|a_u(t)| \in [0, 1]$ for all $u \in V$, $t \in \mathbf{R}^+$, and $\sum_{u \in V} a_u a_u^* = 1$. Again, transitions only occur between adjacent vertices and the evolution of the state vector is given by

$$\frac{d}{dt}|\psi_t\rangle = -i\mu L. \tag{1}$$

Since the evolution of the probability vector of the walk at time t depends on the state vector of the walk (not merely the probability vector), the quantum walk is not a Markov chain. However, given an initial state, $|\psi_0\rangle$, Equation 1 can be solved to give $|\psi_t\rangle = e^{-i\mu Lt}|\psi_0\rangle$ allowing the walk to be analysed using the Laplacian spectrum, as demonstrated in the following section.

3   Commute Times

In this section we will consider the expected time for the quantum walk to travel between each pair of vertices in a graph. For the walk with starting state $|\psi_0\rangle = |u\rangle$, let $X_{(u)}^t$ be the random variable giving the state of the walk at time t. Let $H(u,v)$ be the random variable giving the first hitting time of the vertex v. That is, $H(u,v) = \min\{t|X_{(u)}^t = v\}$. We note that, due to the symmetry of the Laplacian matrix, L, we have $H(u,v) = H(v,u)$. The commute time, $C(u,v)$, between a pair of vertices is defined as the expected time for the walk to travel from the vertex u to v and back to u again. Thus $C(u,v) = O(u,v) + O(v,u)$ where $O(u,v)$ is the expected value of the first hitting time. From the symmetry of the walk this is $Q(u,v) = 2O(u,v)$. The expected hitting time between a pair of vertices can be calculated using

$$O(u,v) = \int_0^\infty P\big(H(u,v) = t\big)t\,dt \tag{2}$$

To calculate $P\big(H(u,v) = t\big)$, we consider the state vector, $|\psi_t\rangle = e^{-iLt}|u\rangle$. Taking as out transition rate, $\mu = 1$, the probability that this walk is at v at time t is given by

$$P(X_{(u)}^t = v) = |\langle v|e^{-iLt}|u\rangle|^2$$

$$= \left(\sum_{j=1}^n \phi_j(u)\phi_j(v)e^{-i\lambda_j t}\right)\left(\sum_{k=1}^n \phi_k(u)\phi_k(v)e^{-i\lambda_k t}\right)^*$$

$$= \sum_{j=1}^n \sum_{k=1}^n \phi_j(u)\phi_j(v)\phi_k(u)\phi_k(v)e^{-i(\lambda_j - \lambda_k)t}$$

where λ_k is an eigenvalue of L and ϕ_k its corresponding eigenvector. As $P(X_{(u)}^t = v)$ is a real number we need only consider the real parts, since the imaginary parts must cancel. Hence, we have that

$$P(X_{(u)}^t = v) = 2\sum_{j=1}^n \left(\sum_{k>j:\lambda_j \neq \lambda_k} \phi_j(u)\phi_j(v)\phi_k(u)\phi_k(v)\cos((\lambda_k - \lambda_j)t)\right.$$

$$\left. \dots + \sum_{k:\lambda_j = \lambda_k} \phi_j^2(u)\phi_j^2(v)\right)$$

$$= F + \sum_{(A,B)\in\mathcal{X}} A\cos(Bt),$$

where the pairs $(A,B) \in \mathcal{X}$, and are defined by

$$\mathcal{X} = \left\{\left(\phi_j(u)\phi_j(v)\phi_k(u)\phi_k(v), \lambda_j - \lambda_k\right)\Big|1 \leq j \leq n,\ 1 \leq k < j,\ \lambda_k \neq \lambda_j\right\}$$

and

$$F = \sum_{j,k:\, \lambda_j = \lambda_k} \phi_j(u)\phi_j(v)\phi_k(u)\phi_k(v).$$

For a particular $u, v \in V$, let $r(t)$ be the probability density function for the first hitting time, $H(u,v)$, and $R(t)$ its cumulative distribution function. The probability function $P(X^t_{(u)} = v)$ is referred to as the hazard function for the walk arriving at v. It is such that the probability of the walk arriving at v in the interval $(\tau, \tau + d\tau)$, given that it has not previously arrived at v, is given by $P(X^\tau_{(u)} = v)d\tau$. For a given hazard function, $P(X^t_{(u)} = v)$, we have that

$$\frac{d}{dt}(1 - R(t)) = -P(X^t_{(u)} = v)(1 - R(t)).$$

Hence,

$$1 - R(t) = (1 - R(0)) \exp\left(-\int_0^t F + \sum_{(A,B)\in \mathcal{X}} A\cos(B\tau)\, d\tau\right)$$

$$= \exp\left(-Ft - \sum_{(A,B)\in \mathcal{X}} \frac{A}{B}\sin(Bt)\right),$$

since $R(0) = 0$. The probability density function for the random variable $H(u,v)$ is then given by $r(t) = (1 - R(t))P(X^t_{(u)} = v)$, hence

$$r(t) = \left\{\exp\left(-Ft - \sum_{(A,B)\in \mathcal{X}} \frac{A}{B}\sin(Bt)\right)\right\}\left(F + \sum_{(A,B)\in \mathcal{X}} B\cos(Bt)\right).$$

Thus, using Equation 2, we can calculate the commute time, $Q(u,v) = 2O(u,v)$, for the vertex v for the walk starting at u.

$$Q(u,v) = 2\int_0^\infty t\left(F + \sum_{(A,B)\in \mathcal{X}} A\cos(Bt)\right)e^{-Ft - \sum_{(A,B)\in \mathcal{X}} \frac{A}{B}\sin(Bt)}\, dt$$

$$= 2\left[-te^{-Ft - \sum_{(A,B)\in \mathcal{X}} \frac{A}{B}\sin(Bt)}\right]_0^\infty + 2\int_0^\infty e^{-Ft - \sum_{(A,B)\in \mathcal{X}} \frac{A}{B}\sin(Bt)}\, dt$$

$$= 2\int_0^\infty e^{-Ft - \sum_{(A,B)\in \mathcal{X}} \frac{A}{B}\sin(Bt)}\, dt$$

In order to calculate this numerically we use Gauss-Laguerre quadrature. Given a function, $f \in L^1(0, \infty)$ we have that

$$\int_0^\infty f(x)dx = \int_0^\infty e^{-x}[e^x f(x)]dx \simeq \sum_{k=1}^n w(x_k)e^{x_k}f(x_k).$$

The abscissas, for x_k, $k = 1, \ldots, n$, for the quadrature order n are given by the roots of the Laguerre polynomial $L_n(x) = \frac{e^x}{n!}\frac{d^n}{dx^n}(e^{-x}x^n)$ and the weights are, $w(x_k) = \frac{x_k}{(n+1)^2(L_{n+1}(x_k))^2}$.

3.1 Embedding Graphs Via Commute Times

Given the commute times, $Q(u,v)$, between all pairs of vertices $u,v \in V$ in a graph, we use Multidimensional Scaling (MDS) to embed the vertices in 2D space. The first step of MDS is to calculate a matrix, M, whose (u,v) entry is given by $M(u,v) = -\frac{1}{2}(Q(u,v)^2 - \hat{Q}_{u.}^2 - \hat{Q}_{.v}^2 + \hat{Q}_{..}^2)$, where $\hat{Q}_{u.} = \frac{1}{n}\sum_{v=1}^{n} Q(u,v)$ is the average distance over the uth row, $\hat{Q}_{.v}$ is the average distance over the vth column and $\hat{Q}_{..} = \frac{1}{n^2}\sum_{u=1}^{n}\sum_{v=1}^{n} Q(u,v)$ is the average distance over all rows and columns of the distance matrix, in this case the matrix of commute times Q.

We subject the matrix M to an eigenvector analysis to obtain a matrix of embedding coordinates Y. If the rank of M is k, where $k < n$, then we will have k non-zero eigenvalues. We arrange these k non-zero eigenvalues in descending order, i.e. $l_1 \geq l_2 \geq \ldots \geq l_k \geq 0$. The corresponding ordered eigenvectors are denoted by $\mathbf{x}_i$ where l_i is the ith eigenvalue. The embedding coordinate system for the graphs is $Y = [\sqrt{l_1}\mathbf{x}_1, \sqrt{l_2}\mathbf{x}_2, \ldots, \sqrt{l_k}\mathbf{x}_k]$. For the vertex u, the vector of coordinates for the embedding, $\mathbf{y}_u$, is a row of matrix Y, i.e. $\mathbf{y}_u = (Y_{u,1}, Y_{u,2}, \ldots, Y_{u,k})^T$.

In the following section we compare the quantum commute times with the classical commute times on a graph, and the embeddings obtained. In [11] Qiu and Hancock show that the commute time for a classical random walk is given by

$$C(u,v) = vol \sum_{j=2}^{|V|} \frac{1}{\lambda_j}(\phi_j(u) - \phi_j(v))^2$$

where the λ_j and ϕ_j are the eigenvalues and corresponding eigenvectors of the Laplacian, as before. They use these classical commute times between vertices to embed graphs. Further, they show that the matrix of coordinates, Θ, for the classical commute time embedding can be calculated directly using $\Theta = \sqrt{vol}D^{-1/2}\Phi^T$ where $\Phi = [\phi_1|\phi_2|\ldots|\phi_n]$.

4   Analysis of Quantum Commute Time Embeddings

In this section we analyse the differences between the quantum and classical commute time embeddings. We use graphs randomly generated according to one of two different models. The first model is the randomly connected graph on n vertices with connection probability p. Here, each pair of vertices is connected with probability p. Our second model is the banded adjacency graph [3]. For the purposes of constructing each graph we take a set of vertices numbered 1 to n. With probability p, the vertex u is then connected to each vertex, $v \in \{u-b, u-b+1, \ldots, u+b\}$. In terms of its adjacency matrix, such a graph only has non-zero entries within a band of width $2b$ centred on the main diagonal. Such graphs approximate many real-world structures, for example, molecules or VLSI circuits. The randomly generated graphs model graphs where no such restriction is obvious, for example, hyperlinks between webpages.

We begin by considering the relationship between commute times and path length. The path length between a pair of vertices is the number of edges that must be traversed along the shortest path between them. It is a less sophisticated measure of distance than commute time since it only takes into account the single shortest path between pair of vertices. The commute time, on the other hand, decreases as the number of alternative paths between the vertices increases. This expresses mathematically the understanding that a pair of vertices with many paths between them of length l should be considered more closely connected than a pair of vertices with just one path of length l connecting them.

Figure 1 shows the mean quantum commute times and mean classical commute times as a function of path length. We see that the functions are both monotonically increasing, however, the quantum commute time has a non-linear relationship with path length. Larger path lengths do not correspond to proportionally larger quantum commute times, as they do in the classical case. Similar behaviour has been observed for the quantum walk on the line [9] and the circle [5] where hitting times were shown to be quadratically faster. We will see this behaviour has important consequences when we consider the treatment of outliers in the embedding of graphs via commute times.

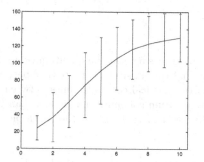 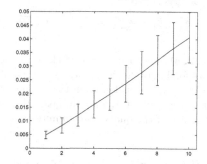

Fig. 1. The average quantum commute time (left) and average classical commute time (right) as a function of path length for a banded adjacency graph on 50 vertices with band width 10 and p=0.3. Note, the error bars show the standard deviation not the standard error.

Although Figure 1 shows that the mean quantum and classical commute times are functions of path length, the standard deviation about these functions are large. Thus, we wish to investigate if there is a relationship between the quantum commute time, $Q(u, v)$, and the classical commute time, $C(u, v)$, for a given pair of vertices, (u, v). Figure 2a shows a scatter plot of the classical commute times between pairs of vertices versus the quantum commute times for a set of 10 randomly connected graphs with $n = 10$, $p = 0.3$. We see that there is very little correlation between the commute times for particular vertices. However, as the classical (or quantum) commute time increases, the lower bound for the corresponding quantum (or classical) commute time does also increase by a significantly smaller fraction.

To investigate any correlation further, let d_k^C be the kth decile for the classical commute times and d_j^Q the jth decile for the quantum commute times for a particular graph. We consider the set of probabilities $P(C(u,v) < d_k^C | Q(u,v) < d_j^Q)$. Figure 2b shows a plot of this function. The figure shows that $P(C(u,v) < d_k^C)$ is almost completely independent of any condition of the form $Q(u,v) < d_j^Q$. This demonstrates that the two commute times emphasise different measures of distance.

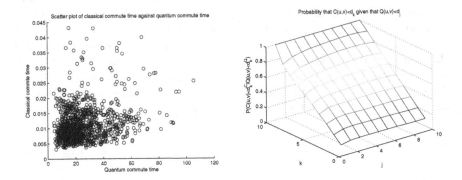

Fig. 2. a) Left: Scatter plot of classical commute times against the quantum commute times for 10 graphs. b) Right: The probability that the classical commute time between a pair of vertices is less than the kth decile of the classical commute times between all pairs of vertices in the graph, given that the quantum commute time is less than the $j^t h$ decile of the quantum commute times. 10 graphs, $n = 35$, $p = 0.3$ for both plots.

We now give some sample embeddings, using both the classical and quantum commute times, for the two different models of randomly generated graphs. Figure 3 shows an example of the quantum and classical commute time embeddings for a randomly connected graph. The most obvious difference between the two embeddings is the classical commute time embedding's tendency to produce a few outliers while confining the majority of the vertices to a small area. The problem of one distance outweighing all others is mush less of a problem for the quantum commute time embedding, and consequently the vertices are distributed more evenly. Embeddings of a banded adjacency graph are given in Figure 4. The classical commute time embedding allows the distance along the band to outweigh the others distances, causing the graph to lie almost entirely along a curve, obscuring almost all of the local structure. The quantum commute time embedding, however, shows both the band structure (principally captured by the x component of the embedding) and the local structure along the band (principally captured by the y component).

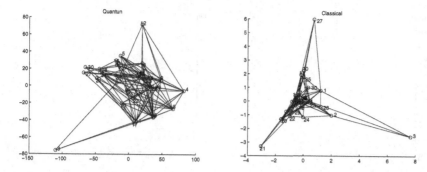

Fig. 3. Graph embeddings using the quantum commute time (left) and the classical commute time (right) for a graph with 30 vertices and probability of each pair of vertices being connected 0.3

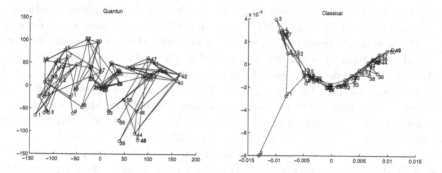

Fig. 4. Graph embeddings using the quantum commute time (left) and the classical commute time (right) for a banded adjacency graph with 50 vertices, band width $b = 5$ and probability of each pair of vertices in the band being connected $p = 0.5$

5   Experiments

We consider the problem of producing a 2D embedding of a set of graphs in order to cluster them. Given a set of graphs we calculate distances between them using the Euclidean distance between their Laplacian spectra. We use the Laplacian spectra as a vehicle to test our approach, its use in this context has been well studied in the literature [7]. We represent the set of graphs using a weighted data similarity graph (WDSG). The vertices of the WDSG graph are the original graphs from the set we wish to cluster. We then use the classical and quantum commute times to embed these graphs. We also embed the graphs directly using Multidimensional Scaling (MDS) on the distances for comparison.

Let S be our set of m graphs. We denote by $\mathbf{x}_k$ the vector of the ordered eigenvalues of the Laplacian matrix, L_k, of the kth graph. We append zeros to the shorter vectors so that all the vectors of eigenvalues are of the same length. The graph, G, representing this data set is the complete graph $G = (V, E, W)$,

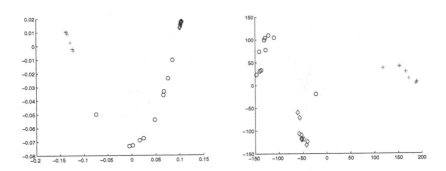

Fig. 5. Embeddings of the house sequences using MDS on the spectral distances directly (left) and the quantum commute times (right)

where $V = \{1, \ldots, k\}$, $E = \{\{u, v\}|u, v \in V, u \neq v\}$ and the weights are given by $W(u, v) = e^{-||\mathbf{x}_u - \mathbf{x}_v||}$, where $||\mathbf{x}_u - \mathbf{x}_v||$ is the standard Euclidean distance.

We use the standard CMU, MOVI and chalet house sequences as our dataset [7]. The houses are viewed at angular intervals and Delaunay triangulations are derived from these to form the set of graphs. The embeddings obtained using MDS on the distances between the spectra, the classical commute times, and the quantum commute times are given in Figures 5 and 6. The embedding using the quantum commute times clearly distinguishes between the 3 clusters. A reasonably good separation of the houses is also obtained using MDS on the spectral distances directly. However, the embedding using MDS on the spectral distances directly is effectively one dimensional, this is a common problem with using MDS on graph distances [7]. For datasets with more than three classes this is likely to prevent this method from successfully clustering different classes separately. The quantum commute time embedding, however, fully utilizes the available dimensions.

The embedding using the quantum commute times distinguishes between the different classes far more clearly than the embedding using the classical commute

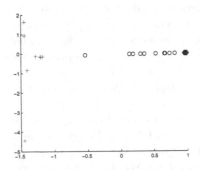

Fig. 6. Embeddings of the house sequences using MDS on the classical commute times

time is able to. One problem with the embedding using the classical commute is that it does not fully utilize the available dimensions. This problem can be seen to be a result of outliers which cause the majority of the nodes to be restricted to submanifolds within the embedding space. In Figure 6 we can see that the majority of the points in the embedding using the classical commute time are restricted to a single line. In comparison, the embedding using the quantum commute time fully utilizes the available dimensions, clearly and unambiguously distinguishing the three clusters.

6 Conclusions

We have shown how the commute time between a pair of vertices in a graph can be calculated for the continuous time quantum walk on a graph. The commute time between vertices can be used as a more robust measure of distance between vertices than the path length between vertices. We analyse how the quantum commute time compares with the classical commute time which was considered in [11]. The commute times between the vertices of a graph can be used to embed that graph in a low dimensional space. We show how the quantum commute time produces enbeddings of graphs that are less prone problems caused by outliers than classical commute time embeddings. In addition, quantum commute time embeddings make full use of both dimensions of the embedding space rather than restricting the graph to a submanifold, as can occur with classical commute time embeddings. We show how quantum commute time embeddings can be used to cluster real-world datasets effectively. In future work we hope to apply the quantum commute time embedding to the problem of image segmentation.

References

1. Borgwardt, K., Schönauer, S., Vishwanathan, S., Smola, A., Kriegel, H.: Protein function prediction via graph kernels. Bioinformatics 21, (June 2005)
2. Childs, A., Farhi, E., Gutmann, S.: An example of the difference between quantum and classical random walks. Quantum Information Processing 1(1-2), 35–43 (2002)
3. DePiero, F., Carlin, J.: Structural matching via optimal basis graphs. In: ICPR '06, pp. 449–452 (2006)
4. Emms, D., Severini, S., Wilson, R.C., Hancock, E.: Coined quantum walks lift the cospectrality of graphs and trees. In: EMMCVPR, pp. 332–345 (2005)
5. Kempe, J.: Quantum random walks – an introductory overview. Contemporary Physics 44(4), 307–327 (2003)
6. László Lovász. Combinatorics, Paul Erdös is Eighty, chapter Random Walks on Graphs: A Survey, János Bolyai Mathematical Society, vol. 2, pp. 353–398., Budapest (1996)
7. Luo, B., Wilson, R.C., Hancock, E.: Spectral embedding of graphs. Pattern Recognition 36(10), 2213–2230 (2003)
8. Meila, M., Shi, J.: A random walks view of spectral segmentation. In: AI and STATISTICS (AISTATS) 2001 (2001)
9. Nayak, A., Vishwanath, A.: Quantum walk on the line (2000)

10. Neuhaus, M., Bunke, H.: A random walk kernel derived from graph edit distance. In: SSPR, pp. 191–199 (2006)
11. Qiu, H., Hancock, E.: Robust multi-body motion tracking using commute time clustering. In: Proc ECCV, pp. 160–173 (2006)
12. Robles-Kelly, A., Hancock, E.: String edit distance, random walks and graph matching. International Journal of Pattern Recognition and Artificial Intelligence 18(3), 315–327 (2004)
13. Zhu, X., Ghahramani, Z., Lafferty, J.: Semi-supervised learning using gaussian fields and harmonic functions. In: ICML, pp. 1561–1566 (2003)

Graph Embedding in Vector Spaces by Means of Prototype Selection

Kaspar Riesen, Michel Neuhaus, and Horst Bunke

Department of Computer Science, University of Bern,
Neubrückstrasse 10, CH-3012 Bern, Switzerland
{riesen,mneuhaus,bunke}@iam.unibe.ch

Abstract. The field of statistical pattern recognition is characterized by the use of feature vectors for pattern representation, while strings or, more generally, graphs are prevailing in structural pattern recognition. In this paper we aim at bridging the gap between the domain of feature based and graph based object representation. We propose a general approach for transforming graphs into n-dimensional real vector spaces by means of prototype selection and graph edit distance computation. This method establishes the access to the wide range of procedures based on feature vectors without loosing the representational power of graphs. Through various experimental results we show that the proposed method, using graph embedding and classification in a vector space, outperforms the tradional approach based on k-nearest neighbor classification in the graph domain.

1 Introduction

The field of pattern recognition can be divided into two sub-fields, namely the statistical and the structural approach. In statistical pattern recognition, patterns are represented by feature vectors $(x_1, \ldots, x_n) \in \mathbb{R}^n$. The recognition process is based on the assumption that patterns of the same class are located in a compact region of $\mathbb{R}^n$. In recent years a huge amount of methods for clustering and classification of patterns represented by feature vectors have been proposed, such as k-means clustering, Bayes classifier, neural network, support vector machine, and many more. Object representations given in terms of feature vectors have a number of useful properties. For example, object similarity, or distance, can easily be computed by means of Euclidean distance. Computing the sum or weighted sum of two objects represented by vectors is straightforward, too. Yet, graph-based representations, which are used in the field of structural pattern recognition, have a number of advantages over feature vectors. Graphs are much more powerful and flexible than vectors, as feature vectors provide no direct possibility to describe structural relations in the patterns under consideration. Furthermore, vectors are constrained to a predefined length, which has to be preserved for all patterns encountered in a particular application. Obviously, graphs have a higher representational power than feature vectors. On the other

F. Escolano and M. Vento (Eds.): GbRPR 2007, LNCS 4538, pp. 383–393, 2007.

hand, a major drawback of graph representations is their lack of suitable methods for clustering and classification. This is mainly due to the fact that some of the basic operations needed in clustering and classification are not available for graphs. Hence, up to a few exceptions [1] classification of patterns represented by graphs is more or less limited to nearest-neighbor classifiers using some graph distance measure.

In this paper we describe a general method that aims at preserving the best of both approaches, that is the high representational power given by graphs and the large amount of algorithms for clustering and classification in feature vector spaces. Our approach is based on graph embedding in an n-dimensional feature vector space by means of prototype selection and edit distance computation. Originally, this idea was proposed in order to map patterns into dissimilarity spaces [2,3]. Later it was extended so as to map string representations into vector spaces [4]. In the current paper we go one step further and generalize the methods described in [4] to the domain of graphs. The key-idea of our approach is to use the distances of an input graph to a number of training graphs as vectorial description of the graph. Consequently, any statistical pattern recognition method will be applicable to such a pattern representation. In the remainder of this paper we will use the term *graph embedding* for the task of mapping graphs from the graph space into a vector space.

2   Graph Edit Distance

In contrast to statistical pattern recognition, where patterns are described by vectors, graphs do not offer a straightforward distance model like the Euclidean distance. However, a common approach to define a distance model for graphs is given by *graph edit distance*, which is one of the most flexible graph distance measures that is applicable to various kinds of graphs [5,6]. The key idea of graph edit distance is to define the dissimilarity, or distance, of graphs by the amount of distortion that is needed to transform one graph into another. These distortions are given by insertions, deletions, and substitutions of nodes and edges. Given two graphs – the source graph g_1 and the target graph g_2 – the idea is to delete some nodes and edges from g_1, relabel some of the remaining nodes and edges (substitutions) and possibly insert some nodes and edges, such that g_1 is finally transformed into g_2. A sequence of edit operations that transforms g_1 into g_2 is called an *edit path* between g_1 and g_2. One can introduce cost functions for each edit operation measuring the strength of the given distortion. The idea of such cost functions is that one can define whether or not an edit operation represents a strong modification of the graph. Hence, between two structurally similar graphs, there exists an inexpensive edit path, representing low cost operations, while for structurally different graphs an edit path with high costs is needed. Consequently, the *edit distance* of two graphs is defined by the minimum cost edit path between two graphs. The edit distance can be computed, for example, by a tree search algorithm [5,7]. Typically, the edit distance is used to classify an input graph by computing its distance to a number of training graphs and feeding the resulting distance values into a nearest-neighbor classifier. In our approach we make use of edit distances to construct a vectorial description of a given graph.

3   Graph Embedding by Means of Prototype Selection

The idea of embedding a population of graphs in an m-dimensional real vector space is motivated through the lack of suitable classification and clustering algorithms in the graph domain. An approach to graph embedding has been proposed in [8]. This method is based on algebraic graph theory and utilizes spectral matrix decomposition. In our approach we will explicitly make use of graph edit distance. Hence, we can easily deal with various kinds of graphs (labelled, unlabelled, directed, undirected, etc.) and utilize domains specific knowledge in defining the dissimilarity of nodes and edges through edit costs. Thus a high degree of robustness against various graph distortions can be achieved. The idea underlying our method was originally developed for the problem of embedding sets of feature vectors in a dissimilarity space [2,3,9,10]. In this paper we introduce an extension of this method to the domain of graphs. Assume we have a labeled set of training graphs, $T = \{g_1, \ldots, g_n\}$, and a dissimilarity measure $d(g_i, g_j)$. After having selected a set $P = \{p_1, \ldots, p_m\}$ of $m < n$ prototypes from T, we compute the dissimilarity of a graph $g \in T$ to each prototype $p \in P$. This leads to m dissimilarities, $d_1 = d(g, p_1), \ldots, d_m = d(g, p_m)$, which can be interpreted as an m-dimensional vector $(d_1, \ldots, d_m)$. In this way we can transform any graph from the training set, as well as any other graph from a validation or testing set, into a vector of real numbers. Note that whenever a graph from the training set, which has been choosen as a prototype before, is transformed into a vector $\mathbf{x} = (x_1, \ldots, x_m)$ one of the vector components is zero. Formally, if $T = \{g_1, \ldots, g_n\}$ is a training set of graphs and $P = \{p_1, \ldots, p_m\} \subseteq T$ is a set of prototypes, the mapping $t_m^P : T \to \mathbb{R}^m$ is defined as a function $t_m^P(g) \mapsto (d(g, p_1), \ldots, d(g, p_m))$ where $d(g, p_i)$ is a dissimilarity measure — in our case graph edit distance — between the graph g and the i-th prototype.

3.1   Prototype Selectors

The first problem to be solved is an appropriate choice of the prototype set $P = \{p_1, \ldots, p_m\}$. A good selection of m prototypes seems to be crucial to succeed with the classification algorithm in the feature vector space. Intuitively, the prototypes should mirror the distribution of the graphs in T as well as possible. That means protoypes should avoid redundancies in terms of selection of similar graphs, and prototypes should include as much information as possible. Hence, they should be uniformly distributed over the whole set of patterns. In this section we discuss five different algorithms for the task of prototype selection, one randomized method and four deterministic algorithms. Assume there are k different classes $c_1, \ldots, c_k$ represented in the training set T. We distinguish between class-wise and class-independent selection, that is to say that the selection can be executed over the whole training set T to get m prototypes, or the selection can be performed individually for each of the k different classes $c_1, \ldots, c_k$. In the latter case, l_i prototypes are selected independently for each class c_i such that $\sum_{i=1}^{k} l_i = m$. Whether the class-wise or class-independent method is more convenient depends on a number of factors, including the size of T, the structure of the graphs in T, whether or not classes are balanced, and the application. The five prototype selectors used in this paper are described below.

- CENTERS. The CENTERS prototype selector selects prototypes situated in the center of the graph set T. Assume that the set median graph of a set S, $median(S)$, is the most central graph. That is, the set median graph $median(S) \in S$ is the graph whose sum of distances to all other graphs $g_i \in S$ is minimal: $median(S) = \operatorname{argmin}_{g_1 \in S} \sum_{g_2 \in S} d(g_1, g_2)$. Then the set of prototypes $P = \{p_1, \ldots, p_m\}$ is iteratively constructed as follows:

$$ P_i = \begin{cases} \varnothing & \text{if } i = 0 \\ P_{i-1} \cup \{g_i\} & \text{if } 0 < i \leqslant m \end{cases} \text{, where } g_i = median(T \setminus P_{i-1}). $$

It seems that the CENTERS prototype selector is not a very appropriate idea for selecting m prototypes, because it neither avoids redundancies nor distributes the prototypes uniformely. Nevertheless, we mention it here for the purpose of completeness. To obtain a better distribution, one can apply the CENTERS prototype selector also class-wise (CENTERSC). In this case, supposably, the prototypes mirror the given distribution better than the class-independent version.

- RANDOM. A random selection of m prototypes from T is performed. Of course, the RANDOM prototype selector can be applied class-independent or class-wise (RANDOMC). RANDOMC provides a random selection of l_i prototypes per class c_i.

- SPANNING. A set of prototypes, P, is selected by the SPANNING prototype selector by means of the following iterative procedure. The first prototype selected is the set median graph. Each additional prototype selected by the spanning prototype selector is the graph the furthest away from the already selected prototype graphs.

$$ P_i = \begin{cases} median(T) & \text{if } i = 1 \\ P_{i-1} \cup \{p_i\} & \text{if } 1 < i \leqslant m \end{cases} \text{, where } p_i = \operatorname*{argmax}_{g \in T \setminus P_{i-1}} \min_{p \in P_{i-1}} d(g, p). $$

The SPANNING prototype selector can be applied class-independent or class-wise (SPANNINGC).

- k-CENTERS. The k-CENTERS prototype selector tries to choose m graphs from T so that they are evenly distributed with respect to the dissimilarity information given by d. The algorithm proceeds as follows:

  1. Select an initial set of m prototypes: $P_0 := \{p_1, \ldots, p_m\}$. One can choose the initial prototypes randomly or by a more sophisticated procedure, for example, the above-mentioned spanning prototype selector.

2. Construct m sets S_i where each set consists of one prototype: $S_1 = \{p_1\}, \ldots, S_m = \{p_m\}$ For each graph $g \in T \setminus P$ find its nearest neighbor $p_i \in P$ and add the graph under consideration to the set S_i corresponding to prototype p_i. This step results in m disjoint sets with $T = \bigcup_{1 \leqslant i \leqslant m} S_i$.
3. For each set S_i find its center c_i, that is the graph for which the maximum distance to all other objects in S_i is minimum.
4. For each center c_i, if $c_i \neq p_i$, replace p_i by c_i in S_i. If any replacements is done, return to step 2, otherwise stop.

The procedure stops when no more changes in the sets S_i occur. The prototypes are given by the centers of the m disjoint sets. The k-CENTERS prototype selector can be applied class-independent as well as class-wise (k-CENTERSC). Note that this prototype selector is similar to k-means clustering [11].

– TARGETSPHERE. The TARGETSPHERE prototype selector first looks for the center graph g_c in T. The center graph is the graph for which the maximum distance to all other graphs in T is minimum. After finding the center graph, the graph the furthest away from g_c, i.e. the graph $g_f \in T$ whose distance to g_c is maximum, is located. Both graphs (g_c and g_f) are selected as prototypes. The distance $d_{max} = d(g_c, g_f)$ is then divided in $m - 1$ partitions with $interval = \frac{d_{max}}{m-1}$. The $m - 2$ graphs that are located the nearest to the interval borders in terms of edit distance are selected as prototypes:

$$P_i = \begin{cases} \{g_c, g_f\} & \text{if } i = 0 \\ P_{i-1} \cup \{g_i\} & \text{if } 0 < i \leqslant m - 2 \end{cases}, \text{ where } g_i = \underset{g \in T \setminus P_{i-1}}{\operatorname{argmin}} |d(g, g_c) - i \cdot interval|.$$

The TARGETSPHERE prototype selector can be applied class-independent as well as class-wise (TARGETSPHEREC).

Of course, one can imagine other techniques and strategies for prototype selection. The intention of all methods remains the same — finding a good selection of m prototypes that lead to a good performance of the resulting classifier in the vector space.

4 The Classification Problem

4.1 Classification in Graph Spaces — k-NN Classifier

The traditional approach to addressing the classification problem in a graph space is to apply a k-nearest-neighbor classifier (k-NN) in conjunction with edit distance. Given a labeled set of training graphs, an unknown graph is assigned to the class that occurs most frequently among the k nearest graphs (in terms of edit distance) from the training set. Formally, let us assume that a pattern space X, a space of class labels Y, and a labeled training set of patterns $\{(x_i, y_i)\}_{i=1,\ldots,m} \subseteq X \times Y$ is given. If $\{(x_{(1)}, y_{(1)}), \ldots, (x_{(k)}, y_{(k)})\} \subseteq \{(x_i, y_i)\}_{i=1,\ldots,m}$ are the k patterns that have the smallest distance $d(x, x_{(i)})$ to a test pattern x, the k-NN classifier $f : X \to Y$ can be defined by

$$f(x) = \underset{y \in Y}{\operatorname{argmax}} |\{(x_{(i)}, y_{(i)}) : y_{(i)} = y\}|$$

If $k = 1$ the k-NN classifiers decision is based on just one element from the training set, no matter if this element is an outlier or a true class representative. Obviously, a choice of parameter $k > 1$ reduces the influence of outliers by evaluating which class occurs most frequently in a neighborhood around the test pattern. This classifier in the graph domain will serve us as a reference system.

4.2   Classification in Vector Spaces — Support Vector Machine Classifier

As already pointed out, in vector spaces a large amount of methods for pattern classification exist. Besides the k-NN classifier, one can choose among more sophisticated algorithms, such as neural network, Bayes classifier, decision tree classifier, support vector machine, and others. Pattern classification by means of support vector machines (SVMs) has become very popular recently. In the present paper we want to compare the classification accuracy achieved by k-NN classifiers in the graph domain with k-NN classifiers and SVM in vector spaces. It is our objective to find out if we can outperform the classification accuracy obtained by k-NN classifiers in the graph space by classifiers relying on vectorial representations after embedding the graphs in an m-dimensional vector space, especially by SVM. For the sake of completeness, we give a brief overview of SVMs below. For a more thorough introduction we refer the reader to [12,13,14]

The basic idea of SVM is to separate classes of patterns by hyperplanes. Assume a pattern space $X = \mathbb{R}^n$, two classes $Y = \{-1, +1\}$ and a labeled training set $\{(x_i, y_i)\}_{i=1,\ldots,m} \subseteq X \times Y$ are given, and the two classes are linearly seperable. Then, there exist infinitely many possible separating hyperplanes that correctly classify the training data. The fact that all patterns are classified correctly can be written as $y_i \cdot (\langle w, x_i \rangle + b) > 0$ for $i = 1,\ldots, m$, with parameters $w \in \mathbb{R}^n$ and $b \in \mathbb{R}$. Intuitively, one would choose a hyperplane such that its distance to the clostest pattern of either class is maximal. Such hyperplanes are commonly called maximum-margin hyperplanes. The sum of the distances from the hyperplane to the closest pattern of each class is commonly termed *margin*. The maximum-margin hyperplane is expected to perform best on an independent test set. Since multiplying the parameters w and b with a constant does not change the hyperplane, one can rescale them such that $\min_{i=1,\ldots,m} |\langle w, x_i \rangle + b| = 1$. This rescaled hyperplane is called to be in *canonical form*. Assume a hyperplane in canonical form is given. Then it is obvious that $margin = \frac{2}{||w||}$. Hence, the smaller the length of the weight vector $||w||$, the larger is the margin. Since we are looking for an optimal hyperplane, which maximizes *margin*, we have to find parameters w and b such that

- $margin = \frac{2}{||w||}$ is maximum
- subject to $y_i \cdot (\langle w, x_i \rangle + b) \geqslant 1$ for $i = 1,\ldots, n$.

The first line corresponds to the maximization of *margin*, while the second line assures that all training samples are classified without error. In a more realistic scenario the two classes would not be linearly seperable anymore. Thus the second condition does not hold. In order to handle linearly non-seperable data, so called *slack variables* ξ are used. Whenever a training element x_i is misclassified by the hyperplane, the corresponding slack variable ξ_i is greater than zero. Thus we have to find parameters w and b such that

- *margin* $= \frac{||w||^2}{2} + C \sum_{i=1}^{m} \xi_i$ is minimum
- subject to $y_i \cdot (\langle w, x_i \rangle + b) \geqslant 1 - \xi_i$ for $i = 1, \ldots, m$.

Minimization of $\frac{||w||^2}{2}$ is equivalent to the maximization of *margin*. Quantity $C \geqslant 0$ denotes a regularization parameter to control whether the maximization of *margin* or the minimization of the sum of errors is more important. Finally, we have to deal with general non-linear classification problems, where a linear hyperplane will not work any longer. Fortunately, one can show that we can transform a linear to a non-linear classifier by only substituting the original dot product with a specific kernel function K [13,14]. Hence, by changing the dot product to a kernel function we can get different non-linear classifiers. In our experiments we used the following kernel functions:

- **Linear:** $K(u, v) = u' \cdot v$
- **Polynomial:** $K(u, v) = (\gamma \cdot u' \cdot v)^d$
- **Radial Basis Function (rbf):** $K(u, v) = exp(-\gamma \cdot ||u - v||^2)$

where $\gamma \in \mathbb{R}$ and $d \in \mathbb{N}$.

5   Experimental Results

As main contribution of this paper, we have introduced a general approach for transforming graphs into n-dimensional real vector spaces by means of prototype selection and graph edit distance. It is furthermore our intention to demonstrate that certain classification tasks can be better solved with methods that use the resulting vectorial patterns rather than the original graph representation. Hence, the reference system in our experiments is given by a k-NN classifier in the graph domain, while the proposed statistical classifiers in real vector spaces are given by different SVMs. Note that k-NN classifiers are the only classifiers that can be directly applied in the original graph domain. In each of our experiments we make use of three disjoint graph sets, viz. *validation set*, *test set* and *training set*. The validation set is used to determine optimal parameter values for graph embedding and classification. The embedding parameters consist of the number of prototypes, i.e. the dimensionality of the resulting feature vector space, and the best performing embedding method, while the parameters for classification consist of parameter k for the nearest neighbor classifier and the different parameters for the SVMs, i.e. C, γ and d. That is, for each embedding method and dimensionality an individual SVM is trained. The parameter values, the embedding method, and the dimensionality that result in the lowest classification error on the validation set are then applied to the independent test set.

5.1 Letter Database

The first database used in the experiments consists of graphs representing distorted letter drawings. In this experiment we consider the 15 capital letters of the Roman alphabet that consists of straight lines only (*A, E, F, ...*). For each class, a prototype line drawing is manually constructed. To obtain aribtrarily large sample sets of drawings with arbitrarily strong distortions, distortion operators are applied to the prototype line drawings. This results in randomly shifted, removed, and added lines. These drawings are then converted into graphs in a simple manner by representing lines by edges and ending points of lines by nodes. Each node is labeled with a two-dimensional attribute giving its position. The graph database used in our experiments is composed of a training set, a validation set, and a test set, each of size 750. In Table 1 we give the classification accuracy of a k-nearest neighbor classifier in the graph space, a k-nearest neighbor classifier in the feature vector space, and the 3 different SVM-classifiers. The best accuracy on the validation set with all SVMs is achieved by the k-CENTERSC prototype selector (Distortions 0.1 and 0.5) and the TARGETSPHEREC prototype selector (Distortions 0.3, 0.7 and 0.9). Therefore, these prototype selectors have been used for all SVMs when classifying the test set. It turns out that classification accuracy can be improved by all considered SVMs on all distortion levels. Note that 9 out of 15 improvements are statistically significant.

Table 1. Letter Database: Classification accuracy in the graph and vector space

	Ref. System	Embedding classifiers			
Distortion	k-NN (graph)	k-NN (vector)	SVM (lin)	SVM (poly)	SVM (rbf)
0.1	98.27	98.53	98.93	98.40	98.53
0.3	97.60	97.47	98.53 ○	98.53 ○	98.80 ○
0.5	94.00	93.60	97.07 ○	97.20 ○	96.93 ○
0.7	94.27	92.53 ●	95.33	95.47	95.47
0.9	90.13	91.20	92.93 ○	92.93 ○	92.93 ○

○ Statistically significantly better than the reference system ($\alpha = 0.05$).
● Statistically significantly worse than the reference system ($\alpha = 0.05$).

5.2 Real World Data

For a more thorough evaluation of the proposed methods we additonally use three real world data sets. First we apply the proposed method to the problem of image classification. Images are converted into graphs by segmenting them into regions, eliminating regions that are irrelevant for classification, and representing the remaining regions by nodes and the adjacency of regions by edges [15]. The Le Saux image database consists of five classes (*city, countryside, people, snowy, streets*) and is split into a training set, a validation set and a test set of size 54 each. The best accuracy on the validation set with all SVMs is achieved by the TARGETSPHERE prototype selector. The classification accuracies obtained by the different methods are given in the first row of Table 2. We note that the SVM

Table 2. Fingerprint-, Image- and Molecules Database: Classification accuracy in the graph and vector space

	Ref. System	Embedding classifiers			
Database	k-NN (graph)	k-NN (vector)	SVM (lin)	SVM (poly)	SVM (rbf)
Le Saux	57.4	48.2	59.3	57.4	64.8
NIST-4	82.6	81.8	85.4 ○	82.4	85.0 ○
Molecules	97.1	95.9 ●	97.7	98.2 ○	98.1 ○

○ Statistically significantly better than the reference system ($\alpha = 0.05$).
● Statistically significantly worse than the reference system ($\alpha = 0.05$).

with polynomial kernel results in the same error rate as the reference system, while the other two kernels lead to lower error rates. Although the SVM with the rbf kernel function improves the accuracy by 7.4%, this improvement is not statistically significant. This is due to the small size of Le Saux database.

The second real world dataset is given by the NIST-4 fingerprint database [16]. We construct graphs from fingerprint images by extracting characteristic regions in fingerprints and converting the results into attributed graphs [17]. We use a validation set of size 300 and a test and training set of size 500 each. In this experiment we address the 4-class problem (*arch, left-loop, right-loop, whorl*). Validation of parameter values needed by linear and polynomial SVM prove to be difficult. With many parameter value combinations, both SVMs are not able to terminate the optimization in a reasonable time. Thus, the optimization of the parameter values is based on a subset of all parameter combinations only. Nevertheless, the linear and rbf kernel SVMs achieve statistically significantly better results than the reference system. These results are achieved by the TAR-GETSPHEREC prototype selector. The number of choosen prototypes per class is proportional to the respective class size.[1]

Finally, we apply the proposed method of graph embedding and subsequent SVM classification to the problem of molecule classification. To this end, we construct graphs from the AIDS Antiviral Screen Database of Active Compounds [18]. Our molecule database consists of two classes (*active, inactive*), which represent molecules with activity against HIV or not. We use a validation set of size 250, a test set of size 1500 and training set of size 250. Thus, there are 2000 elements totally (1600 inactive elements and 400 active elements). The molecules are converted into graphs in a straightforward manner by representing atoms as nodes and the covalent bonds as edges. Nodes are labeled with the number of the corresponding chemical symbol and edges by the valence of the linkage. The results achieved on this database are shown in the third row of Table 2. Although the accuracy of the reference system in the graph domain is quite high, it can be improved by graph embedding and SVM classification. In two out of three cases, the improvement is statistically significant. On this data set the SPANNING prototype selector obtains the best result on the validation set and is therefore used on the test set.

---

[1] In contrast with other databases, the individual classes are of different size.

6   Conclusions

Although graphs have a higher representational power than feature vectors, there is a lack of suitable methods for pattern classification using graph representations. By contrast, a large number of methods for classification have been proposed for object representations given in terms of feature vectors. In this paper, we propose a general approach for bridging the gap between structural and statistical pattern recognition. The idea is to map graphs to an n-dimensional real vector space by means of prototype selection and graph edit distance. To this end, we discuss different prototype selectors with the objective of finding a good distribution of these prototypes. With the proposed graph embedding a large number of different methods from statistical pattern recognition become available to graph representations. It has been our intention in this paper to improve the accuracy achieved by nearest neighbor classifiers in the graph domain by classifiers operating on vectorial representations. We used SVMs as a popular method from statistical pattern recognition and showed that this approach outperforms the nearest neighbor classifiers in the graph domain. From the results of our experiments, one can conclude that the classification accuracy can be statistically significantly enhanced by most SVMs and different prototype selection algorithms. In our future work we will study additional statistical classifiers and try to further improve the recognition accuracy by using classifier ensembles.

Acknowledgements

This work has been supported by the Swiss National Science Foundation (Project 200021-113198/1). Furthermore, we would like to thank R. Duin for valuable discussions and hints regarding our embedding methods. Finally, we thank B. Le Saux for making the Le Saux database available to us.

References

1. Bianchini, M., Gori, M., Sarti, L., Scarselli, F.: Recursive processing of cyclic graphs. IEEE Transactions on Neural Networks 17(1), 10–18 (2006)
2. Pekalska, E., Duin, R., Paclik, P.: Prototype selection for dissimilarity-based classifiers. Pattern Recognition 39(2), 189–208 (2006)
3. Duin, R., Pekalska, E.: The Dissimilarity Representations for Pattern Recognition: Foundations and Applications. World Scientific, Singapore (2005)
4. Spillmann, B., Neuhaus, M., Bunke, H., Pekalska, E., Duin, R.: Transforming strings to vector spaces using prototype selection. In: Yeung, D.-Y., Kwok, J.T., Fred, A., Roli, F., de Ridder, D. (eds.) Structural, Syntactic, and Statistical Pattern Recognition. LNCS, vol. 4109, pp. 287–296. Springer, Heidelberg (2006)
5. Bunke, H., Allermann, G.: Inexact graph matching for structural pattern recognition. Pattern Recognition Letters 1, 245–253 (1983)
6. Sanfeliu, A., Fu, K.S.: A distance measure between attributed relational graphs for pattern recognition. IEEE Transactions on Systems, Man, and Cybernetics (Part B) 13(3), 353–363 (1983)

7. Hart, P.E., Nilsson, N.J., Raphael, B.: A formal basis for the heuristic determination of minimum cost paths. IEEE Transactions of Systems, Science, and Cybernetics 4(2), 100–107 (1968)
8. Wilson, R.C., Hancock, E.R., Luo, B.: Pattern vectors from algebraic graph theory. IEEE Trans. on Pattern Analysis ans Machine Intelligence 27(7), 1112–1124 (2005)
9. Hjaltason, G., Samet, H.: Properties of embedding methods for similarity searching in metric spaces. IEEE Trans. on Pattern Analysis ans Machine Intelligence 25(5), 530–549 (2003)
10. Roth, V., Laubm, J., Kawanabe, M., Buhmann, J.: Optimal cluster preserving embedding of nonmetric proximity data. IEEE Trans. on Pattern Analysis ans Machine Intelligence, vol. 15(12) (2003)
11. MacQueen, J.: Some methods for classification and analysis of multivariant observations. In: Proc. 5th. Berkeley Symp University of California Press 1, pp. 281–297 (1966)
12. Burges, C.: A tutorial on support vector machines for pattern recognition. Data. Mining and Knowledge Discovery 2(2), 121–167 (1998)
13. Shawe-Taylor, J., Cristianini, N.: Kernel Methods for Pattern Analysis. Cambridge University Press, Cambridge (2004)
14. Schölkopf, B., Smola, A.: Learning with Kernels. MIT Press, Cambridge, MA (2002)
15. Le Saux, B., Bunke, H.: Feature selection for graph-based image classifiers. In: Marques, J.S., de la Blanca, N.P., Pina, P. (eds.) IbPRIA 2005. LNCS, vol. 3523, pp. 147–154. Springer, Heidelberg (2005)
16. Watson, C.I., Wilson, C.L.: NIST special database 4, fingerprint database. National Institute of Standards and Technology, March (1992)
17. Neuhaus, M., Bunke, H.: A graph matching based approach to fingerprint classification using directional variance. In: Kanade, T., Jain, A., Ratha, N.K. (eds.) AVBPA 2005. LNCS, vol. 3546, pp. 191–200. Springer, Heidelberg (2005)
18. Development Therapeutics Program DTP. Aids antiviral screen (2004), http://dtp.nci.nih.gov/docs/aids/aids_data.html.

Grouping Using Factor Graphs: An Approach for Finding Text with a Camera Phone

Huiying Shen and James Coughlan

Smith-Kettlewell Eye Research Institute
San Francisco, CA 94115
hshen@ski.org

Abstract. We introduce a new framework for feature grouping based on factor graphs, which are graphical models that encode interactions among arbitrary numbers of random variables. The ability of factor graphs to express interactions higher than pairwise order (the highest order encountered in most graphical models used in computer vision) is useful for modeling a variety of pattern recognition problems. In particular, we show how this property makes factor graphs a natural framework for performing grouping and segmentation, which we apply to the problem of finding text in natural scenes. We demonstrate an implementation of our factor graph-based algorithm for finding text on a Nokia camera phone, which is intended for eventual use in a camera phone system that finds and reads text (such as street signs) in natural environments for blind users.

1 Introduction

The ability to read street signs and other informational signs would be very useful to people who have visual impairments that make it difficult or impossible to find and read signs. A growing body of work in computer vision tackles the problem of finding text in natural scenes [1–4], a task that is especially challenging in highly cluttered environments; once text is located, well-established OCR (optical character recognition) techniques can be used to read it. So far almost all of this work on finding text has been implemented on standard personal (e.g. desktop or laptop) computers. While computers are continually improving in terms of power and portability, they are still too heavy, bulky and expensive to be convenient for most visually impaired users.

An attractive hardware alternative is the camera cell phone (or smart phone), which is lightweight, inexpensive, multi-purpose and nearly ubiquitous. Since most people already carry a cell phone, it has the added benefit of requiring *no additional device to purchase or carry*.

However, an important limitation of the camera phone is that it has substantially less processing power than a standard computer. The camera phone CPU is significantly slower than the kind found in desktop computers; in addition, the camera phone lacks a floating point processing unit (FPU), which means that floating point calculations – a mainstay of most computer vision algorithms – are particularly slow. Integer arithmetic is faster on the camera phone, but it is still up to an order of magnitude slower than on a standard computer.

F. Escolano and M. Vento (Eds.): GbRPR 2007, LNCS 4538, pp. 394–403, 2007.

The need for an algorithm to find text efficiently enough to run on a camera phone has motivated us to develop a new framework for simple, fast text segmentation. To this end we have adapted a graphical model-based framework originally developed for finding pedestrian crosswalks in traffic intersections [5] to the problem of finding text in natural scenes [6]. In this approach, we cast text detection as a problem of segmenting edge-based text features extracted from an image into figure or ground. The signature of a text region is an abundance of text features that are aligned in a fairly regular way. By contrast, text features occur more sparsely outside of text regions, and are spaced less regularly. The purpose of the graphical model framework is to exploit this pattern to segment all the text features in an image into figure or ground, corresponding to text or non-text regions, respectively. The graphical model achieves the desired behavior by expressing appropriate grouping criteria among the text features.

In this paper we describe a new framework for segmentation that is an outgrowth of our previous work, which provides for more expressive grouping criteria, and which is simpler and faster. The framework is based on factor graphs [7], which provide a convenient way of expressing interactions of any order in a graphical model. We have used this framework to develop a text-finding algorithm that relies almost entirely on integer arithmetic calculations, which enables an efficient camera phone implementation. Preliminary experiments demonstrate the ability of the algorithm to segment text regions in an image in several seconds on a camera phone (Nokia 6681, 220 MHz ARM CPU).

2 Past Work on Text Detection

A large body of work addresses the problem of detecting and reading printed text, but so far this problem is considered solved only in the domain of OCR (optical character recognition). This domain is limited to the analysis of high-resolution, high-contrast images of printed text with little background clutter. The broader challenge of detecting and reading text in highly cluttered scenes, such as indoor or outdoor scenes with informational signs, is much more difficult and is a topic of ongoing research. (We focus on the problem of segmentation in this paper, leaving the task of reading segmented text for future research.)

Many text segmentation algorithms employ deterministic, bottom-up processes for grouping text features into candidate text regions using features such as edges, color or texture [1–4]; a recent and comprehensive survey is found in [8]. Statistical methods have recently been developed, such as an Adaboost-based algorithm [9] that uses a cascade of filters trained from a labelled data set of natural scenes containing text.

We build on our recent work [6] casting text detection as a figure-ground segmentation problem represented using a probabilistic graphical model. We now propose to use a novel factor graph grouping technique that permits a more expressive graphical model to be used – specifically, one that allows *higher-order* interactions among several features, rather than being restricted to pairwise interactions (between pairs of features). The advantage of the factor graph grouping framework is that it allows grouping to be performed on very simple features that can be extracted rapidly from an image. The simple features can be grouped according to complex criteria using higher-order factors. As we will see later, the computational complexity that could arise from the use of higher-order factors is avoided because of the particular form of the factor graph.

Recent work related to ours [10, 11] also uses a graphical model framework. Unlike our approach, the former work tackles text detection solely in documents, and the latter work uses color to initiate the segmentation and requires images in which individual letters are clearly visible. By contrast, we have designed our algorithm to process natural grayscale images with letters that may be poorly resolved (e.g. Fig. 5). This allows us to segment text in images in which the letters appear small and/or process the images at coarser scales (which decreases the amount of computation required for segmentation).

3   Grouping with Factors

Most methods devised for clustering or grouping data (such as normalized cuts [12] and graphical-model based typical cuts [13]) rely on affinities defined on *pairs* of data points to express the likelihood that two points should be grouped together. However, many clustering problems necessitate the use of higher-order affinities; for instance, the problem of grouping points on a 2-D plane into lines requires an affinity defined on triplets of points, since every pair of points is trivially collinear. Some recent work [14] has investigated hypergraph partitioning techniques for handling these higher-order affinities.

We propose that a particular form of graphical model known as a *factor graph* [7] provides a natural framework for grouping with higher-order affinities that results in simple and efficient grouping algorithms. Our framework is well suited to modeling object-specific figure-ground segmentation, which is how we cast the problem of text detection. It is inspired by object-specific figure-ground segmentation work by [15] and from work on clustering using graphical models [13].

The factor graph provides a convenient way of representing interactions among arbitrary numbers of variables, generalizing the pairwise interactions often used in graphical models. An important property of factor graphs is that, as for all graphical models, rapid inference can be performed on them using a form of belief propagation (BP).

In the next subsection we introduce factor graphs and factor graph BP and demonstrate how they can be used to implement figure-ground segmentation. The application of figure-ground segmentation to text detection is described subsequently.

3.1   Factor Graphs and Belief Propagation

Factor graphs [7] provide a convenient framework for representing graphical models in a way that shows interactions of any order in a visual format. Fig. 1 shows an example of a factor graph. Each square node represents a factor, or interaction, among one or more variables, depicted by circles, and the topology of the factor graph indicates how the joint distribution of all variables factors.

Belief propagation (BP) can be extended to factor graphs. Here we present a brief overview of factor graph BP, using notation similar to that of [7]. We note that factor graph BP is very similar to standard BP: messages are sent from one node to another, and each message is a function of the state of one node variable. However, a chief

Fig. 1. Factor graph. This graph represents a distribution on four variables w,x,y,z (drawn as circles) using three factors f, g, h (drawn as squares). Edges connect factors with the variables they influence. The joint distribution represented by this factor graph is $P(w,x,y,z) = f(w,x,y)g(x,y,z)h(y,z)$.

difference is that there are *two* types of messages in factor graph BP, those that are sent from variables to factors and those sent from factors to variables. We consider the max-product version of factor BP rather than the sum-product version because the former can be implemented very efficiently in the log domain using only addition and subtraction, without the need for multiplication. (We convert to the log domain by taking the log of the original max-product equations, and renaming the messages and factors to absorb the logs.) As we will see, this arithmetic simplification is key to our ability to implement factor BP efficiently on a camera phone CPU.

The max-product version of factor graph BP is expressed in the log domain according to the following update equations:

$$m_{x \to f}(x) \leftarrow \sum_{h \in n(x)\setminus\{f\}} m_{h \to x}(x) \tag{1}$$

where $n(x)$ is the set of neighbors of x, and

$$m_{f \to x}(x) \leftarrow \max_{\sim\{x\}}\left(f(X) + \sum_{y \in n(f)\setminus\{x\}} m_{y \to f}(y)\right) \tag{2}$$

where $X = n(f)$ is the set of arguments of function f, and $\sim\{x\}$ denotes the set of all arguments of f except for x. (The sums in these two equations correspond to products before converting to the log domain.) Note that, for each factor f and neighboring variable x, updating all messages $m_{f \to x}(x)$ has worst-case complexity $O(|S|^m)$, where m is the number of variables coupled by factor f and $|S|$ is the number of allowed states of each of the m variables (assuming the state spaces are the same size for each variable). This is because Eq. 2 must be iterated for each value of x on the left-hand side, and for each value of x the max must be evaluated over the remaining $m - 1$ variables $\sim\{x\}$.

Once the messages have converged to some value (which, in general, we can only hope happens after enough message updates), we can calculate the "belief function" for each node:

$$b(x) = \sum_{f \in n(x)} m_{f \to x}(x) \tag{3}$$

In sum-product BP, the belief is an estimate of the marginal probability of each node. In max-product BP, the belief is a function with a weaker property: the state that maximizes the belief of a node is an estimate of the node's state in the most likely global configuration of states across the entire graphical model (i.e. the MAP estimate if the graphical model is interpreted as representing a posterior distribution).

3.2   Factor Graphs for Figure-Ground Segmentation

We now specialize our discussion of factor graphs to the problem of figure-ground segmentation. In this context, each node variable x_i in the factor graph is binary-valued: $x_i = 1$ and $x_i = 0$ represent figure and ground states, respectively. A factor of m variables $f(x_1, \ldots, x_m)$ expresses the likelihood of every possible assignment of states to all m variables, irrespective of the other variables in the factor graph. A *unitary* factor of one node variable $f(x)$ enforces a prior bias towards figure or ground, independent of other nodes.

A simple form of this type of factor graph – a graphical model with only unitary factors and pairwise interactions – was used in our previous work on figure-ground segmentation [6]. Generalizing from that form, we now stipulate that each factor in our graph has a special form: $f(x_1, \ldots, x_m)$ is non-zero *only when* $x_1 = \ldots = x_m = 1$, i.e. when all the variables are in figure states, and $f(x_1, \ldots, x_m) = 0$ otherwise. In other words, the factor can only assume two possible values: one value when its arguments (variables that it influences) are in the "figure" state, and zero otherwise.

We make a further non-negativity requirement that $K_f \doteq f(x_1 = 1, \ldots, x_m = 1) \geq f(x_1 = 0, \ldots, x_m = 0) = 0$. This additional requirement implies great computational savings: factor BP for this factor graph will converge in *only one iteration* of message updates. It is straightforward to show this by first initializing all messages $m_{f \to x}$ and $m_{x \to f}$ to zero and noticing that the first update will yield $m_{f \to x}(0) = 0$ and $m_{f \to x}(1) = K_f$. (The messages from variables to factors all remain zero: $m_{x \to f}(x) = 0$ for $x = 0$ and $x = 1$.) Thanks to the non-negativity requirement, subsequent message updates will leave the message values unaltered (provided the messages are "normalized" after each iteration by uniformly shifting them by an appropriate amount – an operation that does not affect the outcome of BP).

The beliefs then have the following simple form: $b_x(x = 1) = \sum_{f \in n(x)} K_f$ and $b_x(x = 0) = 0$. Since $b_x(x = 0) = 0$ is fixed and only the difference $b_x(x = 1) - b_x(x = 0)$ matters, we use a simplified notation to represent the beliefs:

$$B_x = \sum_{f \in n(x)} K_f \tag{4}$$

This result shows that the beliefs can be calculated without the need for any message updates! This computational savings comes at a price, however. First, the non-negativity requirement means that a unitary prior favoring ground over figure (e.g. $f(x = 0) = 0, f(x = 1) < 0$) is not allowed. As a result, the state configuration that maximizes the sum of all the (non-negative) factors in the graph is simply $x = 1$ for all node variables x, which is a degenerate result. A simple way to work around this problem is to omit all unitary factors but to assign each node variable x to figure only if its belief B_x is sufficiently large. In this way, only nodes with sufficient support from other nodes will be assigned to figure, and the rest will be assigned to ground.

Second, the fact that BP converges in only one iteration means that information will not be propagated over long distances in the factor graph. Such propagation can be very useful for "filling in" an image region that has weak evidence with stronger evidence outside the region. However, the benefit of our simple factor graph is that it is easy to include *factors of arbitrarily high order*, whereas in general the computational complexity increases exponentially with the factor order (see the complexity analysis immediately following Eq. 2). Since information is still propagated within overlapping factors – which can be of very high order and thus encompass large regions of an image – this means that information can still be propagated over long distances.

4   Grouping Text Features

We have devised a bottom-up procedure for grouping edges into composite features that are signatures of regions containing text. The next subsection describes how these features are constructed, and the subsequent subsection explains how the features are grouped into factors.

4.1   Constructing Features

Speed is a major consideration, so we used a very simple edge detector to provide the basic elements to be grouped. First, the image is blurred slightly, converted to grayscale and decimated to 640×480 pixels or smaller. Two kinds of edges are detected, corresponding to local maxima or minima of the horizontal and vertical image intensity derivatives. The edges are grouped into line segments, which are approximately straight and fully connected sequences of edge pixels (with between 3 and 20 pixels, which sets an appropriate range of scales for text detection). There are two kinds of line segments, those that are oriented (approximately) vertically and those that are oriented (approximately) horizontally. Vertical segments that are sufficiently close together and have opposite polarity are grouped into "weakly matched vertical edges", shown in Fig. 2(a). "Weakly matched horizontal edges" are determined in a similar way (see Fig. 2(b)). As the figure suggests, weakly matched edges are features designed to be prevalent along the borders of letter strokes.

Next we prune the set of weakly matched vertical segments to obtain our final features, "anchored verticals" (see Fig. 3). An anchored vertical feature is a weakly

Fig. 2. Edge features used to construct text features shown on cropped image of street sign. Left, weakly matched vertical edge segments. Right, weakly matched horizontal edge segments. Edges shown in red and green to indicate opposite polarities.

Fig. 3. Anchored verticals. Left, anchored verticals shown for image in previous figure. Right, shown for an image. Note density and regularity of anchored verticals in text region, where bottoms and tops tend to be at the same level. By contrast, anchored verticals are scattered sparsely and irregularly throughout rest of the image.

matched vertical segment whose topmost or bottommost pixel lies sufficiently close to the leftmost or rightmost pixel of a weakly matched horizontal segment. By "sufficiently close" pixels we mean that they are either identical or one pixel is one of the eight nearest neighbors of the other.

As Fig. 3 shows, anchored verticals have a distribution on text regions that is significantly different from the distribution outside of text regions. Anchored verticals are distributed densely on text regions, and their bottoms and tops tend to be aligned to the same level. By contrast, outside of text regions, anchored verticals are distributed more sparsely and irregularly. We will exploit this differential distribution of anchored verticals to segment out text regions in an image.

4.2 Building Factors

Having constructed a set of useful anchored vertical features that have a distinctive signature in text regions, we now proceed to construct a factor graph based on these features. In the factor graph, each anchored vertical is a variable node. Factor nodes are defined as groups of anchored verticals that may plausibly belong to one text region. As we have shown, our particular factor graph has the advantage that the results of BP can be calculated very simply without any iterative message updating. However, the trade-off is that the search for building suitable factor nodes is computationally intensive, since many of these factors are high-order (i.e. they bind many anchored verticals).

The search for factors is conducted by means of a sliding window of size 5 pixel rows by 30 pixel columns. Rather than having to slide the window pixel by pixel across each column and row of the image, the left side of the window is aligned to the top or bottom of each node (anchored vertical), one after the other. For each node the window is aligned to (which we call the "reference" node), all node tops or bottoms to its right are found that lie in the window. Two kinds of factors are constructed, one linking the tops of the nodes and the other linking the bottoms. For simplicity we will describe the case of linking node tops; the same procedure is followed for linking node bottoms.

For each candidate factor f, we calculate a measure of alignment with respect to a horizontal line at the level of the top of the reference node. This alignment measure is the maximum of the absolute value of the vertical pixel distance between each node top and the horizontal line, where the maximum is taken over all nodes in the window. To accomodate text that may be slightly off-horizontal, we also calculate similar alignment measures for two other lines, one with a slope of 1/10 and the other with slope -1/10. The final error measure E_f corresponding to factor f equals the minimum of the three alignment errors, divided by the number of nodes in the sliding window.

Empirically we find that error measures E_f of 5 or more correspond to very weak factors, so we discard any factor f for which $E_f \geq 5$. Otherwise, we define $K_f = 5 - E_f$, so that a lower error equates to a stronger factor.

One of the strengths of our max-product factor BP approach is that almost all calculations can be done in simple integer arithmetic, which greatly speeds up the algorithm on a camera phone CPU (which lacks a floating-point unit). Rather than using floating-point to calculate and represent the error measure E_f (which is defined above using division), we rescale it by a factor of 100 into a suitable range of integers. Obviously, this rescaling is implicit in all subsequent calculations.

5   Experimental Results

We implemented our algorithm in Symbian C++ on a Nokia 6681 camera phone. Although the built-in camera has a resolution of approximately 1 megapixel, our algorithm decimated the images to 640×480 or smaller to speed execution.

Fig. 4. Experimental results. Left, text segmentation from Fig. 3(b) demonstrating robustness to non-uniform lighting. Right, result demonstrating performance in high clutter.

Fig.'s 4-5 show results of the algorithm for pictures taken by the camera phone of local street scenes. The algorithm took several seconds per image to execute. Note the algorithm's ability to handle considerable amounts of scene clutter. The algorithm's robustness to non-uniform lighting conditions is shown in Fig. 4. However, false positives still occur, especially in image regions characterized by periodic structures such as windows or fences that resemble text at a local scale (Fig. 5).

Fig. 5. More experimental results

6   Discussion

We have demonstrated the feasibility of a novel grouping framework, based on factor graphs, which we have applied to the problem of segmenting text in natural scenes. The algorithm is simple and fast enough to implement on a camera phone. Future work will focus on improving the false positive and false negative rates, which may be accomplished in part by learning the potentials [13] from training samples of manually segmented text. In order for our algorithm to function as part of a practical system for finding and reading text, we will also have to use the text features output by it to determine appropriate bounding boxes to enclose the text, and use OCR to actually read the text. We note that the OCR stage will have the benefit of discarding some false positives which cannot be ruled out by our algorithm alone.

Acknowledgment

The authors would like to acknowledge support from NIH grant NIH EY015187-01A2, and JMC also acknowledges support from NIH grant 1R01EY013875 for preliminary work on computer vision-based wayfinding systems.

References

1. Wu, V., Manmatha, R., Riseman, E.M.: Finding text in images. ACM DL DL, 3–12 (1997)
2. Jain, A., Tu, B.: Automatic text localization in images and video frames. Pattern Recognition 31, 2055–2076 (1998)
3. Li, H., Doermann, D., Kia, O.: Automatic text detection and tracking in digital videos. IEEE Transactions on Image Processing 9, 147–156 (2000)
4. Gao, J., Yang, J.: An adaptive algorithm for text detection from natural scenes (2001)
5. Coughlan, J., Shen, H.: A fast algorithm for finding crosswalks using figure-ground segmentation. In: Leonardis, A., Bischof, H., Pinz, A. (eds.) ECCV 2006. LNCS, vol. 3951, Springer, Heidelberg (2006)
6. Shen, H., Coughlan, J.: Finding text in natural scenes by figure-ground segmentation. In: ICPR (4), pp. 113–118 (2006)
7. Kschischang, F.: Loeliger: Factor graphs and the sum-product algorithm. IEEETIT: IEEE Transactions on Information Theory, vol. 47 (2001)
8. Liang, J., Doermann, D., Li, H.: Camera-based analysis of text and documents: a survey. International Journal on Document Analysis and Recognition 7, 84–104 (2005)
9. Chen, X., Yuille, A.L.: Detecting and reading text in natural scenes. In: Proc. Computer Vision and Pattern Recognition 2004 (2004)
10. Zheng, Y., Li, H., Doermann, D.: Text identification in noisy document images using markov random field. In: International Conference on Document Analysis and Recognition (2003)
11. Zhang, D., Chang, S.: Learning to detect scene text using a higher-order mrf with belief propagation. In: Computer Vision and Pattern Recognition (CVPR) (2004)
12. Shi, J., Malik, J.: Normalized cuts and image segmentation. IEEE Transactions on Pattern Analysis and Machine Intelligence 22, 888–905 (2000)
13. Shental, N., Zomet, A., Hertz, T., Weiss, Y.: Pairwise clustering and graphical models. In: NIPS (2003)
14. Agarwal, S., Lim, J., Zelnik-Manor, L., Perona, P., Kriegman, D., Belongie, S.: Beyond pairwise clustering. In: Proceedings of Computer Vision and Pattern Recognition (CVPR 2005) vol. 2, pp. 838–845 (2005)
15. Yu, S.X., Shi, J.: Object-specific figure-ground segregation. In: Computer Vision and Pattern Recognition (CVPR) (2003)

Generalized vs Set Median Strings for Histogram-Based Distances: Algorithms and Classification Results in the Image Domain

Christine Solnon and Jean-Michel Jolion

LIRIS, UMR 5205 CNRS / Université de Lyon 1 / INSA de Lyon
Nautibus, 43 bd du 11 novembre, 69622 Villeurbanne Cedex, France
{christine.solnon,jean-michel.jolion}@liris.cnrs.fr

Abstract. We compare different statistical characterizations of a set of strings, for three different histogram-based distances. Given a distance, a set of strings may be characterized by its generalized median, i.e., the string —over the set of all possible strings— that minimizes the sum of distances to every string of the set, or by its set median, i.e., the string of the set that minimizes the sum of distances to every other string of the set. For the first two histogram-based distances, we show that the generalized median string can be computed efficiently; for the third one, which biased histograms with individual substitution costs, we conjecture that this is a NP-hard problem, and we introduce two different heuristic algorithms for approximating it. We experimentally compare the relevance of the three histogram-based distances, and the different statistical characterizations of sets of strings, for classifying images that are represented by strings.

1 Motivations

To manage the huge data sets that are now available, and more particularly classify, recognize or search them, one needs statistical measures to characterize them. This statistical characterization is both well defined and easily computed when data are numerical values, or more generally vectors of numerical values. However, many objects are poorly modelized with such vectors of numerical values, that cannot express the sequentiality of attributes. Strings are symbolic structures that allow a richer modelization by integrating a notion of order.

To exploit sets of strings, one needs a statistical characterization of these sets. This characterization depends on a distance measure, that quantifies the dissimilarity of two strings: given a distance, a set of strings may be characterized by its generalized median, i.e., the string —over the set of all possible strings— that minimizes the sum of distances to every string of the set, or by its set median, i.e., the string of the set that minimizes the sum of distances to every other string of the set.

The complexity of the computation of generalized and set median strings depends on the considered distance. For example, for the well known edit distance

F. Escolano and M. Vento (Eds.): GbRPR 2007, LNCS 4538, pp. 404–414, 2007.

of Levenshtein, the set median string may be computed in polynomial time, whereas the computation of the generalized median string is a NP-hard problem [dlHC00, SP01].

In this paper, we focus on three histogram-based distances for strings: the first one, called d_H, considers strings as sets of symbols and is basically defined as a sum of differences of distributions of symbols; the second one, called d_{H_ω} integrates a notion of order by associating weights to positions in strings; the third one, called $d_{H_{\omega,c}}$, biased distances with individual substitution costs of symbols occurring at a same position, in order to express the fact that some symbols are rather similar, whereas some others are very different. These three histogram-based distances have the same computational complexity, which is linear with respect to the size of the strings and the alphabet, and are an order quicker than the edit distance.

A goal of this paper is to study statistical characterizations of sets of strings when considering these histogram-based distances. For the first two distances, we show that the generalized median string can be computed efficiently; for the third one, that biased histograms with individual substitution costs, we conjecture that this is a NP-hard problem, and we introduce two different heuristic algorithms for approximating it.

An application in image classification is proposed as an illustration of these results.

2 Background

2.1 Notations

The alphabet is noted $\mathcal{A}$ and symbols of $\mathcal{A}$ are noted α_i with $1 \leq i \leq |\mathcal{A}|$. Strings are finite length sequences of symbols from $\mathcal{A}$ and are noted s_j with $j \geq 1$. The set of all strings from $\mathcal{A}$ is noted $\mathcal{A}^*$. The length of a string s_j is noted $|s_j|$, and the kth symbol of a string s_j is noted s_j^k.

2.2 Statistical Characterisation of a Set of Strings

Let $d : \mathcal{A}^* \times \mathcal{A}^* \to \mathbb{R}^+$ be a distance or a dissimilarity measure for any pair of strings from $\mathcal{A}$ (see 3). The first moment, also called *generalized median*, of a set of strings $S \subseteq \mathcal{A}^*$ is defined as a string of $\mathcal{A}^*$ that minimizes the sum of distances to every string of S, i.e.,

$$\text{generalized\_median}(S) = arg \min_{s_{j_1} \in \mathcal{A}^*} \sum_{s_{j_2} \in S} d(s_{j_1}, s_{j_2}) \qquad (1)$$

The complexity of the computation of the generalized median string depends on the distance considered. When this complexity is too high, one may approximate the generalized median string by constraining the search to the set S, yielding the *set median* of S as

$$\text{set\_median}(S) = arg \min_{s_{j_1} \in S} \sum_{s_{j_2} \in S} d(s_{j_1}, s_{j_2}) \qquad (2)$$

3   Distances Between Strings

3.1   Edit Distance

The most famous distance between strings has been proposed by Levenshtein, e.g., the edit distance [Lev66]. The edit distance between two strings s_{j_1} and s_{j_2}, denoted by $d_e(s_{j_1}, s_{j_2})$, is defined by the minimum cost set of edit operations required to transform s_{j_1} into s_{j_2}. Three edit operations are allowed (substitution of a symbol by another symbol, deletion of a symbol, and insertion of a symbol); costs are associated with these operations. A simple algorithm using dynamic programming for computing the edit distance can be found in [WF74]. Its computational time complexity is in $\mathcal{O}(|s_{j_1}| \cdot |s_{j_2}|)$.

For this edit distance, the computation of the generalized median string is a NP-hard problem [dlHC00, SP01]. The generalized median string may be approximated by using heuristic algorithms, such as greedy algorithms [MHJC00] or genetic search [JBC04].

3.2   Histogram-Based Distance d_H

An alternative to the edit distance is to consider a sequence not as a string but as a set of symbols. Thus a basic distance between two sets is the comparison of the distributions of symbols defined as:

$$d_H(s_{j_1}, s_{j_2}) = \sum_{\alpha_i \in \mathcal{A}} abs(H(s_{j_1}, \alpha_i) - H(s_{j_2}, \alpha_i)) \tag{3}$$

where $H(s_j, \alpha_i)$ is the number of occurrences of symbol α_i in string s_j, and abs is the function that returns the absolute value[1]. The main advantage of this distance is its computational cost which is in $\mathcal{O}(|s_{j_1}| + |s_{j_2}| + |\mathcal{A}|)^2$. For strings of different sizes, the histograms must be normalized before comparison.

For this histogram-based distance, the generalized median string of a set of strings S can be constructed as follows: starting from an empty string, for each symbol $\alpha_i \in \mathcal{A}$, insert k times the symbol α_i to the string, where k is the median value of the set $\{H(s_j, \alpha_i), s_j \in S\}$. This generalized median string can be computed in $\mathcal{O}(|S| \cdot (l + |\mathcal{A}|))$, where l is the length of the strings of S[3].

---

[1] There exists many other different histogram-based distances such as, e.g., kullback-Leibler or Kolmogorov-Smirnov. Our work, based on a distance defined by means of absolute differences of distributions, could be extended to other histogram-based distances as well.

[2] In case of very large alphabets, one may use a hashing table in order to consider only the symbols of the alphabet that actually occur in the strings, thus computing the distance in $\mathcal{O}(|s_{j_1}| + |s_{j_2}|)$.

[3] Note that the median element of a set can be selected in linear time with respect to the size of the set by using a "divide-and-conquer" approach similar to the one used for the quicksort [CLR90]: the idea is to partition the set in two parts containing elements greater than (resp. lower or equal to) a given element; depending on the cardinalities of these two parts, the search for the median element can be recursively continued in one of the two parts.

3.3 Weighted Histogram-Based Distance d_{H_ω}

The histogram-based distance d_H does not take into account the order the symbols appear in the strings. One could integrate information on the sequentiality of the symbols by using n-grams, thus comparing the distributions of sub-sequences of n symbols. However, in some applications (as the one described in section 5), the order of the symbols in a string may not express a strong sequentiality, but a difference in the importance of the symbols. In this case, the fact that a symbol is just before another symbol is not very significant; the main information contained in the string structuring is the global position of symbols, those at the beginning of a string being more important than those at the end.

In this case of decreasing importance strings, one may associate a weight ω_k with every position k in strings. To emphasize differences at the beginning of the strings, this weight may be defined, e.g., by $\omega_k = 1 + l - k$ where l is the length of the string. To compare strings of different sizes, it is then necessary to complete the shortest string with a new extra symbol until the two strings have the same size.

Hence, we define the weighted histogram associated with a string s_j and a symbol α_i:

$$H_\omega(s_j, \alpha_i) = \sum_{1 \leq k \leq |s_j|, s_j^k = \alpha_i} \omega_k$$

and the weighted histogram-based distance between two strings s_{j_1} and s_{j_2}:

$$d_{H_\omega}(s_{j_1}, s_{j_2}) = \sum_{\alpha_i \in \mathcal{A}} abs(H_\omega(s_{j_1}, \alpha_i) - H_\omega(s_{j_2}, \alpha_i)) \qquad (4)$$

This distance has the same computational cost than the basic histogram-based distance d_H.

For this weighted histogram-based distance, one can construct a "generalized median weighted histogram" of a set of strings S as follows: for each symbol $\alpha_i \in \mathcal{A}$, set the weighted histogram value associated with α_i to the median value of the set $\{H_\omega(s_j, \alpha_i), s_j \in S\}$. This generalized median weighted histogram can be computed within the same time complexity than for the histogram-based distance, i.e., in $\mathcal{O}(|S| \cdot (l + |\mathcal{A}|))$, where l is the length of the strings of S. Note that it may not be possible to construct a string corresponding to this weighted histogram. However, it may be used to statistically characterize a set of strings.

3.4 Weighted Histogram-Based Distance with Substitution Costs $d_{H_{\omega,c}}$

The histogram-based distances d_H and d_{H_ω} assume that all symbols are "equally different". However, some symbols may be considered as rather similar, whereas some others may be very different. Therefore, [RLJS05] has proposed a new distance, which has the same computational complexity as d_H and d_{H_ω}, but which is biased with individual substitution costs of symbols occurring at a same position.

This new distance is based on weighted histograms with substitution costs. Given two strings s_{j_1} and s_{j_2} and a symbol α_i, these histograms are defined as follows

$$H_{\omega,c}(s_{j_1}, \alpha_i) = \sum_{1 \leq k \leq |s_{j_1}|, s_{j_1}^k = \alpha_i} \omega_k \cdot c(\alpha_i, s_{j_2}^k)$$

$$H_{\omega,c}(s_{j_2}, \alpha_i) = \sum_{1 \leq k \leq |s_{j_2}|, s_{j_2}^k = \alpha_i} \omega_k \cdot c(s_{j_1}^k, \alpha_i)$$

where $c : \mathcal{A} \times \mathcal{A} \to \mathbb{R}^+$ is a function which defines the cost of substituting one symbol by another symbol.

Then, the weighted histogram-based distance with substitution costs is defined by

$$d_{H_{\omega,c}}(s_{j_1}, s_{j_2}) = \sum_{\alpha_i \in \mathcal{A}} abs(H_{\omega,c}(s_{j_1}, \alpha_i) - H_{\omega,c}(s_{j_2}, \alpha_i)) \qquad (5)$$

Note that this distance is not a metric, and does not satisfy the triangular inequality property.

The fact that the histogram is biased by the individual substitution cost of every pair of symbols occurring at a same position implies that one cannot construct a "generalized median weighted histogram" of a set of strings S, independently from any candidate median string. Therefore, we conjecture that the computation of the generalized median string of a set of strings is NP-hard.

4   Approximations of the Generalized Median String for $d_{H_{\omega,c}}$

This section describes two algorithms for approximating the generalized median string of a set of strings $S \subseteq \mathcal{A}^*$, when considering the weighted histogram-based distance with substitution costs $d_{H_{\omega,c}}$.

We shall assume that all strings of S have the same length l: if this is not the case, it is always possible to complete every string that is shorter than l with a new extra symbol.

4.1   Greedy Algorithm

The generalized median string of S may be approximated in a greedy way: starting from an empty string s_{greedy}, symbols are iteratively added at the end of s_{greedy} until the length of s_{greedy} is equal to l. At each step, one selects the symbol $\alpha_i \in \mathcal{A}$ that minimizes the sum of distances between $s_{greedy} \cdot \alpha_i$ and every string of S (restricted to the $|s_{greedy}| + 1$ first symbols).

A key point to keep a low time complexity is to incrementally evaluate the sum of distances induced by each candidate symbol. This is done by maintaining, at each iteration $l' \leq l$:

- for every string $s_j \in S$, two arrays H_1^j and H_2^j such that for every symbol $\alpha_i \in \mathcal{A}$:

$$H_1^j[\alpha_i] = \sum_{1 \leq k < l', s_j^k = \alpha_i} \omega_k \cdot c(\alpha_i, s_{greedy}^k)$$

$$H_2^j[\alpha_i] = \sum_{1 \leq k < l', s_{greedy}^k = \alpha_i} \omega_k \cdot c(\alpha_i, s_j^k)$$

- an array sum such that for every string $s_j \in S$,

$$sum[s_j] = \sum_{\alpha_i \in \mathcal{A}} abs(H_1^j[\alpha_i] - H_2^j[\alpha_i])$$

Thanks to these data structures, the choice of the next symbol to add is done in $\mathcal{O}(|\mathcal{A}| \cdot |S|)$. Each time a new symbol is added at the end of the string, these data structures are updated in $\mathcal{O}(|S|)$. As a consequence, the time complexity of the greedy algorithm is in $\mathcal{O}(l \cdot |\mathcal{A}| \cdot |S|)$.

4.2   Local Search

The generalized median string of S may also be approximated by iteratively modifying an initial string of length l: at each iteration, a symbol of the string is replaced by a new symbol such that the sum of distances to every string of S is decreased; these replacements are performed until no more replacement can decrease the sum of distances, thus obtaining a locally optimal string that cannot be improved by a simple replacement.

We have compared different strategies (including meta-heuristics such as tabu search) for selecting the next replacement to perform at each step. On average, the best compromise between solution quality and CPU-time has been reached when considering a "first-improvement" strategy, i.e., when selecting the first found replacement that decreases the sum of distances.

This local search process may be started from different initial strings, e.g., from the set median string of S, from the string constructed by the greedy algorithm, or a string which is randomly generated from $\mathcal{A}^*$.

The same data structures than for the greedy algorithm may be used to evaluate replacements at low cost. With such data structures, given a position k and a new symbol α_i, the replacement of the symbol at position k by α_i may be done in $\mathcal{O}(|S|)$ so that the time complexity of the local search algorithm is in $\mathcal{O}(n \cdot |S|)$ where n is the number of replacements that are evaluated. Of course, n depends on the strategies considered for selecting the replacements to perform and for building the initial string from which starting the local search; it also depends on the length of the string.

4.3   Experimental Results

Table 1 compares the quality of the different approximations of the generalized median string introduced previously, e.g., the set median string, the string

Table 1. Comparison of approximations of the generalized median string: each line first gives the length of the strings and the sum of distances to the set median string, and then the percentage of improvement of this sum of distances when considering strings computed by the greedy and local search algorithms (average results for the 10 classes of the SIMPLIcity base described in 5, each class having 100 strings)

Length	set median	greedy	LS(set median)	LS(greedy)	LS(random)
100	468.87	13.76%	13.63%	14.28%	**14.31%**
200	425.57	16.56%	16.50%	**17.20%**	16.98%
400	381.29	19.41%	19.18%	**20.06%**	19.79%
800	329.93	21.29%	20.93%	**22.09%**	21.63%

Table 2. Comparison of CPU-times: each line displays the length of the strings and the CPU times (in seconds) spent to compute the different approximate generalized median strings (average results for the 10 classes of the SIMPLIcity base described in 5, each class having 100 strings)

Length	set median	greedy	LS(set median)	LS(greedy)	LS(random)
100	0.02	0.29	2.59	1.67	2.32
200	0.03	0.51	5.44	4.40	5.03
400	0.04	1.01	14.62	10.38	13.35
800	0.09	2.13	33.45	32.24	34.47

computed by the greedy algorithm (greedy), and the strings computed by the local search algorithm starting from different initial strings, i.e., from the set median string (LS(set median)), the string computed by the greedy algorithm (LS(greedy)) and a randomly generated string (LS(random)).

This comparison is done on strings of the SIMPLIcity base described in section 5. This base contains 10 classes of 100 strings of 4000 symbols. The table gives average results on the 10 classes, when successively limiting the length of the strings to the 100, 200, 400, and 800 first symbols. For each length, the table first displays the sum of distances to the set median string, and then, for each approximation of the generalized median string, the percentage of improvement with respect to this sum of distances.

Both greedy and local search algorithms significantly better approximate the generalized median string than the set median string. The best improvements are usually obtained by local search, when it is started from the string generated by the greedy algorithm. Surprisingly, starting local search from the set median string often leads to a slightly worse approximation than starting from a randomly generated string. However, all approximations obtain rather close results.

Table 1 also shows that the larger the strings, the better improvements: when strings are limited to the 100 first symbols, the sum of distances to the greedy approximation is 13.76% as small as the sum of distances to the set median string;

when considering the 200 (resp. 400 and 800) first symbols, this percentage of improvement rises to 16.56 (resp. 19.41 and 21.29).

Table 2 compares CPU-times spent to compute the different approximations on a $2.16GHz$ Intel dual core with a 2MB cache. This table shows us that computing the set median string is more than ten times as fast as computing an approximation with the greedy algorithm, which itself is more than ten times as fast as computing an approximation with the local search approaches. Also, when starting local search from the string generated by the greedy algorithm, CPU time is slightly smaller than when starting from a set median or a randomly generated string.

The quality improvement is thus balanced by the CPU-time cost. However, in applications such as classification of unknown strings in already known clusters, the best representative of each cluster, e.g., the generalized median string, is computed off-line, one-for-all.

5 Classification Results in the Image Domain

5.1 Representing Images by Strings

We introduced in [SJ05] a new representation of images based on strings of symbols. This signature, both precise and compact, is based on notions such as interest points, contrast and order. First, a given image is binarized such that it keeps all the contrasts. Then local maxima of the contrast energy are extracted and associated with their local 3×3 binary neighborhood in the binary image. We thus get a 2D map of symbols, e.g. the 3×3 binary patterns. As any local maxima, e.g. interest points, is also characterized by a measure of contrast energy, we use this measure to sort the points, yielding a string of symbols. The contrast energy measure is no longer kept in the final signature. In this application, the alphabet is made of 512 symbols, corresponding to the 2^9 different 3×3 binary neighborhoods.

Note that with this representation of images by strings, the edit distance of Levenshtein is not relevant and gives very disappointing results for classification purposes. Indeed, the edit distance mainly considers the "local" order of symbols —their relative positions— whereas we are more interested in a "global" order of symbols —their global positions in the string, as symbols are sorted with respect to their contrast energy and we consider very long strings: we mainly want to distinguish symbols with high contrast energy, at the beginning of the strings, from symbols with low contrast energy, at the end of the strings. Moreover, the edit distance is an order slowler, which makes it prohibitive on large strings of more than one thousand symbols length.

5.2 Test Suite and Experimental Settings

We have performed experiments on the SIMPLIcity database [WLW01] which contains 1000 images of size 384×256 extracted from the well known old commercial COREL database[4]. The database contains ten clusters representing semantic generalized medianingful categories such as Africa people and villages, beaches, buildings, buses, dinosaurs, elephants, flowers, food, forses and mountains and glaciers. There are 100 images per cluster. Each image of the database is represented by a string of 4000 symbols max, as described in the previous section.

For distances d_{H_ω} and $d_{H_{\omega,c}}$, which associate a weight ω_k with every position k in strings, we have defined $\omega_k = l - k + 1$, where l is the length of the strings, in order to emphasize differences at the beginning of the strings.

For the distance $d_{H_{\omega,c}}$, which biased histograms with individual substitution costs, we have tuned for this database by a basic adaptive process. Let M be a 512×512 matrix initialized to 0. We scan all the possible pairs of symbols $(s_{j_1}^k, s_{j_2}^k)$. If the strings s_{j_1} and s_{j_2} belong to the same cluster, $M(s_{j_1}^k, s_{j_2}^k)$ is decreased by 1 else it is increased by $\frac{1}{1-NC}$ where NC is the number of clusters (in order to take into account the *a priori* probability of two strings to belong to the same cluster). The final cost matrix is then discretized based on the sign of M and we set each cost $c(\alpha_{i_1}, \alpha_{i_2})$ to 1 (resp. 2 and 3) if $M(\alpha_{i_1}, \alpha_{i_2})$ is negative (resp. null and positive).

We have classified the strings extracted from the SIMPLIcity base according to a nearest neighbour approach: to classify a string, we compute the distance between this string and the representative of every class (the set median, or an exact or approximated generalized median); the closest representative determines the class. We have computed representatives of every class according to a "leave-out-one" principle: the string which is classified is removed from its class before computing its representatives.

We have performed experiments with different lengths of strings: strings extracted from images have 4000 symbols (some strings were shorter, but we have completed them with a new extra symbol); to study the influence of the length of the strings, we report experimental results obtained when limiting the number of symbols to different lengths varying from 50 to 2500.

5.3   Experimental Results

We now compare the different histogram-based distances (d_H, d_{H_ω}, and $d_{H_{\omega,c}}$), and the different statistical characterizations (set median string, exact generalized median string for d_H and d_{H_ω}, and approximated generalized median string for $d_{H_{\omega,c}}$), for classifying strings representing images of the SIMPLIcity database.

Table 3 compares global classification rates (GCR), i.e., percentages of strings which have been assigned to the right classes. Let us first compare GCR when classes are characterized by set median strings for the three different histogram-based distances introduced in 3. We note that introducing weights ω_k to emphasize differences at the beginning of the strings improves GCR when strings are long enough (i.e., for lengths greater than a thousand or so symbols), whereas it decreases GCR for shorter strings. Note also that introducing individual substitution costs significantly improves GCR.

---

[4] The SIMPLIcity database can be downloaded on the James Z. Wang web site at http://wang.ist.psu.edu/jwang/test1.tar.

Table 3. Comparison of global classification rates (GCR) (average results for the 10 classes of the SIMPLIcity base, each class having 100 strings). Each line successively displays the length of the strings, the GCR obtained when representatives are set median strings (for distances d_H, d_{H_ω}, and $d_{H_{\omega,c}}$), and the GCR obtained when representatives are generalized median strings (for distances d_H and d_{H_ω}), and approximations computed by greedy and local search algorithms (for distance $d_{H_{\omega,c}}$); GCR obtained with (exact or approximated) generalized median strings are followed in brackets by the improvement with respect to the set median string.

Length	set median strings			(exact or approximated) generalized median strings			
	H	H_ω	$H_{\omega,c}$	H	H_ω	$H_{\omega,c}$	
						Greedy	LS(Greedy)
50	28.4	27.2	35.6	33.2 (+4.8)	33.4 (+6.2)	41.0 (+5.4)	**43.9** (+8.3)
100	35.2	34.3	41.4	43.7 (+8.5)	41.0 (+6.7)	**45.7** (+4.3)	44.6 (+3.2)
300	44.1	45.3	48.6	**60.5** (+16.4)	57.8 (+12.5)	55.0 (+6.4)	56.8 (+8.2)
500	55.3	48.1	52.5	**63.8** (+8.5)	61.8 (+13.7)	61.0 (+8.5)	61.7 (+9.2)
800	57.2	52.5	61.9	**68.1** (+10.9)	66.4 (+13.9)	65.1 (+3.2)	63.9 (+2.0)
1000	59.5	57.8	60.8	**69.6** (+10.1)	67.8 (+10.0)	65.1 (+4.3)	65.3 (+4.5)
1250	62.0	58.4	63.4	**69.7** (+7.7)	68.9 (+10.5)	67.4 (+4.0)	66.8 (+3.4)
1500	60.7	61.9	65.5	70.3 (+9.6)	**70.5** (+8.6)	68.6 (+3.1)	69.4 (+3.9)
1750	62.9	63.3	63.9	68.8 (+5.9)	**71.8** (+8.5)	67.9 (+4.0)	68.7 (+4.8)
2000	57.5	64.3	62.7	63.9 (+6.4)	**71.4** (+7.1)	68.2 (+5.5)	68.0 (+5.3)
2500	53.7	61.0	62.6	63.3 (+9.6)	**70.1** (+9.1)	66.5 (+3.9)	65.9 (+3.3)
avg.	52.4	52.2	56.3	61.4 (+9.0)	61.9 (+9.7)	61.0 (+4.7)	61.4 (+5.1)

Let us now compare GCR when classes are characterized by exact or approximated generalized median strings. We note that these (approximated) generalized median strings are better representatives than set median strings. However, exact generalized median strings, computed for the distances d_H and d_{H_ω}, improve more significantly GCR than approximated ones, computed for the distance $d_{H_{\omega,c}}$: on average, the GCR is improved by 9 (resp. 9.7) points for d_H (resp. d_{H_ω}) when representing classes by generalized instead of set median strings; however, this GCR is only improved by 4.7 (resp. 5.1) points for $d_{H_{\omega,c}}$ when representing classes by approximations computed by the greedy (resp. local search) algorithm.

Note finally that GCR obtained with approximations computed by local search are not significantly better than GCR obtained with approximations computed by the greedy algorithm: on average, the GCR is improved of 0.4 points only.

6   Discussion

We introduced in this paper three histogram-based distances for strings. For the first two ones, the generalized median string can be computed in polynomial time, and we experimentally show that classification is significantly improved when characterizing classes with generalized median strings instead of set median strings. However, we conjecture that the computation of generalized median strings for the third histogram-based distance is a NP-hard problem, so that we have proposed two heuristic algorithms for approximating generalized median strings in this case. Experimental results showed us that, if classification is improved when characterizing classes with these approximations, they are not as relevant as we would like and improvements are twice as small as improvements obtained with exact generalized median strings. Hence, further work will first concern an explanation of these disappointing results: are they due to the fact that our heuristic algorithms build approximations that are far from optimality, or are they due to the distance itself? Actually, we need information on the distribution of strings with respect to distances. We thus are currently working on the definition of a probability density function on such space and algorithms to approximate this function.

Another trend will be to relate our string-based approach to the more usual graph-based representation of images. Of course, we shall investigate the seriation of a graph-based representation of an image but also alternatives such as graphs or trees of strings, each string being related to a localized area in an image.

References

[CLR90]   Cormen, T.H., Leiserson, C.E., Rivest, R.L.: Introduction to Algorithms. MIT Press, Cambridge (1990)

[dlHC00]   de la Higuera, C., Casacuberta, F.: Topology of strings: Median string is np-complete. Theoretical Computer Science 230(1/2), 39–48 (2000)

[JBC04]   Jiang, X., Bunke, H., Csirik, J.: Median strings: A review. In: Last, M., Kandel, A., Bunke, H. (eds) World Scientific, (eds), Data Mining in Time Series Databases, pp. 173–192 (2004)

[Lev66]   Levenstein, A.: Binary codes capable of correcting deletions, insertions and reversals. Sov. Phy. Dohl. 10, 707–710 (1966)

[MHJC00]   Martinez-Hinarejos, C.D., Juan, A., Casacuberta, F.: Use of median string for classification. In: International Conference on Pattern Recognition, vol. 2, pp. 903–906 (2000)

[RLJS05]   Ros, J., Laurent, C., Jolion, J.M., Simand, I.: Comparing string representations and distances in a natural image classification task. In: Brun, L., Vento, M. (eds.) GbRPR 2005. LNCS, vol. 3434, pp. 71–83. Springer, Heidelberg (2005)

[SJ05]   Simand, I., Jolion, J.M.: Représentation d'images par chaînes de symboles: application á la recherche par le contenu. In: Presses universitaires de Louvain, editor, Actes du 20éme colloque GRETSI: Traitement du signal et des images, vol. 2, pp. 925–928 (2005)

[SP01]   Sim, J.S., Park, K.: The consensus string problem for a metric is np-complete. Journal of Discrete Algorithms 2(1), 115–121 (2001)

[WF74]   Wagner, R.A., Fisher, M.J.: The string to string correction problem. Journal of the ACM 21(1), 168–173 (1974)

[WLW01]   Wang, J.Z., Li, J., Wiederhold, G.: Simplicity: Semantics-sensitive integrated matching for picture libraries. IEEE Trans. on Pattern Analysis and Machine Intelligence 23(9), 947–963 (2001)

Author Index

Lecture Notes in Computer Science

For information about Vols. 1–4418

please contact your bookseller or Springer